Springer Series in

SOLID-STATE SCIENCES 144

Springer
Berlin
Heidelberg
New York
Hong Kong
London
Milan
Paris
Tokyo

Springer Series in
SOLID-STATE SCIENCES

The Springer Series in Solid-State Sciences consists of fundamental scientific books prepared by leading researchers in the field. They strive to communicate, in a systematic and comprehensive way, the basic principles as well as new developments in theoretical and experimental solid-state physics.

126 **Physical Properties of Quasicrystals**
Editor: Z.M. Stadnik

127 **Positron Annihilation in Semiconductors**
Defect Studies. By R. Krause-Rehberg and H.S. Leipner

128 **Magneto-Optics**
Editors: S. Sugano and N. Kojima

129 **Computational Materials Science**
From Ab Initio to Monte Carlo Methods. By K. Ohno, K. Esfarjani, and Y. Kawazoe

130 **Contact, Adhesion and Rupture of Elastic Solids**
By D. Maugis

131 **Field Theories for Low-Dimensional Condensed Matter Systems**
Spin Systems and Strongly Correlated Electrons. By G. Morandi, P. Sodano, A. Tagliacozzo, and V. Tognetti

132 **Vortices in Unconventional Superconductors and Superfluids**
Editors: R.P. Huebener, N. Schopohl, and G.E. Volovik

133 **The Quantum Hall Effect**
By D. Yoshioka

134 **Magnetism in the Solid State**
By P. Mohn

135 **Electrodynamics of Magnetoactive Media**
By I. Vagner, B.I. Lembrikov, and P. Wyder

136 **Nanoscale Phase Separation and Colossal Magnetoresistance**
The Physics of Manganites and Related Compounds
By E. Dagotto

137 **Quantum Transport in Submicron Devices**
A Theoretical Introduction
By W. Magnus and W. Schoenmaker

138 **Phase Separation in Soft Matter Physics**
Micellar Solutions, Microemulsions, Critical Phenomena
By P.K. Khabibullaev and A.A. Saidov

139 **Optical Response of Nanostructures**
Microscopic Nonlocal Theory
By K. Cho

140 **Fractal Concepts in Condensed Matter Physics**
By T. Nakayama and K. Yakubo

141 **Excitons in Low-Dimensional Semiconductors**
Theory, Numerical Methods, Applications By S. Glutsch

142 **Two-Dimensional Coulomb Liquids and Solids**
By Y. Monarkha and K. Kono

143 **X-Ray Multiple-Wave Diffraction**
Theory and Application
By S.-L. Chang

144 **Physics of Transition Metal Oxides**
By S. Maekawa, T. Tohyama, S.E. Barnes, S. Ishihara, W. Koshibae, and G. Khaliullin

145 **Point-Contact Spectroscopy**
By Y. Naidyuk and I.K. Yanson

146 **Optics of Semiconductors and Their Nanostructures**
Editors: H. Kalt and M. Hetterich

147 **Solid-State Theory**
An Introduction By U. Rössler

Volumes 1–125 are listed at the end of the book.

S. Maekawa T. Tohyama S.E. Barnes
S. Ishihara W. Koshibae G. Khaliullin

Physics of Transition Metal Oxides

With 137 Figures

Springer

Prof. Sadamichi Maekawa
Prof. Takami Tohyama
Dr. Wataru Koshibae
Institute for Materials Research
Tohoku University
Sendai 980-8577, Japan

Prof. Stewart E. Barnes
Physics Department
University of Miami
Coral Gables
FL 33124, USA

Prof. Sumio Ishihara
Department of Physics
Tohoku University
Sendai 980-8578, Japan

Prof. Giniyat Khaliullin
MPI für Festkörperforschung
Heisenbergstr. 1
70569 Stuttgart, Germany

Series Editors:

ISSN 0171-1873

ISBN 3-540-21293-0 Springer-Verlag Berlin Heidelberg New York

Library of Congress Control Number: 2004102074

Springer-Verlag is a part of Springer Science+Business Media
springeronline.com

Typesetting by the authors
Final processing by Satztechnik Katharina Steingraeber, Heidelberg
Cover concept: eStudio Calamar Steinen
Cover production: *design & production* GmbH, Heidelberg

Printed on acid-free paper SPIN: 11531319 57/3111/tr - 5 4 3 2 1

Preface

The fact that magnetite (Fe_3O_4) was already known in the Greek era as a peculiar mineral is indicative of the long history of transition metal oxides as useful materials. The discovery of high-temperature superconductivity in 1986 has renewed interest in transition metal oxides. High-temperature superconductors are all cuprates. Why is it? To answer to this question, we must understand the electronic states in the cuprates.

Transition metal oxides are also familiar as magnets. They might be found stuck on the door of your kitchen refrigerator. Magnetic materials are valuable not only as magnets but as electronics materials. Manganites have received special attention recently because of their extremely large magnetoresistance, an effect so large that it is called colossal magnetoresistance (CMR). What is the difference between high-temperature superconducting cuprates and CMR manganites?

Elements with incomplete d shells in the periodic table are called transition elements. Among them, the following eight elements with the atomic numbers from 22 to 29, i.e., Ti, V, Cr, Mn, Fe, Co, Ni and Cu are the most important. These elements make compounds with oxygen and present a variety of properties. High-temperature superconductivity and CMR are examples.

Most of the textbooks on magnetism discuss the magnetic properties of transition metal oxides. However, when one studies magnetism using traditional textbooks, one finds that the transport properties are not introduced in the initial stages. On the other hand, when one starts with texts which describe the theory of metals, it is not easy to understand the individual properties of a particular oxide. Everyday, many papers concerning transition metal oxides are published in a wide variety of journals. We would like to understand the most basic physics of transition metal oxides and the recent progress in this field at the same time. This is our aim in writing this book.

This book has been written by six authors; namely, S. Maekawa (Chap. 1), T. Tohyama (Chap. 2 and Appendices A and B), S. E. Barnes (Chap. 3), S. Ishihara (Chaps. 4 and 5 and Appendix C), W. Koshibae (Chap. 6 and Appendix D) and G. Khaliullin (Chap. 7). They intended to make the book coherent. In cases where there exists no real consensus on the understanding of some problems, the author's view is presented and the other views are noted at the same time.

Acknowledgements

The authors would like to thank a number of colleagues in Tohoku University, Nagoya University, University of Tokyo, Max-Planck Institute for Solid State Research in Stuttgart, and University of Miami for the long-standing collaboration and countless discussion of the subject. Among others, they are especially indebted to J. van den Brink, Y. Endoh, O. Fisher, A. Fujimori, H. Fukuyama, B. Giovannini, Z. M. Hasan, T. Hatakeyama, K. Hirota, P. Horsch, J. Inoue, B. Keimer, C. Kim, T. Kimura, R. Kilian, H. Kishida, D. Khomskii, H. Kondo, R. B. Laughlin, M. Mori, H. Matsueda, Y. Mizuno, Y. Moritomo, Y. Murakami, N. Nagaosa, H. Nojiri, Y. Ohta, H. Okamoto, S. Okamoto, A. M. Oleś, H. Onodera, Y. Onose, M. Peter, P. Prelovšek, T. M. Rice, Z.-X. Shen, Y. Shibata, J. Sirker, M. Takahashi, Y. Tokura, K. Tsutsui, S. Uchida, M. Yamanaka, K. Yoshioka, S. Yunoki, in alphabetical order. Finally, they are thankful to Dr. C. Ascheron of Springer Verlag for his cooperation and support.

February 2004

Sadamichi Maekawa
(On behalf of the authors)

Contents

1. Introduction
S. Maekawa 1
1.1 Transition Metal Ions in Crystal Fields 1
1.2 Crystal Structure and Physical Properties 4
1.3 Exchange Interaction 8
1.4 Orbital Degeneracy 15
1.5 Double-Exchange Interaction 19
1.6 Magnetic Anisotropy 22
1.6.1 Single Ion Anisotropy 22
1.6.2 Anisotropic Exchange Interactions 25
References 34

2. Cuprates
T. Tohyama 37
2.1 Underlying Electronic Structure of Cuprates 38
2.1.1 Charge-Transfer-Type Insulators 38
2.1.2 Model Hamiltonian 41
2.1.3 Superexchange Interaction 43
2.1.4 Cyclic Four-Spin Interaction 49
2.1.5 Zhang-Rice Singlet State 50
2.1.6 Optical Excitations 55
2.2 One-Dimensional Cuprates 58
2.2.1 Spin-Charge Separation 58
2.2.2 Realization of Spin-Charge Separation 63
2.2.3 Charge Dynamics in Insulating Cuprates 68
2.2.4 Nonlinear Optical Response 74
2.2.5 Spin Dynamics in Insulating Cuprates 79
2.3 Two-Dimensional Cuprates 80
2.3.1 Single Carrier in Mott Insulator 81
2.3.2 Phase Diagram 86
2.3.3 Optical Conductivity 90
2.3.4 Single-Particle Spectral Function 93
2.3.5 Chemical Potential 94
2.4 Summary 95
References 96

3. Theory of Superconductivity
S.E. Barnes ... 101
3.1 The BCS Pairing Theory ... 103
3.2 Phonons in Solids ... 107
3.3 Phonons as Intermediate Bosons ... 108
3.4 Theory of the Anti-ferromagnetic Parent Compounds ... 111
3.5 The Jordan–Wigner Transformation and Flux Tubes ... 116
3.6 Coherent States, Grassman Variables and Flux Tubes ... 120
3.7 Mean-Field Approximations and Flux States ... 122
3.8 Bogoliubov Theory for a Bose Superfluid ... 124
3.9 Auxiliary Particle Methods ... 126
3.10 Magnetic Exchange Interactions via Intermediate Bosons ... 130
3.11 The (Mean-Field) RVB Slave Boson Theory ... 131
3.12 The Gutzwiller Projection and a U(1) Symmetry ... 136
3.13 Auxiliary Particles and the Introduction of Flux Tubes ... 137
3.14 Spin Pairing ... 140
3.15 Fermionic Excitations in an Anti-ferromagnet ... 141
3.16 SU(3) Approach to Hole Coherent States ... 143
3.17 The Effective Exchange for Coherent Doping ... 147
3.18 SO(5) Theory ... 150
3.19 SO(5) and SU(3) Theories ... 155
3.20 Gossamer Superconductivity ... 158
3.21 Summary ... 163
References ... 165

4. Manganites
S. Ishihara ... 167
4.1 Manganite with Cubic Perovskite Structure ... 167
4.1.1 Introduction ... 167
4.1.2 Electronic Hamiltonian and Exchange Interaction ... 170
4.1.3 Jahn-Teller Effect and Cooperative Jahn-Teller Effect ... 178
4.1.4 Phase Diagram and Orbital Order ... 182
4.1.5 Orbital Liquid State ... 186
4.2 Manganite with Layered Structure ... 190
4.2.1 Phase Diagram ... 190
4.2.2 Stability of Orbital and Magnetic Structure ... 193
4.2.3 Experiments for Spin and Orbital Correlation ... 194
4.3 Resonant X-ray Scattering (RXS) ... 197
4.3.1 Experiments ... 198
4.3.2 Scattering Cross Section ... 200
4.3.3 Azimuthal Angle Dependence ... 203
4.3.4 Mechanism of RXS ... 206
4.3.5 Microscopic Calculations of the RXS Intensity ... 209
4.4 Orbital Excitation ... 211

4.5 Other Orbital-Related Topics ... 216
4.6 Summary ... 219
References ... 219

5. Titanates and Vanadates
S. Ishihara ... 225
5.1 Introduction ... 225
5.2 Orbital States ... 226
5.2.1 Perovskite Titanates ... 226
5.2.2 Perovskite Vanadates ... 229
5.3 Metal–Insulator Transition ... 230
5.4 Electronic State and Model Hamiltonian ... 231
5.5 Summary ... 238
References ... 238

6. Cobaltates
W. Koshibae ... 241
6.1 Low-, High- and Intermediate-Spin States ... 241
6.2 Thermoelectric Materials and Cobalt Oxides ... 245
6.3 Thermoelectric Effect ... 246
6.4 Linear Response Theory for Thermoelectric Systems ... 250
6.5 Thermopower in Strongly Correlated Systems: Approach from High Temperature Side ... 252
6.6 Spin and Orbital States and the Thermopower ... 254
6.7 Thermopower of the Degenerate Electron Gas ... 257
6.8 Summary and Discussion – The Path to New Thermoelectric Materials ... 257
References ... 259

7. Quantum Effects in Orbitally Degenerate Systems
G. Khaliullin ... 261
7.1 Systems with e_g Orbital Degeneracy ... 262
7.1.1 Spin-Orbital Model ... 262
7.1.2 Orbital-Only Model ... 269
7.1.3 Orbital-Charge Coupling, Orbital Polarons ... 272
7.1.4 Orbital Liquids, Anomalous Transport ... 279
7.2 Systems with t_{2g} Orbital Degeneracy ... 282
7.2.1 Spin-Orbital Model ... 283
7.2.2 Orbital-Only Model ... 291
7.3 High Spin Systems with t_{2g} Orbital Degeneracy ... 298
7.3.1 Formation of One-Dimensional Orbital Chains ... 298
7.3.2 Spin-Orbital Dimerization ... 302
7.4 Summary ... 306
References ... 306

Appendices

A. Optical Conductivity
T. Tohyama . . . 311

B. Exact Diagonalization Technique: The Lanczos Method
T. Tohyama . . . 317

C. Projection Method (Memory Function Method, Composite Operator Method)
S. Ishihara . . . 321

D. Thermoelectric Effects
W. Koshibae . . . 323
D.1 Seebeck Effect . . . 323
D.2 Peltier Effect . . . 325
D.3 Thomson Effect . . . 326
D.4 The First and Second Laws of Thermodynamics and Thermoelectric Response in Solids: Thomson's Considerations . . . 328
D.5 The Third Law of Thermodynamics and Thermoelectric Response in Solids . . . 330

References for Appendices . . . 333

Index . . . 335

1. Introduction

An electron is an elementary particle with spin S=1/2 and charge $-e$. However, in materials, in particular, in transition metal oxides, it manifests itself in many different ways. This is because electrons interact strongly not only each other but the crystal potentials. In this chapter, the basic properties of the electronic structure of transition metal oxides are described.

1.1 Transition Metal Ions in Crystal Fields

A highlight of elementary quantum mechanics is the description of the periodic table of elements [1]. Quantum mechanics explains the nature of the elements from H with atomic number 1 up to Ku with 104. In this table, we consider the transition metal elements from Ti with 22 to Cu with 29. In their elemental form, there are $3d$ electrons in addition to the electrons with the configuration $(1s)^2(2s)^2(2p)^6(3s)^2(3p)^6$. Let us examine the ions with 3+. These, for example, are Ti^{3+} which has $(3d)^1$ and V^{3+} with $(3d)^2$. Upon further increasing the atomic number, the number of 3d electrons increases.

The radial quantum number n and azimuthal or orbital angular-momentum quantum number l of $3d$ electrons are n=3 and l=2, respectively. The angular part of the wave function corresponds to the spherical harmonics $Y^2_m(\theta, \phi)$ written here in terms of the spherical polar coordinates. Here, m is the magnetic quantum number and has the value $-2 \leq m \leq 2$. The $3d$ states are five-fold degenerate in a spherical potential since $2l + 1 = 5$, but they are not in a crystal. In many transition metal oxides, transition metal ions are surrounded by six oxygen ions (O^{2-}), making octahedron as in Fig. 1.1. Oxides with the perovskite crystal structure are examples (Fig. 1.2).

Let us consider a transition metal ion (M). It is placed equi-distant from six oxygen ions. In this case, M is in a crystal field potential with cubic symmetry.

Each O^{2-} ion has a charge distribution with spherical symmetry. Neglecting the overlap of the $3d$ wave function with the surrounding ions, we treat each ion as a point charge. So, the potential acting on a 3d electron from the surrounding ions is expressed as

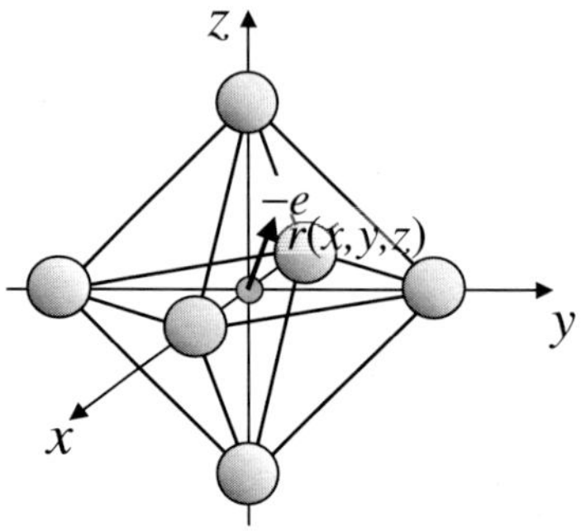

Fig. 1.1. An electron in the crystal field of the octahedron.

$$V(\boldsymbol{r}) = \sum_i \frac{Z_i e^2}{|\boldsymbol{r} - \boldsymbol{R}_i|}, \tag{1.1}$$

where $\boldsymbol{r}$ is the position vector of the $3d$ electron, $\boldsymbol{R}_i$ and $Z_i e$ denote the position vector and charge of the i-th ion, respectively. The expression in (1.1) is called the crystal field potential. In particular, when the $3d$ electron is taken to be a point charge, the quantity,

$$V_0 = \sum_i \frac{Z_i e^2}{|\boldsymbol{R}_i|}, \tag{1.2}$$

is called the Madelung potential. In a cubic crystal field, the potential acting on the $3d$ electron is expressed in the rectangular coordinates as,

$$V_4(\boldsymbol{r}) = \frac{5}{2} V_{40} \left(x^4 + y^4 + z^4 - \frac{3}{5} r^4 \right). \tag{1.3}$$

where V_{40} is a constant determined by the configuration of the surrounding ions.

In the cubic crystal field given by (1.3), the eigen-functions of a $3d$ electron are not the spherical harmonics but are given by the linear combination of them as follows:

$$d(x^2 - y^2) \propto \sqrt{\frac{2\pi}{5}} \left(Y_2^2 + Y_2^{-2} \right) = \frac{\sqrt{3}}{2} \frac{(x^2 - y^2)}{r^2}$$

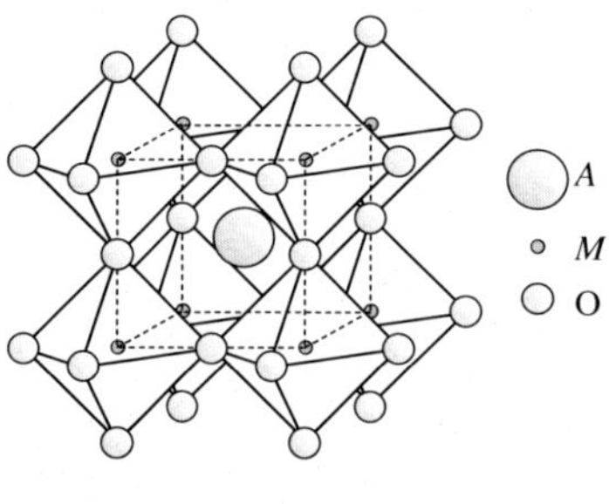

Fig. 1.2. Crystal structure of perovskite AMO_3.

$$
\begin{aligned}
d(3z^2 - r^2) &\propto \sqrt{\frac{4\pi}{5}} Y_2^0 = \frac{1}{2} \frac{(3z^2 - r^2)}{r^2} \\
d(xy) &\propto \frac{1}{i} \sqrt{\frac{2\pi}{5}} \left(Y_2^2 - Y_2^{-2}\right) = \sqrt{3} \frac{xy}{r^2} \\
d(yz) &\propto \sqrt{\frac{2\pi}{5}} \left(Y_2^{-1} + Y_2^1\right) = \sqrt{3} \frac{yz}{r^2} \\
d(zx) &\propto \frac{1}{i} \sqrt{\frac{2\pi}{5}} \left(Y_2^{-1} - Y_2^1\right) = \sqrt{3} \frac{zx}{r^2}
\end{aligned}
\tag{1.4}
$$

In Fig. 1.3, these five eigen-functions are plotted. In Fig. 1.4, the energy level schemes in spherical, cubic and tetragonal (elongated in the z direction) crystal fields are shown. In a cubic crystal field, the two higher energy states are $d(x^2 - y^2)$ and $d(3z^2 - r^2)$, and are often referred to as the e_g orbitals. The three lower energy states, t_{2g} orbitals, comprise $d(xy)$, $d(yz)$,and $d(zx)$. (* The notation $d\gamma$ and $d\varepsilon$, in place of e_g and t_{2g}, is sometimes used). The e_g orbitals extend towards the oxygen ions which have a negative charge so that a $3d$ electron in these orbitals has its energy raised by the Coulomb interaction. On the other hand, the t_{2g} orbitals point away from the oxygen ions so that an electrons in these orbitals have an energy which is less affected by these Coulomb interactions. As a result, the degeneracy is partly removed. The splitting between e_g and t_{2g} orbitals is written as $10Dq$. The value of this splitting is about 2–3 eV in typical oxides with perovskite structure. When the splitting is measured relative to the energy levels in a spherical potential, the e_g level goes up by $6Dq$ and the t_{2g} level comes down by $4Dq$. Thus, the center of gravity of the levels remains constant.

Let us discuss another effect of the crystal field, i.e., the quenching of orbital angular momentum. In a free ion, the magnetic moment is caused by both the spin and orbital angular momenta. In the crystal field, however, the orbital angular momentum vanishes. To see this, let us observe that the time-independent Schrödinger equation is a real equation. Therefore, its non-degenerate eigen-functions are, to within a phase factor, real functions. On the other hand, the orbital angular momentum operator, $\boldsymbol{L}$, is an imaginary operator, i.e.,

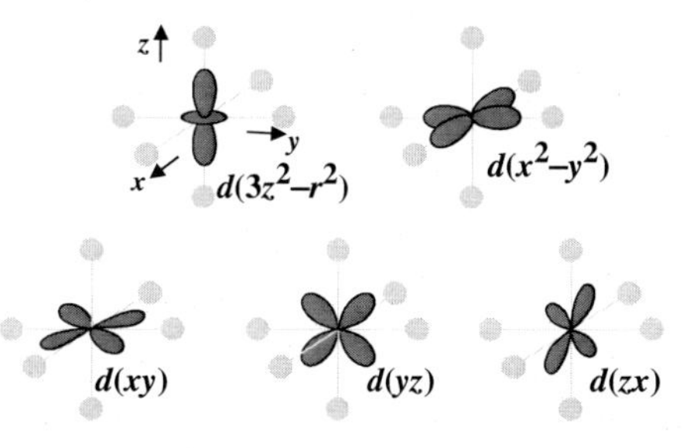

Fig. 1.3. 3*d* orbitals.

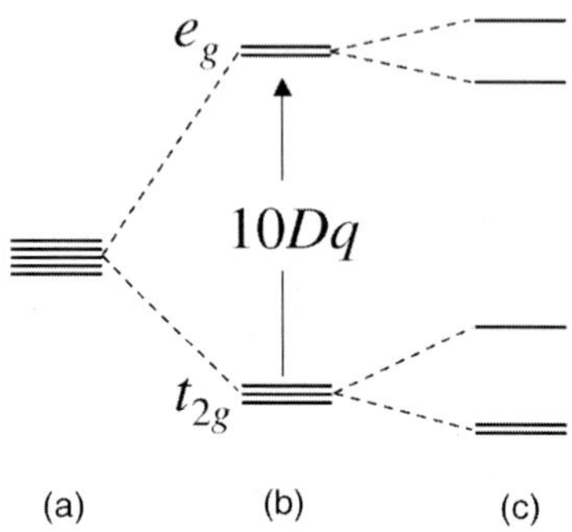

Fig. 1.4. Energy levels in the crystal fields. (**a**) spherical symmetry, (**b**) cubic symmetry, (**c**) tetragonal symmetry where the octahedron is elongated in the z direction.

$$\boldsymbol{L} = -\mathrm{i}\hbar \boldsymbol{r} \times \frac{\partial}{\partial \boldsymbol{r}}. \tag{1.5}$$

The mean angular momentum in the state (Ψ) vanishes, i.e.,

$$\int \Psi^*(\boldsymbol{r})\boldsymbol{L}\Psi(\boldsymbol{r})\mathrm{d}^3r = 0. \tag{1.6}$$

This is because the diagonal part of the Hermite operator ($\boldsymbol{L}$) must be real. The quenching of angular momentum always occurs in the transition metal ions in which the crystal field is much stronger than the spin-orbit interaction. Therefore, the magnetic moment of the transition metal compounds usually reflects that of the spin. When the ground state of a transition metal ion is degenerate, the wave functions can be complex and the angular momentum is not always quenched. However, the degeneracy is often removed by the interactions with lattice and/or between electrons and by the Fermi degeneracy due to the itinerant nature of the d-electrons. These effects will be discussed later.

1.2 Crystal Structure and Physical Properties

Transition metal oxides exist with a great variety of crystal structures, which are reflected in the richness of their physical properties. The orbital states of $3d$ electrons are to a large extent responsible for this complex relationship between the electronic properties and the crystal structure.

Let us consider the perovskite structure which is one of the most common for the transition metal oxides. The basic structure with the chemical formula, AMO_3, is shown in Fig. 1.2. A transition metal ion (M) is surrounded by six oxygen (O) ions which form an octahedron. The ferroelectric titanate, $BaTiO_3$, and a colossal magnetoresistive (CMR) manganite, $La_{1-x}Sr_xMnO_3$, have this crystal structure. The ideal structure of AMO_3 has the relation, $r_A + r_O = \sqrt{2}\,(r_M + r_O)$, where r_A, r_M and r_O are the ionic radii of A, M, and O ions, respectively. However, in the real oxides, the relation,

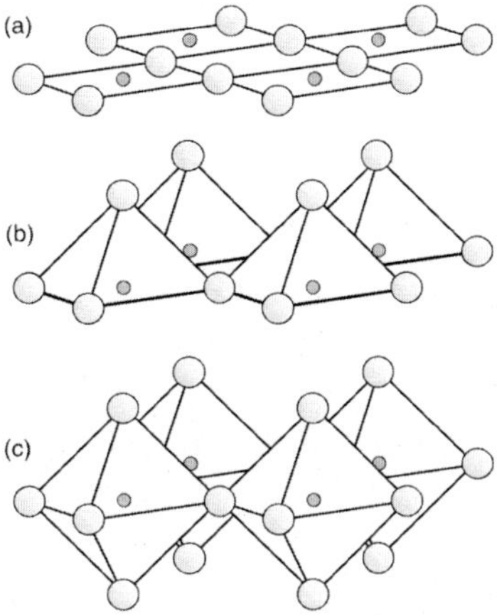

Fig. 1.5. CuO_2 plane structures. Large and small circles are O and Cu ions, respectively.

$$r_A + r_O = t\sqrt{2}\,(r_M + r_O) \tag{1.7}$$

with $t = 0.8 \sim 1.0$ holds.

In addition to AMO_3 with a three-dimensional structure, there exist the related two-dimensional structures written as $A_{n+1}M_nO_{3n+1}$. The layered K_2CuF_4-type structure has $n = 1$, i.e., A_2MO_4. The high temperature superconducting (HTSC) cuprate, $La_{2-x}Sr_xCuO_4$, has this structure. The double-layered CMR compound, $La_{3-x}Sr_xMn_2O_7$ is a manganite with $n = 2$ (M=Mn). The three-dimensional structure shown in Fig. 1.2 corresponds to this lattice with $n = \infty$.

All the HTSC cuprates are layered perovskite oxides. There are not only CuO_6-octahedron units but also CuO_5-pyramid and CuO_4-square units as shown in Fig. 1.5. In Fig. 1.6 is shown the crystal structure of one of the HTSC cuprates with CuO_5 units, $YBa_2Cu_3O_7$. All the HTSC cuprates have stacks of CuO_2 planes which are composed of the above units. The CuO_2 planes with CuO_6 and CuO_5 units are hole-doped HTSC while the planes with CuO_4 units are the electron-doped HTSC. This relation between the units and doping can be understood as reflecting the ionic character of the CuO_2 planes [2–5]: Since a Cu ion is 2+ while an O ion is 2−, the planes with CuO_6 and CuO_5 units are more negative than that with CuO_4 units. When the background charge is accounted for this difference causes the planes in the first two cases to accept hole carriers with positive charge while for the latter case the planes attracts electron carriers with negative charge. The difference in carrier doping is, thus, due to apical oxygens. In fact, the planes with CuO_6 units are sometimes over-doped with hole carriers.

The HTSC cuprates without carriers such as La_2CuO_4 and $YBa_2Cu_3O_{6.5}$ are called the *parent* compounds. Neglecting the covalency for a moment, the compounds are assumed to be ionic with O^{2-} and Cu^{2+}. Even though the covalency *is* strong in the CuO_2 plane, the ionicity dominates the interac-

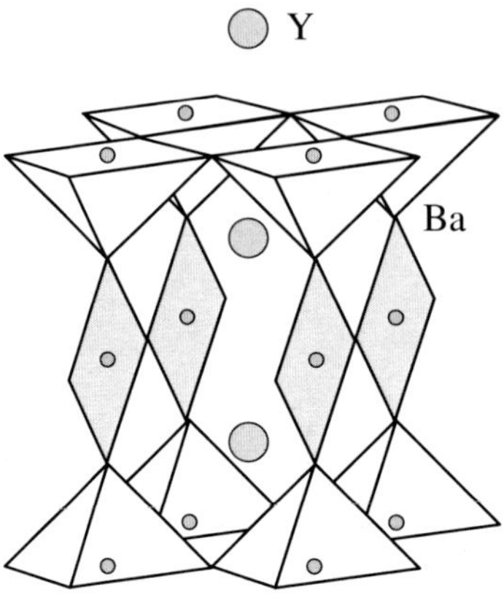

Fig. 1.6. Crystal structure of YBa_2CuO_7. Small circles are Cu ions. Each corner has an O ion.

tion between the layers. Let us consider the hole-doped HTSC cuprates. As discussed in the previous Sect. 1.1, in the ionic compounds, a central role is played by the electrostatic potential which is defined for a hole at the i-site as

$$V_i = \sum_{(i \neq j)} \frac{Z_j e^2}{|\boldsymbol{R}_j - \boldsymbol{R}_i|} \tag{1.8}$$

where $\boldsymbol{R}_j$ is the position of the j-th ion with charge $Z_j e$. The electrostatic energy, V_i, is that required to bring a hole from infinity to the i-site (Madelung site potential). Assuming lattice periodicity and total charge neutrality, V_i is calculated using the standard Ewald method for lattice summation. Let us introduce the Madelung site potentials for a hole on an oxygen in the CuO_2 plane (O(P)) and for an apex oxygen (O(A)), $V_{\mathrm{O(P)}}$ and $\mathrm{V}_{\mathrm{O(A)}}$, respectively. The difference between $\mathrm{V}_{\mathrm{O(A)}}$ and $V_{\mathrm{O(P)}}$ is written as

$$\Delta V_A = V_{\mathrm{O(A)}} - V_{\mathrm{O(P)}}, \tag{1.9}$$

which is a measure of the stability of a hole introduced into O(P).

The crystal-field splitting of the d-orbitals of a Cu ion in the CuO_2 plane (Cu(P)) due to the electrostatic potential may also be calculated from the spatial variation of the Madelung potential around the Cu(P) ion. There are six points along the Cu(P)-O(P) and Cu(P)-O(A) bond directions at which the atomic wave function of the Cu 3d orbital is maximum (or at points 0.32 Å apart from the Cu(P) site). The Madelung potentials are evaluated at these points. Then, the level splitting between $d(x^2 - y^2)$ and $d(3z^2 - r^2)$ orbitals due to the electrostatic potentials, ΔV_d, is defined to be an average of these potentials. In Figs. 1.7 and 1.8, the superconducting transition temperature (T_c) in HTSC cuprates is plotted as a function of ΔV_A and ΔV_d evaluated in parent compounds, respectively [4]. All the compounds are located in the clear area in the T_c vs. ΔV_A curve, indicating that T_c scales very well with ΔV_A. A correlation between T_c and ΔV_d is also seen as the shaded area in

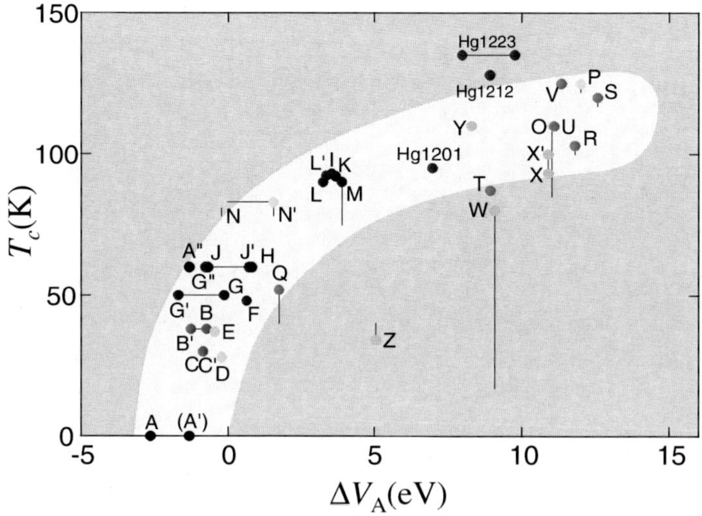

Fig. 1.7. The T_c versus ΔV_A correlation in hole-doped HTSC cuprates [4]. The Madelung potential ΔV_A was evaluated for the corresponding parent compounds. A: $La_2SrCu_2O_{6.2}$, A': $La_{1.9}Ca_{1.1}Cu_2O_6$, B and B': $La_{1.85}Sr_{0.15}CuO_4$, C and C': $La_{1.85}Ba_{0.15}CuO_4$, D: $Nd_{0.66}Sr_{0.205}Ce_{0.135}CuO_4$, E: $Sm_1La_{0.75}Sr_{0.25}CuO_{3.95}$, F: $(Ba_{0.67}Eu_{0.33})_2(Eu_{0.67}Ce_{0.33})_2Cu_3O_{8+x}$, G, G' and G": $Y_{0.8}Ca_{0.2}Ba_2Cu_3O_{6.11}$, H: $YBa_2Cu_3O_{6.5}$, I: $YBa_2Cu_3O_7$, J and J': $ErBa_2Cu_3O_{6.53}$, K: $ErBa_2Cu_3O_7$, L and L': $Y_{0.9}Ca_{0.1}Ba_2Cu_4O_8$, M: $(Ca_{0.5}La_{0.5})(Ba_{1.25}La_{0.75})Cu_3O_{6+\delta}$, N and N': $Pb_2Sr_2Y_{0.5}$Ca0.5Cu_3O_8, O: $Pb_{0.5}Tl_{0.5}Sr_2CaCu_2O_7$, P: $Pb_{0.5}Tl_{0.5}Sr_2Ca_2Cu_3O_9$, Q: $Tl(Ba_{0.6}La_0.4)CuO_{5-\delta}$, R: $TlBa_2CaCu_2O_{7-\delta}$, S: $TlBa_2Ca_2Cu_3O_{9-\delta}$, T: $Tl_2Ba_2CuO_{6+\delta}$, U: $Tl_2Ba_2CaCu_2O_{8+\delta}$, V: $Tl_2Ba_2Ca_2Cu_3O_{10+\delta}$, W: $Bi_2Sr_2CuO_{6+\delta}$, X: $Bi_2Sr_2Ca_{0.9}Y_{0.1}Cu_2O_{8.24}$, X': $Bi_{2+x}Sr_{2+y}Ca_{1+x}Cu_2O_{8-\delta}$ ($x + y + z \simeq 1$), Y: $Bi_2Sr_2Ca_2Cu_3O_{10+\delta}$, and Z: $Bi_2Sr_2(Gd_{0.82}Ce_{0.18})_2Cu_2O_{10.24}$, Hg1201: $HgBa_2CuO_4$, Hg1212: $HgBa_2CaCu_2O_6$, Hg1223: $HgBa_2CaCu_3O_8$.

Fig. 1.9. However, the fact that T_c scales better with ΔV_A than with ΔV_d means that doped holes extend to the oxygen sites. The correlation between T_c and ΔV_A is caused by the stability of the Zhand-Rice singlet (see Sect. 2.1.5). This is discussed in [4]. The electronic properties in the CuO_2 planes will be discussed in Chap. 2 in detail.

The relation between crystal structure and superconductivity in HTSC cuprates has been also discussed from several other points of view. In Fig. 1.9, T_c is plotted as a function of the distances between Cu(P) and O(P), d_P and between Cu(P) and O(A), d_A. For anisotropic CuO_2 planes, the average value of d_P is given in Fig. 1.9(a). In the figures, each dot corresponds to a HTSC cuprate defined in the figure caption of Fig. 1.7. Although there is no clear correlation between T_c and d_P or d_A, it is found that in HTSC cuprates, d_A is much larger than d_P. This is one of their characteristic features when compared with other transition metal oxides [4].

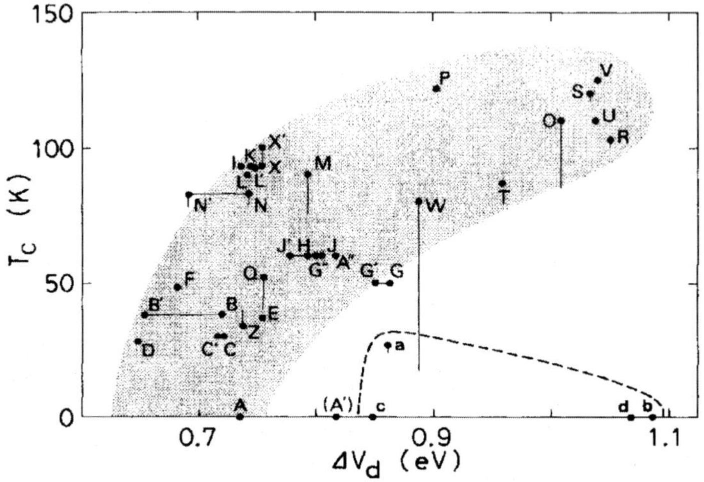

Fig. 1.8. The T_c versus ΔV_d correlation [4]. For the materials, see Fig. 1.7.

The physical properties of the CMR manganites also depend on the crystal structure. The crystal structure of the double-layered manganite, $La_{1.2}Sr_{1.8}Mn_2O_7$, is shown in Fig. 1.10. In this compound, there are three kinds of oxygen ions, O(1), O(2) and O(3). However, the difference in the Mn-O bond lengths is small. When a hydrostatic pressure is applied to this compound, the electric resistivity in the c-direction, i.e., perpendicular to the MnO_2 plane decreases by one order of magnitude at low temperatures. This is surprising, considering the fact that the c axis spacing decreases with pressure. This is considered to be due to the change of the e_g levels with pressure (Fig. 1.11). The splitting of e_g orbitals due to the electrostatic potential, ΔV_d, is calculated using the crystal structure measured by the X-ray scattering experiments under pressure up to 6 kbar. The results of ΔV_d vs. pressure are shown in Fig. 1.12. As seen in this figure, the $d(3z^2 - r^2)$ orbital, for an electron, becomes unstable relative to the $d(x^2 - y^2)$ equivalent, i.e., there is a level crossing at around 6 kbar. Once, under high pressure, the $d(x^2 - y^2)$ level is lowest and the hopping of electrons between the MnO_2 planes weakens. As a result, the resistivity in the c-direction increases.

As discussed in this sub-section, the dimensionality of electronic properties of the oxides is determined not only by the crystal structure but also by the nature of d electron orbitals.

1.3 Exchange Interaction

The Coulomb repulsion between electrons is the origin of interactions between spins. Consider two $3d$ electrons, 1 and 2. The individual wave functions for

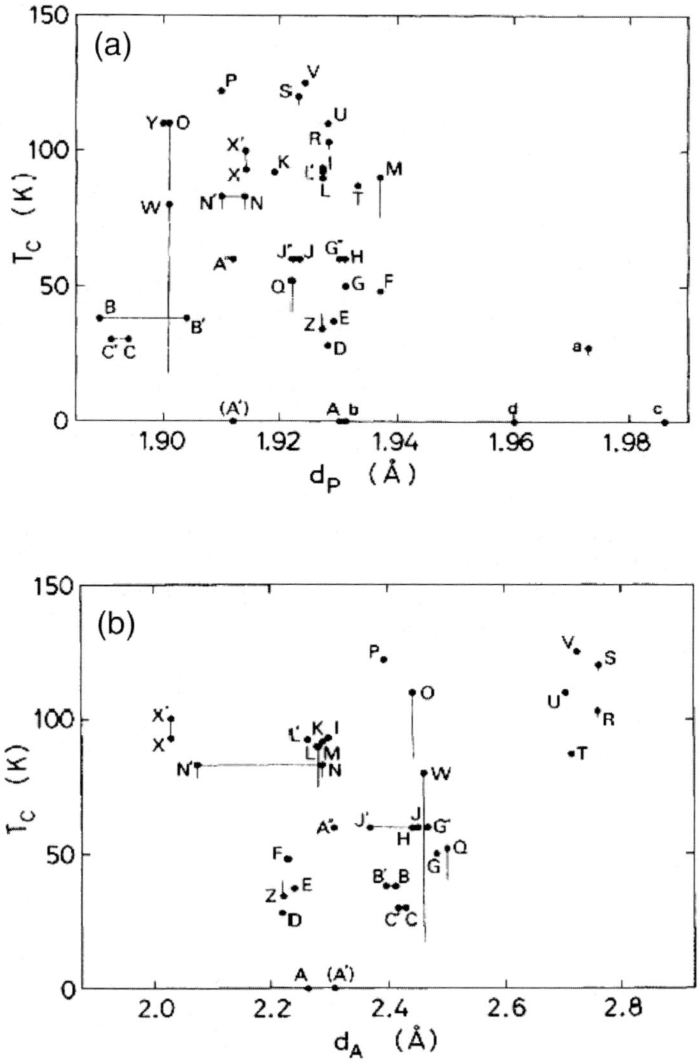

Fig. 1.9. (**a**) The T_c versus d_P and (**b**) the T_c versus d_A correlations [4]. d_P is one-half of the nearest Cu(P)-Cu(P) bond length and d_A is the Cu(P)-O(A) bond length. The average distance is shown for orthorhombic compounds. For the materials, see Fig. 1.7.

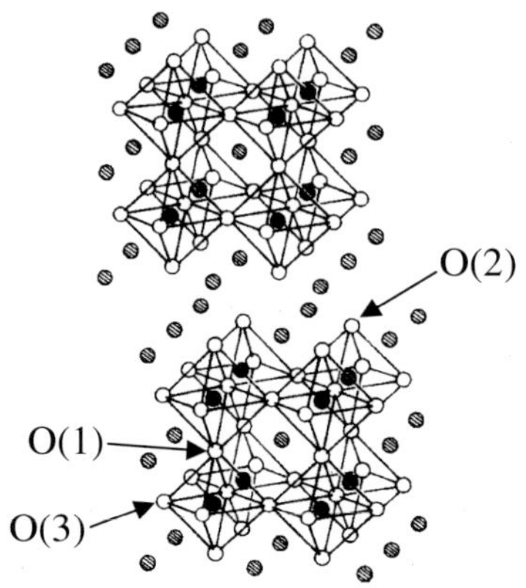

Fig. 1.10. Crystal structure of $(La,Sr)_3Mn_2O_7$.

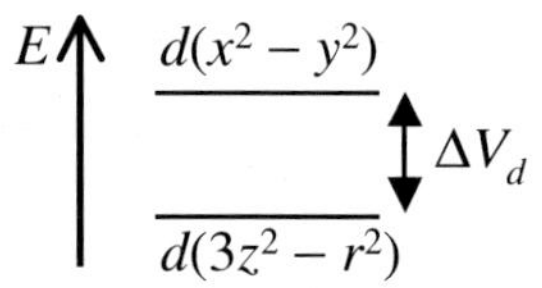

Fig. 1.11. e_g level splitting due to the Madelung potential, ΔV_d, for an electron.

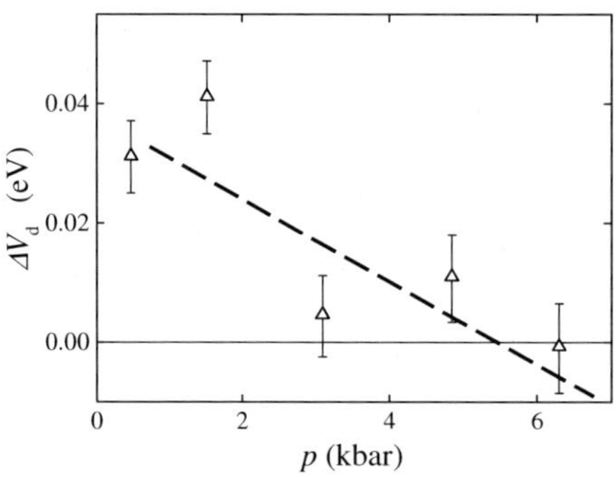

Fig. 1.12. The ΔV_d versus pressure [6]. ΔV_d was evaluated from the Madelung potential.

these electrons with positions r_1 and r_2, are written as $\Psi(r_1)$ and $\Psi(r_2)$.The Coulomb interaction is then reflected by

$$H = \frac{1}{2} \int \int \Psi^\dagger(r_1) \Psi^\dagger(r_2) \frac{e^2}{|r_1 - r_2|} \Psi(r_2) \Psi(r_1) \mathrm{d}\tau_1 \tau_2, \tag{1.10}$$

where the integrals run over the spatial variable and there is a sum over the spin degrees of freedom. The second-quantized wave function $\Psi(r)$ is expressed in terms of the ortho-normal states as

$$\Psi(r) = \sum_{n,m,\sigma} c_{nm\sigma} \psi_{nm}(r) \theta_\sigma. \tag{1.11}$$

Where n, m, and σ denote the ionic site, orbital and spin quantum numbers, respectively, ψ_{nm} and θ_σ are the spatial and spin wave functions. The electron has S=1/2, and $c_{nm,\sigma}$ is the annihilation operator for such an electron. Integrating over the spatial part, (1.10) is written as

$$H = \frac{1}{2} \sum_{n,m,\sigma} \langle n_1 m_1, n_2 m_2 | \frac{e^2}{|r_1 - r_2|} | n_3 m_3, n_4 m_4 \rangle \\ c^\dagger_{n_1 m_1 \sigma_1} c^\dagger_{n_2 m_2 \sigma_2} c_{n_4 m_4 \sigma_2} c_{n_3 m_3 \sigma_1} \tag{1.12}$$

where $\langle \cdots | \, | \cdots \rangle$ denotes a matrix element. Equation (1.12) includes several processes. In the following, they are examined separately.

i) *Hubbard interaction.*
That part of (1.12) with $n_1 = n_2 = n_3 = n_4 = n$ and $m_1 = m_2 = m_3 = m_4$ corresponds to the electron-electron interaction on the same orbital at the same site. This part is expressed as

$$H_U = \frac{1}{2} \sum_\sigma \langle n, n | \frac{e^2}{|r_1 - r_2|} | n, n \rangle c^\dagger_{n\sigma} c^\dagger_{n-\sigma} c_{n-\sigma} c_{n\sigma} = U n_{n\uparrow} n_{n\downarrow} \tag{1.13}$$

with $n_{n\sigma} = c^\dagger_{n\sigma} c_{n\sigma}$. It is usual to abbreviate the matrix element as U. For a free ion, U is estimated to be 15~20 eV. However, in transition metal oxides, because of the mixing between $3d$ orbital in a transition metal ion and $2p$ orbitals in surrounding oxygens and because of the screening due to the electrons in the ion, U is reduced to 6~8 eV. Irrespective of such effects which reduce U, this remains the largest interaction between electrons, as emphasized by Hubbard in his derivation of the so-called Hubbard Hamiltonian [7]. Although this is not a direct exchange interaction, the Hubbard interaction causes the antiferromagnetic interactions between ions, as discussed in iv) later.

ii) *Hund coupling.*
For two electrons in different orbitals but on the same ion ($n_1 = n_2 = n_3 = n_4 = n$), the relevant part of (1.12) is given by

$$\begin{aligned} H_H &= \frac{1}{2} \langle m_1, m_2 | \frac{e^2}{|r_1 - r_2|} | m_1, m_2 \rangle \sum_{\sigma_1, \sigma_2} c^\dagger_{m_1 \sigma_1} c_{m_1 \sigma_1} c^\dagger_{m_2 \sigma_2} c_{m_2 \sigma_2} \\ &- \frac{1}{2} \langle m_1, m_2 | \frac{e^2}{|r_1 - r_2|} | m_2, m_1 \rangle \sum_{\sigma_1, \sigma_2} c^\dagger_{m_1 \sigma_1} c_{m_1 \sigma_2} c^\dagger_{m_2 \sigma_2} c_{m_2 \sigma_1} \\ &= K^0_{m_1 m_2} n_{m_1} n_{m_2} - 2 J^0_{m_1 m_2} \left(\boldsymbol{S}_{m_1} \cdot \boldsymbol{S}_{m_2} + \frac{1}{4} n_{m_1} n_{m_2} \right) . \end{aligned} \tag{1.14}$$

The following relations were introduced with (1.14),

$$\begin{aligned} n_m &= n_{m\uparrow} + n_{m\downarrow} \\ S_{mz} &= \frac{1}{2} \left(c^\dagger_{m\uparrow} c_{m\uparrow} - c^\dagger_{m\downarrow} c_{m\downarrow} \right) \\ S_{m+} &= c^\dagger_{m\uparrow} c_{m\downarrow}, \; S_{m-} = c^\dagger_{m\downarrow} c_{m\uparrow} \, . \end{aligned} \tag{1.15}$$

The first term on the right hand side of (1.14) is the Coulomb interaction with a positive $K^0_{m_1m_2}$. The second term reflects the Hund rule coupling and $J^0_{m_1m_2}$ ~1 eV in transition metal oxides and favors parallel spins. The Hund coupling is responsible not only the multiplet-structure of atoms or ions but also the ferromagnetic states of elementary metallic transition metal compounds. This will be discussed in Sect. 1.5.

iii) *Direct exchange interaction.*
Consider two ions ($n_1 = n_4, n_2 = n_3$ and $n_1 = n_3, n_2 = n_4$), each of which is assumed to have a single orbital. In this case, by exchanging the indices n and m in (1.14), we have

$$\begin{aligned} H_J &= \frac{1}{2}\langle n_1, n_2| \frac{e^2}{|r_1 - r_2|} |n_1, n_2\rangle \sum_{\sigma_1,\sigma_2} c^\dagger_{n_1\sigma_1} c_{n_1\sigma_1} c^\dagger_{n_2\sigma_2} c_{n_2\sigma_2} \\ &\quad - \frac{1}{2}\langle n_1, n_2| \frac{e^2}{|r_1 - r_2|} |n_2, n_1\rangle \sum_{\sigma_1,\sigma_2} c^\dagger_{n_1\sigma_1} c_{n_1\sigma_2} c^\dagger_{n_2\sigma_2} c_{n_2\sigma_1} \\ &= K_{n_1n_2} n_{n_1} n_{n_2} - 2J_{n_1n_2}\left(\boldsymbol{S}_{n_1} \cdot \boldsymbol{S}_{n_2} + \frac{1}{4} n_{n_1} n_{n_2} \right). \end{aligned} \tag{1.16}$$

Here, in the same way as for (1.14), $K_{n_1n_2}$ and $J_{n_1n_2}$ are called the Coulomb and exchange integrals, respectively. The exchange integral $J_{n_1n_2}$ causes the ferromagnetic interactions. Since this term reflects the exchange of electrons between ions, it is called the *direct exchange interaction*. An estimate of $J_{n_1n_2}$ involves the overlap between $3d$ and $2p$ orbitals on the neighboring oxygens. The detailed calculations of this direct ion-ion exchange interaction have been made for the cuprates and will be discussed in Chap. 2.

iv) *Superexchange interaction.*
Many of the transition metal oxides are antiferromagnetic insulators. The simplest model is given by the Hubbard Hamiltonian [7];

$$H = -t \sum_{\langle ij\rangle\sigma} \left(c^\dagger_{i\sigma} c_{j\sigma} + \text{h.c.} \right) + U \sum_i n_{i\uparrow} n_{i\downarrow}. \tag{1.17}$$

where each ion is assumed to have a single orbital. Here, t is the hopping integral of electron between neighboring sites and $\langle i, j\rangle$ runs once of such pairs of sites. In (1.17), U is considered to be much larger than t.

When the number of electrons is the same as the number of ions (the half-filled case), in the ground state, each electron localizes on an ion. When an electron moves to the neighboring site, there are two electrons on the same site and the energy increases by U. Because of this large energy cost electrons cannot move and the system is insulating. Such *Mott insulators* are common among the many transition metal oxides.

Starting with the Mott insulating state, the t term is taken as a perturbation. The second order processes in involving t are shown in Fig. 1.13. The

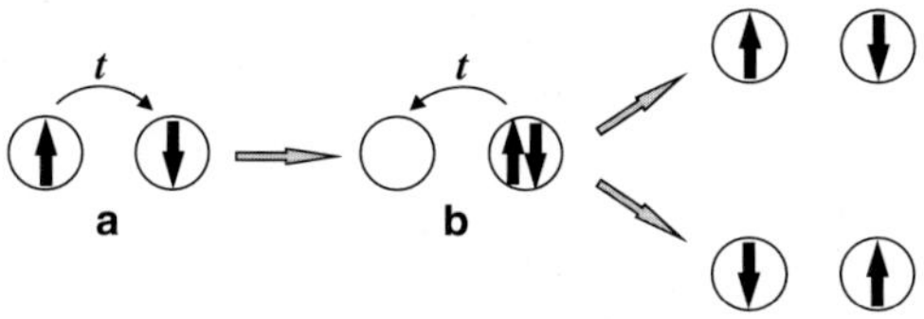

Fig. 1.13. The processes of the superexchange interaction.

intermediate state (Fig. 1.13(b)) has an energy larger by U than the ground state (Fig. 1.13(a)). Note that the intermediate state with parallel spin configuration is not allowed because of the Pauli exclusion principle. There are two possible final states depending on which electron goes to which ion. The energy due to the second order processes corresponds to

$$H' = -\sum_{\langle ij \rangle} \sum_{\sigma,\sigma'} \frac{2t^2}{U} c^\dagger_{i\sigma'} c_{j\sigma'} c^\dagger_{j\sigma} c_{i\sigma} . \tag{1.18}$$

Taking the summation over the spin in (1.18) and re-writing electron operators in terms of spin operators, (1.18) becomes

$$H' = \sum_{\langle i,j \rangle} \frac{4t^2}{U} \left(\boldsymbol{S}_i \cdot \boldsymbol{S}_j - \frac{1}{4} n_i n_j \right) . \tag{1.19}$$

In the half-filled case, $n_i = n_j = 1$ so that the energy goes to zero for the parallel spin configuration. This directly reflects the fact that an electron cannot hop when their spins are parallel. Equation (1.19) is the antiferromagnetic interaction between spins caused by the motion of electrons [8]. For this reason this interaction is sometimes called the kinetic exchange interaction. In the transition metal oxides, this interaction between spins is mediated by oxygens which sit between transition metal ions and is often called the *superexchange interaction.*

In 1959, P. W. Anderson [8] argued that the superexchange interaction ((1.19)) is much larger than the direct exchange interaction ((1.16)), unless the superexchange interaction disappears due to the orthogonality of wave functions. Although it is not easy to calculate the exchange interaction for a given ionic configuration, there exists a rule for the sign of the exchange interaction based on the orthogonality of wave functions. This is called the Kanamori-Goodenough rule [9, 10].

For transition metal ions on the right hand side of the periodic table, e.g., Cu and Ni, the Hubbard model ((1.17)) is not always sufficient. The oxygen $2p$ orbitals must be taken into account explicitly. In Fig. 1.14, the energy level scheme appropriate to the high temperature superconducting cuprates is shown. Here, although the energy of the $3d(x^2 - y^2)$ orbital of the Cu^{2+} ion is lower than that of the O^{2-} $2p$ orbital, there exists only one electron in the $3d(x^2 - y^2)$, this reflecting the strong Coulomb interaction (U). The excitation energy of an electron from oxygen to copper is called the charge-transfer energy (Δ). In the cuprates, Δ is 2∼3 eV and is considerably smaller

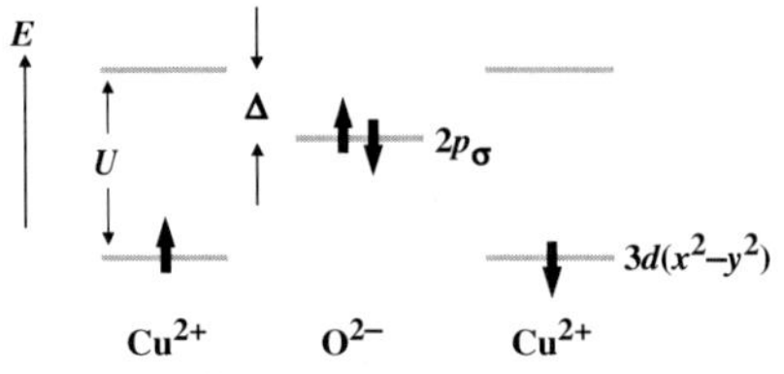

Fig. 1.14. The energy level scheme of Cu^{2+} and O^{2-} ions in HTSC cuprates. Arrows denote electron spins.

than U. Materials are classified as being charge-transfer (CT) [11], $\Delta < U$, or Mott-Hubbard (MH), $\Delta > U$, insulators depending upon the relative magnitude of these two energies. For the cuprates which are CT insulators, the antiferromagnetic exchange interaction between Cu^{2+} is fourth order in hopping integral between copper and oxygen ions, t_{pd}, i.e., the effective exchange is

$$J = \frac{4t_{pd}^4}{\Delta^2}\left(\frac{1}{\Delta} + \frac{1}{U}\right). \tag{1.20}$$

For the HTSC cuprates, Δ is small and the mixing is large between Cu $3d(x^2 - y^2)$ and O $2p$ orbitals (t_{pd}). As a result, the J value is large; $J \sim$0.1-0.2 eV, compared with that for other transition metal oxides.

As will be discussed in detail in Sect. II, since the crystal structures of HTSC cuprates are similar it is possible to estimate the value of J [12]. In Fig. 1.15, the calculated J is plotted together with the peak value ω_s of the

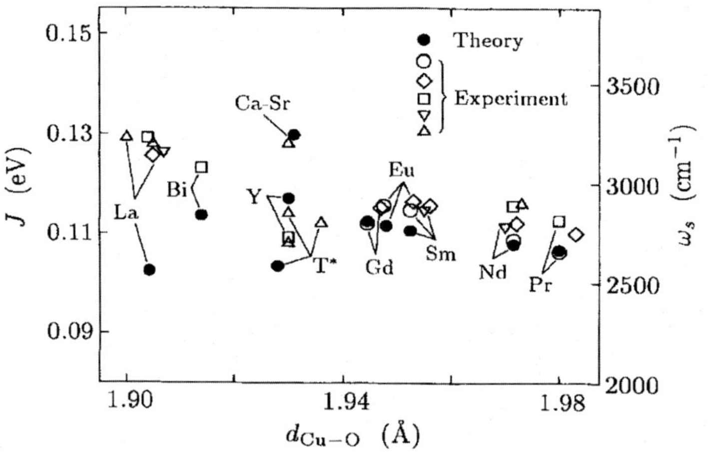

Fig. 1.15. The calculated superexchange interaction J compared with the two-magnon Raman peak position ω_s [12]. Two scales on the ordinate are given by assuming an approximate relation $\hbar\omega_s = 3.1J$. The distance d_{Cu-O} is that between nearest neighboring Cu and O ions in the CuO_2 plane. La; T phase La_2CuO_4, Bi; $Bi_2Sr_2YCu_2O_8$, Ca-Sr; (Ca-Sr)CuO_2, Y; $YBa_2Cu_3O_6$, T*; T* phase (La-Gd-Sr)$_2CuO_4$, L(=Gd, Eu, Sm, Nd and Pr); T' phase L_2CuO_4.

magnetic Raman scattering spectra which is assumed to have the relation $\omega_s = 3.1J$. In the figure, $d_{\mathrm{Cu-O}}$ denotes the distance between neighboring Cu(P) and O(P). Only a weak material dependence of J is found. This characteristic of the CT materials reflects a cancellation of the material dependence of Δ against that of t_{pd} in the expression for J. This contrasts with the typical MH materials, e.g., V_2O_3, in which J increases strongly with decreasing the lattice constant [13].

1.4 Orbital Degeneracy

As discussed in the previous subsections, the fact that the crystal structure is reflected in the ordering of t_{2g} and e_g orbitals is one of the origins of a variety of phenomena in transition metal oxides. In this sub-section, the superexchange interaction caused by orbital degeneracy is discussed with particular interest in manganese oxides with perovskite structure. For an octahedron consisting of six oxygen irons, the crystal field leads to three degenerate t_{2g} and two degenerate e_g orbitals. Consider a Mn^{3+} ion in the cubic crystal field, three electrons go into the t_{2g} orbitals and one occupies the e_g orbital. The spins of these four electrons align due to the Hund coupling as shown in Fig. 1.16. As was shown in Fig. 1.3, because of their orientation, the t_{2g} orbitals are orthogonal to the oxygen $2p_\sigma$ orbitals.

As a consequence, the t_{2g} electrons localize on the Mn ion and exhibit a weak superexchange interactions with neighboring Mn ions via oxygen $2p_\pi$ orbitals. The compound, $CaMnO_3$, is an antiferromagnetic insulator with Néel temperature T_N=120K. In this compound, there are three electrons in the t_{2g} orbitals of an Mn^{4+} ion. The magnetism reflects the Hamiltonian,

$$H_{t_{2g}} = J_{t_{2g}} \sum_{\langle ij \rangle} \boldsymbol{S}_i \cdot \boldsymbol{S}_j, \tag{1.21}$$

with $J_{t_{2g}} \sim 1.5$meV and S=3/2, where the summation runs over the nearest neighbor sites. On the other hand, the e_g orbitals on a Mn^{3+} ion are differently orientated and have a large overlap with oxygen $2p_\sigma$ orbitals. This fact results in a strong superexchange interaction. In this case, the degeneracy of e_g orbitals has a dramatic effect. Let us write the Hamiltonian for the $3d$ electrons in the e_g orbitals as

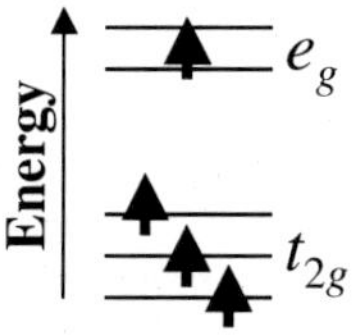

Fig. 1.16. Electronic state of a Mn^{3+} ion in the cubic crystal field. Arrows denote spins.

$$H_{e_g} = -\sum_{\langle ij\rangle}\sum_{\gamma\gamma'}\sum_{\sigma} t_{ij}^{\gamma\gamma'}\left(d_{i\gamma\sigma}^{\dagger}d_{j\gamma'\sigma}+\text{h.c.}\right)+U\sum_{i}\sum_{\gamma}n_{i\gamma\uparrow}n_{i\gamma\downarrow}$$
$$+U'\sum_{i}n_{ia}n_{ib}+J_H\sum_{i}\sum_{\sigma\sigma'}d_{ia\sigma}^{\dagger}d_{ib\sigma'}^{\dagger}d_{ia\sigma'}d_{ib\sigma}, \quad (1.22)$$

where the orbitals are considered as mixtures of e_g orbitals of Mn^{3+} with the $2p_\sigma$ levels of oxygen ions. In the first term on the right hand side of (1.22), $d_{i\gamma\sigma}^{\dagger}$ ($d_{i\gamma\sigma}$) is the creation (annihilation) operator for an electron with spin σ in orbital γ at site i, $t_{ij}^{\gamma\gamma'}$ is the hopping integral between γ at i and γ' at j. The Coulomb U and U' correspond to interactions within and between orbitals on a given site, respectively, and the Hund coupling for the e_g orbitals is J_H. Here, we neglect the so-called pair hopping interaction, for simplisity (See Chap. 4). Realistic values are $U \sim 7\text{eV}$, $U' \sim 5\text{eV}$ and $J_H \sim 2\text{eV}$, with a relationship $U = U' + 2J_H$. In (1.22), a and b denote two different e_g orbitals, and $n_{i\gamma\uparrow}$ and n_{ia} are defined as

$$n_{i\gamma\uparrow} = d_{i\gamma\uparrow}^{\dagger}d_{i\gamma\uparrow}, \quad n_{ia} = n_{ia\uparrow} + n_{ia\downarrow}. \quad (1.23)$$

As discussed in the Sect. 1.3, the superexchange interaction is second order in the hopping integral. We note that the hopping integral depends on the nature of the orbitals. Let us examine $t_{ij}^{\gamma\gamma'}$. When the hopping integral is between a $3d_\gamma$ orbital of a Mn ion and a $2p_\sigma$ orbital of an O ion, t_{pd}^{γ}, is written as $t_{pd}^{\gamma} = \alpha(\gamma)V_{pd}$, $t_{ij}^{\gamma\gamma'}$ is given by

$$t_{ij}^{\gamma\gamma'} = \alpha(\gamma)\,\alpha(\gamma')\,t_0, \quad t_0 \propto V_{pd}^2, \quad (1.24)$$

where $\alpha(\gamma)$ is obtained by integrating the angular part of the wave functions. As an example, we calculate the hopping integral between e_g orbitals via $O2p_\sigma$ orbital in the z direction as shown in Fig. 1.17.

We define $\alpha(\gamma = d(3z^2 - r^2)) = 1$ and write $t_{ij}^{\gamma\gamma} = t_0$ for $\gamma = d(3z^2 - r^2)$. From Fig. 1.17, we find $\alpha(\gamma' = d(x^2 - y^2)) = 0$. Therefore, the relation, $t_{ij}^{\gamma'\gamma'} = t_{ij}^{\gamma\gamma'} = t_{ij}^{\gamma'\gamma} = 0$ for $\gamma = d(3z^2 - r^2)$ and $\gamma' = d(x^2 - y^2)$.

The hopping integral in the x direction shown in Fig. 1.18 is determined as follows: Upon integrating the angular part of the wave functions, the relations, $\alpha(\gamma = d(3z^2 - r^2)) = \pm 1/2$ and $\alpha(\gamma' = d(x^2 - y^2)) = \pm\sqrt{3/2}$ are obtained,

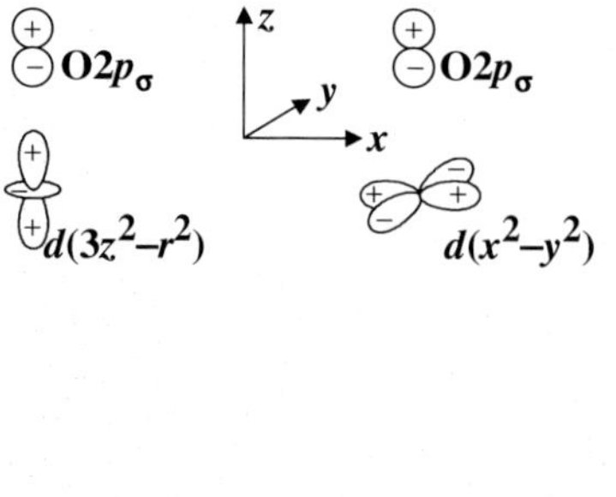

Fig. 1.17. Hopping integral between e_g orbitals via $O2p\sigma$ orbital in the z direction.

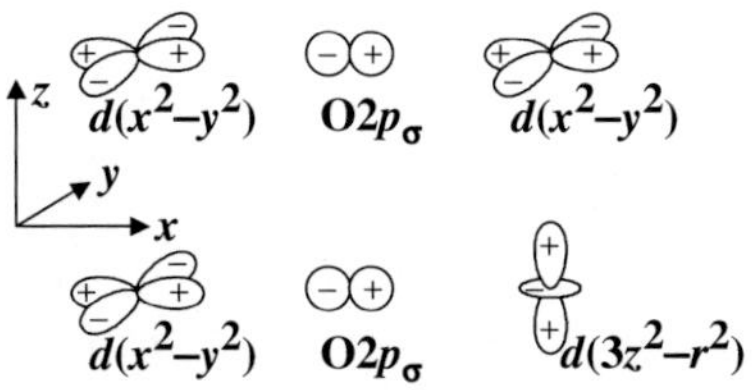

Fig. 1.18. Hopping integral between e_g orbitals via O2p_σ orbital in the x direction.

where the sign depends on the phase of the wave functions. Inserting these relations into (1.24), we have $t_{ij}^{\gamma\gamma} = -t_0/4$, $t_{ij}^{\gamma\gamma'} = t_{ij}^{\gamma'\gamma} = \pm\sqrt{3}t_0/4$ and $t_{ij}^{\gamma'\gamma'} = -3t_0/4$ for $\gamma = d(3z^2-r^2)$ and $\gamma' = d(x^2-y^2)$. The hopping integrals in the y direction may be obtained in the same way.

The superexchange interaction is a second order process involving $t_{ij}^{\gamma\gamma'}$. Here, depending on which orbitals are involved, there are several processes for this interaction. Some of them are shown in Fig. 1.19. In the process (a), the orbital degeneracy does not affect the interaction, so that the interaction is antiferromagnetic as shown in the Sect. 1.3 and has the value $J = 2\xi t_0^2/U$, where ξ is a constant reflecting the nature of the orbitals. In process (b), two degenerate orbitals are used in the intermediate states with energy $U' + J_H$. However, this process is essentially the same as process (a) and results in the antiferromagnetic interaction with $J = 2\xi t_0^2/(U + J_H)$. On the other hand, process (c) is quite different. Since the Hund coupling affects the intermediate state in this process, a ferromagnetic interaction with $J = 2\xi t_0^2/(U' - J_H)$ occurs. Among (a), (b) and (c), the process (c) is the most effective since the intermediate states have the lowest energy. Therefore, collecting all the process (c), we have the Hamiltonian,

$$\tilde{H}_{e_g} = -2J_1 \sum_{\langle ij \rangle} \left(\frac{3}{4} + \boldsymbol{s}_i \cdot \boldsymbol{s}_j \right) \left(\frac{1}{4} - \Psi_i^\dagger \hat{\tau}_{ij} \Psi_j \right),$$

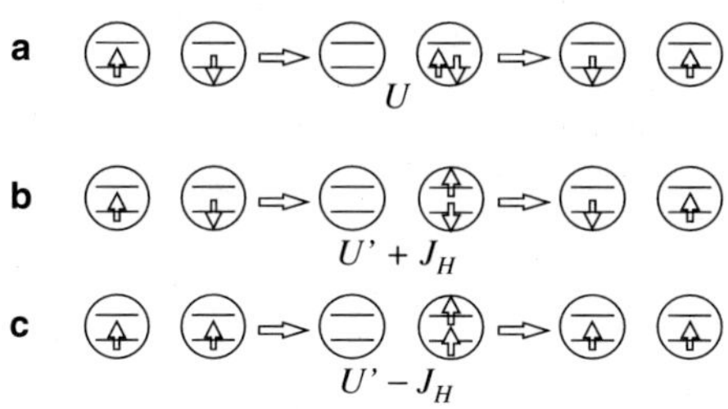

Fig. 1.19. Superexchange processes between e_g orbitals. Arrows denote spins. The energy in the intermediate states of the processes (a), (b) and (c) are U, $U' + J_H$, $U' - J_H$, respectively.

$$\begin{cases} \hat{\tau}_{i,i\pm x} = \frac{1}{4}\begin{pmatrix} 1 & \sqrt{3} \\ \sqrt{3} & 1 \end{pmatrix}, \\ \hat{\tau}_{i,i\pm y} = \frac{1}{4}\begin{pmatrix} 1 & -\sqrt{3} \\ -\sqrt{3} & 1 \end{pmatrix}, \\ \hat{\tau}_{i,i\pm z} = \begin{pmatrix} 1 & 0 \\ 0 & 0 \end{pmatrix}, \end{cases}$$

$$\Psi_i^\dagger = (T_i^z, T_i^x), \tag{1.25}$$

with $J_1 = t_0^2/(U' - J_H)$, where $\boldsymbol{T}_i$ is the pseudo-spin describing the e_g orbitals and is given by

$$\boldsymbol{T}_i = \frac{1}{2}\sum_{\gamma\gamma'}\sum_{\alpha\beta} d_{i\gamma\alpha}^\dagger \boldsymbol{\sigma}_{\alpha\beta} d_{i\gamma'\beta}, \tag{1.26}$$

where σ are the Pauli matrices. The eigen-states of $T^z = 1/2$ and $-1/2$ are $d(3z^2 - r^2)$ and $d(x^2 - y^2)$, respectively. The following is true of (1.25): (i)The sign of the interaction between spins is different from that between orbitals, (ii) only T^x and T^z but not T^y appear in the interaction, and (iii) the interaction between orbitals depends on the direction in the crystal. For example, for the $d(x^2 - y^2)$ orbitals, there is no superexchange interaction in the z direction, since the hopping integral disappears.

The compound, $LaMnO_3$, has a so-called A-type antiferromagnetic spin structure. The spin and orbital orientations are presented in Fig. 1.20, where the $d(3x^2 - r^2)$ and $d(3y^2 - r^2)$ align alternately and the spins order ferromagnetically in the xy plane. Since the orbital states are expressed as

$$\begin{cases} |3x^2 - r^2\rangle = -\frac{1}{2}|3z^2 - r^2\rangle + \frac{\sqrt{3}}{2}|x^2 - y^2\rangle, \\ |3y^2 - r^2\rangle = -\frac{1}{2}|3z^2 - r^2\rangle - \frac{\sqrt{3}}{2}|x^2 - y^2\rangle, \end{cases} \tag{1.27}$$

it is found from (1.25) that the ferromagnetic superexchange interaction is finite in the xy plane, but that there is no ferromagnetic interaction in the z direction. As a result, the remaining, weak, antiferromagnetic interaction acts on spins in e_g orbitals as well as between t_{2g} orbitals. It can be appreciated from (1.25), that the orbitals can be expected to have their own dynamics analogous to those of the spins. In fact, it has been shown that there exist elementary excitations of the orbitals, which are called *orbital waves* or *orbitons*, in $LaMnO_3$. These excitations as well as the more detailed expression of the interactions will be discussed in Chap. 4.

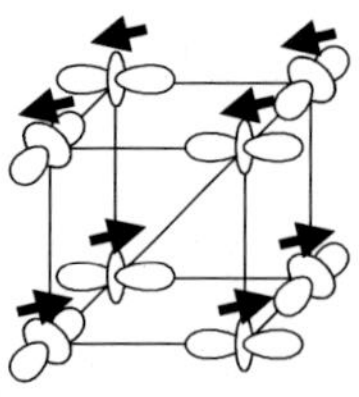

Fig. 1.20. Spin and orbital structures in $LaMnO_3$. Arrows denote spins.

1.5 Double-Exchange Interaction

Many of transition metal oxides are in the mixed valance states and exhibit metallic properties. High temperature superconductivity in cuprates and colossal magnetoresistance (CMR) in manganites are examples of the properties of these mixed valance states. In this sub-section, the relation between metallic nature and ferromagnetism in the mixed valance states of the manganites is discussed.

The magnetism and electrical transport in manganites have been studied since the 1950's. However the discovery of CMR in manganites attracted the renewed interest with an emphasis on strong electron correlations.

The manganites, $LaMnO_3$ and $CaMnO_3$, are both antiferromagnetic insulators with perovskite crystal structure. In $LaMnO_3$ and $CaMnO_3$, Mn ions are 3+ with spin $S = 2$ and 4+ with $S = 3/2$, respectively. Jonker and van Santen [16] prepared the mixed crystal, $La_{1-x}Ca_x(Mn^{+3}_{1-x}, Mn^{4+}_x)O_3$, in 1950 and discovered the metallic ferromagnetic state in the range $0.2 < x < 0.4$. Following this discovery, Zener [17], Anderson and Hasegawa [18] and de Gennes [19] theoretically examined the relation between ferromagnetism and their metallic nature. Since then, the study has continued to the present day.

Let us consider $La_{1-x}Ca_x(Mn^{+3}_{1-x}Mn^{+4}_x)O_3$, where there exist three $3d$ electrons in the t_{2g} states of a Mn^{4+} ion, and three $3d$ electrons in the same states and one in the e_g states for the ion Mn^{3+}. The spins of these electrons are parallel with each other because of the Hund coupling. Suppose there are neighboring Mn^{3+} and Mn^{4+} ions. An e_g electron on the Mn^{3+} can move to an empty e_g state of the Mn^{4+} ion and, thus, these ions exchange valence. Note that the electron hopping depends on the spin states: If the spins of the ions are parallel, the resonance, $Mn^{3+}- Mn^{4+} \Leftrightarrow Mn^{4+}- Mn^{3+}$ occurs. On the other hand, if the spins are antiparallel, the resonance is inhibited by the Hund coupling. This means that the motion of electrons is allowed only in the ferromagnetic state. In other words, ferromagnetism and the metallic state occur cooperatively. This phenomenon was named the "double-exchange interaction" by Zener [17]. In the following, we examine the double-exchange interaction in some detail.

Upon further consideration of the above mixed valent crystal, the following assumptions are introduced: (i) The degeneracy of e_g states is neglected. (ii) The spin of three $3d$ electrons localized in the t_{2g} states are treated as a classical spin ($S = 3/2$), neglecting quantum effects. (iii) The electronic states in the surrounding oxygens are considered to be included in those of Mn ions. (iv) The electron-lattice interaction is neglected. Although it is not clear how these assumptions work in the real materials, some discussion of (ii) and of (i) $\sim$ (iv) are found in references [18] and [19], respectively.

Based on these assumptions, the Hamiltonian is written as

$$
\begin{aligned}
H = & -\sum_{\langle ij\rangle}\sum_{\sigma} t_{ij}\left(c_{i\sigma}^{\dagger}c_{j\sigma} + \text{h.c.}\right) + U\sum_{i} n_{i\uparrow}n_{i\downarrow} \\
& -\frac{1}{2}J_H\sum_{i}\sum_{\sigma\sigma'} \boldsymbol{S}_i\cdot c_{i\sigma}^{\dagger}\boldsymbol{\sigma}_{\sigma\sigma'}c_{i\sigma'} + J\sum_{\langle ij\rangle}\boldsymbol{S}_i\cdot\boldsymbol{S}_j,
\end{aligned}
\tag{1.28}
$$

where $c_{i\sigma}^{\dagger}$ ($c_{i\sigma}$) denotes the creation (annihilation) operator of an e_g electron with spin σ at site i, $n_{i\sigma} = c_{i\sigma}^{\dagger}c_{i\sigma}$, $\boldsymbol{S}_i$ is the spin operator of t_{2g} electrons ($S = 3/2$), and $\boldsymbol{\sigma}$ are the Pauli spin matrices. The hopping integral of an e_g electron between sites i and j, t_{ij}, is expressed as

$$
t_{ij} = \int \varphi_i(r)H'\varphi_j(r)\mathrm{d}r^3,
\tag{1.29}
$$

where $\varphi_i(r)$ denotes the wave function at site i, which is given by the Fourier transform of the Bloch wave function and is orthogonal to $\varphi_j(r)$ with $i \neq j$, and H' includes the kinetic energy and the ionic potential. Since H' does not depend on spin, the electron spin is conserved by hopping. In (1.28), U is the Coulomb interaction between e_g electrons and J_H is the Hund coupling between e_g and t_{2g} spins. The superexchange interaction between neighboring t_{2g} spins, $J(>0)$, is estimated to be $J \simeq 20K$ from the Neel temperature $T_N = 6JS(S+1)/3k_B$ of $CaMnO_3$.

For the manganites, we have $U \simeq 5 \sim 8$eV and $J_H \simeq 1 \sim 2$eV. Therefore, the limit of large U and J_H is taken. Then, the spin of an e_g electron is always parallel to the t_{2g} spin at each site. It is convenient to introduce the polar coordinate (θ_i, ϕ_i) at site i and express the electron operator as

$$
\begin{aligned}
c_{i\sigma} &= c_i z_{i\sigma} \\
z_{i\sigma} &= \begin{pmatrix} \cos\frac{\theta_i}{2} \\ \mathrm{e}^{\mathrm{i}\phi_i}\sin\frac{\theta_i}{2} \end{pmatrix},
\end{aligned}
\tag{1.30}
$$

where c_i denotes the spinless fermion at site i and $z_{i\sigma}$ is a spinor which reflects the spin state with SU(2) symmetry. Since the spin is conserved by the electron hopping, the hopping integral is given by

$$
\begin{aligned}
t_{ij}^{\text{eff}} &= t z_{i\sigma}^{\dagger}z_{j\sigma} \\
&= t\left[\cos\frac{\theta_i}{2}\cos\frac{\theta_j}{2} + \mathrm{e}^{\mathrm{i}(-\phi_i+\phi_j)}\sin\frac{\theta_i}{2}\sin\frac{\theta_j}{2}\right],
\end{aligned}
\tag{1.31}
$$

and the Hamiltonian, (1.28), is reduced to

$$
H = -\sum_{\langle ij\rangle} t_{ij}^{\text{eff}}\left(c_i^{\dagger}c_j + \text{h.c.}\right) + J\sum_{\langle ij\rangle}\boldsymbol{S}_i\cdot\boldsymbol{S}_j.
\tag{1.32}
$$

Note that t_{ij}^{eff} includes a phase part which originates with the spin structure and acts as a gauge field. Thus, an anomalous Hall effect may occur due to the spin texture [20–23].

Let us examine (1.32) in the two-sublattice model following de Gennes [19]. In this case, the angles ϕ_i are irrelevant and are ignored. The Hamiltonian, (1.32), is re-written as

$$H = -t\cos\frac{\theta}{2}\sum_{\langle ij\rangle}\left(c_i^\dagger c_j + \text{h.c.}\right) + \frac{1}{2}\tilde{z}JS^2N_0\cos\theta, \tag{1.33}$$

with $\theta = \theta_i - \theta_j$, $S = 3/2$, and $\tilde{z}$ and N_0 being the number of the nearest neighbor sites and the total number of Mn ions, respectively.

In $LaMnO_3$, the layered antiferromagnetic A-type structure occurs. Therefore, a one-dimensioned lattice with $\tilde{z} = 2$ is considered, for simplicity, since the electronic state is uniform in the other two directions. Performing the Fourier-transform on the first term of (1.33), the band energy of electrons is written as

$$E_k = -2t\cos\frac{\theta}{2}\cos k, \tag{1.34}$$

where the momentum k is measured in units of the inverse lattice constant. Note that the electron hopping prefers $\theta = 0$, whereas the superexchange interaction prefers $\theta = \pi$.

For the small carrier concentration ($x << 1$), carriers exist near the bottom of the energy band, with $k \sim 0$. Therefore, the total energy is given as a function of θ by

$$E(\theta) = -2txN_0\cos\frac{\theta}{2} + JS^2N_0\cos\theta, \tag{1.35}$$

Taking $\partial E(\theta)/\partial\theta = 0$, the angle between spins is given by

$$\cos\frac{\theta}{2} = \begin{cases} \frac{tx}{2J^2}, & tx < 2JS^2, \\ 1, & tx > 2JS^2. \end{cases} \tag{1.36}$$

It was proposed by de Gennes [19] that this canted spin structure (see Fig. 1.21) occurs in $La_{1-x}Ca_xMnO_3$. However, in the real materials, the phase separation between ferromagnetic metallic and antiferromagnetic insulating states also occurs. The phase separation will be discussed in Chap. 4.

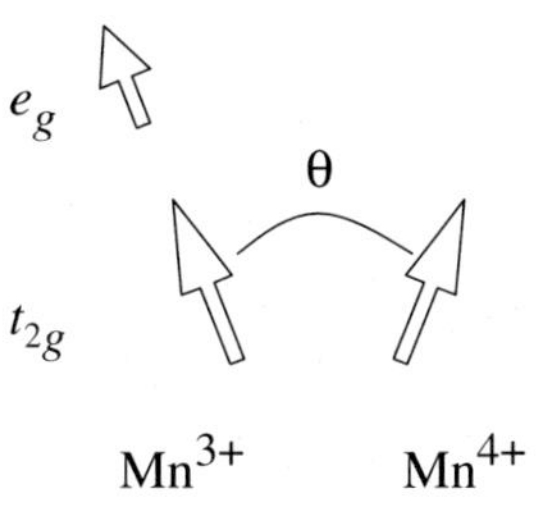

Fig. 1.21. Canted spin structure in a manganite.

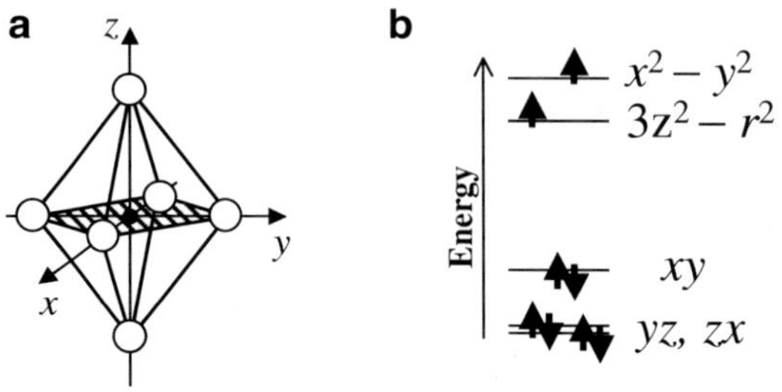

Fig. 1.22. **(a)** An elongated NiO_6 octahedron along the z-axis. Open circles indicate O ions. A Ni ion is located at the center of the octahedron. The hatched area is a cross section of the octahedron in the xy-plane. **(b)** The crystal-field-level structure. The arrows indicate the $3d$ electrons. Due to the distortion, the e_g and t_{2g} levels split, as shown in the figure.

1.6 Magnetic Anisotropy

Transition metal Mott insulators often exhibit antiferromagnetism with magnetic anisotropy. Although the exchange interactions discussed above determines the arrangements of neighboring spins, they do not fix the direction of the magnetic moments. The orbitals see the crystal axis through the electronic potential of the ions. Then, the spin-orbit interaction transfers the crystal environment to the spins and causes the magnetic anisotropy.

1.6.1 Single Ion Anisotropy

Many antiferromagnets have a single-ion anisotropy which determines the easy axis of the magnetic moment. Let us examine the origin of the single-ion anisotropy, using an antiferromagnet of Ni compound as an example.

A distorted NiO_6 octahedron is shown in Fig. 1.22. This is elongated along the z-axis. This distortion causes an energy-level splitting of the $3d$ orbitals.

The Hamiltonian of the spin-orbit interaction is expressed as

$$\begin{aligned} H_{so} &= \lambda \sum_i \boldsymbol{L}^i \cdot \boldsymbol{s}^i \\ &= \frac{\lambda}{2} \sum_i \sum_{nm} \sum_{\alpha\beta} \boldsymbol{L}_{nm} \cdot \boldsymbol{\sigma}_{\alpha\beta} {c^i}^\dagger_{n\alpha} {c^i}_{m\beta}, \end{aligned} \tag{1.37}$$

where λ is the coupling constant and i is the index for the $3d$ electrons. The vector $\boldsymbol{L}^i$ ($\boldsymbol{s}^i$) is the orbital (spin) angular momentum with $L = 2$ ($S = 1/2$). n and m (α, β) are indices for the orbital (spin). The vector $\boldsymbol{L}_{nm}$ ($\boldsymbol{\sigma}_{\alpha\beta}$) denotes the matrix elements of the orbital angular momentum (the Pauli matrices). In second order perturbation theory, the following is obtained for the single-ion anisotropy,

$$H_A = -\sum_\nu A^\nu (S_\nu)^2, \tag{1.38}$$

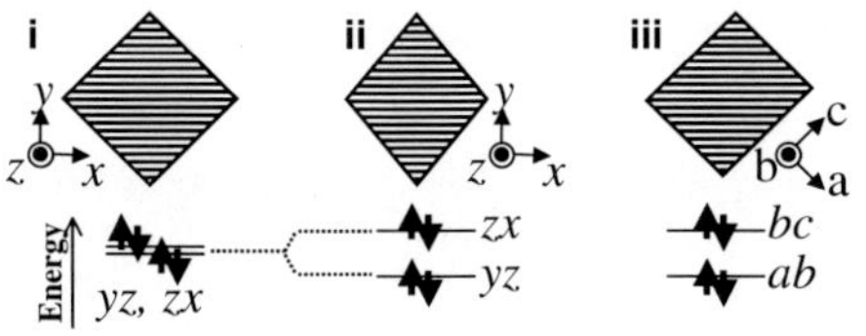

Fig. 1.23. Distortion of the cross section in the xy-plane shown in Fig. 1.22 and the energy-level splitting. (i) In the case that the cross section is a square. The yz and zx levels are degenerate. (ii) A rhombic distortion and the energy-level splitting. (iii) A rectangular distortion and the energy-level splitting.

with

$$A^x = \left(\frac{\lambda}{2}\right)^2 \left(\frac{1}{\varepsilon_{x^2-y^2} - \varepsilon_{yz}} + \frac{3}{\varepsilon_{3z^2-r^2} - \varepsilon_{yz}}\right), \tag{1.39}$$

$$A^y = \left(\frac{\lambda}{2}\right)^2 \left(\frac{1}{\varepsilon_{x^2-y^2} - \varepsilon_{zx}} + \frac{3}{\varepsilon_{3z^2-r^2} - \varepsilon_{zx}}\right), \tag{1.40}$$

$$A^z = \left(\frac{\lambda}{2}\right)^2 \left(\frac{4}{\varepsilon_{x^2-y^2} - \varepsilon_{xy}}\right), \tag{1.41}$$

where the energies ε are those of the $3d$ orbitals and S_ν is a spin operator with $S = 1$ in magnitude. Since $\varepsilon_{xy} > \varepsilon_{yz} = \varepsilon_{zx}$, it follows that $A^x = A^y > A^z$. A distortion in the xy-plane lifts the degeneracy of ε_{yz} and ε_{zx} (see Fig. 1.23). With a rhombic distortion as shown in Fig. 1.23(ii), the yz is lower in energy than the zx-orbital since the negative ligand ions are closer to the electron cloud of the zx-orbital. In the same way, the ab is lower than bc-level for a rectangular distortion of the type shown in Fig. 1.23(iii). The lifting of the degeneracy determines the easy axis due to the single ion anisotropy; (1.39)$\sim$(1.41) give the relation $A^y > A^x > A^z$ for the example illustrated in Fig. 1.23(ii) while $A^a > A^c > A^b$ when Fig. 1.23(iii) is relevant.

There exist experimental data which will illustrate the relation between the distortion of the NiO_6 octahedron and magnetic structure [24–28]. The crystal structure of La_2NiO_4 is shown in Fig. 1.24. In the temperature region 770K $\sim$ 70K, the crystal is orthorhombic. A phase transition occurs at $\sim$ 70K and the crystal structure becomes tetragonal below this temperature. The

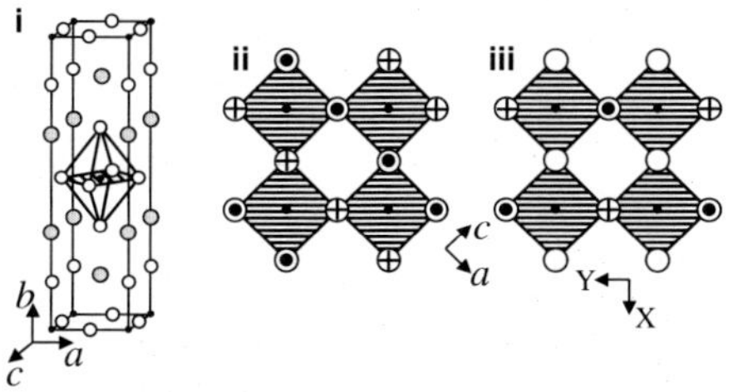

Fig. 1.24. The crystal structure of La_2NiO_4. (i) Solid, open and shaded circles indicate Ni, O and La ions, respectively. (ii) NiO_2 plane in the orthorhombic phase. The symbol $\odot$ ($\oplus$) indicates the O ions which are tilted up (down) from the plane. (iii) NiO_2 plane in the tetragonal phase.

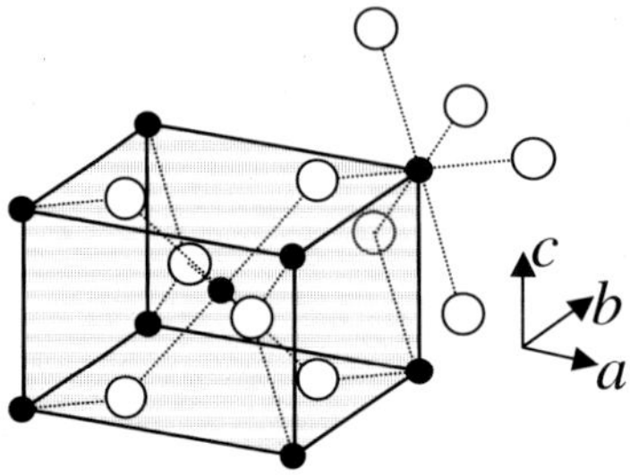

Fig. 1.25. The rutile structure of NiF_2. Solid and open circles indicate Ni and F ions, respectively.

difference in the crystal structure is characterized by a rotation of the NiO_6 octahedra, i.e., in the orthorhombic phase, the octahedra rotate along the a-axis as shown in Fig. 1.24(ii). On the other hand, in the tetragonal phase the axis of rotation is the X-axis as shown in Fig. 1.24(iii). This rotation causes a distortion of the NiO_6 octahedra. This distortion in the orthorhombic phase is that shown in Fig. 1.23(iii) while for the tetragonal phase, Fig. 1.23(ii) is relevant. This difference causes the easy axis to be in the a direction in the orthorhombic phase whereas it is in the y direction for the tetragonal phase.

In some antiferromagnets, the single ion anisotropy results in a weak ferromagnetic moment. The compound NiF_2 is an example. As shown in Fig. 1.25, the antiferromagnet has a rutile crystal structure. A Ni ion is surrounded by an octahedron of F ions. The octahedron has an orthorhombic distortion. In a unit cell, two Ni ions are involved; there is a 90 degrees difference in the rotation around the c-axis for the two orientations of the NiF_6 octahedra. Consequently, the single ion anisotropies of the two Ni ions reflect this 90 degrees difference. In the antiferromagnetic state, each Ni spin is in its own magnetic sub-lattice. The magnetization is in the ab-plane, and a tiny net moment is observed along the a-axis. In the ab plane, the directions of easy axes due to the single ion anisotropy between the sub-lattices differ from each other by 90 degrees (see Fig. 1.26). The spatial structure of the single ion anisotropy causes a canting of the moments. This is the origin of the weak ferromagnetism in NiF_2.

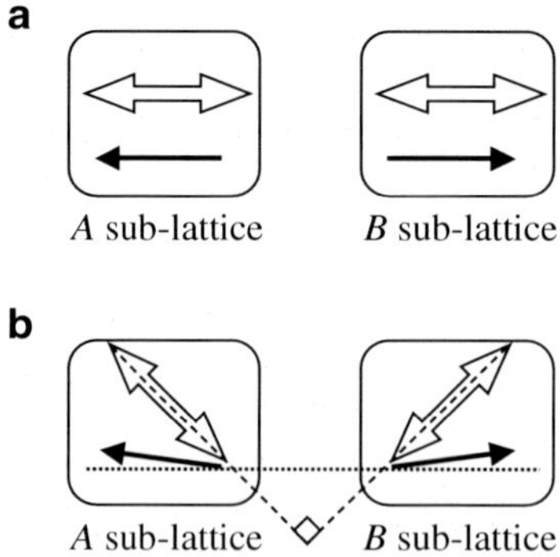

Fig. 1.26. Weak ferromagnetism induced by the single ion anisotropy. The directions of magnetic moment (solid arrow) and easy axis (open arrow) due to the single ion anisotropy in a sub-lattice are shown in a rounded box. **(a)** The antiferromagnetic state without weak ferromagnetism. The A and B sub-lattices have the same easy axis. **(b)** The antiferromagnetic state with weak ferromagnetism. The easy axes of the A and B sub-lattices form an angle of 90 degrees.

In La_2NiO_4 with the tetragonal structure shown in Fig. 1.24(iii), weak ferromagnetism also has been observed. Although the configuration of octahedron of O ions in each sub-lattice is different from each other, the rotation angle of the octahedron is only $\sim$ 6 degrees. This is too small an angle for the mechanism based on the single ion anisotropy to be relevant. Here, weak ferromagnetism is due to another mechanism, which reflects the crystal symmetry. Let us discuss this weak ferromagnetism in the next sub-section.

1.6.2 Anisotropic Exchange Interactions

An antiferromagnet with low crystal-symmetry exhibits weak ferromagnetism in some cases. Based on a phenomenological study, Dzyaloshinski showed that the following interaction exists between sub-lattice magnetizations $\boldsymbol{M}_A$ and $\boldsymbol{M}_B$;

$$\boldsymbol{D} \cdot (\boldsymbol{M}_A \times \boldsymbol{M}_B) . \tag{1.42}$$

This interaction is anti-symmetric, since it involves the vector product $\boldsymbol{M}_A \times \boldsymbol{M}_B$. Because of this, the existence of the anti-symmetric interaction requires a sufficiently low crystal symmetry. This is zero when there exists a center of inversion between adjacent magnetic moments. Crystal symmetry considerations imply that an anti-symmetric interaction cannot exist in the NiF_2. The anti-symmetric and antiferromagnetic interactions, $\boldsymbol{D} \cdot (\boldsymbol{M}_A \times \boldsymbol{M}_B)$ and $J\boldsymbol{M}_A \cdot \boldsymbol{M}_B$, explain the weak ferromagnetism (see Fig. 1.27). The canting angle θ of the magnetic moment is given by

$$\theta = \frac{1}{2} \tan^{-1} \frac{|\boldsymbol{D}|}{J} . \tag{1.43}$$

Moriya has discussed the microscopic origin of the anti-symmetric interaction in terms of the superexchange mechanism accounting for the spin-orbit interaction and derived the following Hamiltonian;

$$H = \sum_{ij} J_{ij} \boldsymbol{S}_i \cdot \boldsymbol{S}_j + \sum_{ij} \boldsymbol{D}_{ij} \cdot (\boldsymbol{S}_i \times \boldsymbol{S}_j) + \sum_{ij} \boldsymbol{S}_i \overleftrightarrow{\Gamma}_{ij} \boldsymbol{S}_j . \tag{1.44}$$

The first term is the superexchange interaction which is isotropic. The second and third terms are the anisotropic superexchange interactions; the former is anti-symmetric and the latter is symmetric. The symmetric exchange-interaction $\overleftrightarrow{\Gamma}_{ij}$ is called the pseudodipolar interaction. The anti-symmetric exchange-interaction is usually called the Dzyaloshinski-Moriya interaction. Moriya's perturbation theory shows that the coupling constant $\boldsymbol{D}_{ij}$ is of the first order in spin-orbit interaction whereas $\overleftrightarrow{\Gamma}_{ij}$ is of second order. Because of this, it might be imagined that the Dzyaloshinski-Moriya dominates the pseudo-dipolar interaction. However, the pseudo-dipolar interaction gives an easy-axis anisotropy along the principal axis of the tensor $\overleftrightarrow{\Gamma}_{ij}$, whereas the

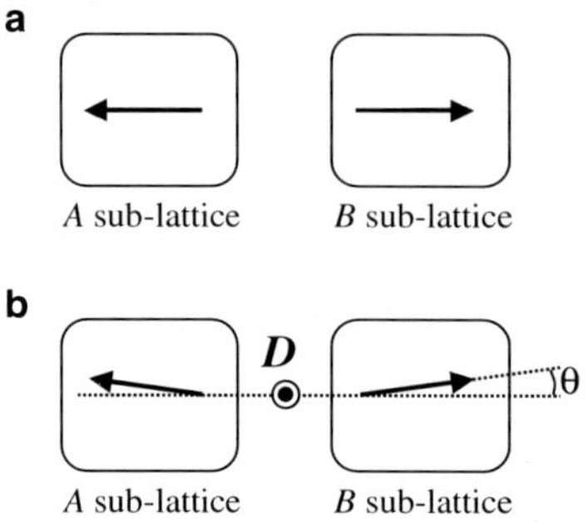

Fig. 1.27. Weak ferromagnetism induced by the Dzyaloshinski-Moriya interaction. The direction of magnetic moment in a sub-lattice is shown by an arrow in a rounded box. The symbol $\odot$ indicates the direction of the vector $\boldsymbol{D}$ which is perpendicular to this page. **(a)** The antiferromagnetic state without weak ferromagnetism. **(b)** The antiferromagnetic state with weak ferromagnetism.

Dzyalonshinski-Moriya interaction gives an easy-plane anisotropy perpendicular to the vector $\boldsymbol{D}_{ij}$ with a canting angle $\theta = \frac{1}{2}\tan^{-1}\frac{|\boldsymbol{D}_{ij}|}{J}$. As seen in Fig. 1.27(b), the energy gain is obtained from the relation,

$$\boldsymbol{D} \cdot (\boldsymbol{M}_A \times \boldsymbol{M}_B) = -|\boldsymbol{D}||\boldsymbol{M}|^2 \sin 2\theta \simeq -\frac{|\boldsymbol{D}|^2}{J}|\boldsymbol{M}|^2, \tag{1.45}$$

and is of second order in the spin-orbit interaction, i.e., is of the same order as that due to the pseudo-dipolar interaction.

Moriya's perturbation theory might be applied to the distorted NiO_2 plane shown in Fig. 1.24. A Ni spin is composed of two electrons in the $x^2 - y^2$ and $3z^2 - r^2$ orbitals of the e_g doublet. These orbitals have the hopping integrals with $2p_\sigma$ orbitals on O ions even when there is no distortion of the NiO_2 plane. The arrangement of the orbitals is shown in Fig. 1.28(a); the $2p_Y$ orbital has a hopping integral with the $3z^2 - r^2$ orbital. The isotropic

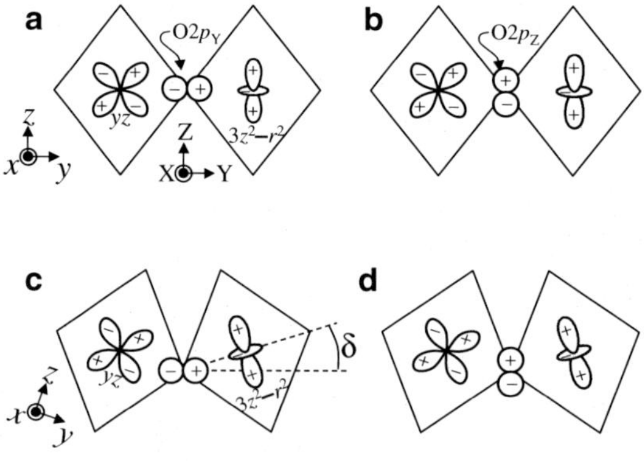

Fig. 1.28. Configuration of orbitals in the distorted NiO_2 plane. The rhombic box indicates a view of a NiO_6 octahedron on the YZ-plane. The xyz-coordinate system is on the NiO_6 octahedron. **(a)** and **(b)** indicate the arrangements of orbitals without the distortion of NiO_2 plane. In these cases, the xyz and XYZ coincide. **(c)** and **(d)** show the arrangements of orbitals in the distorted NiO_2 plane. Here, δ is the tilt angle of the NiO_6 octahedron.

superexchange interaction $J_{ij}\boldsymbol{S}_i \cdot \boldsymbol{S}_j$ results from these hopping integrals. Spin-dependent hopping integrals appear when the NiO_6 plane is distorted caused by the spin-orbit interaction. The relevant Hamiltonian is written as

$$H_t = \sum_{kq,jn}\sum_{\alpha\beta}\left[t_{kq,jn}\mathbf{1}_{\alpha\beta} - \frac{\lambda}{2}\sum_m t_{kq,jn}\frac{(\boldsymbol{L}_{nm})^*}{\varepsilon_n - \varepsilon_m}\cdot\boldsymbol{\sigma}_{\alpha\beta}\right] p^\dagger_{kq\alpha} d_{jn\beta} + \text{h.c.}, \tag{1.46}$$

where $\mathbf{1}$ is a 2×2 unit matrix, n and m are the indices for the Ni $3d$ orbitals, respectively, and $p^\dagger_{kq\alpha}$ ($d_{jn\beta}$) is the creation (annihilation) operator of an electron with spin α (β) for the O $2p_q$ (Ni $3d_n$) orbital, q being X, Y and Z at site k. The hybridization ($t_{kq,jn}$) vanishes for the NiO_2 plane without the distortion: In Fig. 1.28(c), the $2p_{\mathrm{Y}}$ orbital is expressed as

$$p^\dagger_{\mathrm{Y}} = p^\dagger_y + (\pm)\delta\ p^\dagger_z \tag{1.47}$$

in the xyz-coordinate system of the NiO_6 octahedron (see the caption of Fig. 1.28), where δ is the tilt angle of a NiO_6 octahedron. The sign of the tilt angle depends on which NiO_6 octahedron is involved. The yz orbital on the Ni ion hybridizes with the $2p_{\mathrm{Y}}$ equivalent because it contains the $2p_z$ component, (1.47). The strength of this hybridization is

$$t_{yz,p_{\mathrm{Y}}} = (\pm)\delta\ t_\pi \tag{1.48}$$

where t_π is the hopping integral for a π bond. There are matrix elements of the orbital angular momentum between the yz and $3z^2 - r^2$ orbitals of a Ni ion given by the following

$$\boldsymbol{L}_{yz,3z^2-r^2} = -\mathrm{i}\sqrt{3}\ \hat{x}, \tag{1.49}$$

where $\hat{x}$ denotes the unit vector along the x-axis. Using the matrix elements (1.48) and (1.49), a contribution to the spin-dependent hopping-matrix-element in (1.46) is found. Another contribution to (1.46) follows from the considerations illustrated in Fig. 1.28(d). Due to the spin-orbit interaction, there exists a hopping-matrix-element between the O$2p_{\mathrm{Z}}$ and $3z^2-r^2$ orbitals for the left Ni ion in Fig. 1.28(d). The neighboring O $2p_{\mathrm{Z}}$ and Ni $3z^2 - r^2$ orbitals on the right hybridize via,

$$t_{p_{\mathrm{Z}},3z^2-r^2} = (\mp)\delta t_{p_y,3z^2-r^2} \tag{1.50}$$

in the xyz-coordinate system of the right NiO_6 octahedron. This matrix element (1.50) is also responsible for part of the anisotropic superexchange interaction in (1.46).

The anisotropic superexchange interactions arise from (1.46) in fourth order perturbation theory. The Dzyaloshinski-Moriya interaction $\boldsymbol{D}_{ij}$ is of first order and the pseudodipolar interaction is of second order in $\delta\lambda$. The detailed expressions for $\boldsymbol{D}_{ij}$ and $\overleftrightarrow{\Gamma}_{ij}$ are given in [38,39,41] and will not be reproduced here. It is emphasized that the anisotropic superexchange disappears in the NiO_2 plane without the distortion (δ).

There exists another mechanism which leads to an anisotropic exchange interaction. This occurs for the system, even without the distortions, shown in Figs. 1.28(c) and (d). The spin-orbit interaction causes the excitation of an electron from yz orbital to one of the e_g states. Although there is no exchange path for the superexchange interaction between the yz orbital on the left Ni ion and the $3z^2-r^2$ wave function on the right in Figs. 1.28(a) and (b), there does exist the direct exchange ferromagnetic interaction between the spins in these orbitals. Perturbation theory leads to the following expression for the energy:

$$\frac{\lambda^2}{4}\left\{\sum_m \frac{J_{m,l}}{(\varepsilon_n-\varepsilon_m)^2}[(\boldsymbol{s}_i\cdot\boldsymbol{L}_{nm})(\boldsymbol{L}_{mn}\cdot\boldsymbol{s}_j)+(\boldsymbol{s}_i\cdot\boldsymbol{L}_{mn})(\boldsymbol{L}_{nm}\cdot\boldsymbol{s}_j)\right.$$
$$\left.-(\boldsymbol{s}_i\cdot\boldsymbol{s}_j)(\boldsymbol{L}_{nm}\cdot\boldsymbol{L}_{mn})]\right\}, \tag{1.51}$$

where n (l) indicates a x^2-y^2 or $3z^2-r^2$ orbital on the left (right) Ni ion, and $\boldsymbol{s}_i$ and $\boldsymbol{s}_j$ are the $S=\frac{1}{2}$ spin operators for the orbital n and l, respectively. The $J_{m,3z^2-r^2}$ are the direct exchange interactions between the spins on the orbital m on the left Ni ion and $3z^2-r^2$ orbital on the right Ni ion. Finally, $\boldsymbol{L}$ is the orbital angular momentum on the left Ni ion. The anisotropic exchange interaction contributes to the pseudodipolar interaction. Note that the anisotropy due to the superexchange mechanism is of order in $\lambda^2\delta^2$, i.e. reflects a lattice distortion. On the other hand, the direct exchange mechanism gives an anisotropy of the order of $\lambda^2 J_{m,l}/(\varepsilon_n-\varepsilon_m)^2$ where l is x^2-y^2 or $3z^2-r^2$. Therefore, although the direct exchange interaction $J_{l,m}$ may not be large compared with the superexchange interaction J_{ij} in (1.44), the direct exchange mechanism must be accounted for in both $\boldsymbol{D}_{ij}$ and $\overleftrightarrow{\Gamma}_{ij}$ in (1.44).

Crystal symmetry considerations show that the Dzyaloshinski-Moriya interaction $\boldsymbol{D}_{ij}$ is along the X-axis. This is also confirmed in the perturbation theory, because only the x-component of the orbital angular momentum arises in the perturbation process (cf. (1.49)). Then, since the single ion anisotropy gives the easy axis along the y-axis, the antiferromagnetic state in the tetragonal phase of La_2NiO_4 shows weak ferromagnetism where the net moment is perpendicular to the NiO_2 plane. On the other hand, the orthorhombic phase of La_2NiO_4 does not show the weak ferromagnetism although there exists the vector $\boldsymbol{D}_{ij}$ for each adjacent Ni spin. In this phase, however, the easy axis determined by the single ion anisotropy is along the a-axis. As a result, the Dzyaloshinski-Moriya interaction is not effective.

Usually, the effect of single ion anisotropy is stronger than that of anisotropic exchange interactions. However, in a spin-1/2 system, e.g. cuprates and titanates, the single ion anisotropy does not exist. In this case, the only sources of magnetic anisotropy are the exchange interactions, i.e., the Dzyaloshinski-Moriya and pseudo-dipolar terms. Shekhtman, Entin-Wohlman and Aharony have studied the magnetism of La_2CuO_4 and shown that the the Dzyaloshinski-Moriya and pseudo-dipolar interactions compete with each other. Let us examine the interactions in this compound.

The anisotropic superexchange interaction arises via the hopping integral when the spin-orbit interaction is accounted for, and is reflected as the spin-dependent hopping term in (1.46). Due to this term, the orbitals on the neighboring Cu ions are connected by spin-dependent hopping integrals. For simplicity, let us eliminate the O ion orbitals on and ignore the anisotropy in the direct exchange interaction (see (1.51)). The simplest model Hamiltonian for the neighboring sites i and j with the spin-orbit interaction is

$$H_{ij} = \sum_{\alpha\beta} \left(t\mathbf{1}_{\alpha\beta} + \boldsymbol{C}_{ij} \cdot \boldsymbol{\sigma}_{\alpha\beta}\right) d^{\dagger}_{i\alpha} d_{j\beta} + \text{h.c.} + U\left(n_{i\uparrow}n_{i\downarrow} + n_{j\uparrow}n_{j\downarrow}\right), \tag{1.52}$$

where the components of the vector $\boldsymbol{C}_{ij}$ originate from the spin-orbit interaction and are pure imaginary. When the z-axis is taken along the vector $\boldsymbol{C}_{ij}$, the Hamiltonian becomes,

$$\begin{aligned} H_{ij} = & \sum_{\sigma} \left\{ \left[t + \text{sign}\,(\sigma)\,\text{i}C\right] d^{\dagger}_{i\sigma} d_{j\sigma} + \text{h.c.} \right\} \\ & + U\left(n_{i\uparrow}n_{i\downarrow} + n_{j\uparrow}n_{j\downarrow}\right), \end{aligned} \tag{1.53}$$

where $C = |\boldsymbol{C}|$ and sign(σ) gives $+$ $(-)$ for $\sigma =\uparrow$ $(\downarrow)$. With an electron at each site, the second order perturbation in the hopping integrals t and C is equivalent to the spin Hamiltonian,

$$H_{\text{spin}} = 4\frac{t^2 - C^2}{U}\left(\boldsymbol{S}_i \cdot \boldsymbol{S}_j\right) + 8\frac{tC}{U}\left(S_i^x S_j^y - S_i^y S_j^x\right) + 8\frac{C^2}{U} S_i^z S_j^z. \tag{1.54}$$

The coupling constants of the first, second and third terms in (1.54) correspond to the superexchange interactions, J_{ij}, the z component of $\boldsymbol{D}_{ij}$ and zz element of $\overleftrightarrow{\Gamma}_{ij}$ in (1.44), respectively. Let us deal with the spins $\boldsymbol{S}_i$ and $\boldsymbol{S}_j$ as the classical vectors with magnitude S. It is seen that $\overleftrightarrow{\Gamma}_{ij}$ gives an easy axis anisotropy along z axis and the energy of the antiferromagnetic state along the easy axis is $S^2 4(t^2 + C^2)/U$. On the other hand, $\boldsymbol{D}_{ij}$ corresponds to a planar anisotropy in the xy plane and a canting angle $\theta = \tan^{-1}(C/t)$ for the spins in the plane (cf. Fig. 1.27). The energy of the antiferromagnetic state with the canting angle θ is also $S^2 4(t^2 + C^2)/U$. This equals the energy of the antiferromagnetic state along z axis, i.e. the Hamiltonian (1.54) has no magnetic anisotropy. This is a property of (1.52) rather than being a result of perturbation theory. [32,35] In order to see this, let us define the following rotation of the spins $\boldsymbol{S}_i$ and $\boldsymbol{S}_j$,

$$\begin{pmatrix} S_i^x \\ S_i^y \\ S_i^z \end{pmatrix} = \begin{pmatrix} \cos\theta & \sin\theta & 0 \\ -\sin\theta & \cos\theta & 0 \\ 0 & 0 & 1 \end{pmatrix} \begin{pmatrix} S_i^\xi \\ S_i^\psi \\ S_i^\varsigma \end{pmatrix} \tag{1.55}$$

$$\begin{pmatrix} S_j^x \\ S_j^y \\ S_j^z \end{pmatrix} = \begin{pmatrix} \cos\theta & -\sin\theta & 0 \\ \sin\theta & \cos\theta & 0 \\ 0 & 0 & 1 \end{pmatrix} \begin{pmatrix} S_j^\xi \\ S_j^\psi \\ S_j^\zeta \end{pmatrix}. \tag{1.56}$$

The rotations are by $+\theta$ or $-\theta$ as the spin index is i or j. In the $\xi\psi\zeta$ coordinate system, the operators are expressed as

$$d_{i\sigma}^\dagger = \exp\left[-\text{sign}\,(\sigma)\,\mathrm{i}\,(\theta/2)\right] c_{i\sigma}^\dagger, \tag{1.57}$$
$$d_{j\sigma} = \exp\left[-\text{sign}\,(\sigma)\,\mathrm{i}\,(\theta/2)\right] c_{j\sigma}, \tag{1.58}$$

while the spins operators are

$$\boldsymbol{S}_i = \frac{1}{2}\sum_{\alpha\beta} c_{i\alpha}^\dagger \boldsymbol{\sigma}_{\alpha\beta} c_{i\beta} \tag{1.59}$$

and

$$\boldsymbol{S}_j = \frac{1}{2}\sum_{\alpha\beta} c_{j\alpha}^\dagger \boldsymbol{\sigma}_{\alpha\beta} c_{j\beta}, \tag{1.60}$$

in the $\xi\psi\zeta$ coordinate system. In this coordinate system, the Hamiltonian (1.53) becomes

$$H_{ij} = \sqrt{t^2 + C^2}\sum_\sigma \left(c_{i\sigma}^\dagger c_{j\sigma} + \text{h.c.}\right) + U\left(n_{i\uparrow}n_{i\downarrow} + n_{j\uparrow}n_{j\downarrow}\right), \tag{1.61}$$

and is manifestly isotropic. Perturbation theory then gives the spin Hamiltonian,

$$H_{\text{spin}} = \frac{4\left(t^2 + C^2\right)}{U}\boldsymbol{S}_i \cdot \boldsymbol{S}_j, \tag{1.62}$$

in the $\xi\psi\zeta$ coordinate system.

In reality, the magnetic anisotropies caused by the direct exchange mechanism and the superexchange mechanism associated with the arrangements of orbitals of transition metal and ligand ions destroy the isotropy of (1.62). The magnetic dipolar interaction may also result in a magnetic anisotropy.

The magnetic anisotropy in CuO_2 plane with distortion shown in Fig. 1.24(iii) (the readers should interpret the solid circles as Cu instead of Ni hereafter) provides an example for which the pseudodipolar interaction suppresses the weak ferromagnetism caused by Dzyaloshinski-Moriya interaction.

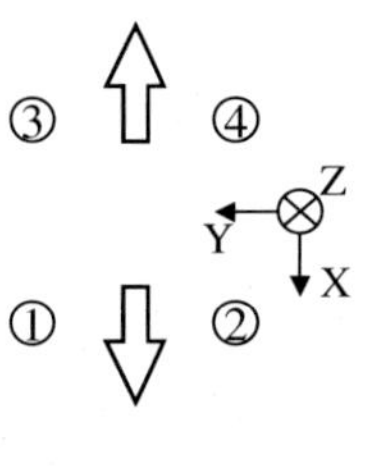

Fig. 1.29. Magnetic anisotropy of a distorted CuO_2 plane where the distortion is the same in Fig. 1.24(iii). The circles 1 $\sim$ 4 denote the copper site with the number, respectively. The open arrows indicate the Dzyaloshinski-Moriya interaction.

In the CuO_2 plane, the distortion of Cu-O-Cu bond angle from 180° is absent on the neighboring bonds 1-3 and 2-4 of Fig. 1.29. Therefore, there exist Dzyaloshinski-Moriya and pseudo-dipolar interactions caused by superexchange mechanism on the neighboring bonds 1-2 and 3-4 only. Since the sign of the tilt angle δ of the bonds is different, the Dzyaloshinski-Moriya interaction may be expressed in the XYZ-coordinate system as follows:

$$\boldsymbol{D}_{12} \cdot (\boldsymbol{S}_1 \times \boldsymbol{S}_2) = \left(S_1^X, S_1^Y, S_1^Z\right) \begin{pmatrix} 0 & 0 & 0 \\ 0 & 0 & d \\ 0 & -d & 0 \end{pmatrix} \begin{pmatrix} S_2^X \\ S_2^Y \\ S_2^Z \end{pmatrix}, \tag{1.63}$$

and

$$\boldsymbol{D}_{34} \cdot (\boldsymbol{S}_3 \times \boldsymbol{S}_4) = \left(S_3^X, S_3^Y, S_3^Z\right) \begin{pmatrix} 0 & 0 & 0 \\ 0 & 0 & -d \\ 0 & d & 0 \end{pmatrix} \begin{pmatrix} S_4^X \\ S_4^Y \\ S_4^Z \end{pmatrix}, \tag{1.64}$$

with $d = |\boldsymbol{D}_{12}| = |\boldsymbol{D}_{34}|$. On the other hand, the pseudo-dipolar interaction corresponds to

$$\boldsymbol{S}_1 \overleftrightarrow{\Gamma}_{12} \boldsymbol{S}_2 = \left(S_1^X, S_1^Y, S_1^Z\right) \begin{pmatrix} \Gamma_s & 0 & 0 \\ 0 & -\Gamma_s & 0 \\ 0 & 0 & -\Gamma_s \end{pmatrix} \begin{pmatrix} S_2^X \\ S_2^Y \\ S_2^Z \end{pmatrix}, \tag{1.65}$$

and with $\overleftrightarrow{\Gamma}_{34} = \overleftrightarrow{\Gamma}_{12}$ in the XYZ-coordinate system.

The anisotropies d and Γ_s require a tilting of the CuO_6 octahedron with an angle δ. However, the direct exchange mechanism (1.51) (n and l are both $x^2 - y^2$ orbitals in this case) gives an anisotropy without distortion:

$$\overleftrightarrow{\Gamma}_{12}^{\text{direct}} = \overleftrightarrow{\Gamma}_{34}^{\text{direct}} = \begin{pmatrix} \Gamma_{yz} - \Gamma_{xy} & 0 & 0 \\ 0 & -\Gamma_{yz} - \Gamma_{xy} & 0 \\ 0 & 0 & \Gamma_{xy} - \Gamma_{yz} \end{pmatrix}, \tag{1.66}$$

and

$$\overleftrightarrow{\Gamma}_{13}^{\text{direct}} = \overleftrightarrow{\Gamma}_{24}^{\text{direct}} = \begin{pmatrix} -\Gamma_{zx} - \Gamma_{xy} & 0 & 0 \\ 0 & \Gamma_{zx} - \Gamma_{xy} & 0 \\ 0 & 0 & \Gamma_{xy} - \Gamma_{zx} \end{pmatrix}, \tag{1.67}$$

where

$$\Gamma_{yz} = \frac{\lambda^2 J_{yz,x^2-y^2}}{4\left(\varepsilon_{x^2-y^2} - \varepsilon_{yz}\right)^2}, \quad \Gamma_{zx} = \frac{\lambda^2 J_{zx,x^2-y^2}}{4\left(\varepsilon_{x^2-y^2} - \varepsilon_{zx}\right)^2} \tag{1.68}$$

and

$$\Gamma_{xy} = \frac{\lambda^2 J_{xy,x^2-y^2}}{4\left(\varepsilon_{x^2-y^2} - \varepsilon_{xy}\right)^2}, \tag{1.69}$$

respectively. Due to the symmetry, $\Gamma_{yz} = \Gamma_{zx}$, and Γ_{yz} is replaced by Γ_{zx} in the following.

As a result, the Hamiltonian for a plaquette (see Fig. 1.29) in the distorted CuO_2 plane takes the form:

$$H = \boldsymbol{S}_1 \overleftrightarrow{C} \boldsymbol{S}_2 + \boldsymbol{S}_4 \overleftrightarrow{C} \boldsymbol{S}_3 + \boldsymbol{S}_1 \overleftrightarrow{D} \boldsymbol{S}_3 + \boldsymbol{S}_2 \overleftrightarrow{D} \boldsymbol{S}_4, \tag{1.70}$$

where

$$\overleftrightarrow{C} = \begin{pmatrix} J+\Gamma_{zx}-\Gamma_{xy}+\Gamma^s & 0 & 0 \\ 0 & J-\Gamma_{zx}-\Gamma_{xy}-\Gamma^s & d \\ 0 & -d & J+\Gamma_{xy}-\Gamma_{zx}-\Gamma^s \end{pmatrix}, \tag{1.71}$$

and

$$\overleftrightarrow{D} = \begin{pmatrix} J-\Gamma_{zx}-\Gamma_{xy} & 0 & 0 \\ 0 & J+\Gamma_{zx}-\Gamma_{xy} & 0 \\ 0 & 0 & J+\Gamma_{xy}-\Gamma_{zx} \end{pmatrix}. \tag{1.72}$$

Let us examine the classical ground state of the Hamiltonian (1.70), i.e., the spins $\boldsymbol{S}_1 \sim \boldsymbol{S}_4$ are taken to be the classical vectors. The Hamiltonian (1.70) can be rewritten as:

$$\begin{aligned} H = \; & \boldsymbol{S}_1 \left(\overleftrightarrow{E} + \overleftrightarrow{F}\right) \boldsymbol{S}_2 + \boldsymbol{S}_4 \left(\overleftrightarrow{E} + \overleftrightarrow{F}\right) \boldsymbol{S}_3 \\ & + \boldsymbol{S}_1 \left(\overleftrightarrow{E} - \overleftrightarrow{F}\right) \boldsymbol{S}_3 + \boldsymbol{S}_2 \left(\overleftrightarrow{E} - \overleftrightarrow{F}\right) \boldsymbol{S}_4, \end{aligned} \tag{1.73}$$

where

$$\overleftrightarrow{E} = \begin{pmatrix} J-\Gamma_{xy}+\frac{1}{2}\Gamma^s & 0 & 0 \\ 0 & J-\Gamma_{xy}-\frac{1}{2}\Gamma^s & \frac{1}{2}d \\ 0 & -\frac{1}{2}d & J+\Gamma_{xy}-\Gamma_{zx}-\frac{1}{2}\Gamma^s \end{pmatrix}, \tag{1.74}$$

and

$$\overleftrightarrow{F} = \begin{pmatrix} \Gamma_{zx}+\frac{1}{2}\Gamma^s & 0 & 0 \\ 0 & -\Gamma_{zx}-\frac{1}{2}\Gamma^s & \frac{1}{2}d \\ 0 & -\frac{1}{2}d & -\frac{1}{2}\Gamma^s \end{pmatrix}. \tag{1.75}$$

Since the eigenvalues of $\overleftrightarrow{F}$ are much smaller than those of $\overleftrightarrow{E}$, the ground state is determined by diagonalizing the interaction tensor $\overleftrightarrow{E}$.

In the ground state, the lattice points are divided into two-sublattices A and B according to the direction of the magnetizations, i.e., $\boldsymbol{S}_1$ and $\boldsymbol{S}_4$ ($\boldsymbol{S}_2$ and $\boldsymbol{S}_3$) belong A (B) sublattice. Using the sub-lattice magnetization vectors $\boldsymbol{M}_A$ and $\boldsymbol{M}_B$, the diagonalized form of $\overleftrightarrow{E}$ is:

$$\left(M_A^{X'}, M_A^{Y'}, M_A^{Z'}\right) \begin{pmatrix} J_1 & 0 & 0 \\ 0 & J_2 & 0 \\ 0 & 0 & J_3 \end{pmatrix} \begin{pmatrix} M_B^{X'} \\ M_B^{Y'} \\ M_B^{Z'} \end{pmatrix}, \tag{1.76}$$

where

$$J_1 = J - \Gamma_{xy} + \frac{1}{2}\Gamma^s, \tag{1.77}$$

$$J_2 = \frac{1}{2}\Gamma_{zx} - \Gamma_{xy} + \sqrt{\frac{1}{4}d^2 + \left[J - \frac{1}{2}(\Gamma_{zx} + \Gamma^s)\right]^2}, \tag{1.78}$$

$$\tag{1.79}$$

and

$$J_3 = -\frac{1}{2}\Gamma_{zx} + \Gamma_{xy} + \sqrt{\frac{1}{4}d^2 + \left[J - \frac{1}{2}(\Gamma_{zx} + \Gamma^s)\right]^2}. \tag{1.80}$$

The $X'Y'Z'$-coordinate systems depending on A and B are defined as follows:

$$\begin{pmatrix} M_A^{X'} \\ M_A^{Y'} \\ M_A^{Z'} \end{pmatrix} = \begin{pmatrix} 1 & 0 & 0 \\ 0 & \cos\theta & -\sin\theta \\ 0 & \sin\theta & \cos\theta \end{pmatrix} \begin{pmatrix} M_A^X \\ M_A^Y \\ M_A^Z \end{pmatrix}, \tag{1.81}$$

and

$$\begin{pmatrix} M_B^{X'} \\ M_B^{Y'} \\ M_B^{Z'} \end{pmatrix} = \begin{pmatrix} 1 & 0 & 0 \\ 0 & \cos\theta & \sin\theta \\ 0 & -\sin\theta & \cos\theta \end{pmatrix} \begin{pmatrix} M_B^X \\ M_B^Y \\ M_B^Z \end{pmatrix}, \tag{1.82}$$

respectively, and

$$\tan 2\theta = \frac{\frac{1}{2}d}{J - \frac{1}{2}(\Gamma_{zx} + \Gamma^s)}. \tag{1.83}$$

Let us examine the role of the magnetic anisotropies. When the tilt angle δ is zero, $d = \Gamma^s = 0$. Then, $\theta = 0$ is obtained. In this case, $J_1 = J_2 = J - \Gamma_{xy}$ and $J_3 = J + \Gamma_{xy} - \Gamma_{zx}$. The direct exchange leads Γ_{xy}, $\Gamma_{zx} < 0$, and $|\Gamma_{xy}| > |\Gamma_{zx}|$ due to the distortion of the CuO_6 octahedron. Therefore, $J_1 = J_2 > J_3$, i.e., the anisotropy due to the direct exchange mechanism confines the magnetizations M_A and M_B to the XY plane. In that plane, however, the mechanism does not give any anisotropy.

The magnetic anisotropy in the XY plane is determined by d and Γ^s via the superexchange mechanism. When $J_2 > J_1$, the magnetizations M_A and M_B point along the Y' axis, i.e. the weak ferromagnetism appears with canting angle θ in (1.83). On the other hand, when $J_2 < J_1$ the magnetizations are along X $(= X')$ axis with uniaxial anisotropy and $\theta = 0$. With the limitations discussed in connection with (1.52), the anisotropy is expressed as $\Gamma_s = d^2/(4J)$, and implies that $J_1 - J_2 \simeq \Gamma^s/2$, so that the magnetizations M_A and M_B point along the X axis.

Reality is even more complicated. Detailed studies of the magnetic anisotropy of a distorted CuO_2 plane have been made by several authors [33–49]. Stein et al. [45] have shown that antiferromagnetism with uniaxial anisotropy along the X axis is stabilized in the distorted CuO_2 plane shown in Fig. 1.24(iii) due to the pseudo-dipolar interaction. Tsukada et al. [49] have investigated the magnetic susceptibility of $(La_{1.8}Eu_{0.2})CuO_4$ and found that the experimental data are explained by the theory of Stein et al. [45]. These studies illustrate the importance of the pseudo-dipolar interaction.

Zheludev et al. [50,51] have suggested both the Dzyaloshinski-Moriya and the pseudo-dipolar interactions are required to explain the neutron scattering experiments on the spin-1/2 system $Ba_2CuGe_2O_7$.

It is important to remember, as was discussed in connection with (1.44), that the *both* the Dzyaloshinski-Moriya *and* pseudo-dipolar interactions must be accounted for in any discussion of the anisotropy arising from the exchange.

References

1. For example, L. T. Schiff: *Quantum Mechanics* (Mcgraw Hill Book Co. Inc., 1955)
2. Y. Tokura: *Physics of High Temperature Superconductors*, eds. S. Maekawa, M. Sato (Springer Series in Solid State Sciences 106, 1992) p.191.
3. J. Kondo: J. Phys. Soc. Jpn. **58**, 2884 (1989)
4. Y. Ohta, T. Tohyama, S. Maekawa: Phys. Rev. B**43**, 2968 (1991)
5. S. Maekawa, Y. Ohta, T. Tohyama: *Physics of High Temperature Superconductors*, eds. S. Maekawa, M. Sato (Springer Series in Solid State Sciences 106, 1992) P.29.
6. S. Ishihara, S. Okamoto, S. Maekawa: J. Phys. Soc. Jpn. **66**, 2965 (1997)
7. J. Hubbard: Proc. Roy. Soc. A276, 238 (1963) and A**277**, 237 (1964)
8. P. W. Anderson: Phys. Rev. **115**, 2 (1959) and Adv. in Phys. **46**, 3 (1997)
9. J. Kanamori: J. Phys. Chem. Solids **10**, 87 (1959)
10. J. B. Goodenough: Phys. Rev. **100**, 564 (1955)
11. J.Zaanen, G. Sawatzky, J. W. Allen: Phys. Rev. Lett. **55**, 418 (1985)
12. Y. Ohta, T. Tohyama, S. Maekawa: Phys Rev. Lett. **66**, 1228 (1991)
13. D. B. McWhan, A. Menth, J. P. Remeika, W. F. Brinkman, T. M. Rice: Phys. Rev. B**7**, 1920 (1973)
14. K. I. Kugel, D. I. Khomskii: Sov. Phys. JETP **37**, 725 (1973)
15. S. Ishihara, J. Inoue, S. Maekawa: Physica C **263**, 130 (1996) and Phys Rev. B**55**, 8280 (1997)
16. G. H. Jonker, J. H. Santen: Physica **16**, 337 (1950)
17. C. Zener: Phys. Rev. **82**, 403 (1951)
18. P.W. Anderson, H. Hasegawa: Phys. Rev. **100**, 675 (1955)
19. P.G. de Gennes: Phys. Rev. **118**, 141 (1960)
20. M. Yamanaka, W. Koshibae, S. Maekawa: Phys. Rev. Lett. **81**, 5604 (1998)
21. J. Ye, Y.B. Kim, A.J. Millis, B.I. Shraiman, P. Majumdar, Z. Tesanovic: Phys. Rev. Lett. **83**, 3737 (1999)
22. K. Ohgushi, S. Murakami, N. Nagaosa: Phys. Rev. B**62**, R6065 (2000)
23. Y. Taguchi, Y. Oohara, H. Yoshizawa, N. Nagaosa, Y. Tokura: Science **291**, 2573 (2001)
24. J. Rodríguez-Carvajal, J. L. Martínez, J. Pannetier R. Saez-Puche: Phys. Rev. B **38**, 7148 (1988)
25. J. Rodríguez-Carvajal, M. T. Fernández-Dáz J. L. Martínez: J. Phys.: Condens. Mattter **3**, 3215 (1991)

26. K. Yamada, M. Arai, Y. Endoh, S. Hosoya, K. Nakajima, T. Perring, A. Taylor: J. Phys. Soc. Jpn. **60** 1197 (1991)
27. K. Yamada, T. Omata, K. Nakajima, S. Hosoya, T. Sumida Y. Endoh: Physica C **191** 15 (1992); S. Hosoya, T. Omata, K. Nakajima, K. Yamada, Y. Endoh: *ibid* **202**, 188 (1992)
28. K. Nakajima, K. Yamada, S. Hosoya, T. Omata, Y. Endoh: J. Phys. Soc. Jpn. **62**, 4438 (1993); K. Nakajima: Thesis, Tohoku University (1992)
29. I. Dzyaloshinsky: J. Phys. Chem. Solids **4**, 241 (1958)
30. T. Moriya: Phys. Rev. **120**, 91 (1960)
31. T. Moriya: Magnetism Vol. I, pp. 85, Weak Ferromagnetism Eds. G. T. Rado, H. Suhl Academic Press (1963)
32. T. A. Kaplan: Z. Phys. B**49**, 313 (1983)
33. D. Coffey, K.S. Bedell, S.A. Trugman: Phys. Rev. B **42**, 6509 (1990)
34. D. Coffey, T.M. Rice, F.C. Zhang: Phys. Rev. B **44**, 10 112 (1991); *ibid.* **46**, 5884(E) (1992)
35. L. Shekhtman, O. Entin-Wohlman, A. Aharony: Phys. Rev. Lett. **69**, 836 (1992)
36. L. Shekhtman, O. Entin-Wohlman, A. Aharony: Phys. Rev. B**47**, 174 (1993)
37. N.E. Bonesteel: Phys. Rev. B **47**, 11 302 (1993)
38. W. Koshibae, Y. Ohta, S. Maekawa: Phys. Rev. B**47**, 3391 (1994); and *ibid.* **48**, 3580(E) (1993)
39. W. Koshibae, Y. Ohta, S. Maekawa: Phys. Rev. Lett. **71**, 467 (1993)
40. L. Shekhtman, O. Entin-Wohlman, A. Aharony: Phys. Rev. Lett. **71**, 468 (1993)
41. W. Koshibae, Y. Ohta, S. Maekawa: Phys. Rev. B **50**, 3767 (1994)
42. T. Yildirim, A.B. Harris, O. Entin-Wohlman, A. Aharony: Phys. Rev. Lett. **72**, 3710 (1994)
43. H.E. Viertio, N.E. Bonesteel: Phys. Rev. B **49**, 6088 (1994)
44. T. Yildirim, A.B. Harris, A. Aharony, O. Entin-Wohlman: Phys. Rev. B **52**, 10 239 (1995)
45. J. Stein, O. Entin-Wohlman, A. Aharony: Phys. Rev. B **53**, 775 (1996)
46. S. Shamoto, T. Kiyokura, M. Sato, K. Kakurai, Y. Nakamura, S. Uchida: Physica C **203**, 7 (1992)
47. B. Keimer, R.J. Birgeneau, A. Cassanho, Y. Endoh, M. Greven, M.A. Kastner, G. Shirane: Z. Phys. B: Condens. Matter **91**, 373 (1993)
48. M.K. Crawford, R.L. Harlow, E.M. McCarron, W.E. Farneth, N. Herron, H. Chou, D.E. Cox: Phys. Rev. B **47**, 11 623 (1993)
49. I. Tsukada, X. F. Sun, Seiki Komiya, A. N. Lavrov, Yoichi Ando: Phys. Rev. B **67**, 224401 (2003)
50. A. Zheludev, S. Maslov, I. Tsukada, I. Zaliznyak, L.P. Regnault, T. Masuda, K. Uchinokura, R. Erwin, G. Shirane: Phys. Rev. Lett. **81**, 5410 (1998)
51. A. Zheludev, S. Maslov, G. Shirane, I. Tsukada, T. Masuda, K. Uchinokura, I. Zaliznyak, R. Erwin, L.P. Regnault: Phys. Rev. B **59**, 11 432 (1999)

2. Cuprates

In 1986, Bednorz and Müller [1] discovered high-temperature (high-T_c) superconductivity in a copper oxide La-Ba-Cu-O. The superconductivity occurred on a two-dimensional plane composed of CuO_2 units when a small number of hole carriers was introduced. Soon after this discovery, the importance of strong electron correlation in the CuO_2 plane was pointed out [2]. These correlations are manifest by the fact that the CuO_2 planes without extra carriers are a Mott insulator. The emerging physics which has to be elucidated is therefore that of a carrier-doped Mott insulator. In order to solve the problem and clarify the mechanism of high-T_c superconductivity, we must, first of all, know the details of the electronic structures of the cuprates. This will be examined in this chapter. In particular, the electronic structures and magnetic properties of the parent insulating cuprates are discussed on the basis of a multiband model that contains relevant orbitals on Cu and O atoms. Furthermore, a procedure that transforms the multiband model to a single band model will be shown in detail. The single band model is found to be adequate for describing the low-energy physics of carrier-doped cuprates as well as the insulating equivalents.

In strongly correlated one-dimensional electron systems, it is known that the low-energy excitations are not quasi-particles with charge $-e$ and spin $1/2$, as in an ordinary Fermi liquid, but are rather decoupled collective modes of spin and charge. In other words, an electron or a hole introduced into the one-dimensional system decays into spin and charge degrees of freedom (the spin-charge separation). The appearance of this separation and its effects on various physical properties in one-dimensional cuprates will be discussed in relation to existing experimental results.

High-T_c superconductivity is obtained by introducing not only holes but also electrons into the two-dimensional insulating cuprates. The differences in the phase diagram, as well as various physical quantities, for both doping directions are discussed on the basis of the electronic states of the insulating cuprates. In the discussion, the importance of long-range hopping of the carriers beyond the nearest-neighbor lattice sites is emphasized. A discussion of the mechanism for high-T_c superconductivity will be given in Chap. 3.

2.1 Underlying Electronic Structure of Cuprates

The copper oxides discussed in this chapter invariably contain CuO_4 units that are linked to form a variety of Cu-O networks, where the Cu-O distance is usually in the range of 1.89 to 1.99 Å. The cuprates are usefully classified as being either one-dimensional (1D), two-dimensional (2D), or ladder systems. Representative materials are shown in Fig. 2.1. The material Sr_2CuO_3 is an example of the 1D compounds, where the corners of the CuO_4 units are shared forming a straight Cu-O-Cu bond (corner-sharing). The 1D corner-sharing chain compounds, Sr_2CuO_3 and Ca_2CuO_3, are known to be magnetically ideal 1D quantum spin chains based on the magnetic susceptibility measurements [3]. In $SrCuO_2$, two corner-sharing chains are combined, forming a zigzag-type double chain where the CuO_4 units in the two chains share their edges. In spite of the double chain structure, electronic and magnetic properties are very similar to those of the single chain compounds.

Shown in Fig. 2.1b is Li_2CuO_2 which is also an 1D material but for which the edges of the CuO_4 units are shared with almost 90-degree Cu-O-Cu bond angle (edge-sharing). This type of network is seen in other materials such as $CuGeO_3$ (the first known inorganic spin-Peierls compound [4]), $Ca_2Y_2Cu_5O_{10}$ and $La_6Ca_8Cu_{24}O_{41}$.

A 2D network of corner-sharing CuO_4 is realized in high-T_c cuprate superconductors to form a CuO_2 plane. An example of the high-T_c cuprate, $La_{2-x}Sr_xCuO_4$, is shown in Fig. 2.1c. Here the CuO_4 units are accompanied by additional two oxygens below and above (apical oxygens) forming CuO_6 octahedra elongated along the apical oxygens (Jahn-Teller distortion). The CuO_2 planes with only one apical oxygen (pyramidal coordinate) are seen in $YBa_2Cu_3O_{7-\delta}$. The electron-doped high-T_c cuprate, $Nd_{2-x}Ce_xCuO_4$, has no apical oxygen.

The ladder cuprates comprise of more than a single corner-sharing chain. For two-leg ladder compounds, the two chains are linked sharing the corner of the CuO_4 units. With increasing numbers of chains, the ladder approaches a 2D system. Representative two-leg ladder compounds are $SrCu_2O_3$ shown in Fig. 2.1d, $LaCuO_{2.5}$, and $(Sr,Ca)_{14}Cu_{24}O_{41}$. Except for $LaCuO_{2.5}$, adjacent two-leg ladders are connected, sharing the edges of the CuO_4 units, similar to the zigzag double chain structure. Of these materials $(Sr,Ca)_{14}Cu_{24}O_{41}$ becomes superconducting under high pressure [5].

2.1.1 Charge-Transfer-Type Insulators

Let us consider the electronic structure of cuprates which is independent of the dimensionality as well as the structure of the Cu-O network. For the moment we neglect the hybridization effect among the copper and oxygen orbitals, that is, we consider a simplified ionic model where the direct overlap between ions of the same kind as well as the covalency between Cu and O ions is not included. In the undoped cuprates which are insulators, the

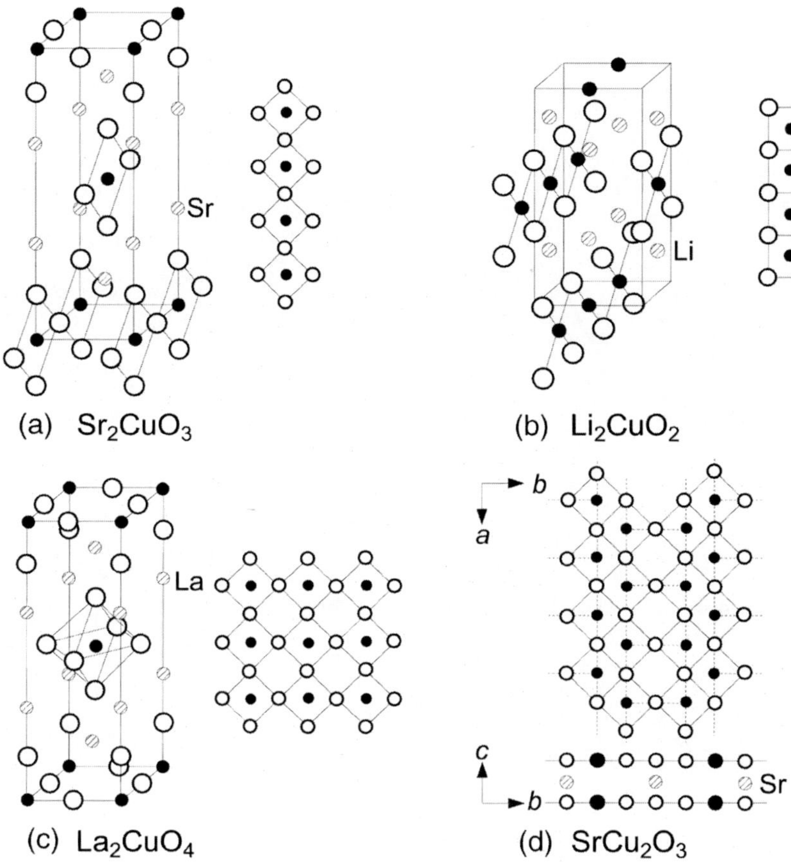

Fig. 2.1. Crystal structures of copper oxides with different dimensionality. (**a**) Sr_2CuO_3 and (**b**) Li_2CuO_2 for 1D, (**c**) La_2CuO_4 for 2D, and (**d**) $SrCu_2O_3$ for ladder. Schematics of the Cu-O networks are also shown in (**a**), (**b**), and (**c**).

formal valence of Cu atom is 2+ with a single hole in $3d$ orbitals, i.e., a $3d^9$ configuration. Due to the crystal field, the hole is located in $3d_{x^2-y^2}$ orbital (see Chap. 1). The hole is localized on the Cu ion because of strong on-site Coulomb interaction (U_d) that comes from localized nature of Cu $3d$ orbitals. There is thereby a localized spin on each Cu ion. The oxygen ion has, of course, a closed shell with a $2p^6$ configuration.

Here, we consider two processes that are associated with insulator gap formation [6]: The first possibility is that one of the localized holes on Cu ion is moved to one of the other Cu ions, while other process has the localized hole moving to one of the oxygen ions. Both are schematically shown in Fig. 2.2. In the former case, two $3d^9$ configurations change into $3d^{10}$ and $3d^8$. The resulting total energy change, which is given by the sum of the energy

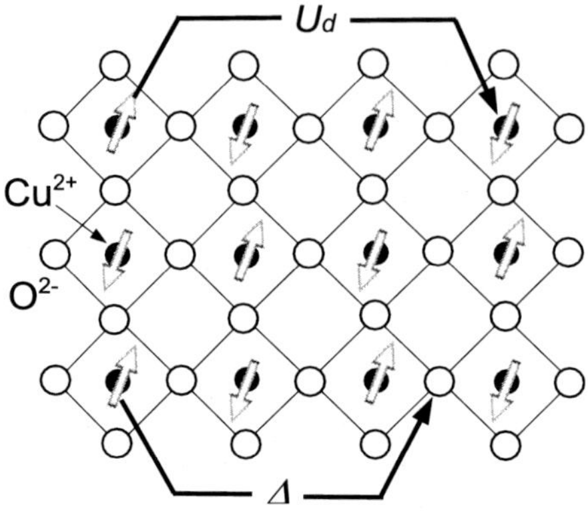

Fig. 2.2. A schematic representation of CuO_2 plane consisting of Cu^{2+} ion ($3d^9$) and closed shell O^{2-} ions. The arrows denote spin of a localized hole in a Cu ion. The two processes to transfer the localized hole to other sites are indicated: one is related to the Coulomb interaction energy on Cu ion U_d, and the other is the charge transfer energy Δ.

required to change $3d^9$ to $3d^{10}$ and that to $3d^8$, is nothing but the definition of the Coulomb interaction U_d. The latter process is of a charge transfer type from cation to anion and the energy for the hole to be transferred is called the charge transfer energy Δ. It was found from the analysis of high-energy spectroscopy spectra such as photoemission that in most of cuprates and nickelates $U_d > \Delta$, while $U_d < \Delta$ for titanates and vanadates [7, 8]. These are respectively called the charge-transfer or the Mott-Hubbard insulators.

An explicit form of U_d in cuprates may be written within the ionic picture as

$$\begin{aligned} U_d &= \left(-I_{\mathrm{Cu}}(2) - V_{\mathrm{Cu}} - E_{\mathrm{Cu}}^{\mathrm{pol}}\right) + \left(I_{\mathrm{Cu}}(3) + V_{\mathrm{Cu}} - E_{\mathrm{Cu}}^{\mathrm{pol}}\right) \\ &= I_{\mathrm{Cu}}(3) - I_{\mathrm{Cu}}(2) - 2E_{\mathrm{Cu}}^{\mathrm{pol}}, \end{aligned} \tag{2.1}$$

where $I_{\mathrm{Cu}}(3)$ and $I_{\mathrm{Cu}}(2)$ are the third and second ionization energies of copper atom, respectively, and V_{Cu} is the Madelung potential for hole at the Cu site and which eventually cancels out. We note that the atomic Coulomb energy $U_{\mathrm{atom}} = I_{\mathrm{Cu}}(3) - I_{\mathrm{Cu}}(2)$ is reduced by the ionic polarization energy $E_{\mathrm{Cu}}^{\mathrm{pol}}$ induced by the removed or added hole at the Cu site. The value of U_{atom} is 16.5 eV, taking the experimental values of $I_{\mathrm{Cu}}(3) = 36.8$ eV and $I_{\mathrm{Cu}}(2) = 20.3$ eV for isolated atoms. So the value of $U_d \sim 8$ eV, estimated from the analysis of the photoemission data, may be reproduced if $E_{\mathrm{Cu}}^{\mathrm{pol}} \sim$ 4 eV. However, the precise estimate of the polarization energy and thus the value of U_d is a very difficult task and remains a problem to be solved.

According to the estimation of U_d, the charge transfer energy

$$\Delta = V_{\mathrm{O}} - V_{\mathrm{Cu}} + A_{\mathrm{O}}(2) - I_{\mathrm{Cu}}(2) - E_{\mathrm{O}}^{\mathrm{pol}} - E_{\mathrm{Cu}}^{\mathrm{pol}}, \tag{2.2}$$

where $A_{\mathrm{O}}(2)$ is the second electron affinity of an oxygen atom. In contrast with U_d, Δ is dependent on the Madelung potential. A linear relationship is evident in Fig. 2.3, where the energy gaps E_{g} determined by the optical absorption measurements for high-T_{c} parent compounds are plotted against the Madelung potential difference $\Delta V_{\mathrm{M}} = V_{\mathrm{O}} - V_{\mathrm{Cu}}$. Since E_{g} is predominantly

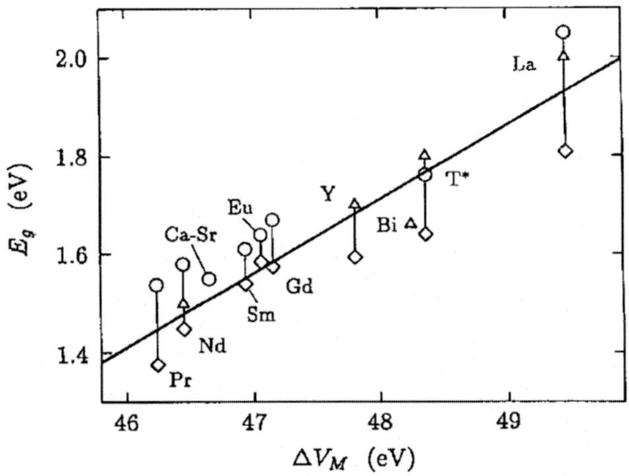

Fig. 2.3. Correlation between the observed energy gap E_g for 2D insulating cuprates and the Madelung potential difference ΔV_M calculated for various compounds. Collected are the data for L_2CuO_4 (L=Pr, Nd, Sm, Eu, and Gd), La_2CuO_4, T^* phase $(La\text{-}Ga\text{-}Sr)_2CuO_4$ and $(La\text{-}Tb\text{-}Sr)_2CuO_4$, $(Ca\text{-}Sr)CuO_2$, $Bi_2Sr_2YCu_2O_8$, and $YBa_2Cu_3O_6$. The solid line is a guide to the eye. (From [9])

governed by Δ, this relationship is reasonable. Although the expression for Δ in (2.2) looks simple, the precise determination of Δ from simple electronic structure arguments is as difficult as for U_d. Since ΔV_M is around 48 eV, a roughly estimate of Δ is 10 eV, this taking the standard value of $A_O(2) \sim -10$ eV (the precise value is unknown) and $E^{pol}_{Cu,O} \sim 4$ eV obtained above. As the value of Δ is experimentally known to be around 3 to 4 eV, this model overestimates Δ by a factor of three. This may be an indication of the overestimate of the Madelung potential due to the exclusion of the effect of covalency. However, if one is interested in systematics such as that illustrated in Fig. 2.3, the use of the Madelung potential with the formal valence may still be useful, provided that phenomenological corrections are made in the determination of Δ [9–11].

2.1.2 Model Hamiltonian

Up to now we have neglected hybridization effects. In order to define a valid model Hamiltonian for the electronic states of cuprates we must introduce hybridization terms. The most important orbital on oxygen relevant to the electronic states are the $2p_\sigma$ orbitals whose wave function has the largest overlap with Cu $3d_{x^2-y^2}$ orbitals. These have either $2p_x$ or $2p_y$ character. Figure 2.4 shows the orbitals of the CuO_4 unit. We remark that the orbitals are not simple atomic orbitals but are constructed to be orthogonal to each other, i.e., are Wannier orbitals [12]. Taking $3d^{10}$ and $2p^6$ configurations as

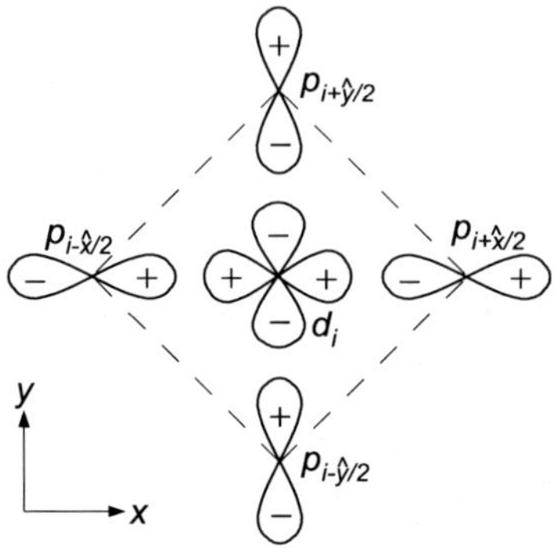

Fig. 2.4. Schematic diagram of the $3d_{x^2-y^2}$ (center), $2p_x$, and $2p_y$ orbitals for the CuO_4 unit. The signs + and − represent the phase of the orbitals.

the vacuums of the system, we introduce a minimum model for the CuO_4 network, including the energy levels of $3d_{x^2-y^2}$ and $2p_\sigma$ holes (ε_d and $\varepsilon_d+\Delta$, respectively), the Coulomb repulsion energy of holes on $3d_{x^2-y^2}$ (U_d), and on $2p_\sigma$ (U_p), the nearest-neighbor intersite Coulomb energy (U_{pd}), and the hopping energies of holes between nearest-neighbor $3d_{x^2-y^2}$ and $2p_\sigma$ orbitals (T_{pd}), and between nearest-neighbor $2p_\sigma$ orbitals (T_{pp}). The model is called the three-band Hubbard model (named for three orbitals in the CuO_2 unit), or the *d-p* model (named for the inclusion of *d* and *p* orbitals), otherwise the Emery model [13].

The Hamiltonian reads

$$H_{dp} = H_\varepsilon + H_U + H_T \tag{2.3}$$

with

$$H_\varepsilon = \varepsilon_d \sum_{i,\sigma} n^d_{i,\sigma} + (\varepsilon_d + \Delta) \sum_{i,\boldsymbol{\delta},\sigma} n^p_{i+\boldsymbol{\delta}/2,\sigma}, \tag{2.4}$$

$$\begin{aligned} H_U = {} & U_d \sum_i n^d_{i,\uparrow} n^d_{i,\downarrow} + U_p \sum_{i,\boldsymbol{\delta}} n^p_{i+\boldsymbol{\delta}/2,\uparrow} n^p_{i+\boldsymbol{\delta}/2,\downarrow} \\ & + U_{pd} \sum_{i,\boldsymbol{\delta},\sigma,\sigma'} n^d_{i,\sigma} n^p_{i+\boldsymbol{\delta}/2,\sigma'}, \end{aligned} \tag{2.5}$$

$$\begin{aligned} H_T = \sum_{i,\sigma} \Big[& T_{pd} d^\dagger_{i,\sigma} \left(p_{i+\boldsymbol{y}/2,\sigma} - p_{i+\boldsymbol{x}/2,\sigma} - p_{i-\boldsymbol{y}/2,\sigma} + p_{i-\boldsymbol{x}/2,\sigma} \right) \\ & + T_{pp} (p^\dagger_{i+\boldsymbol{y}/2,\sigma} p_{i+\boldsymbol{x}/2,\sigma} - p^\dagger_{i-\boldsymbol{x}/2,\sigma} p_{i+\boldsymbol{y}/2,\sigma} \\ & + p^\dagger_{i-\boldsymbol{y}/2,\sigma} p_{i-\boldsymbol{x}/2,\sigma} - p^\dagger_{i+\boldsymbol{x}/2,\sigma} p_{i-\boldsymbol{y}/2,\sigma}) + \text{h.c.} \Big], \end{aligned} \tag{2.6}$$

where the operator $d_{i,\sigma}$ annihilates holes with spin σ in $3d_{x^2-y^2}$ orbital at site i, and $n^d_{i,\sigma} = d^\dagger_{i,\sigma} d_{i,\sigma}$. Here the $p_{i\pm\boldsymbol{x}(\boldsymbol{y})/2,\sigma}$ are the hole annihilation operators for the $2p_x$ ($2p_y$) orbitals at site $i \pm \boldsymbol{x}(\boldsymbol{y})/2$, $\boldsymbol{x}$ ($\boldsymbol{y}$) being the vector connecting neighboring Cu ions along the x (y) directions, and $\boldsymbol{\delta}$ in (2.4) and (2.5) is either $\boldsymbol{x}$ or $\boldsymbol{y}$ and where $n^p_{i+\boldsymbol{\delta}/2,\sigma} = p^\dagger_{i+\boldsymbol{\delta}/2,\sigma} p_{i+\boldsymbol{\delta}/2,\sigma}$. The signs of the hopping terms in (2.6) are determined by the phase relations between two orbitals represented in Fig. 2.4: For hole hopping, the sign in front of T_{pd} and T_{pp} becomes + (−) when the phases of the orbitals facing each other are same (different). Hereafter the Cu-Cu distance is taken to be unity, i.e., $|\boldsymbol{x}| = |\boldsymbol{y}| = 1$.

If all of the Coulomb interactions are neglected in (2.5), i.e., H_U=0, the model is reduced to a non-interacting system with the tight-binding bands. The resulting Hamiltonian $H_\varepsilon + H_T$ can be easily diagonalized within the momentum space representation, after performing the following Fourier transformations of the orbitals

$$d_{\boldsymbol{k},\sigma} = \frac{1}{\sqrt{N}}\sum_i d_{i,\sigma}\mathrm{e}^{-\mathrm{i}\boldsymbol{k}\cdot\boldsymbol{R}_i}, \quad p_{x,\boldsymbol{k},\sigma} = \frac{1}{\sqrt{N}}\sum_i p_{i+\boldsymbol{x}/2,\sigma}\mathrm{e}^{-\mathrm{i}\boldsymbol{k}\cdot\boldsymbol{R}_i+\boldsymbol{x}/2},$$
$$p_{y,\boldsymbol{k},\sigma} = \frac{1}{\sqrt{N}}\sum_i p_{i+\boldsymbol{y}/2,\sigma}\mathrm{e}^{-\mathrm{i}\boldsymbol{k}\cdot\boldsymbol{R}_i+\boldsymbol{y}/2}, \tag{2.7}$$

where N is the total number of the Cu sites. For a 2D system with the CuO_2 plane, the tight-binding Hamiltonian $H_\varepsilon + H_T$ reads

$$\sum_{\boldsymbol{k},\sigma}\left(d^\dagger_{\boldsymbol{k},\sigma}, p^\dagger_{x,\boldsymbol{k},\sigma}, p^\dagger_{y,\boldsymbol{k},\sigma}\right)$$
$$\times\begin{pmatrix} 0 & -2\mathrm{i}T_{pd}\sin(k_x/2) & 2\mathrm{i}T_{pd}\sin(k_y/2) \\ 2\mathrm{i}T_{pd}\sin(k_x/2) & \Delta & 4T_{pp}\sin(k_x/2)\sin(k_y/2) \\ -2\mathrm{i}T_{pd}\sin(k_y/2) & 4T_{pp}\sin(k_x/2)\sin(k_y/2) & \Delta \end{pmatrix}$$
$$\times {}^t\left(d_{\boldsymbol{k},\sigma}, p_{x,\boldsymbol{k},\sigma}, p_{y,\boldsymbol{k},\sigma}\right) \tag{2.8}$$

where we take $\varepsilon_d = 0$. The dispersion relation is obtained by diagonalizing the 3×3 matrix in (2.8). Since there is one hole per copper site, the bonding band (in the electron picture this is an antibonding band) is half-filled by holes. The resulting dispersion relations are the starting point for approaches starting from the weak correlation limit.

In real materials with the CuO_2 planes, the approximate parameter values for the Hamiltonian (2.3) are believed to be $U_d \sim 8$, $U_p \sim 4$, $U_{pd} \sim 1$, $\Delta \sim 3$, $T_{pd} \sim 1$, and $T_{pp} \sim 0.5$. (For more precise values, refer to the original references [14–16].) In order to capture the essential physics of the model (2.3), it is necessary to retain non-zero values of *at least* U_d, Δ, and T_{pd}. Note that a perturbation expansion in terms of T_{pd}, whose value is smaller than U_d, U_p, and Δ, is usually used to clarify the low-energy physics of the model, as will be discussed in Sect. 2.1.5.

2.1.3 Superexchange Interaction

Corner-Sharing Cuprates

When the localized spins on the Cu $3d_{x^2-y^2}$ orbitals are anti-parallel, they interact with each other through the oxygens located between the Cu ions. This

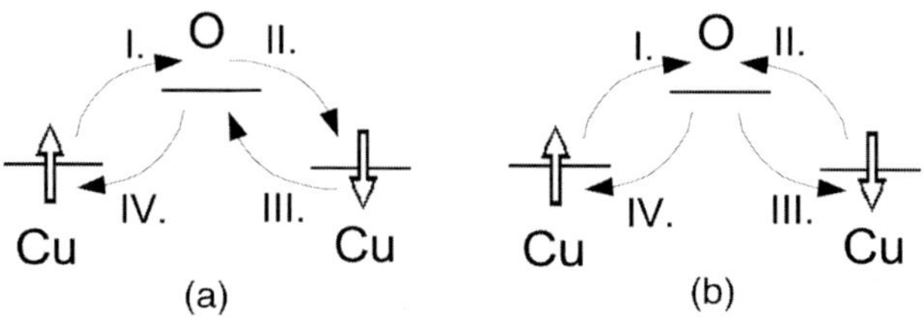

Fig. 2.5. Schematic diagrams of the fourth-order processes in terms of hole hopping T_{pd}, which give rise to the superexchange interaction between nearest-neighbor Cu spins. The process (**a**) contains an intermediate state with two holes on the copper atoms, while two holes are on oxygen in an intermediate state in the process (**b**).

interaction is called the *superexchange* or *kinetic exchange* interaction [17]. Two possible superexchange processes are schematically shown in Fig. 2.5. The process (a) involves an intermediate state with two holes on Cu and is the process considered in the original superexchange theory of Anderson [17]. The role of the process (b), where the two holes are on oxygen in the intermediate state, has been emphasized in connection with the Néel temperature of the transition-metal monoxides [6].

As shown in Fig. 2.5, a Cu-O-Cu (i.e., $3d_{x^2-y^2}$-$2p_x$-$3d_{x^2-y^2}$) cluster with two holes is sufficient to understand the fundamental physics of the superexchange. Here, we derive a fourth-order in T_{pd} expression for the superexchange interaction by constructing the Hamiltonian matrices for the singlet and triplet spaces. Starting from a state with two holes localized on the two coppers (d_1 and d_2 for the left and right site, respectively), one can generate other states by applying T_{pd}. For the singlet space, the Hamiltonian matrix is given by

$$\begin{pmatrix} 0 & \sqrt{2}T_{pd} & 0 & 0 \\ \sqrt{2}T_{pd} & \Delta + U_{pd} & -2T_{pd} & -\sqrt{2}T_{pd} \\ 0 & -2T_{pd} & 2\Delta + U_p & 0 \\ 0 & -\sqrt{2}T_{pd} & 0 & U_d \end{pmatrix}, \tag{2.9}$$

where the corresponding states are

$$|s_1\rangle = \frac{1}{\sqrt{2}}\left(d^\dagger_{2,\uparrow}d^\dagger_{1,\downarrow} + d^\dagger_{1,\uparrow}d^\dagger_{2,\downarrow}\right)|\text{vac}\rangle, \tag{2.10}$$

$$|s_2\rangle = \frac{1}{2}\left(p^\dagger_{\uparrow}d^\dagger_{1,\downarrow} + d^\dagger_{1,\uparrow}p^\dagger_{\downarrow} - d^\dagger_{2,\uparrow}p^\dagger_{\downarrow} - p^\dagger_{\uparrow}d^\dagger_{2,\downarrow}\right)|\text{vac}\rangle, \tag{2.11}$$

$$|s_3\rangle = p^\dagger_{\uparrow}p^\dagger_{\downarrow}|\text{vac}\rangle, \tag{2.12}$$

$$|s_4\rangle = \frac{1}{\sqrt{2}}\left(d^\dagger_{1,\uparrow}d^\dagger_{1,\downarrow} + d^\dagger_{2,\uparrow}d^\dagger_{2,\downarrow}\right)|\text{vac}\rangle, \tag{2.13}$$

while for the triplet space

$$\begin{pmatrix} 0 & \sqrt{2}T_{pd} \\ \sqrt{2}T_{pd} & \Delta + U_{pd} \end{pmatrix}, \tag{2.14}$$

with

$$|t_1\rangle = \frac{1}{\sqrt{2}} \left(d_{2,\uparrow}^\dagger d_{1,\downarrow}^\dagger - d_{1,\uparrow}^\dagger d_{2,\downarrow}^\dagger \right) |\text{vac}\rangle \,, \tag{2.15}$$

$$|t_2\rangle = \frac{1}{2} \left(p_\uparrow^\dagger d_{1,\downarrow}^\dagger - d_{1,\uparrow}^\dagger p_\downarrow^\dagger - d_{2,\uparrow}^\dagger p_\downarrow^\dagger + p_\uparrow^\dagger d_{2,\downarrow}^\dagger \right) |\text{vac}\rangle \,. \tag{2.16}$$

The superexchange interaction J for two nearest-neighbor spins is the energy difference between the triplet and singlet states, as is easily understood from a two-spin problem with the Hamiltonian $J\boldsymbol{S}_1 \cdot \boldsymbol{S}_2$ $(J > 0)$. Therefore, the superexchange interaction, J, in the Cu-O-Cu cluster is just the difference of the ground state energies in the singlet and triplet spaces. Since the first 2×2 matrix elements in the singlet space (2.9) are the same as the triplet matrix (2.14), the difference only comes from virtual processes via $|s_3\rangle$ and $|s_4\rangle$. The expression for J up to the forth order in T_{pd} is thus easily derived from the matrix (2.9), i.e.,

$$J = \frac{2T_{pd}^2}{(\varDelta + U_{pd})^2} \left(\frac{2T_{pd}^2}{U_d} + \frac{4T_{pd}^2}{2\varDelta + U_p} \right) . \tag{2.17}$$

This expression is very useful for a qualitative understanding of the correspondence between the value of J and the electronic structure parameters. Of course, if one diagonalizes the matrices numerically, one can find the *exact* value of J up to infinite order within the Cu-O-Cu cluster. Although it is possible to obtain higher-order perturbation expressions for J in T_{pd} and T_{pp} [18], getting convergence is actually a difficult for the realistic parameter region where $T_{pd}/\varDelta$ is not so small. In such a situation, numerical estimations of J for larger Cu-O clusters may be the most reliable approach.

The values of J for the corner-sharing cuprates are known to be strongly dependent on dimensionality. Their values can be determined from various experiments. In 1D corner-sharing cuprates, the values of J have been estimated to be ~ 0.19 eV from the analysis of the temperature dependence of the magnetic susceptibilities [3]. It is also possible to determine J from optical conductivity measurements [19], where phonon-assisted multimagnon absorption appears in the mid infrared region [20]. From the analysis of the absorption spectrum measured at $T = 32$ K, J for Sr_2CuO_3 has been reported to be 0.246 eV [21]. The discrepancy between the values estimated from the susceptibility and optical absorption may be understood if one takes into account the thermal lattice expansion which leads to a temperature dependence of J [22]. Actually, T_{pd} is proportional to the Cu-O distance with a power of about -3.5 [23], and so J is expected to be very sensitive to the thermal expansion. For 2D insulating cuprates such as La_2CuO_4, the values of J are smaller than those for the 1D corner-sharing compounds. There is no strong materials dependence [9], with values of J determined from the two-magnon Raman [24] and neutron scattering experiments [25] in the range 0.1$\sim$0.16 eV.

Here let us consider the origin of the dependence of J on dimensionality. Supposing that the Coulomb energies U_d and U_p are material independent, so that any material and dimensionality dependence of J must come from T_{pd} and/or Δ according to (2.17). However, a change in Δ cannot be the origin of the larger J in 1D, because the charge-transfer gap experimentally observed in the 1D cuprates is almost the same as, or slightly larger, than in 2D. Furthermore, if T_{pd} and T_{pp} are assumed to be only dependent on the bond-length as usual [23], then one cannot obtain large J values in 1D because there is no systematic change of the bond-length between 1D and 2D cuprates. Therefore, the different magnitude of J for the 1D and 2D cuprates was puzzling. This has been studied by Mizuno et al. [26]. They have found that the changes in hopping matrix elements T_{pd} and T_{pp} reflect changes in the Madelung potential around Cu-O and O-O bonds. As a result, J depends on the crystal structure through the modification of T_{pd} and T_{pp}. These findings explain the larger value of J in 1D. We will examine this in some detail.

Given a Cu $3d$ Wannier orbital, ϕ_d, at $\boldsymbol{R}_{\mathrm{Cu}}$ and an adjacent O $2p$ Wannier orbital, ϕ_p, at $\boldsymbol{R}_{\mathrm{O}}$, T_{pd} is defined as

$$T_{pd} = \int \phi_p^*(\boldsymbol{r} - \boldsymbol{R}_{\mathrm{O}}) H \phi_d(\boldsymbol{r} - \boldsymbol{R}_{\mathrm{Cu}}) \mathrm{d}\boldsymbol{r}, \tag{2.18}$$

where $H = -\hbar^2/(2m)\nabla^2 + U(\boldsymbol{r})$, and where U is the periodic potential in the crystal. This may be expressed as $U = U_{\mathrm{atom}} + U_{\mathrm{M}}$, where U_{atom} is given by the linear combination of the atomic potentials at the positions of ions, while U_{M} is the Madelung potential determined by the spatial distribution of ions around the position $\boldsymbol{r}$: $U_{\mathrm{M}}(\boldsymbol{r}) = \sum_i V_i(\boldsymbol{r} - \boldsymbol{R}_i)$ with $V_i(\boldsymbol{r} - \boldsymbol{R}_i) = Z_i e^2 / |\boldsymbol{r} - \boldsymbol{R}_i|$ and where Z_i is the valence of an ion at site i. Often solid-state texts do not give and explicit expression for $U_{\mathrm{M}}(\boldsymbol{r})$. Using these expressions, we obtain

$$T_{pd} = T_{pd}^0 + T_{pd}^{\mathrm{M}}, \tag{2.19}$$

with

$$T_{pd}^0 = \int \phi_p^*(\boldsymbol{r} - \boldsymbol{R}_{\mathrm{O}}) \left(-\frac{\hbar^2}{2m}\nabla^2 + U_{\mathrm{atom}}(\boldsymbol{r}) \right) \phi_d(\boldsymbol{r} - \boldsymbol{R}_{\mathrm{Cu}}) \mathrm{d}\boldsymbol{r}, \tag{2.20}$$

$$T_{pd}^{\mathrm{M}} = \int \phi_p^*(\boldsymbol{r} - \boldsymbol{R}_{\mathrm{O}}) U'_{\mathrm{M}}(\boldsymbol{r}) \phi_d(\boldsymbol{r} - \boldsymbol{R}_{\mathrm{Cu}}) \mathrm{d}\boldsymbol{r}, \tag{2.21}$$

where in $U'_{\mathrm{M}}(\boldsymbol{r}) = \sum_i' V_i(\boldsymbol{r} - \boldsymbol{R}_i)$ the summation excludes the two sites between which T_{pd} acts. The exclusion of these sites in the summation removes a divergence due to the Coulomb potential of there associated point charges in the integration in (2.21). The excluded terms are almost independent of the dimensionality of Cu-O network because of their local nature and are thus considered to give a constant shift to the hoppings. For the right hand side of (2.20), we use the two-center approximation as usual [27]. Therefore, the T_{pd}^0 term depends only on the distance $d_{\mathrm{Cu-O}} = |\boldsymbol{R}_{\mathrm{Cu}} - \boldsymbol{R}_{\mathrm{O}}|$ and thus an empirical estimate of its value is basically possible, this assuming a functional form, e.g., $d_{\mathrm{Cu-O}}^{-3.5}$ [23], for the distance dependence. In contrast, T_{pd}^{M} is dependent not only on $d_{\mathrm{Cu-O}}$ but also on the crystal structure via U'_{M}. The dimensionality dependence of J in the corner-sharing cuprates is thus predominantly caused by the variation of U'_{M} reflecting the spatial distribution of ions around the CuO_4 units.

The precise determination of T^{M}_{pd} is not easy because we have to know the actual wave functions of the Wannier orbitals, ϕ_d and ϕ_p, and the absolute magnitude of the Madelung potentials U'_{M} in which the effect of covalency should be taken into account. In the reference [26], an attempt to evaluate T^{M}_{pd} has been made. The values of T^{M}_{pd} as well as T^{M}_{pp} in 1D cuprates are found to be different from those in 2D ones, resulting in larger T_{pd} and T_{pp} and thus larger J in 1D. This difference mainly comes from the difference of the stacking patterns of the Cu-O network: in 2D the CuO_2 planes stack along the c axis with block layers inserted between the planes, whereas in 1D the Cu-O chains stack along the b-axis without any layers between the chains. Such structural differences give rise to significant differences in the spatial distribution of the Madelung potentials U'_{M} around the Cu-O and O-O bonds. For the ladder cuprates this mechanism leads to anisotropic exchange interactions along the leg and rung directions which are consistent with the inelastic neutron data for $La_6Ca_8Cu_{24}O_{41}$ [28]. It is found that J along a leg is larger that along a rung. (In the analysis, the importance of a four-spin interaction was also emphasized, which will be discussed in Sect. 2.1.4).

Edge-Sharing Cuprates

In contrast to the corner-sharing cuprates, the angle of the Cu-O-Cu bond in edge-sharing cuprates is nearly 90 degree. However, the angle is actually material dependent. This angle θ in Li_2CuO_2 is 94°, which is smaller than that of $CuGeO_3$ ($\theta = 99°$) but larger than that of $La_8Ca_6Cu_{24}O_{41}$ ($\theta = 91°$). Let us consider how such differences in θ affect the magnetic interactions in the edge-sharing systems whose orbital configurations are shown in Fig. 2.6c.

In the edge-sharing systems, the hopping processes I and III shown in Fig. 2.5a between Cu $3d$, left, and O $2p$-orbitals are as strong as those in the corner-sharing structure, but the processes II and IV are severely suppressed due to almost orthogonal geometry of the $2p$ and Cu $3d$ orbitals on the right-hand side (see Fig. 2.6c). If we assume $\theta = 90°$ and ignore O-O hoppings, the superexchange interaction via oxygen orbitals completely disappears. Therefore, the value of J is in general significantly reduced in edge-sharing cuprates.

As a result of the suppression of the superexchange processes, ferromagnetic contributions become relatively important and dominate the net magnetic interactions. There are two main mechanisms for ferromagnetic interactions: (i) the direct exchange interactions, of the order of 0.1 eV, between the Cu $3d$ and O $2p$ orbitals in the intermediate state of the second-order process of d-p hybridization (Fig. 2.6a), and (ii) a Hund's coupling process where the Hund coupling at O ions arranges two spins ferromagnetically in

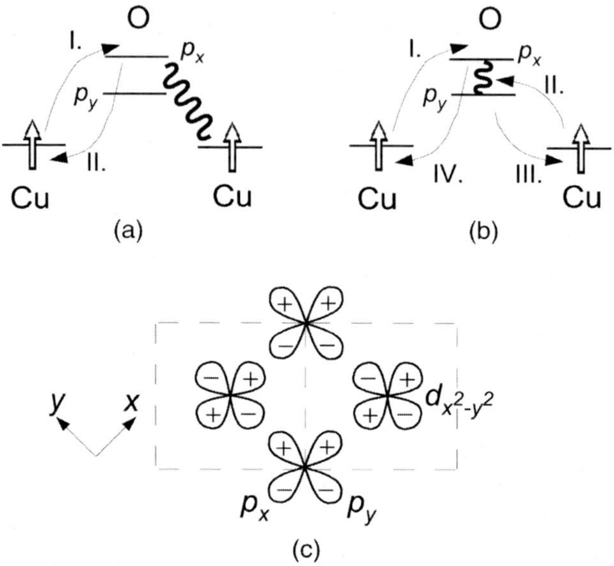

Fig. 2.6. Schematic diagrams of the processes that give rise to ferromagnetic interactions between nearest-neighbor Cu spins in edge-sharing cuprates. Two oxygen orbitals are located in the middle of the Cu ions. The process (**a**) is of the second order and contains the direct exchange interaction represented by the wavy line between O and Cu in the intermediate state, while the process (**b**) shows the fourth-order process with the Hund coupling represented by the wavy line at O ions in the intermediate state. (**c**) Schematic diagram of orbitals in the edge-sharing Cu_2O_2 unit: $3d_{x^2-y^2}$ at Cu site and two $2p_\sigma$ orbitals at O site. The signs + and − represent the phase of the orbitals.

the intermediate state of a fourth-order processes (Fig. 2.6b). Reflecting the order of *d*-*p* hybridization, the mechanism (i) is stronger than (ii). Including all the processes mentioned above, the θ dependence of the nearest-neighbor magnetic interaction J for the edge-sharing insulating cuprates has been examined and the critical angle where J changes from antiferromagnetic to ferromagnetic has been estimated to be $\theta_c \sim 95°$ [11].

Another magnetic interaction we must take into account is a next nearest-neighboring magnetic interaction J'. This is a superexchange interaction coming from Cu-O-O-Cu paths and its magnitude is the same order of J [11]. The effects of J' on the magnetic properties are discussed in connection with the spin-Peierls transition in $CuGeO_3$ [29]. Therefore, an appropriate model describing the edge-sharing chain is a spin 1/2 Heisenberg J-J' model. In real materials the interchain interactions may also play a crucial role in the magnetic properties [30].

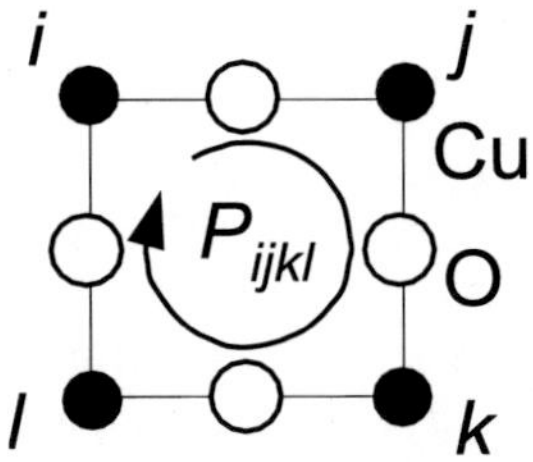

Fig. 2.7. A schematic diagram of the four-spin cyclic permutation.

2.1.4 Cyclic Four-Spin Interaction

The magnetic coupling we have discussed so far is two-body Heisenberg-type interactions coming from spin exchanges between two Cu ions. Many-body interactions also exist [31]. Here we will introduce cyclic or ring exchange of the spins. These higher order processes involving at least of eight powers of the d-p hybridization interaction T_{pd}. Such higher-order interactions might be expected to be small. However, the presence of four-spin cyclic interactions has been confirmed in inelastic neutron scattering measurements [28, 32].

Let us start with the three-spin cyclic interaction corresponding to the cyclic permutation of spins on three Cu ions forming a triangle. The Hamiltonian for the three spins at the i, j, and k sites is proportional to $\left(P_{ijk} + P_{ijk}^{-1}\right)$, where P_{ijk} represents the three-spin cyclic permutation operator, which can be written as the product of two two-spin permutation operators: $P_{ijk} = P_{ij}P_{ik} = 1/4\,(1 + 4\boldsymbol{S}_i \cdot \boldsymbol{S}_j)\,(1 + 4\boldsymbol{S}_i \cdot \boldsymbol{S}_k)$. Using the identity $4\,(\boldsymbol{S}_i \cdot \boldsymbol{S}_j)\,(\boldsymbol{S}_i \cdot \boldsymbol{S}_k) = 2\boldsymbol{S}_j \cdot \boldsymbol{S}_k + 3\mathrm{i}\boldsymbol{S}_i \cdot (\boldsymbol{S}_j \times \boldsymbol{S}_k)$, we obtain $P_{ijk}+P_{ijk}^{-1} = 1/2+2\,(\boldsymbol{S}_i \cdot \boldsymbol{S}_j + \boldsymbol{S}_j \cdot \boldsymbol{S}_k + \boldsymbol{S}_k \cdot \boldsymbol{S}_i)$, thereby reducing the three-spin exchange to a sum of two-spin exchanges. Thus the three-spin contributions can be simply absorbed into the usual two-spin exchange interactions upon adjusting the exchange parameters.

In contrast to the three-spin term, the four-spin cyclic interaction can not be reduced to the summation of two-spin terms. The four-spin interaction is given by the cyclic permutation of four Cu spins at the i, j, k, and l sites forming a plaquette as schematically depicted in Fig. 2.7. The Hamiltonian for the four spins is given by $P_{ijkl} + P_{ijkl}^{-1}$, where P_{ijkl} represents the four-spin cyclic permutation operator. Using $P_{ijkl} = P_{ijk}P_{il}$ and operator relations mentioned above, we find, after some algebra,

$$
\begin{aligned}
P_{ijkl} + P_{ijkl}^{-1} = 4[&(\boldsymbol{S}_i \cdot \boldsymbol{S}_j)(\boldsymbol{S}_k \cdot \boldsymbol{S}_l) + (\boldsymbol{S}_i \cdot \boldsymbol{S}_l)(\boldsymbol{S}_j \cdot \boldsymbol{S}_k) \\
&- (\boldsymbol{S}_i \cdot \boldsymbol{S}_k)(\boldsymbol{S}_j \cdot \boldsymbol{S}_l)] \\
&+ [(\boldsymbol{S}_i \cdot \boldsymbol{S}_j) + (\boldsymbol{S}_j \cdot \boldsymbol{S}_k) + (\boldsymbol{S}_k \cdot \boldsymbol{S}_l) + (\boldsymbol{S}_l \cdot \boldsymbol{S}_i)] \\
&+ [(\boldsymbol{S}_i \cdot \boldsymbol{S}_k) + (\boldsymbol{S}_j \cdot \boldsymbol{S}_l)] + \frac{1}{4}.
\end{aligned}
\tag{2.22}
$$

Adding the cyclic four-spin interaction to the usual Heisenberg-type interactions, we come to a spin Hamiltonian that can describe the spin degrees of freedom in 2D insulating cuprates:

$$H_{\text{spin}} = J_1 \sum_{\langle i,j \rangle} \boldsymbol{S}_i \cdot \boldsymbol{S}_j + J_2 \sum_{\langle\langle i,j \rangle\rangle} \boldsymbol{S}_i \cdot \boldsymbol{S}_j + J_c \sum_{\langle i,j,k,l \rangle} \Big[(\boldsymbol{S}_i \cdot \boldsymbol{S}_j)(\boldsymbol{S}_k \cdot \boldsymbol{S}_l) + (\boldsymbol{S}_i \cdot \boldsymbol{S}_l)(\boldsymbol{S}_j \cdot \boldsymbol{S}_k) - (\boldsymbol{S}_i \cdot \boldsymbol{S}_k)(\boldsymbol{S}_j \cdot \boldsymbol{S}_l) \Big], \tag{2.23}$$

where $\langle i,j \rangle$ and $\langle\langle i,j \rangle\rangle$ represent the first- and second-nearest-neighbor pairs, and the sum of $\langle i,j,k,l \rangle$ is taken over all distinct four-spin cycles on square plaquettes. The value of J_1 is given by a combination dominated by the nearest-neighbor superexchange. The three-spin, and four-spin terms contributions are expected to be one order smaller. Here, note that the three spin contribution is ferromagnetic because the three-spin cyclic permutation is given by the product of the even number of two-spin permutations [31]. The value of J_2 is also given by the three contributions, i.e., the second-nearest-neighbor superexchange, three-spin, and four-spin exchange couplings, but in this case all of them are of the same order and the antiferromagnetic contributions may be partly canceled by the ferromagnetic contribution coming from the three-spin exchange. For J_c, the perturbation expression is eight order in T_{pd}, and given by [33]

$$J_c = \frac{80 T_{pd}^8}{U_d^3 \Delta^7} (U_d + \Delta) \left(U_d^2 + U_d \Delta + \Delta^2 \right). \tag{2.24}$$

Numerical estimates of the parameters in (2.23) involve small clusters of Cu and O described by the three-band Hubbard Hamiltonian [34], with the result: J_2 is almost zero, and $J_c/J_1 \sim 0.29$. The analysis of the magnetic excitations observed by neutron scattering give estimates for these parameters. Figure 2.8 shows the spin-wave dispersions for La_2CuO_4 [32]. The experimental data are well reproduced by classical linear spin-wave theory where $J_1 = 138$ (146) meV and $J_c = 38$ (61) meV at $T = 295$ (10) K . The evidence for the presence of the four-spin interaction is a clearly noticeable dispersion along the antiferromagnetic zone boundary, i.e., along the $(\pi, 0) - (\pi/2, \pi/2)$ direction, which can not arise in the Heisenberg model with J_1 only. A similar high-energy dispersion anomaly due to the four-spin interaction is observed in a ladder cuprate $La_6Ca_8Cu_{24}O_{41}$ [28]. Thus the insulating cuprates joins important systems such as a nuclear magnet ^{3}He [35] and a weak ferromagnet NiS_2 [36], where cyclic exchange interactions play important roles in their magnetic properties.

2.1.5 Zhang-Rice Singlet State

As mentioned in Sect. 2.1.1, insulating cuprates are charge-transfer-type insulators, where the on-site Coulomb interaction U_d at Cu site is larger than the charge-transfer energy Δ between the Cu and O ions. Such a situation is schematically drawn in Fig. 2.9. The density of states (DOS) reflects the

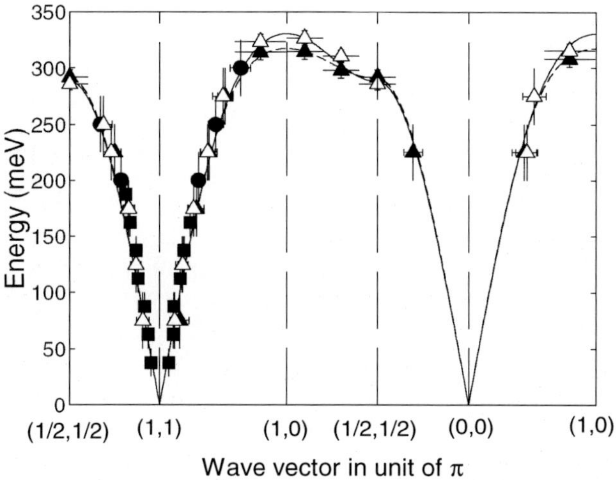

Fig. 2.8. The spin-wave dispersions for La_2CuO_4. Neutron scattering data are taken at T = 10 K (open symbols) and 295 K (solid symbols). Solid (dashed) line is a fit by using classical linear spin-wave theory. The wave vector notation corresponds to a square lattice in two dimension. (From [32])

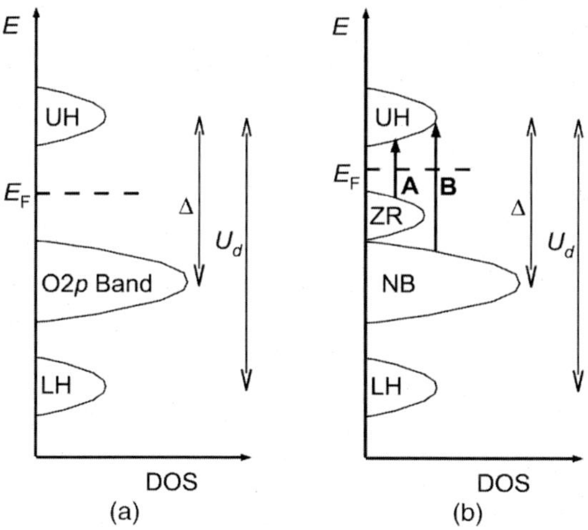

Fig. 2.9. Schematic density of state (DOS) for insulating cuprates. (**a**) Basic energy level structure with the lower Hubbard (LH), upper Hubbard (UH), and O $2p$ bands. (**b**) The O $2p$ band in (**a**) is split into two parts: non-bonding (NB) and Zhang-Rice (ZR) singlet bands. The labels **A** and **B** denote the optical interband excitations. The Fermi energy is E_F.

lower and upper Hubbard bands, which are separated by U_d. In between is the O $2p$ band. This view of the DOS is, however, of very primitive one, since the DOS is intended to represent single-particle excitations. That part which lies below the Fermi level corresponds to electron-removal while DOS above this level is probed by electron-addition. Since this is a many body system one has to take into account interactions related to the created hole and electron carriers. In reality, holes are introduced predominantly into O $2p_\sigma$ orbitals which are strongly coupled with a localized hole on the Cu $3d_{x^2-y^2}$ orbital, leading to a bound state which splits off from the O $2p$ band. This bound state is called the Zhang-Rice (ZR) singlet named after the seminal work by Zhang and Rice [37]. It is a key ingredient to description of the cuprate electronic states. The details of the bound state will be given below. The resulting schematic DOS is shown in Fig. 2.9b, where the ZR singlet bound state is reflected by a new band split-off from oxygen non-bonding band. Therefore, the lowest-energy charge excitation from occupied to unoccupied states corresponds to the interband excitation from the ZR singlet band to the upper Hubbard band denoted by process **A**. This represents the Mott gap in the insulating cuprates.

Let us derive the ZR singlet state following the original paper [37]. Essential to the existence of a ZR singlet state are Δ, U_d, and T_{pd} in the three-band Hamiltonian (2.3). Start with H_ε and H_T [(2.4) and (2.5)] with $\varepsilon_d = 0$ and $T_{pp} = 0$. Since there are two O $2p$ orbitals in the CuO_2 unit, Zhang and Rice introduced two O $2p$ Wannier orbitals centered at the Cu site: the phase symmetry of one such orbital is such that it *can* hybridize with the $3d_{x^2-y^2}$ orbital while the other *cannot*. In the momentum space representation, the Wannier operators for the two orbitals are found from (2.8)

$$\phi^{\mathrm{s}}_{\boldsymbol{k},\sigma} = \mathrm{i}\beta_{\boldsymbol{k}} \left(-\sin(k_x/2)p_{x,\boldsymbol{k},\sigma} + \sin(k_y/2)p_{y,\boldsymbol{k},\sigma}\right), \tag{2.25}$$

$$\phi^{\mathrm{a}}_{\boldsymbol{k},\sigma} = \mathrm{i}\beta_{\boldsymbol{k}} \left(-\sin(k_y/2)p_{x,\boldsymbol{k},\sigma} - \sin(k_x/2)p_{y,\boldsymbol{k},\sigma}\right), \tag{2.26}$$

where $\beta_{\boldsymbol{k}}$ is a normalization factor

$$\beta_{\boldsymbol{k}} = \left[1 - \frac{1}{2}\left(\cos k_x + \cos k_y\right)\right]^{-1/2}. \tag{2.27}$$

Thus,

$$H_\varepsilon + H_T = \sum_{\boldsymbol{k},\sigma} \left[\Delta\left(\phi^{\mathrm{s}\dagger}_{\boldsymbol{k},\sigma}\phi^{\mathrm{s}}_{\boldsymbol{k},\sigma} + \phi^{\mathrm{a}\dagger}_{\boldsymbol{k},\sigma}\phi^{\mathrm{a}}_{\boldsymbol{k},\sigma}\right) + 2T_{pd}\beta^{-1}_{\boldsymbol{k}}\left(d^\dagger_{\boldsymbol{k},\sigma}\phi^{\mathrm{s}}_{\boldsymbol{k},\sigma} + \mathrm{h.c.}\right)\right]. \tag{2.28}$$

Performing the Fourier transform to the Wannier operators (2.25) and (2.26), real space, Wannier orbitals that are centered at the Cu site i are obtained. The resulting total Hamiltonian $H_{dp} = H_\varepsilon + H_U + H_T$ reads, in the real space representation,

$$H_\varepsilon + H_U = \Delta \sum_{i,\sigma} \left[\left(\phi^{\mathrm{s}\dagger}_{i,\sigma}\phi^{\mathrm{s}}_{i,\sigma} + \phi^{\mathrm{a}\dagger}_{i,\sigma}\phi^{\mathrm{a}}_{i,\sigma}\right) + U_d \sum_i n^d_{i,\uparrow} n^d_{i,\downarrow}\right], \tag{2.29}$$

$$H_T = 2T_{pd} \sum_{i,j,\sigma} \tau_{i,j} \left(d^\dagger_{i,\sigma} \phi^{\rm s}_{j,\sigma} + \text{h.c.} \right), \tag{2.30}$$

where $\tau_{i,j} = N^{-1} \sum_{\boldsymbol{k}} \beta_{\boldsymbol{k}}^{-1} \exp\left[\mathrm{i}\boldsymbol{k} \cdot (\boldsymbol{R}_i - \boldsymbol{R}_j)\right]$, is a measure of the spatial extent of the $2p$ Wannier orbitals. The values of $\tau_{i,j}$ are obtained by replacing $N^{-1}\sum_{\boldsymbol{k}}$ by the integral over the first Brillouin zone $\int_{\rm BZ}(2\pi)^{-2}\mathrm{d}\boldsymbol{k}$. They are $\tau_{i,j} = 0.958$, -0.140, and -0.024 for $\boldsymbol{R}_i - \boldsymbol{R}_j = (0,0)$, $(\pm1,0)$ $(0,\pm1)$, and $(\pm1,\pm1)$, respectively, and smaller for longer distances.

Since T_{pd} is the smallest parameter in H_{dp}, in order to derive an effective Hamiltonian, it is reasonable to use perturbation theory up to the second order in T_{pd}. Taking into account the fact that, in the zeroth order, the holes introduced as carriers are on the oxygen ions with the d holes frozen, let us introduce the projection operator $P_1 = \prod_i (n^d_{i,\uparrow} - n^d_{i,\downarrow})^2$, which projects H_{dp} on to the subspace with one hole located on each Cu site but which does not restrict the number of $2p$ holes. To describe the intermediate state in the second order processes, we further introduce two projection operators P_2 and P_0, which represent subspaces with one Cu site doubly occupied by holes and with one unoccupied Cu site, respectively, with all the other Cu sites being singly occupied. Using these projection operators, the effective Hamiltonian is given by

$$H_{\rm eff} = P_1 H_{dp} P_1 - \frac{1}{\Delta} P_1 H_T P_0 H_T P_1 - \frac{1}{U_d - \Delta} P_1 H_T P_2 H_T P_1 \,. \tag{2.31}$$

The $\tau_{i,j}$ are such that, e.g., $\sum_i \tau_{i,j}\tau_{i,j'} = 1$, $-1/4$, and 0 for $\boldsymbol{R}_j - \boldsymbol{R}'_j = (0,0)$, $(\pm1,0)$ or $(0,\pm1)$. These and similar relationships are used to write,

$$\begin{aligned} H_{\rm eff} = &\sum_{i,\sigma} \left[\Delta \left(\phi^{\rm s\dagger}_{i,\sigma} \phi^{\rm s}_{i,\sigma} + \phi^{\rm a\dagger}_{i,\sigma} \phi^{\rm a}_{i,\sigma} \right) - 4t_1 d^\dagger_{i,\sigma} d_{i,\sigma} \right] \\ &- 4t_2 \sum_{j,j',\sigma} \left(\delta_{j,j'} - \frac{1}{4}\delta_{\langle j,j'\rangle} \right) \phi^{\rm s\dagger}_{j,\sigma} \phi^{\rm s}_{j',\sigma} \\ &+ 4(t_1 + t_2) \sum_{i,j,j',\sigma,\sigma'} \tau_{i,j}\tau_{i,j'} \phi^{\rm s\dagger}_{j,\sigma} \phi^{\rm s}_{j',\sigma'} d^\dagger_{i,\sigma'} d_{i,\sigma}, \end{aligned} \tag{2.32}$$

where $t_1 = T_{pd}^2/\Delta$, $t_2 = T_{pd}^2/(U_d - \Delta)$, and $\delta_{\langle j,j'\rangle} = 1$ if j and j' are nearest neighbors.

Since the Kondo-type interaction between the localized hole spin and the oxygen holes involved in (2.32) shows a maximum value of $\tau_{i,j}\tau_{i,j'}$ when $i = j = j'$, it is reasonable, in the zeroth order, to consider the local Hamiltonian at site i. Zhang and Rice considered two types of local states. These are a spin singlet and triplet composed of the Cu hole $d_{i,\sigma}$ and the hybridizing oxygen hole $\phi^{\rm s}_{i,\sigma}$:

$$\left| \Psi_i^{\rm S,T} \right\rangle = \frac{1}{\sqrt{2}} \left(\phi^{\rm s\dagger}_{i,\uparrow} d^\dagger_{i,\downarrow} \pm d^\dagger_{i,\uparrow} \phi^{\rm s\dagger}_{i,\downarrow} \right) |\text{vac}\rangle \,. \tag{2.33}$$

The energies of the local singlet and triplet states for $H_{\rm eff}$ are

$$E_{\mathrm{S,T}} = \left\langle \Psi_i^{\mathrm{S,T}} \right| H_{\mathrm{eff}} \left| \Psi_i^{\mathrm{S,T}} \right\rangle = \Delta - 4\,(t_1 + t_2)\left(1 \pm \tau_0^2\right), \tag{2.34}$$

where τ_0 denotes $\tau_{i,j}$ for $\boldsymbol{R}_i - \boldsymbol{R}_j = (0,0)$. Since $\tau_0^2 = 0.918$, the triplet state is closest in energy to $2p$ level with energy Δ, while the singlet state is split off from this level with the binding energy of $7.672\,(t_1 + t_2)$. The bound state in the schematic representation of DOS (Fig. 2.9b) is actually this ZR singlet state. By using parameters mentioned in Sect. 2.1.2, one can obtain a rough value ~ 4 eV for this binding energy which, however, seems to be too large this reflecting the crude approximations used in the derivation.

Up to now we have concentrated on the ZR singlet state in the CuO_2 plane. Such a bound state is also clearly defined in the CuO_3 1D chain, because the values of $\tau_{i,j}$ in the CuO_3 chain are very similar to those of the CuO_2 plane: $\tau_{i,j} = N^{-1}\sum_k (1 - \frac{1}{2}\cos k)\exp\left[ik(R_i - R_j)\right]$ and $\tau_{i,j} = 0.983$ and -0.128 for $R_i - R_j = 0$ and ± 1, respectively. Note that in the CuO chain $\tau_{i,j}$ is more delocalized and thus the local singlet picture becomes worse.

The band width of the ZR singlet state depends on the hopping amplitude of the singlet. Let us evaluate this amplitude. The O $2p$ holes are created in the background of singly occupied Cu holes. When the ZR singlet state moves from site i to j, a Cu hole moves simultaneously from site j to i. This motion is represented by an effective hopping Hamiltonian

$$H_t = \sum_{i \neq j,\sigma} t_{ij}\left(\Psi_j^{\mathrm{S}} d_{i,\sigma}\right)^{\dagger}\left(\Psi_i^{\mathrm{S}} d_{j,\sigma}\right). \tag{2.35}$$

Using the definition that

$$\left|\Psi_i^{\mathrm{S}} d_{j,\sigma}\right\rangle = \frac{1}{\sqrt{2}}\left(\phi_{i,\uparrow}^{\mathrm{s}\dagger} d_{i,\downarrow}^{\dagger} + d_{i,\uparrow}^{\dagger}\phi_{i,\downarrow}^{\mathrm{s}\dagger}\right) d_{j,\sigma}^{\dagger}\left|\mathrm{vac}\right\rangle, \tag{2.36}$$

the effective hopping matrix element t_{ij} is given by

$$\begin{aligned} t_{ij} &= \left\langle \Psi_j^{\mathrm{S}} d_{i,\sigma}\right| H_{\mathrm{eff}} \left|\Psi_i^{\mathrm{S}} d_{j,\sigma}\right\rangle \\ &= 4(t_1 + t_2)\tau_0\tau_{i,j} - \frac{1}{2}t_2\delta_{\langle i,j\rangle}. \end{aligned} \tag{2.37}$$

The second term in the right-hand side of (2.37) comes from the effective hopping of ϕ_i^{s} to the nearest-neighbor site j. Using the values of $\tau_{i,j}$, one can obtain the hopping parameter t_{ij} for the nearest neighbor i and j: $t_{\langle ij\rangle} = -0.536t_1 - 1.036t_2$. The matrix elements for other than nearest-neighbor hopping are smaller by one order in magnitude and thus negligible. After dropping the singlet operator Ψ_i^{S} which acts as the empty state of the d hole at site i and redefining $t_{\langle ij\rangle}$ as $-t$, one obtains an effective Hamiltonian that describes the physics of the ZR singlet band,

$$\begin{aligned} H_{tJ} = &-t\sum_{\langle i,j\rangle,\sigma}\left[\left(1 - n_{i,-\sigma}^{d}\right) d_{i,\sigma}^{\dagger} d_{j,\sigma}\left(1 - n_{j,-\sigma}^{d}\right) + \mathrm{h.c.}\right] \\ &+ J\sum_{\langle i,j\rangle}\boldsymbol{S}_i\cdot\boldsymbol{S}_j, \end{aligned} \tag{2.38}$$

where $\langle i,j\rangle$ represents the nearest-neighbor pairs, and $t > 0$. The exchange interaction term in (2.38) is introduced to describe the antiferromagnetic interaction between the d holes. The effective Hamiltonian (2.38) is called the t-J model. The ratio of J to t in cuprates is $J/t \sim 0.4$.

We can also derive the t-J model from the the single-band Hubbard model

$$H_{\text{Hub}} = -t \sum_{\langle i,j\rangle,\sigma} \left(c^{\dagger}_{i,\sigma} c_{j,\sigma} + \text{h.c.} \right) + U \sum_{i} n_{i,\uparrow} n_{i,\downarrow}, \tag{2.39}$$

if we apply the second-order perturbation theory with respect to t in the large-U limit and neglect correlated hopping terms with the order of t^2/U. Here, $c_{i,\sigma}$ is the annihilation operator for an electron at site i with spin σ. The exchange interaction is given by $J = 4t^2/U$. In this sense, the Hubbard model (2.39) with large U is equivalent to (2.38) which was derived from the three-band Hubbard model (2.3). Physically, the ZR band maps to the lower Hubbard band of the single-band Hubbard model.

As discussed above, hopping between other than nearest-neighbor ZR singlets (identified with the Cu sites) has a small amplitude. Actually the second-neighbor hopping t' along the diagonal direction in 2D CuO_2 plane is given by $-t' = -0.092(t_1 + t_2) \sim -0.13t$, i.e., t and t' have the same sign. However the dominant contribution to t' is known to come from the oxygen-oxygen hopping T_{pp}, which is neglected in the above discussion. Including T_{pp}, the sign of t' changes from positive to negative [38]. The importance of the longer-range hopping on the electronic states of 2D cuprates will be discussed in Sect. 2.3.1.

2.1.6 Optical Excitations

In this subsection, we consider the charge excitations of a charge-transfer insulator. The transitions are from occupied to unoccupied bands in the DOS shown in Fig. 2.9. The excitation with minimum energy, ~ 2 eV, is the process **A** which defines the Mott gap in cuprates. In real space this process is a nonlocal excitation in the sense that a localized hole on a Cu $3d_{x^2-y^2}$ orbital is transferred to O $2p_\sigma$ orbital and forms a ZR singlet state which is coupled to an adjacent localized Cu$3d_{x^2-y^2}$ hole. Therefore the process **A** involves at least two CuO_4 units. In contrast to **A**, the excitation denoted as **B**, from the O $2p$ non-bonding band to the upper Hubbard band with energy scale Δ, is a local excitation within a single CuO_4 unit. It does *not* involve the formation of a ZR singlet. This difference will be seen in the experimental data as discussed below. The excitation from the lower to upper Hubbard band has a large energy of the order of U_d, the largest energy scale and is hard to be detected because many other processes involving the Cu-O planes occur in the corresponding energy region.

The charge excitations couple to the external electric field. The response is given by the real part of the optical conductivity, $\boldsymbol{\sigma}'(\omega)$, where ω is the frequency of the applied electric field (see Appendix A). The optical conductivity is in general a tensor. Here we deal with the diagonal part, $\boldsymbol{\sigma}'_{\alpha\alpha}(\omega)$, of this tensor with the α direction for both the electric field and induced current, and use for this the symbol $\sigma(\omega)$. Then $\sigma(\omega)$ is split into singular and regular parts

$$\sigma(\omega) = 2\pi e^2 D\delta(\omega) + \sigma_{\text{reg}}(\omega), \tag{2.40}$$

where e is the unit of the electric charge ($e > 0$), and D is the charge stiffness, which distinguishes an insulator ($D = 0$) from a metal ($D \neq 0$) at zero temperature $T = 0$ as discussed originally by Kohn [39]. The regular part $\sigma_{\text{reg}}(\omega)$ at temperature T is expressed as

$$\begin{aligned}\sigma_{\text{reg}}(\omega) = {} & \frac{\pi e^2}{NZ}\frac{1-\mathrm{e}^{-\beta\hbar\omega}}{\hbar\omega} \\ & \times \sum_{\substack{n,m \\ E_n \neq E_m}} \mathrm{e}^{-\beta E_n} \left|\langle n\left|j_\alpha\right|m\rangle\right|^2 \delta\left(\omega - (E_m - E_n)/\hbar\right),\end{aligned} \tag{2.41}$$

where $|n\rangle$ is the eigenstate with the eigenvalue E_n, Z is the partition function defined as $Z = \sum_n \mathrm{e}^{-\beta E_n}$, and $\beta = 1/(k_\mathrm{B}T)$, k_B being the Boltzmann constant. The current operator j_α for the three-band model (2.3) with $\alpha = x$ is given by (see Appendix A)

$$\begin{aligned}j_x = \frac{\mathrm{i}d_{\text{Cu-O}}}{\hbar}\sum_{i,\sigma}\Big[& T_{pd}d^\dagger_{i,\sigma}\left(p_{i+\boldsymbol{x}/2,\sigma} + p_{i-\boldsymbol{x}/2,\sigma}\right) \\ & + T_{pp}(-p^\dagger_{i+\boldsymbol{y}/2,\sigma}p_{i+\boldsymbol{x}/2,\sigma} + p^\dagger_{i-\boldsymbol{x}/2,\sigma}p_{i+\boldsymbol{y}/2,\sigma} \\ & \quad + p^\dagger_{i-\boldsymbol{y}/2,\sigma}p_{i-\boldsymbol{x}/2,\sigma} - p^\dagger_{i+\boldsymbol{x}/2,\sigma}p_{i-\boldsymbol{y}/2,\sigma}) - \text{h.c.}\Big],\end{aligned} \tag{2.42}$$

where $d_{\text{Cu-O}}$ is the Cu-O distance. Note that the phases of the linear combination of the $2p$ orbitals are different from those in (2.6) because the dipole transitions induced by the current operators connect states with different parities. The optical conductivity is experimentally determined from the reflectivity measurements using a Kramers-Kronig transformation [40]. The experimental data for a 2D insulating cuprate Pr_2CuO_4 [41] is shown in Fig. 2.10. The excitation **A** corresponds to a peak at around $\hbar\omega$=1.5 eV with large oscillator strength. Such a peak coming from the process **A** is also seen in a corner-sharing 1D insulator Sr_2CuO_3 at around 2 eV (see Fig. 2.11a).

In contrast, an edge-sharing 1D insulator Li_2CuO_2 exhibits no pronounced structure at around 2 eV [11] (see Fig. 2.11b). The oscillator strength of the excitation **A** for the edge-sharing systems is evidently small compared to that for the corner-sharing equivalent. This is because the nonlocal excitation involving neighboring CuO_4 units is suppressed due to the near 90° Cu-O bond angle. On the other hand, Li_2CuO_2 shows a peak with large weight at around $\hbar\omega = 4$ eV. This is attributed to the process **B** in Fig. 2.9 since in the edge-sharing insulator the optical response is expected to be dominated by the local excitation within a CuO_4 unit. In corner-sharing compounds,

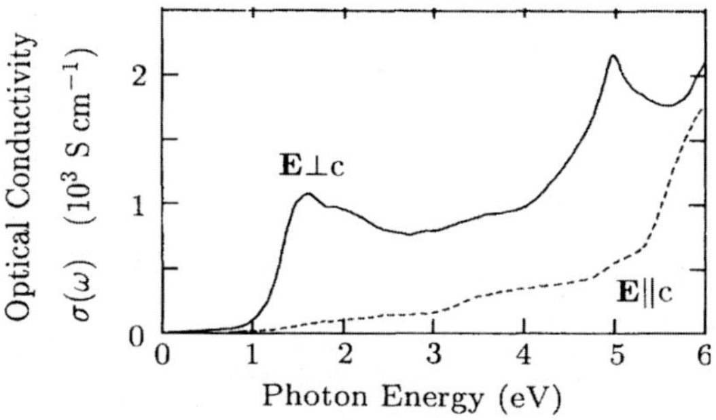

Fig. 2.10. The optical conductivity at room temperature for Pr_2CuO_4 with 2D corner-sharing CuO_2 plane. Solid and dashed lines represent the polarization parallel and perpendicular to the plane, respectively. (From [41])

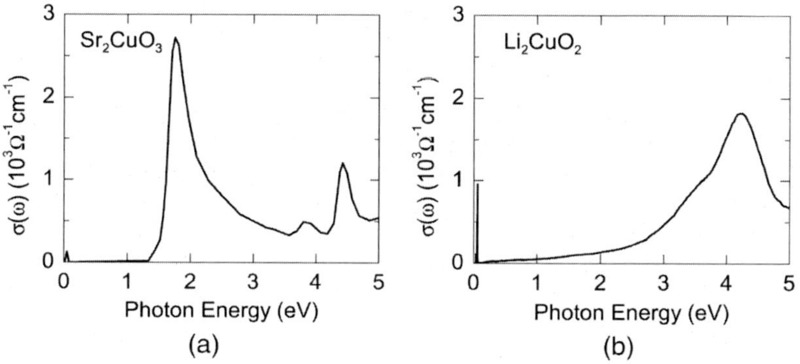

Fig. 2.11. The optical conductivity at room temperature for 1D insulating cuprates with polarization parallel to the Cu-O chain. (**a**) Corner-sharing chain compound Sr_2CuO_3 reproduced from [8] and (**b**) edge-sharing chain compound Li_2CuO_2 from [11].

process **B** may be identified with peak structures in the energy region of 4-5 eV. Note that a different mechanism for the 4-5 eV peaks (excitations from O $2p$ to Cu $4s$) has been discussed [8, 41].

The dramatic contrast between the $\sigma(\omega)$ functions, for the corner- and edge-sharing insulators, can be demonstrated in theoretical calculations. Figure 2.12 shows such calculations at $T = 0$, which are obtained by the numerical exact diagonalization (the Lanczos method: see Appendix B) for small clusters with four CuO_4 units described by the Hamiltonian (2.3). The qualitative features of experiment are reproduced by the calculations. This fact demonstrates the validity of the current description of the electronic states of the cuprates. However, the clusters which can be studied by exact diagonalization are limited in size. More precise calculations of $\sigma(\omega)$ will require other techniques.

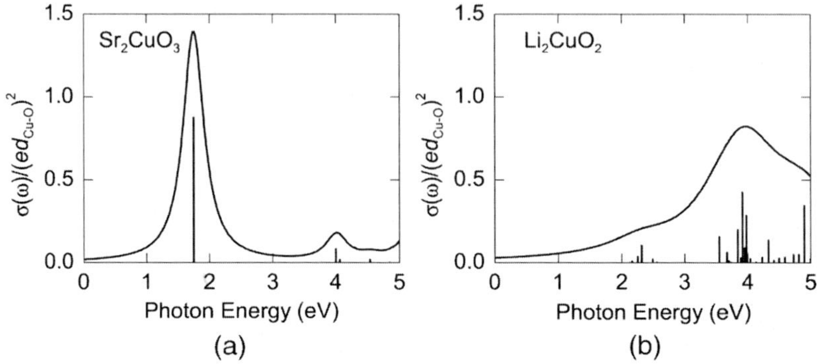

Fig. 2.12. The theoretical results for the optical conductivity at $T = 0$ for 1D insulating cuprates with polarization parallel to the Cu-O chain. The delta functions (vertical bars) are broadened by a Lorentzian with a width of 0.6 eV (solid lines). (**a**) The corner-sharing chain compound Sr_2CuO_3 described by a Cu_4O_{13} cluster with open boundary conditions. Parameters for Hamiltonian (2.3) are taken from [26]. (**b**) The edge-sharing chain compound Li_2CuO_2 described by a Cu_4O_{10} cluster with open boundary conditions [11]).

2.2 One-Dimensional Cuprates

In one-dimensional (1D) systems, the Coulomb interaction dramatically changes the electronic states from a Fermi liquid to a non-Fermi liquid called a Tomonaga-Luttinger liquid [42,43]. Since the interaction is strongly enhanced due to the restricted motion of electrons along one direction, a perturbation approach does not work, and anomalous electronic states emerge in 1D systems, which have no equivalent in 3D systems and which correspond to a Fermi liquid [44]. In this section, we first discuss the theoretical concepts in an effort to understand the Tomonaga-Luttinger liquid behavior, in particular, for a system with strong Coulomb interactions. The discussion is then extended to the spin and charge dynamics in the 1D cuprates which exhibit the Tomonaga-Luttinger liquid behaviors.

2.2.1 Spin-Charge Separation

There are two key factors that lead to the breakdown of the Fermi liquid and four parameters that are necessary for the description of the low-energy physics of the Tomonaga-Luttinger liquid [45–48]:

1. The internal degrees of freedom of the electron, spin and charge, are decoupled and thus move independently with different velocities, v_σ and v_ρ (σ and ρ represent spin and charge, respectively). Thus, elementary excitations of the Tomonaga-Luttinger liquid are not quasi-particles but collective modes for the spin and charge degrees of freedom, i.e., the

spinon with spin 1/2 and the *holon* with charge e. This phenomenon is called *spin-charge separation*.
2. Various physical quantities obey non-universal power-law behaviors as functions of length, energy, and temperature. The power-law scaling is described by the coupling constants K_ρ and K_σ for charge and spin fluctuations, respectively. Since the spin-rotation invariance requires K_σ=1 [45], K_σ has no significant effect. Therefore, K_ρ plays a crucial role in scaling laws.

The low-energy Hamiltonian of the interacting 1D electron systems can be decoupled into spin and change parts using the bosonization technique [49–51]. The representative models are the Tomonaga model [42] and the Luttinger model [43], which are exactly solvable due to the linear dispersion near the Fermi momentum. In this treatment, the spin and charge degrees of freedom are represented by bosonic operators which describe the excitation of spin-density waves and charge-density waves with the velocities v_σ and v_ρ, respectively.

For the single-band Hubbard model (2.39), the dispersion relation for which is $E_k = -2t\cos k$, one can find two such kinds of elementary excitation from the low-energy region by examining the exact solution based on the Bethe ansatz [52]. In the exact solution there appear two types of quantum numbers. One of the quantum numbers represents the spin occupation, while the other reflects the occupied charge states. In the limit of $U \to \infty$, the wave function itself can be decomposed into the product of the two wave functions representing the spin and charge parts [53].

The velocities v_σ and v_ρ are the key to understand the nature of the 1D Hubbard model. Figure 2.13a shows v_σ and v_ρ for the Hubbard model

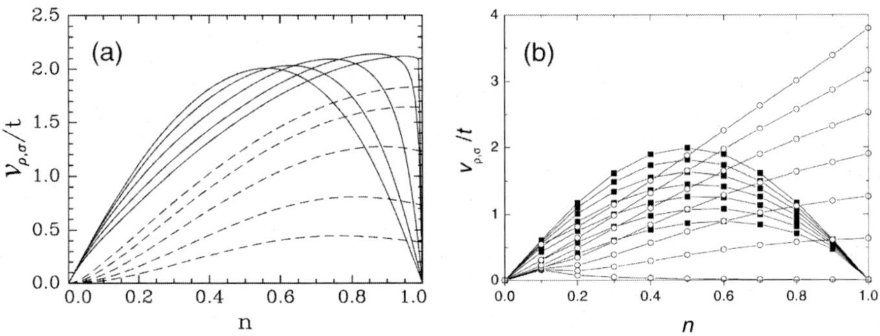

Fig. 2.13. The charge and spin velocities v_ρ (solid lines) and v_σ (dash-dotted lines) of (**a**) the 1D Hubbard model and (**b**) the t-J model (20-site ring), as functions of the electron density n. In (**a**), U/t = 1, 2, 4, 8, and 16 from top to bottom for v_σ and from bottom to top for v_ρ in the left-hand part of the figure. (From [48]) In (**b**), J/t = 0, 0.4, 0.8, 1.2, 1.6, 2, and 2.4 from bottom to top for v_σ and from top to bottom for v_ρ. (From [55])

as functions of the electron density n [48]. Both of the velocities approach the Fermi velocity $v_F = \hbar^{-1}\mathrm{d}E_k/\mathrm{d}k|_{k=k_F} = 2\hbar^{-1}t\sin(\pi n/2)$ for $U \to 0$ as expected. For $U \to \infty$, $v_\rho = 2\hbar^{-1}t\sin(\pi n)$, being equivalent to the velocity of the spinless fermions, and $v_\sigma = \hbar^{-1}(2\pi t^2/U)\{1 - \sin(2\pi n)/(2\pi n)\}$ [53]. At half-filling ($n = 1$), v_ρ is zero irrespective of the value of U, while v_σ for $U \to \infty$ coincides with the spin-wave velocity of the 1D Heisenberg model. For the t-J model, the velocities can be evaluated numerically for finite-size clusters [54,55]. Figure 2.13b shows the velocities obtained by using a 20-site t-J ring. The n dependence in the small J region is similar to that in the large U region of the Hubbard model. At $J/t = 2$, the Bethe-ansatz exact solution is available [56] and gives the exact velocities with values similar to those for the 20-site results. For larger J, there is a phase-separation with hole-free and hole-rich regions. This appears for $2.8 < J/t < 3.6$ [54]. Note that, in realistic parameter regions for cuprates, i.e., close to half-filling and for $U/t \sim 10$ or $J/t \sim 0.4$, v_σ is larger than v_ρ.

Another important quantity that characterizes the Tomonaga-Luttinger liquid behavior is K_ρ as mentioned above. The value of K_ρ in the 1D Hubbard model has been evaluated through the bosonization and conformal-field theories [57]. In the limit of $U/t \to \infty$, K_ρ is 1/2. With decreasing U/t, K_ρ increases and reaches unity at $U/t = 0$. On the other hand, K_ρ in the t-J model [54] increases from 1/2 at $J/t = 0$ to values greater than one, where the effective interactions are attractive, and for which superconducting correlations are dominant at large distances. With further increasing J/t, phase-separation occurs as mentioned above.

Spin and Charge Correlation Functions

One of the methods used to see the spin-charge separation is to investigate the spin and charge correlation functions at zero temperature. These are defined as

$$S(\boldsymbol{q},\omega) = \sum_m \left|\langle m|\, S^z_{\boldsymbol{q}}\, |0\rangle\right|^2 \delta\left(\omega - (E_m - E_0)/\hbar\right) \tag{2.43}$$

and

$$N(\boldsymbol{q},\omega) = \sum_m \left|\langle m|\, N_{\boldsymbol{q}}\, |0\rangle\right|^2 \delta\left(\omega - (E_m - E_0)/\hbar\right), \tag{2.44}$$

with

$$S^z_{\boldsymbol{q}} = \frac{1}{\sqrt{N}}\sum_i \mathrm{e}^{\mathrm{i}\boldsymbol{q}\cdot\boldsymbol{R}_i} S^z_i \qquad N_{\boldsymbol{q}} = \frac{1}{\sqrt{N}}\sum_i \mathrm{e}^{\mathrm{i}\boldsymbol{q}\cdot\boldsymbol{R}_i}(n_i - n), \tag{2.45}$$

where $S^z_i = (n_{i\uparrow} - n_{i\downarrow})/2$, $n_i = n_{i\uparrow} + n_{i\downarrow}$, and $|0\rangle$ represents the ground state with energy E_0. In the small q limit in 1D, these are given by [45]

$$S(q,\omega) = \frac{2}{\pi}\frac{v_\sigma q^2}{\omega^2 - v_\sigma^2 q^2} \qquad N(q,\omega) = \frac{2}{\pi}K_\rho\frac{v_\rho q^2}{\omega^2 - v_\rho^2 q^2}. \tag{2.46}$$

The spectral weight appears at different frequencies, $\omega = v_\sigma|q|$ and $\omega = v_\rho|q|$, for the spin and charge correlations, respectively. This is a manifestation of the spin-charge separation. The positions of the two poles, $v_\sigma|q|$ and $v_\rho|q|$, correspond to the energies of a spinon and holon with momentum q, respectively. On the other hand, each correlation function near $q = 2k_{\mathrm{F}}$, k_{F} being the Fermi momentum, has spectral weight at both $\omega = v_\sigma|q - 2k_{\mathrm{F}}|$ and $\omega = v_\rho|q - 2k_{\mathrm{F}}|$ [45]. Therefore, when one uses a finite-size system with N sites in order to investigate the spin-charge separation, one has to calculate the correlation functions at $q = 2\pi/N$ which is minimum momentum defined in the system.

Single-Particle Spectral Function

Another quantity which manifests the spin-charge separation is the single-particle excitation or the spectral function $A(\boldsymbol{k},\omega)$ at zero temperature, defined by

$$A(\boldsymbol{k},\omega) = A_-(\boldsymbol{k},\omega) + A_+(\boldsymbol{k},\omega) \tag{2.47}$$

and

$$A_\pm(\boldsymbol{k},\omega) = \sum_{m,\sigma} |\langle m\,|a_{\boldsymbol{k},\sigma}|\,0\rangle|^2\, \delta(\omega \mp (E_m - E_0)/\hbar), \tag{2.48}$$

where A_- (A_+) is the electron-removal (electron-addition) spectral function, $a_{\boldsymbol{k},\sigma} = c^\dagger_{\boldsymbol{k},\sigma}$ and $c_{\boldsymbol{k},\sigma}$ for A_+ and A_-, respectively, $c^\dagger_{\boldsymbol{k}\sigma}$ ($c_{\boldsymbol{k}\sigma}$) being the creation (annihilation) operator for an electron with momentum $\boldsymbol{k}$ and spin σ. The chemical potential is set to be zero in energy. Since an electron has both spin and charge degrees of freedom, the excitation of an electron in a 1D system contains information on both the spinon and holon: $A(k,\omega)$ behaves like a convoluted spectrum of spinons and holons, although it is not a simple convolution of the two components because of the presence of non-local phase string through which the spinon and holon interact with each other [58, 59] (see Sect. 2.2.2).

In the large U region of the 1D Hubbard model relevant to 1D cuprates, there is a simplified intuitive picture which describes the decay of an electron or hole into a spinon and holon. Figure 2.14 shows the propagation of a

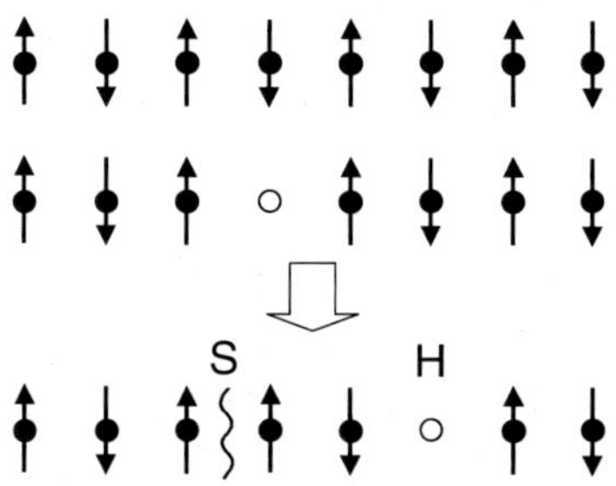

Fig. 2.14. A simplified picture of the spin-charge separation in the large U region. A hole created in a chain with short-range antiferromagnetic ordering decays into a spin excitation (a spinon, labeled as S) and a charge excitation (a holon, labeled as H). The spinon and holon propagate independently with different velocities, v_σ and v_ρ.

hole introduced into a half-filled 1D antiferromagnetic insulator with short-range antiferromagnetic order. Hopping of the hole with a matrix element t to a neighboring site, or equivalently the hopping of a neighboring electron into the hole site, creates a magnetic excitation or a domain wall (a spinon, marked as wavy line in the figure). However, additional hopping in the same direction does not create magnetic excitations. The motion of the charge vacancy (holon) is free from magnetic interactions apart from just after the first step. Note that the spinon can propagate by using the transverse component of the exchange interaction J. Thus, even without going into details, we can see that the quasiparticle dispersion of single particles in the sense of the Fermi liquid does not occur, but the dispersion should be a convolution of two branches of dispersive bands due to two particles, i.e., the spinon and holon, each with its own energy scale.

Analytical expressions for $A(k,\omega)$ near the Fermi momentum have been derived for the Tomonaga-Luttinger model [60]:

$$\begin{aligned} A(k,\omega) &\approx (\omega - v_\sigma p)^{\delta-1/2}\,|\omega - v_\rho p|^{(\delta-1)/2} \qquad (v_\rho > v_\sigma) \\ A(k,\omega) &\approx (\omega - v_\rho p)^{(\delta-1)/2}\,|\omega - v_\sigma p|^{\delta-1/2} \qquad (v_\rho < v_\sigma) \end{aligned} \tag{2.49}$$

with $p = k - k_{\mathrm{F}}$ and

$$\delta = (K_\rho + 1/K_\rho - 2)/4. \tag{2.50}$$

These expressions imply that the spectrum is composed of a continuum, not a quasiparticle, with branch-cut singularities at $\omega = v_\sigma p$ or $v_\rho p$. In addition to the Tomonaga-Luttinger model, the spectral function of the Luther-Emery model [61], which describes 1D systems with gapless charge and gaped spin degrees of freedom, has been constructed [62]. For example, for $v_\rho < v_\sigma$, K_ρ <1 and $p < p_c$, being $p_c = \mathrm{sign}(p)(v_\rho/v_\sigma)\hbar^{-1}\Delta_\sigma(v_\sigma^2 - v_\rho^2)^{-1/2}$ (Δ_σ is the magnitude of the spin gap),

$$A(k,\omega) \approx [-\omega - \varepsilon_\sigma(p)/\hbar]^{\delta-1/2} + \Theta[-\omega - \tilde{\varepsilon}_\rho(p)/\hbar][-\omega - \tilde{\varepsilon}_\rho(p)/\hbar]^{\frac{\delta-1}{2}}, \tag{2.51}$$

where $\Theta(x)$ is the Heaviside function, $\varepsilon_\sigma(p) = \pm\sqrt{(\hbar v_\sigma p)^2 + \Delta_\sigma^2}$, and $\tilde{\varepsilon}_\rho(p) = \varepsilon_\sigma(p_c) + \hbar v_\rho(p - p_c)$. This expression is applicable to the spectral function in the 1D Mott insulators by exchanging σ for ρ and vise versa, as will be discussed in Sect. 2.2.2.

For the t-J model, which is a realistic model of the cuprates, it is hard to obtain analytic expressions for $A(k,\omega)$. Instead, it is instructive to calculate $A(k,\omega)$ numerically for finite-size systems and to investigate the nature of the spin-charge separation. Figure 2.15 represents $A(k,\omega)$ of a 16-site t-J ring with eight electrons ($n = 0.5$) [63] obtained by the exact diagonalization based on the Lanczos method (see Appendix B). The momentum at which an electron can be removed (can be added) with the minimum cost in energy is $k_{\mathrm{F}}^- = 3\pi/16$ ($k_{\mathrm{F}}^+ = 5\pi/16$). Since the spectra are a collection of δ-functions, it is impossible to obtain the exponent of the singularities, as was possible with the analytical expression (2.49) for the Tomonaga-Luttinger model. However,

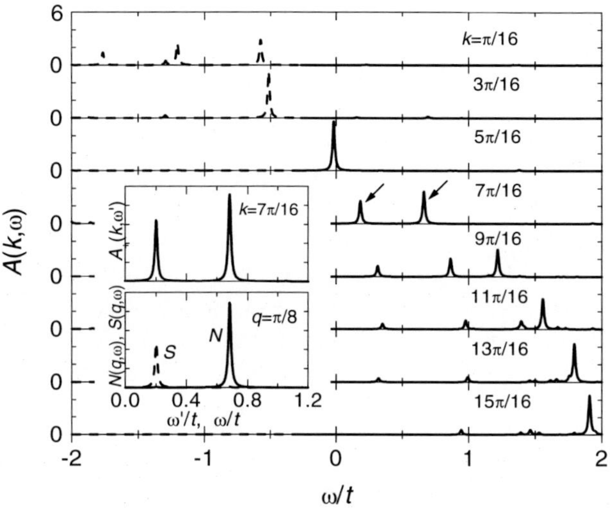

Fig. 2.15. The single-particle spectral function $A(k, \omega)$ for a 1D 16-site t-J model ($J = 0.4$) with an antiperiodic boundary condition at n=0.5 (eight electrons). The dashed and solid lines correspond to A_- and A_+, respectively. k_F^- and k_F^+ are located at $3\pi/16$ and $5\pi/16$, respectively. $\hbar = 1$. Inset shows the two-peak structure of $A_+(7\pi/16, \omega')$ denoted by the arrows, ω' being measured from the nine-electron ground state, and $N(q, \omega)$ (solid line) and $S(q, \omega)$ (dashed line) at $q = \pi/8$ calculated for the nine-electron system. The low-energy (high-energy) peak carries the spin (charge) degree of freedom. (From [63])

the two peaks seen at the momentum of $5\pi/16$ close to k_{F}^+ are easily found to correspond to the branch cuts at $\omega = v_\sigma p$ and $v_\rho p$, by examining $S(q, \omega)$ and $N(q, \omega)$ in the Hilbert space to which the final states of $A(k, \omega)$ belong. This procedure is based on the fact that $S(q, \omega)$ ($N(q, \omega)$) in the small q limit has spectral weight at $\omega = v_\sigma q$ ($v_\rho q$) as shown in (2.46). The inset of Fig. 2.15 shows the two-peak structure of $A_+(7\pi/16, \omega')$, ω' being measured from the nine-electron ground state, and $S(\pi/8, \omega)$ and $N(\pi/8, \omega)$ for the system with nine electrons to which the final states of $A_+(7\pi/16, \omega')$ belong. It is clearly seen that the eigenstate of the low-energy peak at $\omega' = 0.2t$ in $A_+(7\pi/16, \omega')$ is coupled to the spin excitation, i.e., only $S(\pi/8, \omega)$ has weight for this eigenstate. In contrast, the high-energy peak at $\omega' = 0.7t$ in $A_+(7\pi/16, \omega')$ is only related to the charge excitation represented by $N(\pi/8, \omega)$. These are fully consistent with an the behavior expected from the spin-charge separation.

2.2.2 Realization of Spin-Charge Separation

As illustrated in Fig. 2.14, the 1D insulating cuprates constitute a good example of a system showing spin-charge separation. In fact, the development of the technique of angle-resolved photoemission spectroscopy (ARPES) has enabled spinons and holons to be observed in insulating chain compounds. The observations of these new *elementary particles* has been possible for $SrCuO_2$ [65], Sr_2CuO_3 [66], and a vanadium oxide NaV_2O_5 [67]

ARPES

The intensity measured in an ARPES experiment on a low-dimensional system is expressed as

$$I(\boldsymbol{k},\omega) = I_0(\boldsymbol{k},\nu,\omega)\,A_-(\boldsymbol{k},\omega)\,, \tag{2.52}$$

where $\boldsymbol{k}$ is the component of the electron crystal momentum parallel to the surface, ω is the electron energy with respect to the Fermi level. $I_0(\boldsymbol{k},\nu,\omega)$ is proportional to a squared matrix element of an electron-photon interaction between a one-electron orbital in the initial state $\phi_i^{\boldsymbol{k}}$ (from which a photoelectron is emitted) and the wave function of the photoelectron in the final state $\phi_f^{\boldsymbol{k}}$, i.e., $|<\phi_f^{\boldsymbol{k}}|\boldsymbol{A}\cdot\boldsymbol{p}|\phi_i^{\boldsymbol{k}}>|^2$, $\boldsymbol{A}$ and $\boldsymbol{p}$ being the electromagnetic vector potential and the electronic momentum operator, respectively. Therefore, I_0 depends on the frequency ν and polarization of the incoming photon. (A concise explanation of (2.52) and a list of text books about ARPES are found in a recent review article discussing ARPES for high-T_c cuprates [64].) The electron-removal spectral function $A_-(\boldsymbol{k},\omega)$ is related to the retarded Green's function $G^R(\boldsymbol{k},\omega)$ via the Fermi function $f(\omega) = (\exp(\beta\omega)+1)^{-1}$ [50]:

$$A_-(\boldsymbol{k},\omega) = f(\omega)\,A(\boldsymbol{k},\omega) = -\frac{1}{\pi}f(\omega)\,\mathrm{Im}G^R(\boldsymbol{k},\omega)\,. \tag{2.53}$$

Figure 2.16a shows the energy versus k relationship determined from ARPES data at different k points along the chain direction in $SrCuO_2$. Note that the data corresponding to the filled states just below the Mott gap, i.e., to the ZR singlet band discussed in Sect. 2.1.5. The dispersion shows a maximum at $\pi/2$. The width of the lower bound of the broad feature from $k=0$ to $\pi/2$ is almost the same as the width of the dispersion from $\pi/2$ to π, while there is no dispersion that corresponds to the upper bound of the broad feature from $k=0$ to $\pi/2$. The dispersion below 1 eV is difficult to be detected because of the presence of a non-bonding oxygen band.

The ARPES data can be compared with the single-hole spectral function of the t-J model calculated by the Lanczos method (see Appendix B) [65]. Figure 2.16b shows the spectra for $J=0.2$ eV and $t=0.6$ eV. The spectra are collections of the δ-functions due to the finite size, a continuous broad spectrum is expected in the thermodynamic limit. There is a clear similarity between the t-J calculation and the experimental data: (i) between $k=\pi/2$ and π a sharp edge is seen, (ii) between $k=0$ and $\pi/2$ the spectra have a two-edge structure, and (iii) there exist dense spectra between these edges.

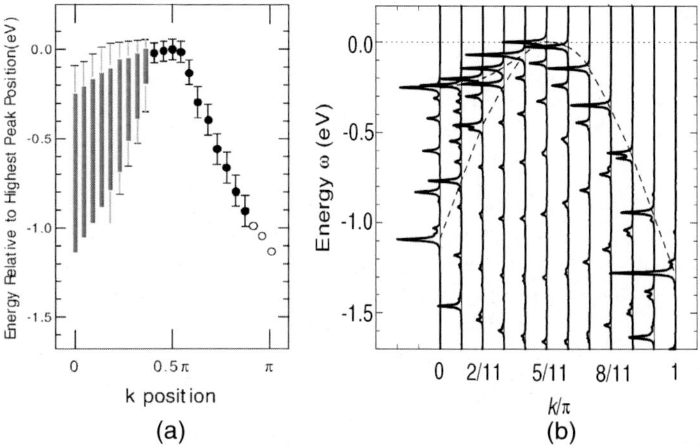

Fig. 2.16. (**a**) The energy versus k relationship of the ZR singlet band in a 1D Mott insulator $SrCuO_2$ determined from ARPES spectra. There is a single band between $\pi/2$ and π that corresponds to holon dispersion. Between 0 and $\pi/2$, the thick lines are the upper and lower bounds of the dispersive features, the former of which is the spinon dispersion and the latter the holon one. (**b**) The spectral function $A(k, \omega)$ for a 1D 22-site t-J model with $J = 0.2$ eV and $t = 0.6$ eV. The δ-functions are convoluted with a Lorentzian broadening of $0.01t$. The broken lines guide the eye. (From [65])

Convolution of Spinon and Holon

These features can be explained as a simple convolution of spinon and holon excitations based on the slave boson formalism [59], where the electron operator $c_{i,\sigma}$ is given by the product of the spinon $f_{i,\sigma}$ regarded as a fermion and a boson holon $b_i^\dagger$, i.e., $c_{i,\sigma} = f_{i,\sigma} b_i^\dagger$ (see Sect. 3.11). The spinon and holon are decoupled in the mean-field treatment, and the Hamiltonian of the 1D t-J model in the momentum-space representation reads

$$H_{\mathrm{sb}} = \sum_k E^{\mathrm{h}}_{k_{\mathrm{h}}} b_k^\dagger b_k + \sum_{k,\sigma} E^{\mathrm{s}}_{k_{\mathrm{s}}} f_{k,\sigma}^\dagger f_{k,\sigma}, \tag{2.54}$$

where the holon dispersion $E^{\mathrm{h}}_{k_{\mathrm{h}}} = -2t_{\mathrm{h}} \cos k_{\mathrm{h}}$ and the spinon dispersion $E^{\mathrm{s}}_{k_{\mathrm{s}}} = -2J_{\mathrm{s}} \cos k_{\mathrm{s}}$, k_{h} and k_{s} being the momentum of holon and spinon, respectively. The hopping amplitude t_{h} and J_{s} are of the order of t and J, respectively. (The precisely numbers are estimated to be $J_{\mathrm{s}} \sim \pi J/4$ and $t_{\mathrm{h}} \sim t$, reflecting the spinon dispersion of the 1D Heisenberg model and a spinless fermionic behavior of the charge degree of freedom in the limit of $J/t \to 0$.)

In the present situation, the initial state, before photoexcitation, contains no holon but has spinons with a half-filled Fermi sea, i.e., the Fermi momentum of spinon is $k_{\mathrm{s}}^{\mathrm{F}} = \pm\pi/2$. Upon photoexcitation, one photohole is created in the sample, resulting in the creation of one holon and annihilation of one spinon. Since the spinon and holon are decoupled, the physical electron Green's function $G(r,t)$ in real space-time is given by the product of the Green's functions of spinon and holon [68]: $G(r,t) = G_{\text{spinon}}(r,t)\,G_{\text{holon}}(-r,-t)$. Note that G_{spinon} and G_{holon} correspond to free-particle Green's functions for fermion and boson, respectively. For instance,

$$G_{\text{spinon}}(r,t) = -\mathrm{i}\sum_{\sigma}\left\langle T f_{\sigma}(r,t) f_{\sigma}(0,0)^{\dagger}\right\rangle, \tag{2.55}$$

where T is the time ordering operator, $f(r,t)$ is the Heisenberg representation of the operator $f_{r,\sigma}$, and $\langle\cdots\rangle$ represents the thermal average. The spectral function $A(k,\omega)$ is obtained by using the relation (2.53). Shifting the energy by a half of the on-site Coulomb energy, we find at zero temperature, after some algebra,

$$\begin{aligned} A(k,\omega) &= \frac{1}{\pi}\int_{-\pi}^{\pi}\mathrm{d}k_{\mathrm{h}}\int_{-\pi/2}^{\pi/2}\mathrm{d}k_{\mathrm{s}}\delta\left(k-k_{\mathrm{s}}+k_{\mathrm{h}}\right)\delta\left(\omega - E^{\mathrm{s}}_{k_{\mathrm{s}}}/\hbar + E^{\mathrm{h}}_{k_{\mathrm{h}}}/\hbar\right) \\ &= \frac{1}{\pi}\frac{1}{\sqrt{F_k^2-(\hbar\omega)^2}} f(k,\omega), \end{aligned} \tag{2.56}$$

with

$$F_k = 2\sqrt{t_{\mathrm{h}}^2 + J_{\mathrm{s}}^2 - 2t_{\mathrm{h}}J_{\mathrm{s}}\cos k} \tag{2.57}$$

$$f(k,\omega) = \left[\Theta\left(\pm\hbar\omega + 2t_{\mathrm{h}}\sin k\right) + \Theta\left(\pm\hbar\omega - 2t_{\mathrm{h}}\sin k\right)\right]\Theta\left(F_k \pm \hbar\omega\right), \tag{2.58}$$

where the upper and lower signs in the right-hand side of (2.58) should be used in the momentum regions of $0 \le k \le k_0$ and $k_0 \le k \le \pi$, respectively. The quantity $k_0 = \arccos(J_{\mathrm{s}}/t_{\mathrm{h}})$ becomes $\pi/2$ when $J_{\mathrm{s}} \to 0$. From (2.56), we find that the momentum and energy conservation are satisfied via the relations that $k = k_{\mathrm{s}} - k_{\mathrm{h}}$ and $\hbar\omega = E^{\mathrm{s}}_{k_{\mathrm{s}}} - E^{\mathrm{h}}_{k_{\mathrm{h}}}$, respectively.

Figure 2.17 illustrates the dispersion relations of the spectral function (2.56). The energy region above $\omega = 0$ corresponds to the region experimentally observed in ARPES for $SrCuO_2$ (Fig. 2.16a). In the region between 0 and $\pi/2$, there are two boundary bands: the band $\hbar\omega = F_k$ with width $2J_{\mathrm{s}}$ is predominantly determined by spinon dispersion for $J_{\mathrm{s}}/t_{\mathrm{h}} \ll 1$, while the band $\hbar\omega = 2t_{\mathrm{h}}\sin k$ is due to holon dispersion. The reason there is only one band, with holon character between $\pi/2$ and π is that the spinon band is half-filled with $k_{\mathrm{F}}^{\mathrm{s}} = \pi/2$. In the meshed regions that are bounded by the spinon and holon branches, the spectral weight is stronger than that in the shaded region. This result is in accord with the numerical results presented in Fig. 2.16b. One finds from (2.56) and (2.58) that a square-root-type divergence occurs at the edge of the spinon dispersions satisfying $\hbar\omega = \pm F_k$. This is also consistent with the numerical data where spectral weight at the spinon edges is enhanced. The numerical data in Fig. 2.16b also shows enhancements of the weight on the holon dispersions from 0 to π. This is, however, not reproduced in (2.56), where the spectral weight on the holon dispersion, $\hbar\omega = \pm 2t_{\mathrm{h}}\sin k$, does not show any singularity. This inconsistency is recovered when the effect

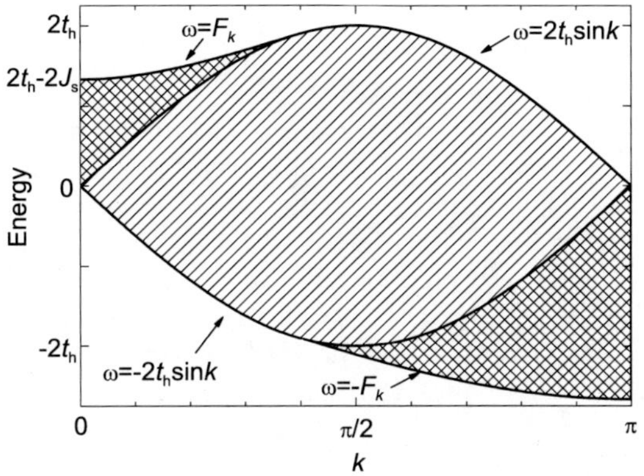

Fig. 2.17. The dispersion relations of the electron-removal spectral function in the 1D Mott insulator obtained from the convolution of spinon and holon (2.56), based on the slave boson mean-field formalism.

of phases acting on spinon and holons is taken into account in evaluating the electron Green's function [59]. This *phase string* effect [58], originates from the Marshall sign rule for the Heisenberg model [69]. Here, we are not going to enter into the details of the theory, but just mention that the spinon Green's function (2.55) is changed to

$$G_{\text{spinon}}(r,t) = -\mathrm{i}\sum_{\sigma}\left\langle T f_\sigma(r,t)\mathrm{e}^{\mathrm{i}\theta_r^{\mathrm{s}}}\mathrm{e}^{-\mathrm{i}\theta_0^{\mathrm{s}}} f_\sigma^\dagger(0,0)\right\rangle \tag{2.59}$$

with the phase θ_r^{s} acting on the spinon. After some algebra, the spectral function is found to show inverse-square-root singularities for the holon dispersion except at $k = \pi/2$ where an exponent of $-3/4$ is obtained. The readers who are interested in the details are referred to the original paper [59] as well as [58].

It should be noted that such inverse-square-root singularities are also reproduced by the Tomonaga-Luttinger liquid picture. As mentioned in Sect. 2.2.1, the spectral line shape contains information on the exponent α, which appears in the branch cut singularities (see (2.49) and (2.51)). In the Tomonaga-Luttinger liquid picture, the Mott insulator is the consequence of instability in the charge channel, caused by Umklapp scattering due to commensurate band fillings. The Umklapp scattering opens a charge gap, while the spin excitation spectrum is gapless. Therefore, the spectral function of the Mott insulator is given by simply exchanging σ and ρ in the spectral function for the Luther-Emery model (2.51). Since $K_\rho = 1$ due to spin-rotational invariance, the exponent α is zero according to (2.50), leading to inverse-square-root singularities in the spinon and holon bands in (2.51). Although we have discussed the exponents of the singularities in the spectral function, the experimental resolution of ARPES, at present, is unfortunately not sufficient to determine the exponents. There is hope that in the near future high-resolution ARPES experiments at low temperature will change this situation.

Metallic Chains

So far we have discussed spin-charge separation seen by a photo-doped carrier in 1D insulating cuprates. Particularly interesting is the metallic state of 1D cuprates. However, to date there has been no success in introducing carriers into 1D corner-sharing insulating cuprates such as $SrCuO_2$ and Sr_2CuO_3. However, there are doped Cu-O chains inside high-T_c materials, e.g., in $YBa_2Cu_3O_7$ (Y123), $YBa_2Cu_4O_8$ (Y124) family. These contain both CuO_2 plane and corner-sharing CuO_3 single chains as in Sr_2CuO_3 (for Y123) or zigzag-type double chain as in $SrCuO_2$ (for Y124). The resistivity along the Cu-O chains is reported to be metallic in Y123 and Y124. However, since highly metallic CuO_2 planes are involved, these two materials do not really represent suitable subjects if one wishes to observe the spinon and holon branches separately [70]. Among the family of cuprates, Pr123 and Pr124 are more suitable choices, since superconductivity is suppressed and there are no carriers in the CuO_2 plane. In fact ARPES experiments have been performed on Pr123 as well as Zn-doped Pr124 and observed are indeed 1D dispersive spectral weights with two structures [71]. These are interpreted in terms of holons and spinons.

2.2.3 Charge Dynamics in Insulating Cuprates

In the previous subsection, Sect. 2.2.2, we discussed the electron-removal spectral function that can describe the occupied states (the Zhang-Rice band) in the corner-sharing 1D insulating cuprates. The charge dynamics in the Mott insulators is, however, controlled by not only the occupied states but also the unoccupied states which corresponds to the upper Hubbard band of Cu $3d$. This corresponds to the excitation of electrons across the Mott gap, as discussed in Sect. 2.1.6 in terms of optical excitations.

Holon-Doublon Model

By mapping the three-band model to the single-band model, the Mott gap excitation in the large U limit is described as a transition which changes two single occupied electron sites to one empty and one doubly occupied sites. The energy cost of the transition is, needless to say, of the order of U. The empty site possesses positive charge as compared with other sites, while the doubly occupied site has negative charge. Since the two sites can change their positions using the hopping processes of electrons, the two carriers with different charges propagate in the system: one is a hole carrier propagating through an unoccupied site and the other is an electron carrier passing through doubly occupied sites. In 1D system with large U, each carrier is similar to a single carrier in the antiferromagnetic background, as shown in Fig. 2.14. Therefore, even in the two-carrier case, the concept of the spin-charge separation may manifest itself, and the spin degree of freedom is not involved in the charge dynamics. More precisely this is based on the fact that in the infinite-U limit of the 1D single-band Hubbard model, the ground-state wave function is explicitly given by the product of a charge part and a spin part [53]. The charge part is expressed in the form of the spinless fermion system, and the spin part is just the Bethe-ansatz solution to a 1D Heisenberg model on a lattice where all hole sites are squeezed out. Thus, naively the correlation functions are just simple products of the spinless fermion charge part and the Heisenberg spin part, and for charge dynamics only the spinless fermion part can contribute.

Based on the above considerations, in the large-U limit, an effective two-particle model with only the charge degrees of freedom may derived:

$$\begin{aligned} H = & \, t \sum_j \left(h_j^\dagger h_{j+1} + d_j^\dagger d_{j+1} + \text{h.c.} \right) + \frac{U}{2} \sum_j \left(h_j^\dagger h_j + d_j^\dagger d_j \right) \\ & - V \sum_j \left(h_{j+1}^\dagger h_{j+1} + h_{j-1}^\dagger h_{j-1} \right) d_j^\dagger d_j . \end{aligned} \tag{2.60}$$

The operators h and d represent the annihilation of the charge degree of freedom of the unoccupied site and the doubly occupied site, respectively. Here these charge carriers will be called a *holon* and *doublon*. (The definition of holon is somehow different from that discussed in Sect. 2.2.2.) Since the holon and doublon have the positive and negative charges, respectively, as compared with the background charge, they may attract each other as expressed by $-V$ in (2.60). The value of V should satisfy $V \ll U$, since the ground state changes from the spin-density wave state (Heisenberg-type) to the charge-density one as V approaches to $U/2$. We also need the constraint that $(h_j^\dagger h_j + d_j^\dagger d_j) \leq 1$, because it is impossible for the site j to be occupied by both holon and doublon at the same time. It is important to notice that this constraint differentiates the holon-doublon model (2.60) from the standard excitonic model in the semiconductors with two particles, a *hole* and a *particle* in the valence and conduction bands, respectively (see, for instance, [72]).

Bound State in Holon-Doublon Model

Since the holon-doublon model is the two-particle problem with hard-core, which is equivalent to the problem of two spinless fermions, the distribution of the eigenstates for the model is easily evaluated by using the wave function,

$$|\phi\rangle = \sum_{j_1 < j_2} a(j_1, j_2) \, |j_1, j_2\rangle , \tag{2.61}$$

where two particles are located at the sites j_1 and j_2. When $j_2 \neq j_1 + 1$, the Schrödinger equation reads

$$\begin{aligned} & t \left[a(j_1+1, j_2) + a(j_1-1, j_2) + a(j_1, j_2+1) + a(j_1, j_2-1) \right] \\ & \quad + U a(j_1, j_2) = E a(j_1, j_2) . \end{aligned} \tag{2.62}$$

When $j_2 = j_1 + 1$,

$$t\left[a\left(j_1-1,j_1+1\right)+a\left(j_1,j_1+2\right)\right]+Ua\left(j_1,j_1+1\right)$$
$$-Va\left(j_1,j_1+1\right)=Ea\left(j_1,j_1+1\right), \tag{2.63}$$

which is different from (2.62). However, if one adds a condition

$$t\left[a\left(j_1+1,j_1+1\right)+a\left(j_1,j_1\right)\right]+Va\left(j_1,j_1+1\right)=0 \tag{2.64}$$

to (2.63), one finds that (2.62) is satisfied even for $j_2 = j_1 + 1$. The Bethe-ansatz tells us that the coefficient $a\left(j_1,j_2\right)$ is given by

$$a\left(j_1,j_2\right)=\mathrm{e}^{\mathrm{i}(k_1j_1+k_2j_2)}+\mathrm{e}^{\mathrm{i}\theta(k_1,k_2)}\mathrm{e}^{\mathrm{i}(k_1j_2+k_2j_1)}, \tag{2.65}$$

where k_1 and k_2 are the momenta of the two particles, and θ is the phase shift due to the presence of interactions. Introducing the total and relative momenta, $q = k_1 + k_2$ and $p = (k_1 - k_2)/2$,

$$a\left(j_1,j_2\right)=\mathrm{e}^{\mathrm{i}(j_1+j_2)q/2}\left(\mathrm{e}^{\mathrm{i}p(j_1-j_2)}+\mathrm{e}^{\mathrm{i}\theta(q,p)}\mathrm{e}^{\mathrm{i}p(j_2-j_1)}\right). \tag{2.66}$$

The constraint (2.64) gives

$$\mathrm{e}^{\mathrm{i}\theta(q,p)}=-\frac{V\mathrm{e}^{-\mathrm{i}p}+2t\cos(q/2)}{V\mathrm{e}^{\mathrm{i}p}+2t\cos(q/2)}, \tag{2.67}$$

and thus the phase shift satisfies

$$\tan\frac{\theta(q,p)}{2}=\frac{V\cos p+2t\cos(q/2)}{V\sin p}. \tag{2.68}$$

From (2.62) and (2.66) the energy

$$E=U+4t\cos\frac{q}{2}\cos p. \tag{2.69}$$

The anti-periodic boundary condition $a(0,j) = -a(j,L)$ results in the quantization conditions $L(q/2+p)+\theta = 2\pi(\lambda_1+1/2)$ and $L(q/2-p)+\theta = 2\pi(\lambda_2+1/2)$ ($\lambda_1, \lambda_2 = 0,1,2,\cdots,L-1$). When $p = 0$, there is no solution for a, i.e., $a(j_1,j_2) = 0$. It is found that there are $L-1$ null solutions with real wavevectors. It is also known that there are $L-1$ solutions exist for which k_1 and k_2 are the complex conjugate each other, i.e., $q = 2k$ and $p = \mathrm{i}k'$. These solutions represent bound states. Using (2.68), if $V > 2t\cos(q/2)$, then $\mathrm{e}^{-k'L} \to 0$ for $L \to \infty$, and the energy of the bound state is given by

$$E_{\mathrm{b}}=U-V-\frac{4t^2}{V}\cos^2\frac{q}{2}. \tag{2.70}$$

The distribution of the eigenenergies (2.69) and (2.70) for $V/t = 2$ is plotted in Fig. 2.18, as a function of the total momentum q. The width is $8t$ at $q = 0$, and it decreases with increasing q. The bound state of a holon and doublon emerges at $q = 0$ if $V > 2t$, and at $q = \pi$ for any finite V. The inequality implies that the optical absorption spectra (no change of the total momentum) shows the bound state split off from the continuum if V is large.

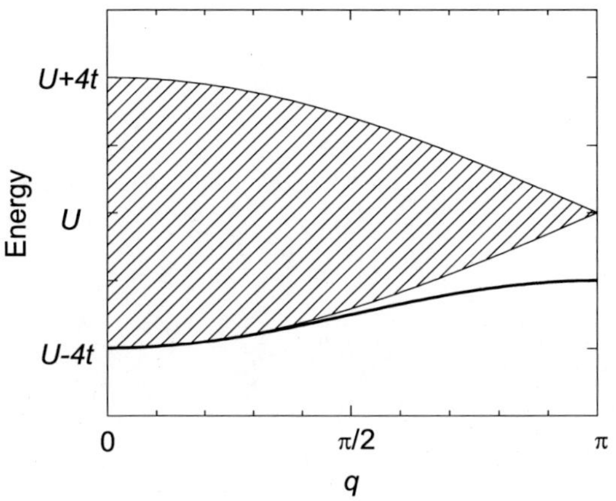

Fig. 2.18. The energy range of the eigenstate on the holon-doublon model as a function of q (shaded area) and the exciton dispersion (solid line) induced by an attractive interaction $-V$ between a holon and a doublon with $V = 2t$.

The expression for the dynamical charge correlation function (2.44) was given in [73]. Leaving the detailed derivation to the original paper, (neglecting small contributions) the spectral density function

$$N(q,\omega) = \frac{16t^2}{U^2} \ln 2 \sin^2 \frac{q}{2} g(q, \omega - U/\hbar) \tag{2.71}$$

with

$$\begin{aligned} g(q,\omega') = {} & \frac{1}{2\pi} \frac{\sqrt{16t^2 \cos^2(q/2) - \hbar^2 \omega'^2}}{4t^2 \cos^2(q/2) + \hbar V \omega' + V^2} \Theta\left(4t \cos(q/2) - |\hbar\omega'|\right) \\ & + \left(1 - 4t^2/V^2 \cos^2(q/2)\right) \delta\left(\omega' - \left(E_\mathrm{b} - U\right)/\hbar\right) . \end{aligned} \tag{2.72}$$

Figure 2.19a shows $g(q, \omega - U/\hbar)$ for various values of q with $V = 2t$. With increasing q, the weight decreases, and at finite q a bound state emerges with large weight. At $q = 0$, $g(q, \omega - U/\hbar)$ is proportional to the imaginary part of the dielectric function, ϵ_2, i.e., the absorption spectrum. Figure 2.19b shows $g(0, \omega - U/\hbar)$ for $V = 0$, t, and $3t$. At $V = 0$, there is no singularity at the edge of the spectrum. This differs from the simple picture of the absorption spectrum as just the joint density of states of the valence and conduction bands. With increasing V, the spectral weight begins to concentrate on the absorption edge, and for $V > 2t$ a bound state splits from the continuum band. Here, since it is assumed that there is no double occupation in the ground state, the enhancement of the spectral weight at the absorption edge reflects the interaction V. In the real 1D cuprates, however, an enhancement of the edge weight may arise not only as a consequence of this interaction but also because of the finite value of U which allows the double occupation in the ground state. For the optical absorption in 1D Hubbard model, please refer to [74].

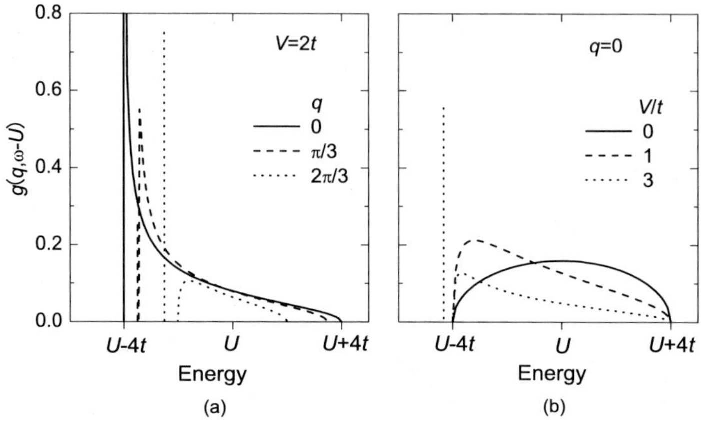

Fig. 2.19. $g(q, \omega - U/\hbar)$. (**a**) $V = 2t$ and $q = 0$ (solid), $\pi/3$ (dashed), and $2\pi/3$ (dotted line). (**b**) $q = 0$ and $V = 0$ (solid), t (dashed), and $3t$ (dotted line). The spectra correspond to the optical absorption.

Electron Energy-Loss Spectroscopy

The electron energy-loss spectroscopy (EELS) is one of the most powerful tools by which to investigate the charge dynamics of Mott insulators. EELS measures the scattering probability of incident electrons that lose their energy $\hbar\omega$ and momentum $\boldsymbol{q}$. Since the electric field induced by the incident electrons in a sample is proportional to $e/|\epsilon|$, $\epsilon = \epsilon_1 + \mathrm{i}\epsilon_2$ being the dielectric function, the energy absorbed into the sample is proportional to $\omega\epsilon_2(\boldsymbol{q},\omega)/|\epsilon(\boldsymbol{q},\omega)|^2$ per second. Thus the scattering probability is proportional to $\epsilon_2(\boldsymbol{q},\omega)/|\epsilon(\boldsymbol{q},\omega)|^2 = \mathrm{Im}\,(-1/\epsilon(\boldsymbol{q},\omega))$, which is called the loss function.

Neudert et al. [75] have carried out EELS measurements on a 1D insulating cuprate Sr_2CuO_3. Figure 2.20a shows the loss function with momentum transfer q parallel to the Cu-O chain direction, where q ranges from $q \sim 0$ to $\sim \pi$. At $q \sim 0$, a broad spectrum at around 2 eV is similar to that of the optical conductivity shown in Fig. 2.11a. The broad spectrum narrows with increasing q, evolving into a single sharp peak centered at 2.8 eV for q=0.4 Å^{-1}. With further increasing q, the peak, at 3.2 eV, becomes slightly broader, and it is located at the zone boundary. It has been suggested [75] that the momentum dependence of the data is consistent with the loss function for the holon-doublon model, especially the sharp peak at large momentum transfer which may correspond to the bound state of holon and doublon as discussed above.

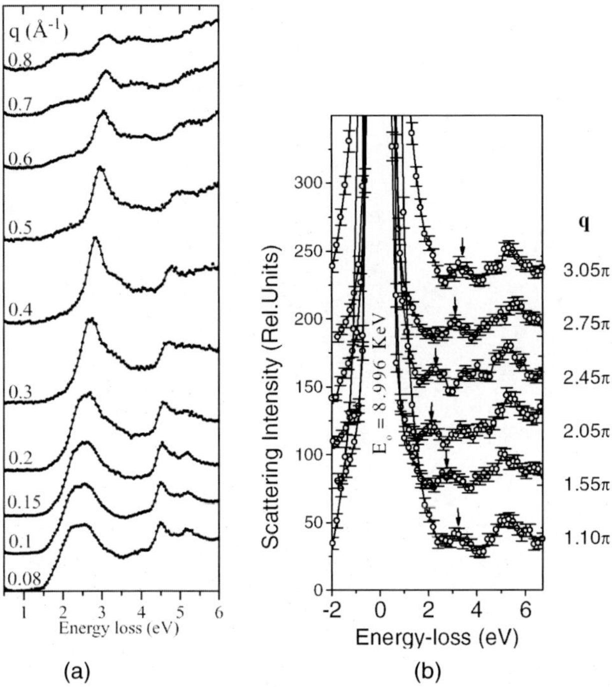

Fig. 2.20. The charge dynamics in 1D insulating cuprates. (**a**) The loss function for Sr_2CuO_3 measured by EELS with the momentum transfer q parallel to the Cu-O chain, where q=0.08 Å^{-1} and 0.8 Å^{-1} correspond to the zone center and boundary, i.e., $q \sim 0$ and $\sim \pi$, respectively. (From [75]). (**b**) Resonant inelastic x-ray scattering spectra for $SrCuO_2$ with the momentum transfer q parallel to the Cu-O chain over the entire Brillouin zone. $q = 2\pi$ corresponds to $q = 0$. The peaks denoted by arrows correspond to those in EELS data. Huge zero-energy peaks come from elastic scattering. (From [76])

Resonant Inelastic X-Ray Scattering

The observation of the momentum-dependent excitations across the gap is possible not only by means of EELS but also by means of resonant inelastic x-ray scattering (RIXS) in which the incident photon energy is tuned through the Cu K absorption edge. Since the technique determines directly the excitation at Cu sites, it is expected to be a powerful tool for studying the momentum dependence in the Cu-O network. Figure 2.21 shows the schematic process involved in the Cu K-edge RIXS for a Mott insulator. An absorption of an incident photon with frequency ω_i, momentum $\boldsymbol{K}_i$, and polarization $\boldsymbol{\epsilon}_i$ causes a dipole transition in which an electron excited from the Cu 1s to 4p orbital [process (a) in Fig. 2.21]. In the intermediate states, 3d electrons (more precisely, electrons in the Zhang-Rice singlet band) interact with a 1s-core hole and a photo-excited 4p electron via the Coulomb interac-

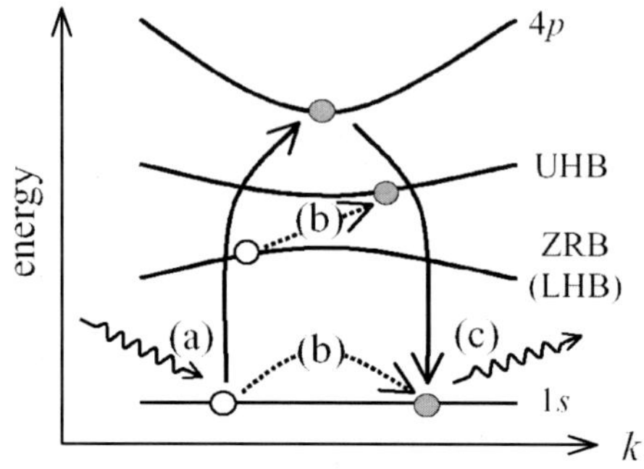

Fig. 2.21. A schematic picture of the Cu K-edge RIXS process. An incident photon is absorbed and dipole transition $1s \to 4p$ is brought about [process (**a**)], and through the intermediate state [process (**b**)] where an electron in the occupied band (the Zhang-Rice singlet band or lower Hubbard band) is excited to the unoccupied band (the upper Hubbard band), the photo-excited $4p$ electron goes to $1s$ again and a photon is emitted [process (**c**)].

tions so that the excitations in the $3d$ electron system are involved [process (b)]. The $4p$ electron returns to the $1s$ orbital and a photon with frequency ω_f, momentum $\boldsymbol{K}_\mathrm{f}$, and polarization $\boldsymbol{\epsilon}_\mathrm{f}$ is emitted [process (c)]. The differences in the energies and the momenta between incident and emitted photons, $\hbar\Delta\omega = \hbar\omega_\mathrm{i} - \hbar\omega_\mathrm{f}$ and $\Delta\boldsymbol{K} = \boldsymbol{K}_\mathrm{i} - \boldsymbol{K}_\mathrm{f}$, are transferred to the $3d$ electron system. (See also Sect. 4.3 for Mn K-edge RXS.) The excitation from the lower to upper Hubbard bands are principally caused by a local attractive interaction between the $1s$-core hole $s_{i,\sigma}$ and $3d$ electron $d_{i,\sigma}$

$$
\begin{aligned}
&- V_{1s\text{-}3d} \sum_{i,\sigma,\sigma'} d^\dagger_{i,\sigma} d_{i,\sigma} s^\dagger_{i,\sigma'} s_{i,\sigma'} \\
&= -\frac{V_{1s\text{-}3d}}{N} \sum_{\substack{\boldsymbol{k}_1,\boldsymbol{k}_2,\boldsymbol{k}_3,\boldsymbol{k}_4,\sigma,\sigma' \\ \boldsymbol{k}_1+\boldsymbol{k}_2=\boldsymbol{k}_3+\boldsymbol{k}_4}} d^\dagger_{\boldsymbol{k}_1,\sigma} d_{\boldsymbol{k}_3,\sigma} s^\dagger_{\boldsymbol{k}_2,\sigma'} s_{\boldsymbol{k}_4,\sigma'} , \qquad (2.73)
\end{aligned}
$$

where the $3d$ electron can be excited from a state with momentum $\boldsymbol{k}_3$ in the lower Hubbard band to that with $\boldsymbol{k}_1$ in the upper Hubbard band.

In Fig. 2.20b, the RIXS spectra for an 1D insulating cuprate $SrCuO_2$ are shown [76]. In addition to huge elastic peaks at zero energy, one can see peak structures, with small weight, in the energy range 2 to 3 eV. These correspond to the excitations across the Mott gap and clearly show momentum dependence similar to those of the EELS data. Therefore, RIXS as well as EELS can capture the physics of charge dynamics in 1D insulating cuprates, which originates from the spin-charge separation.

2.2.4 Nonlinear Optical Response

Since the gap in Mott insulators is a consequence of strong electron correlation, the nature of charge excitation across the gap is expected to be different from that of more usual insulators in which the gap basically originates from band effects. Actually, the measurements of the optical nonlinear susceptibilities in 1D Mott insulators, including a cuprate Sr_2CuO_3, illustrates the remarkable difference between the two types of insulators [77–79], which will be discussed in this section.

Formulation

In the nonlinear optical response, the polarization $\boldsymbol{P}$ of a medium under the influence of an applied electric filed $\boldsymbol{E}$ is described in terms of a power series in the field. In the case of conventional optics, the induced polarization depends linearly on the field as described by the relationship

$$P_j^{(1)}(\omega) = \chi_{jk}^{(1)}(-\omega;\omega)E_k(\omega), \tag{2.74}$$

where the indices j, k refer to the Cartesian components of the field, ω is the frequency of the field, and the coefficient $\chi^{(1)}$ is known as the linear susceptibility. Note that the linear susceptibility is related to the dielectric function, i.e., $\epsilon = 1 + 4\pi\chi^{(1)}$. The second-order nonlinear susceptibility, however, vanishes, when the medium has inversion symmetry, i.e., the system is centrosymmetric: $\boldsymbol{P} \to -\boldsymbol{P}$ and $\boldsymbol{E} \to -\boldsymbol{E}$ for the inversion $\boldsymbol{r} \to -\boldsymbol{r}$, but $\chi^{(2)}$ is unchanged, leading to $\chi^{(2)} = -\chi^{(2)}$ and thus $\chi^{(2)} = 0$. Since the 1D cuprates display this symmetry, the third-order nonlinear optical response dominates nonlinear optics in these materials. The corresponding polarization reads

$$P_j^{(3)}(\omega_\sigma) = K\sum_{klm}\chi_{jklm}^{(3)}(-\omega_\sigma;\omega_1,\omega_2,\omega_3)E_k(\omega_1)E_l(\omega_2)E_m(\omega_3), \tag{2.75}$$

where $\omega_\sigma = \omega_1 + \omega_2 + \omega_3$ and K is the degeneracy factor which is equal to the number of distinct permutations of the frequencies ω_1, ω_2, and ω_3. Introducing the dipole moment of two states, $e\boldsymbol{r}_{ab} \equiv e\,\langle a\,|\boldsymbol{r}|\,b\rangle$, $\boldsymbol{r}$ being the displacement operator, and following textbooks on nonlinear optics [80, 81], the linear and third-order nonlinear susceptibilities at zero temperature may be expressed as

$$\chi_{jk}^{(1)}(-\omega;\omega) = \frac{n_c e^2}{\hbar}\sum_a\left(\frac{r_{0a}^j r_{a0}^k}{\Omega_a - \mathrm{i}\Gamma_a - \omega} + \frac{r_{0a}^k r_{a0}^j}{\Omega_a + \mathrm{i}\Gamma_a + \omega}\right), \tag{2.76}$$

and

$$\begin{aligned}
&\chi_{jklm}^{(3)}(-\omega_\sigma;\omega_1,\omega_2,\omega_3) = \frac{n_c e^4}{3!\hbar^3}\mathcal{P} \\
&\times\sum_{a,b,c}\Bigg(\frac{r_{0a}^j r_{ab}^k r_{bc}^l r_{c0}^m}{(\Omega_a - \mathrm{i}\Gamma_a - \omega_\sigma)(\Omega_b - \mathrm{i}\Gamma_b - \omega_2 - \omega_3)(\Omega_c - \mathrm{i}\Gamma_c - \omega_3)} \\
&+ \frac{r_{0a}^k r_{ab}^j r_{bc}^l r_{c0}^m}{(\Omega_a + \mathrm{i}\Gamma_a + \omega_1)(\Omega_b - \mathrm{i}\Gamma_b - \omega_2 - \omega_3)(\Omega_c - \mathrm{i}\Gamma_c - \omega_3)} \\
&+ \frac{r_{0a}^k r_{ab}^l r_{bc}^j r_{c0}^m}{(\Omega_a + \mathrm{i}\Gamma_a + \omega_1)(\Omega_b + \mathrm{i}\Gamma_b + \omega_1 + \omega_2)(\Omega_c - \mathrm{i}\Gamma_c - \omega_3)}
\end{aligned}$$

$$+\frac{r^k_{0a}r^l_{ab}r^m_{bc}r^j_{c0}}{(\Omega_a+\mathrm{i}\Gamma_a+\omega_1)(\Omega_b+\mathrm{i}\Gamma_b+\omega_1+\omega_2)(\Omega_c+\mathrm{i}\Gamma_c+\omega_\sigma)}\Bigg), \tag{2.77}$$

where n_c is the number density of atoms, and $\mathcal{P}$ in (2.77) represents the sum of the permutations on $(j,\, \omega_1)$, $(k,\, \omega_2)$, and $(l,\, \omega_3)$. The quantity Ω_a is the energy difference between the ground state $|0\rangle$ and excited state $|a\rangle$, and Γ_a is the damping factor. It should be noted that, for the dipole transition due to the electric field, the state $|a\rangle$ as well as $|c\rangle$ have odd parity in terms of the inversion operation, while $|b\rangle$ is an even-parity state. Therefore, combining $\chi^{(1)}$ and $\chi^{(3)}$ measurements, one can obtain information about the distribution of both the odd- and even-parity excited states.

The susceptibility $\chi^{(3)}$ can be obtained in the following experimental methods: (i) *electro-modulation spectroscopy*, in which the change of the linear response due to the static electric filed is observed, leading to $\chi^{(3)}(-\omega;0,0,\omega)$, (ii) *pump and probe spectroscopy*, in which the probe photon with frequency ω_1 detects the change of the linear response due to the pump photon ω_2, leading to two photon absorption (TPA) $\chi^{(3)}(-\omega_1;-\omega_2,\omega_2,\omega_1)$, and (iii) *third harmonic generation spectroscopy* (THG), in which 3ω photons result from an incident photon with frequency ω, leading to $\chi^{(3)}(-3\omega;\omega,\omega,\omega)$. Note that TPA can directly detect the even-parity states.

Experiment and Theory for Sr_2CuO_3

Kishida et al. [77] have reported anomalously enhanced $\chi^{(3)}$-values in the range 10^{-8} to 10^{-5} esu (electrostatic system of units) obtained by analyzing the electroreflectance spectra of 1D Mott insulators such as Ni halides and Sr_2CuO_3. These values are larger than those, 10^{-12} to 10^{-7} esu, for silicon polymer band insulators and Peierls insulators, e.g., the π-conjugated polymers and platinum halides. The anomalous enhancement of $\chi^{(3)}$ in 1D Mott insulators was attributed to a large dipole moment for the nearly degenerate odd- and even-parity states. The data for Sr_2CuO_3 is shown in Fig. 2.22a, where the polarization of the electric field is along the Cu-O chain. In addition, by analyzing the photoinduced absorption of time-resolved femtosecond pump-probe measurements, Ogasawara et al. [78] have shown that an even-parity two-photon state occurs just above linear absorption edge, i.e., an odd-parity state. Figure 2.22b shows the photoinduced absorption efficiency $\beta(\omega_{\text{probe}},\omega_{\text{pump}})$, which is proportional to $\mathrm{Im}\chi^{(3)}(-\omega_1;-\omega_2,\omega_2,\omega_1)$ $(\omega_1=\omega_{\text{probe}},\ \omega_2=\omega_{\text{pump}})$, as functions of $\omega_{\text{probe}}+\omega_{\text{pump}}$. It is found that the maximum of the pump-induced absorption at 2.1 eV is very close to that of the linear absorption, and the two absorptions nearly overlap. The THG spectrum is also shown in Fig. 2.22c [79], where there is a peak denoted A and a shoulder B with an energy relation $3\hbar\omega_{\mathrm{A}} \sim 2\hbar\omega_{\mathrm{B}}$. This implies that, taking into account the nearly degenerate odd- and even-parity states, A and B can be assigned to the three- and two-photon resonances, respectively.

The holon-doublon model (2.60) can explain the experimental data of $\chi^{(3)}$ in Sr_2CuO_3. Figure 2.23 shows the linear and nonlinear susceptibilities

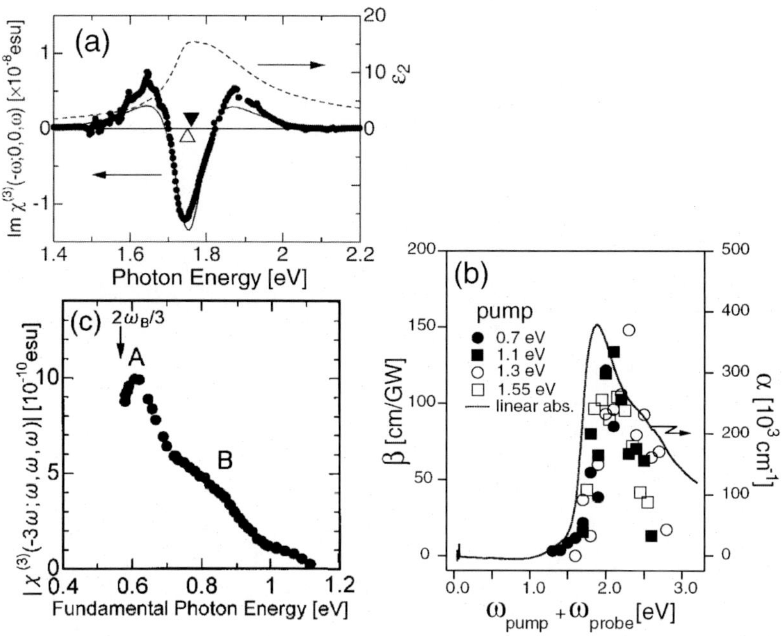

Fig. 2.22. The third-order nonlinear optical responses in Sr_2CuO_3. (**a**) ϵ_2 and $\text{Im}\chi^{(3)}(-\omega;0,0,\omega)$. (From [77]). (**b**) The photoinduced absorption efficiency β that is proportional to $\text{Im}\chi^{(3)}(-\omega_{\text{probe}};-\omega_{\text{pump}},\omega_{\text{pump}},\omega_{\text{probe}})$ for four different pump energies. The dotted line is the linear absorption α that is proportional to ϵ_2. (From [78]) (**c**) THG spectrum. (From [79])

for a 100-site chain. The results are very similar to the experimental data in the sense that (i) ϵ_2 and $\text{Im}\chi^{(3)}(-\omega;-\omega,\omega,\omega)$ are concentrated in the same energy region when the energy in $\text{Im}\chi^{(3)}(-\omega;-\omega,\omega,\omega)$ is doubled, (ii) $\text{Im}\chi^{(3)}(-\omega;0,0,\omega)$ shows an oscillating structure at the spectral edge of ϵ_2 ($\omega \sim 2$), and (iii) there are a peak and a broad hump in $|\chi^{(3)}(-3\omega;\omega,\omega,\omega)|$. These identifications suggest that this model is appropriate for the description of the nonlinear optical response of the 1D Mott insulators. Therefore it might be possible to say that the spin-charge separation and strong correlation are manifest in the nonlinear optics of 1D Mott insulators.

The presence of the nearly degenerate odd- and even-parity states is a characteristic feature of 1D Mott insulators as emphasized above. This is in contrast to the standard exciton model of 1D band and Peierls insulators. Let us discuss this difference very simply. Electron correlations are, of course, the origin of these differences. In the holon-doublon model (2.60), the two particles, the holon and doublon, cannot occupy the same site at the same time because of strong correlations, while the band insulators do permit the double occupation of a particle and a hole. The prohibition on

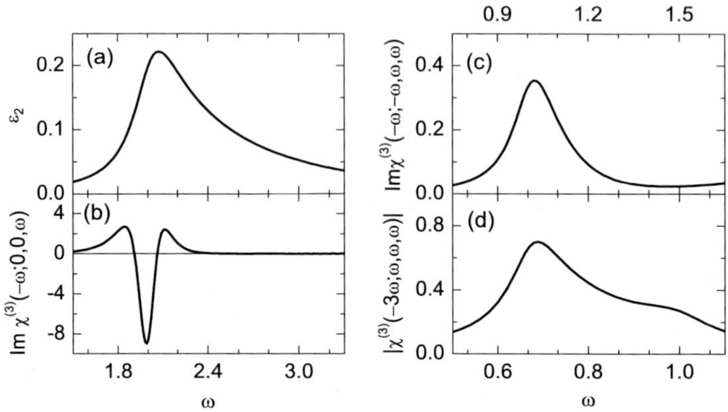

Fig. 2.23. (**a**) ϵ_2, (**b**) $\text{Im}\chi^{(3)}(-\omega; 0, 0, \omega)$, (**c**) $\text{Im}\chi^{(3)}(-\omega; -\omega, \omega, \omega)$, and (**d**) $|\chi^{(3)}(-3\omega; \omega, \omega, \omega)|$ in the holon-doublon model for a 100-site chain with t=0.5, V=1, and U=4. The damping factor Γ is assumed to be independent of the excited states and is taken to be 0.15.

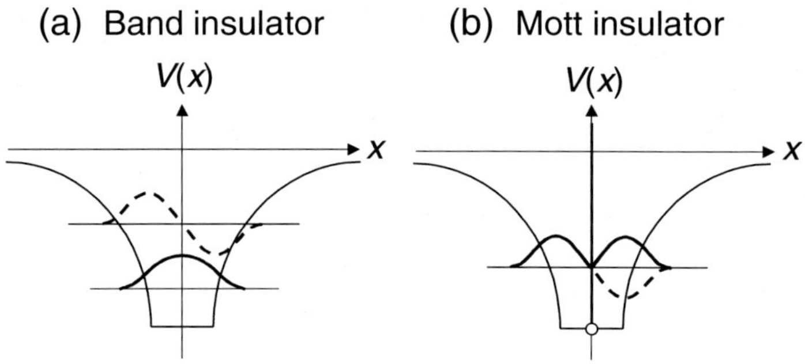

Fig. 2.24. A schematic description of the 1D excitonic envelope wave function of two particles created by photoexcitation in band insulator (**a**) and Mott insulator (**b**) with long-range attractive Coulomb interaction $V(x)$. Usual 1D exciton wave functions without node (solid) and with node (broken line) are shown in the band insulator. In the Mott insulator, irrespective of the symmetry of the wave functions, nodes appear at the origin of the relative coordinate x because of strong on-site Coulomb interaction. The odd- and even-parity states are thus degenerate.

double occupancy necessarily induces a node in the envelope wave function at the origin of the relative coordinates, and the odd- and even-parity states are always degenerate due to the presence of this node [82]. This is schematically represented in Fig. 2.24, where, for band and Mott insulators, the 1D envelop wave functions, under the influence of long-range attractive Coulomb interactions $V(x)$, are compared.

More importantly, Sr_2CuO_3 exhibits not only strong optical nonlinearity but also picosecond recovery of the optical transparency at room temperature, which indicates the great potential of 1D Mott insulators for use as

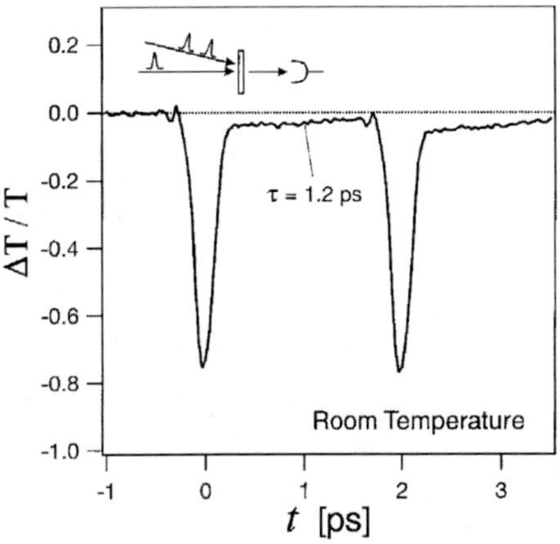

Fig. 2.25. The transmission change induced by two temporally separated equivalent pump pulses in Sr_2CuO_3. The pump and probe photonenergies are 0.88 eV (1.4 μm) and 1.03 eV (1.2 μm), respectively. The relaxation time of the tail part is 1.2 ps. (From [78])

high-bit-rate all-optical switching devices [78]. Such a demonstration is shown in Fig. 2.25. For such an ultrashort response, a fast nonradiative relaxation channel is necessary. A spin channel may be one of the possible channels for such a relaxation [78]; the ultrafast recovery to the ground state occurs in the following two steps, (i) the relaxation of the photoexcited states to the highest energy spin excited states, and (ii) subsequent relaxation of the spin excited states through the emission of multiple phonons and spinons. However, the relationship of this relaxation mechanism to the spin-charge separation picture is not clear at present. Clarifying the mechanism is an urgent problem which remains to be solved. Although a complete understanding of the nonlinear optical response in the 1D Mott insulators has not been obtained, it is possible to say that the concept of the spin-charge separation is not just of purely academic interest but that it is essential to an understanding of the underlying physics of optoelectronic devices based on 1D cuprates.

2.2.5 Spin Dynamics in Insulating Cuprates

Below the charge gap of insulating cuprates, only spin degrees of freedom remain and they are described by the spin 1/2 1D Heisenberg model. For the corner-sharing chains, only the nearest-neighbor antiferromagnetic interaction J is needed in the model as discussed in Sect. 2.1.3. In fact, Sr_2CuO_3 is an almost ideal 1D antiferromagnetic Heisenberg system and thus this compound is the best candidate for use in a comparison of the dynamical spin properties at finite temperature with theoretical predictions obtained by field-theoretical and numerical techniques.

Neutron scattering experiments can measure the spin dynamics over a wide frequency and momentum range. However, few detailed experiments on the spin dynamics have been reported for Sr_2CuO_3. In order to fully clarify the picture of the spin dynamics of 1D insulating cuprates, i.e., the dynamics of spinon, inelastic neutron scattering experiments are highly desirable [128].

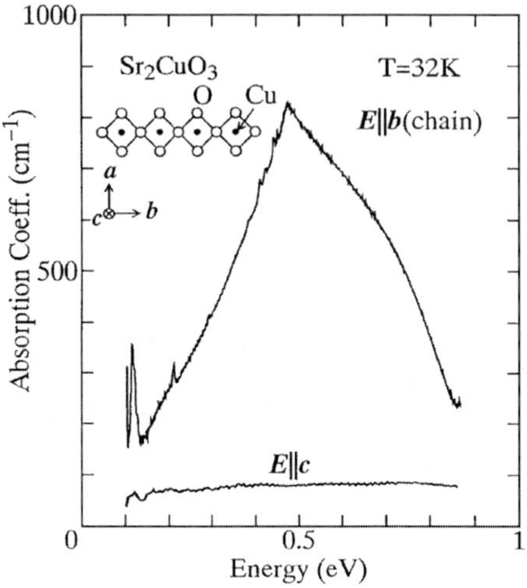

Fig. 2.26. The mid-infrared absorption spectra of Sr_2CuO_3 for the electric field parallel and perpendicular to the chain. The absorption comes from the photon-assisted multimagnon process. (From [19]).

Infrared optical spectra can also offer detailed information on the spin dynamics of insulating cuprates via multimagnon absorption. But this statement sounds somehow strange if one considers the fact that the ordinary optical transition due to a dipole interaction between an electron and a photon does not couple with the spin degrees of freedom of electrons. In particular, in a typical two-magnon excitation, the presence of an inversion center inhibits any asymmetric displacement of charge, and thus the associated dipole moment vanishes. However, the situation changes if the coupling of phonons with electrons is taken into account. In a process where one phonon and two magnons are absorbed, a two-magnon absorption is allowed due to the effective lowering of symmetry of the lattice reflecting the presence of the phonon. This is called the phonon-assisted multimagnon absorption [20].

Suzuura et al. have observed the phonon-assisted magnon absorption for Sr_2CuO_3 [19]. As shown in Fig. 2.26a, the spectrum has an asymmetric cups-like structure at 0.48 eV. This spectrum was analyzed in terms of the 1D Heisenberg model and the agreement was very good [21]. This gives the value of J=0.246 eV, which is directly determined from the position of the cusp located exactly at $\pi J/2$ coming from the saddle point of the des Cloizeaux-Pearson dispersion relation [83].

2.3 Two-Dimensional Cuprates

In 1986, Bednorz and Müller discovered high-T_c superconductivity in a copper oxide La-Ba-Cu-O [1], which has soon been confirmed by Uchida et al. [84]. Even now, almost twenty years after the discovery, we do not have a final answer for the mechanism of superconductivity in spite of intensive studies by many researchers around the world. However, it is also true that there have been much progresses toward the understanding of the unusual properties of the normal and superconducting states of the high-T_c cuprates.

The most dramatic features of high-T_c cuprates are that the superconductivity emerges as a consequence of carrier doping into a two-dimensional (2D) Mott insulator composed of CuO_2 planes and that the superconductivity exists in a certain finite range of carrier density. This naturally suggests the importance of strong electron correlations in high-T_c superconductivity. Although many authors have discussed the electronic properties in 2D cuprates from the view point of strong electron correlation, the problems still remain controversial. In this section, we first discuss the single-carrier motion in 2D antiferromagnetic Mott insulators. The understanding of this is an important initial step in developing a theory for the mechanism of high-T_c superconductivity. We also discuss the change of the electronic states induced by carrier doping with particular emphasis on the asymmetry of hole and electron dopings. The mechanism of superconductivity will be discussed in Chap. 3.

2.3.1 Single Carrier in Mott Insulator

As discussed in Sect. 2.1.3, 2D insulating cuprates, which are the parent compounds of high-T_c superconductors, are antiferromagnetic Mott insulators with Néel temperature $T_N \sim 300$ K. In passing from above to below T_N, there are no remarkable differences in the electronic states except for the occurrence of the magnetic order. This is due to the fact that the antiferromagnetic spin correlations in the CuO_2 plane begin to develop at around $T \sim J/k_B \sim 1000$ K well above T_N and thus the electronic states in the plane is qualitatively similar below and above T_N. For the magnetic excitations from the ordered state, please see the spin-wave excitation in La_2CuO_4 shown in Fig. 2.8.

ARPES

Since the superconductivity occurs when carriers are introduced into the CuO_2 plane, the nature of a single carrier introduced into the 2D Mott insulator was extensively studied soon after the discovery of high-T_c superconductivity. This motion can be described by the single-particle spectral function (2.47). The electron-removal spectral function $A_-(\boldsymbol{k}, \omega)$ is extracted from angle-resolved photoemission spectroscopy (ARPES) for the parent compounds. The first ARPES experiment on undoped compounds was performed by Wells et al. [85] for $Sr_2CuO_2Cl_2$. The data were taken at the temperature of 350 K above T_N of 256 K. Kim et al. [86] found that the dispersion is similar to that at 150 K well below T_N, although the spectral intensity depends slightly on temperature. The data show that along the (0,0)-(π,π) direction, the low energy peak, referred to as the quasiparticle peak, moves by about

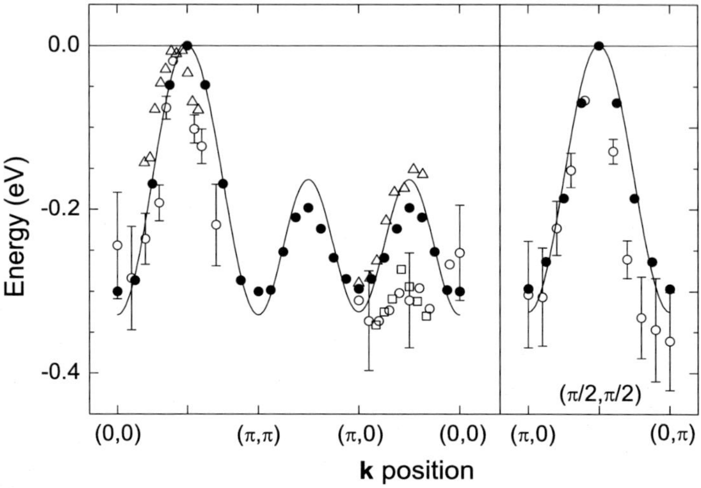

Fig. 2.27. The energy dispersions of quasiparticles for insulating $Sr_2CuO_2Cl_2$ measured from the top of the band. Experimental data are taken from [85] (open circles), [86] (open squares), and [87] (open triangles). Solid circles: self-consistent Born approximation (SCBA) results for the t-t'-t''-J model. The solid lines are obtained by fitting the SCBA data to a dispersion relation given by $E(\mathbf{k})$ in (2.80).

0.3 eV toward the low binding energy side, reaches its minimum at $(\pi/2,\pi/2)$, and then folds back with a rapid reduction of the intensity. Along the (0,0)-$(\pi,0)$ cut, the spectral weight has its maximum at around $(\pi/2,0)$ [87], and as $\boldsymbol{k}$ is increased further toward $(\pi,0)$, the low energy features are suppressed. The dispersion relations experimentally observed are illustrated in Fig. 2.27.

Singel-Hole Motion

Since the ARPES data directly detect the Zhang-Rice singlet bands, their dispersion is expected to be described by the t-J model (2.38). The spectral function of a single hole in the t-J model was examined at $T = 0$ using the numerically exact diagonalization technique for small clusters or the self-consistent Born approximation (SCBA) approach [88]. The latter gives a simple physical picture in which a hole is propagating in a fluctuating spin background as described by the Heisenberg Hamiltonian H_J in (2.38). In this approach H_J is approximately diagonalized, which leads to spin waves, and the kinetic part H_t in (2.38) is rewritten in terms of spinless hole (holons) and spin-wave operators, giving rise to a coupling between holes and spin waves that induces a bound state. This is very similar to the standard polaron problem [50]. Therefore we sometimes calls this bound state a spin-polaron. We also call it a quasiparticle. An interesting aspect of these quasiparticles is the absence of a free kinetic energy for the holon. The coherent propagation of the quasiparticle shows a strongly reduced bandwidth, of the order of J

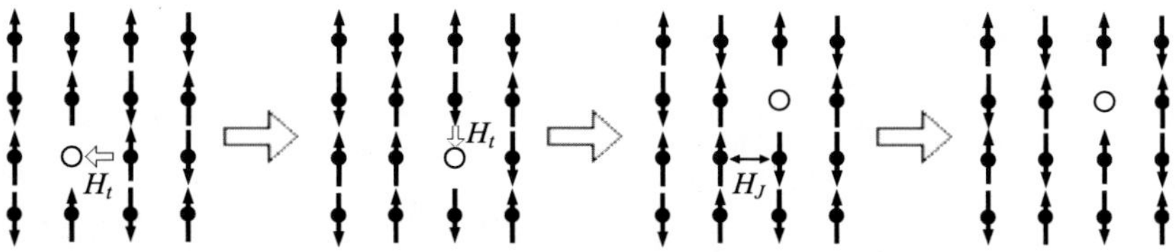

Fig. 2.28. The simplified view of the diagonal motion of a hole in the two-dimensional antiferromagnet. The up and down arrows represent up and down spins, respectively, on Cu sites (black circles), and the white circle exhibits the hole.

for $J < t$, as a result of the coupling to spin waves. This is different from the standard polaron problem, where the free band mass is simply renormalized due to the coupling to phonons. Readers who are interested in the SCBA method are referred to the original papers [88].

A very simple description of hole propagation in the t-J model is schematically shown in Fig. 2.28. Let us consider a configuration where a hole is located in the Néel state in which there are two sublattices with up spins and down spins. An adjacent up spin hops to the hole site by the H_t. The spin moved to the original hole site is now on a different sublattice site, i.e., in the down-spin sublattice, inducing three ferromagnetic aligned bonds. With the next hop of the hole, a down spin comes to the up-spin-sublattice site and the number of ferromagnetic aligned bonds increases to six. Such ferromagnetic aligned bonds disturb the free motion of the hole and the hole tends to be localized. However, the exchange term H_J can now reverse the two wrongly directed spins thereby recovering a fully anti-ferromagnetic configuration. This results in a coherent motion of a hole which effectively hops on just one of the two sublattices. The coherent motion of the hole is controlled by H_J and the band width of the quasiparticle becomes of the order of J. This is a very simplified point of view, and in reality the hole moves coherently but accompanied by a spin cloud. We call this a spin polaron.

Wells et al. [85] found that the t-J model can quantitatively explain the observed ARPES dispersion from (0,0) to (π,π) with a width of $2.2J$. However, the model does not explain the dispersion with a width of about $2J$ along the (0,π)-(π,0) direction (see Fig. 2.27): it predicts a nearly flat dispersion along this direction given approximately by,

$$E_0(\boldsymbol{k}) = -0.55J\left(\cos k_x + \cos k_y\right)^2 , \tag{2.78}$$

where the energy is measured from the ($\pi/2$,$\pi/2$) point, and the coefficient $0.55J$ is determined so as to give the band width of $2.2J$. Note that this dispersion corresponds to the sum of the second-neighbor hopping (diagonal direction) and the third-nearest neighbor hopping (one-lattice spacing) with effective hopping amplitude of $J/4$ and $J/8$, respectively.

The discrepancy between the experiments and t-J results has led to an intense theoretical activity. It was found that the discrepancy may be resolved by adding long-range hopping terms to the t-J model. Needed, at the most, are couplings to the third nearest neighbor with one-lattice spacing:

$$H_{t't''} = -t' \sum_{\langle i,j\rangle_{2\text{nd}},\sigma} \left(\tilde{d}^{\dagger}_{i\sigma}\tilde{d}_{j\sigma} + \text{h.c.}\right) - t'' \sum_{\langle i,j\rangle_{3\text{rd}},\sigma} \left(\tilde{d}^{\dagger}_{i\sigma}\tilde{d}_{j\sigma} + \text{h.c.}\right), \tag{2.79}$$

where $\langle i,j\rangle_{2\text{nd}}$ and $\langle i,j\rangle_{3\text{rd}}$ run over second and third nearest neighbor pairs, respectively, and $\tilde{d}^{\dagger}_{i\sigma} = (1 - n^{d}_{i,-\sigma})d^{\dagger}_{i,\sigma}$. Note that such hoppings are also known to be necessary for the description of the Fermi surface topologies obtained by ARPES as well as from band structure calculations. The solid circles in Fig. 2.27 show the SCBA results for the t-t'-t''-J Hamiltonian, $H_t + H_{t't''} + H_J$, with parameters $t' = -0.12$ eV and $t'' = 0.08$ eV together with $t = 0.35$ eV and $J = 0.14$ eV, which reproduces the experimental data very well not only along the (0,0)-(π,π) direction but also the (π,0)-(0,π) direction. Taking into account t' and t'', the single-hole dispersion of the t-t'-t''-J model may be approximately written as

$$E(\boldsymbol{k}) = E_0(\boldsymbol{k}) + E_1(\boldsymbol{k}) \tag{2.80}$$

with

$$E_1(\boldsymbol{k}) = -4t'_{\text{eff}}\cos k_x \cos k_y - 2t''_{\text{eff}}\left(\cos 2k_x + \cos 2k_y\right), \tag{2.81}$$

where t'_{eff} and t''_{eff} are effective values roughly proportional to t' and t'', respectively. Fitting the SCBA results to this effective dispersion relationship gives $t'_{\text{eff}} = -0.038$ eV and $t''_{\text{eff}} = 0.022$ eV. It is important to notice that the t' and t'' terms strongly effect the energy at (π,0), lowering it by $4(t'_{\text{eff}} - t''_{\text{eff}}) = -0.24$ eV.

Single-Particle Spectral Function

While the dispersion of the lowest energy peak (quasiparticle peak) helps to characterize the motion of a hole, the spectral function itself contains important information. Shown in Fig. 2.29a is the spectral function $A(\boldsymbol{k},\omega)$ at T=0 for a twenty-site cluster of the t-t'-t''-J model. These are exact results obtained using the exact diagonalization (Lanczos) method (see Appendix B). The parameters are $t = 1$, $J = 0.4$, $t' = -0.34$, and $t'' = 0.29$, which have the same ratio as the parameters mentioned above. The electron-removal spectral function $A_-(\boldsymbol{k},\omega)$ (the single-hole spectrum) discussed above corresponds to the spectra below zero energy which is identified with the chemical potential. (The on-site Coulomb interaction U is taken to be $U = 4t^2/J = 10t$.) One finds that the quasiparticle peaks at around $(\pi/2,\pi/2)$ have a large spectral weight, while the peak at the $(\pi,0)$ is strongly suppressed. This is consistent with experiments [86]. The t' and t'' terms are responsible for this suppression. This can be understood if one considers the weight of the Néel configuration $|\Psi^{\text{Néel}}_{\boldsymbol{k}} >$ with a hole (schematically shown in the left-hand

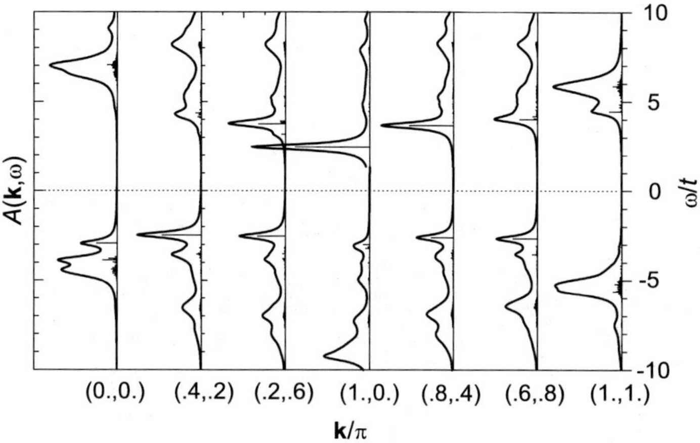

Fig. 2.29. The single-particle spectral function $A(\boldsymbol{k},\omega)$ of the half-filled t-t'-t''-J model for a $\sqrt{20}\times\sqrt{20}$ cluster. $J/|t| = 0.4$, $t'/t = -0.34$ and $t''/t = 0.29$. The curves below and above the chemical potential ($\omega = 0$) represent the electron-removal and addition spectra, respectively, where δ functions are convoluted with a Lorentzian broadening of $0.2t$. The momentum is measured in units of π.

panel of Fig. 2.28) in the final states $|f\rangle$. Since the initial state is antiferromagnetically ordered, the spectral weight is expected to be proportional to the one-hole-Néel-configuration character of the final state:

$$|\langle f\,|c_{\boldsymbol{k},\sigma}|\,i\rangle|^2 = |\langle f\,|c_{\boldsymbol{k},\sigma}|\,\mathrm{AF}\rangle|^2 \sim \left|\langle f|\Psi_{\boldsymbol{k}}^{\text{Néel}}\rangle\right|^2 . \tag{2.82}$$

The weight of $|\Psi_{\boldsymbol{k}}^{\text{Néel}}\rangle$ in the eigenstate for each $\boldsymbol{k}$ has been examined [86,89] and found to be large at around $\boldsymbol{k} = (\pi/2, \pi/2)$ where the quasiparticle weight is large while at $(\pi, 0)$ the weight is small and a spin-liquid-like spin-background is formed. Therefore, the small spectral weight at $(\pi, 0)$ is a consequence of the weakening of the antiferromagnetic spin correlations caused by the presence of t' and t''. At first sight, this may sound strange, because t' and t'' involve the same sublattice hoppings and have no influence on the spin background. The essential point is that, for the chosen values of t' and t'', the self-energy of the one-hole-Néel configuration is increased. The increase makes the $(\pi, 0)$ quasiparticle state shift to higher binding energy (see (2.81)). This is accompanied by a reduction in the weight of the one-hole-Néel configuration. This induces an increase of the other spin configurations, i.e., antiferromagnetic correlations are reduced bringing the system closer to a spin-liquid. The effect of t' and t'' may become clearer when we consider *one-electron motion* in the antiferromagnets.

Above the zero energy in Fig. 2.29, the electron-addition spectral function $A_+(\boldsymbol{k},\omega)$ is shown. This corresponds to the one-electron motion. The difference between hole and electron carriers is included in the calculation through the sign difference of the hopping parameters: $t < 0$, $t' > 0$, and $t'' < 0$

for the electron carriers. The ratio of these are assumed to be the same as those for the hole carriers [90], although they may be carrier-dependent since holes reside in the Zhang-Rice singlet band while the electrons are associated with a true upper Hubbard band. In contrast to the hole carriers, the largest quasiparticle weight is located at $\boldsymbol{k} = (\pi, 0)$. This is easily understood since the sign difference of t' and t'' leads to the stabilization of the one-hole Néel configuration and thus results in the huge quasiparticle weight. Such a stabilization is the source of antiferromagnetism in electron-doped high-T_c cuprates.

Finally, let us comment on the charge excitation across the Mott gap. Form Fig. 2.29 it is found that the excitations are indirect in contrast to the case of 1D Mott insulator (see Sect. 2.2.3). The indirect gap is defined as an excitation from the top of the occupied band at $\boldsymbol{k} = (\pm\pi/2, \pm\pi/2)$ to the bottom of the unoccupied band at $(\pi, 0)$ and $(0, \pi)$. Evidence for such an indirect gap has been obtained in the K-edge resonant inelastic X-ray scattering (RIXS) experiments on $Ca_2CuO_2Cl_2$ [90] (for RIXS see also Sect. 2.2.3).

2.3.2 Phase Diagram

Figure 2.30 shows a schematic phase diagram for high-T_c cuprates, where x represents carrier concentration in the CuO_2 plane for both hole and electron dopings. (For instance, $x = 0.1$ means that on average 0.1 holes (electrons) are found on the CuO_2 unit cell.) As discussed above, at $x = 0$ (undoping), the system is a Mott insulator with antiferromagnetic (AF) order below $T \sim$ 300 K.

Upon hole doping, the system begins to show metallic behavior in the resistivity at high temperatures [91], and the AF order is rapidly destroyed so that the AF phase is confined to very small region. In the region between the AF phase and superconducting phase, a disordered phase showing a spin glass behavior is observed in $La_{2-x}Sr_xCuO_4$ (LSCO). The superconducting transition temperature (T_c) monotonically increases with hole doping (underdoped region), reaching a maximum at $x \sim 0.15$ (optimal doping). Then, T_c decreases and becomes zero (overdoped region). This doping dependence forms a dome-like behavior for T_c. The normal state above the pseudogap temperature (see below) in the underdoped region and well above T_c in the overdoped region exhibits metallic behavior which is different from that of a simple Fermi liquid. For instance, non-T^2 dependence of the resistivity is observed. The anomalous metallic phase may originate from strong correlation in the CuO_2 plane. In the overdoped region, beyond the superconducting dome, a Fermi liquid picture emerges although materials that can reach this region are rare.

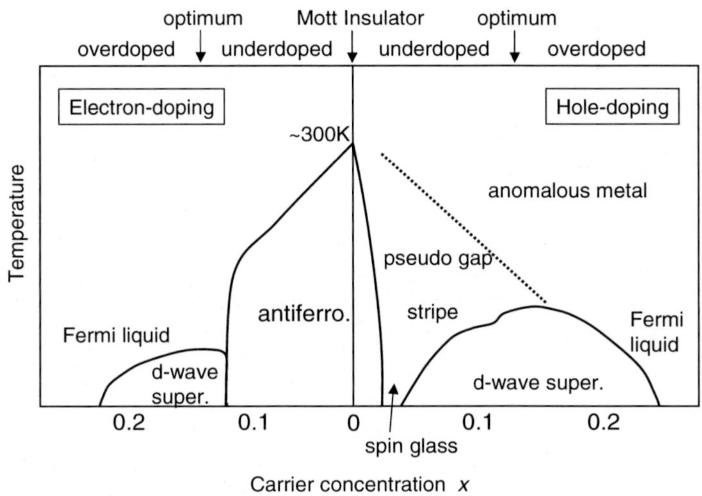

Fig. 2.30. A schematic phase diagram of high-T_c cuprates for both hole and electron dopings. The dotted line is a crossover line between the metallic phase and pseudogap phase.

Pseudogap

In the underdoped region above T_c, there is evidence for the existence of a pseudogap. Both spin and charge properties suggest the existence of such a gap. In fact, many experiments for underdoped region are consistent with a pseudogap. Early measurements of the nuclear magnetic resonance (NMR) relaxation rate [92] and neutron scattering experiments [93] on underdoped $YBa_2Cu_3O_{6+x}$ (YBCO) indicate a strong suppressions of spin excitations below a certain temperature $T^* > T_c$. This was associated with a "spin-gap". After the discovery of this spin-gap, it was shown that a gap is relevant not only in spin channel but also in charge channel. This is evident from specific heat measurements [94], optical absorption along the c axis [95] for YBCO, and ARPES [96] and tunneling spectroscopy [97] for underdoped $Bi_2Sr_2CaCu_2O_{8+\delta}$ (Bi2212). The transition from the normal state to the pseudogap state is considered to be crossover and not a thermodynamic transition. The pseudogap temperature T^* is known to decrease with increasing the hole concentration, but the beginning and ending point of T^* is not at all clear and is controversial. For instance, there is an argument that T^* falls to zero at a critical doping $x = 0.19$ (quantum critical point) [98]. A review of the physics of the pseudogap is found elsewhere [99], in which experiments and theories before 1999 are summarized.

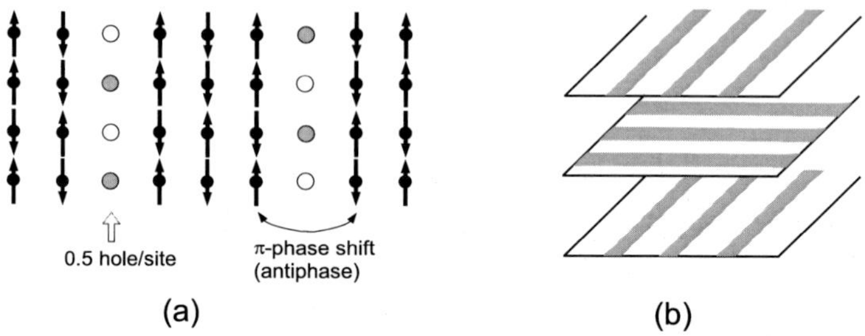

Fig. 2.31. A schematic view of the static stripe structure. (**a**) The charge stripes and spin domains within a CuO_2 plane. (**b**) Stacking structure of the striped CuO_2 planes.

Stripes

In the underdoped region of LSCO, a slight depression of the superconducting dome exits at $x \sim 1/8$, but in a sample in which La is partly substituted by Nd, $La_{1.6-x}Nd_{0.4}Sr_xCuO_4$ (LNSCO), T_c is strongly suppressed. Furthermore, in $La_{2-x}Ba_xCuO_4$ (LBCO) T_c completely vanishes at this concentration. At the same time the antiferromagnetic order is recovered. This is called "1/8" anomaly. One of the remarkable differences between superconducting LSCO and non- (less-) superconducting LBCO (LNSCO) at around $x = 1/8$ is the crystal structure: In the latter samples the structural phase transition from orthogonal to tetragonal phases occurs at 60-80 K above T_c in the underdoped region, while there is no such transition in the former sample and the structure remains orthogonal down to $T = 0$ K. This indicates a dramatic change of the electronic states below the structural transition temperature.

The physics of the 1/8 anomaly has been unveiled through neutron scattering experiments. Tranquada et al. [100] have examined LNSCO at $x = 0.12$ and found that dynamical incommensurate spin fluctuations observed LSCO at wavevectors $(\pi \pm 2\pi x, \pi)$ and $(\pi, \pi \pm 2\pi x)$ [101] condense out and become static with long-range order. More importantly, they have observed evidence for a charge-density wave at $(0, 2\pi \pm 4\pi x)$, which is established at slightly higher temperature than the incommensurate spin-density wave condensation temperature. Reflecting these spin and charge density waves, they propose a charge stripe structure with spin domains, as schematically shown in Fig. 2.31a. In this stripe structure, doped holes (1/8 doping) are not distributed uniformly but form one-dimensional arrays with one hole per two sites, which corresponds to quarter filling in the usual band picture. The spin domains between the charge stripes are antiferromagnetically ordered with a shift of spin-ordering phase by π. This stripe structure in the CuO_2 plane stacks alternately changing the stripe direction (Fig. 2.31b).

The presence of the quarter-filled charge river has been confirmed by ARPES experiments for LNSCO [102], where one-dimensional flat Fermi surfaces are seen. In contrast to many other systems showing spin-density wave and charge-density wave order at low temperatures, the stripe structures in cuprates appear to be metallic. This is also explained by the model since the charge carriers are mobile along the charge stripes. An anomalous transport behavior appears in the Hall constant R_{H}: R_{H} becomes strongly temperature dependent below the structural transition temperature to the tetragonal phase and eventually almost vanishes at temperature $T \sim 0$ [103]. The vanishing R_H is strong evidence for the presence of quarter-filling static charge stripes, since the formation of the one-dimensional charge stripe with an equal concentration of hole and electron carriers (due to the strong on-site Coulomb interaction) leads to the electron-hole symmetry and thus the vanishing of the off-diagonal conductivity σ_{xy} [104]. The behavior of R_{H} implies the static stripe structures set in below the structural transition temperature. As mentioned above, there is no structural transition to the tetragonal phase in LSCO but structural fluctuations have been reported which may represent fluctuating stripes analogous to fluctuations in liquid crystals (see a review article [105]). At present, the static stripe order is only observed in La-based monolayer cuprates with the tetragonal structure at low temperature. It is clear that such a static stripes destroy superconductivity, but it is still controversial whether the fluctuating stripes are helpful for superconductivity. (The physical properties in La-based monolayer cuprates, before stripes became popular, has been reviewed by Kastner et al. [106].)

The origin of the stripe formation is not yet clear. Suggested mechanisms involve electronically intrinsic effects, long-rang Coulomb interactions, and coupling with lattice fluctuations. Theoretically it seems to be difficult to identify a unique origin, but it is fair to say that the opinion, that stripes are intrinsic to the physics of doped Mott insulators, is shared among many researchers. In fact, for the t-J model, one finds that the parameter region for high-T_{c} cuprates, i.e. $J/t \sim 0.4$ and $x \leq 0.2$, is very close to the boundary to the phase-separated state (see, for instance, [55]). Thus two phases, the phase-separated state and d-wave superconducting state, may compete each other. Therefore one can expect that the competition induces unusual phases, e.g., stripes, when small (external) perturbations are applied to the system.

Charge Inhomogeneity

As one of unusual states in high-T_{c} cuprates, atomic-resolution scanning tunneling microscopy measurements have revealed large charge density modulation on Bi2212 surfaces cleaved at low temperatures [107], showing spatially cross-correlated variations in the local density of states and the superconducting energy gap on a remarkably short length of about 14Å. The superconductivity particularly in the underdoped region has been proposed to be granular with nanoscale superconducting grains separated by non-superconducting regions [108]. The non-superconducting region shows a pseudogap with magnitude larger than 50 meV. This suggests that a mixture of superconducting and (non-superconducting) pseudogap phases occurs in, at least, the surface of underdoped Bi2212. It has been conjectured that the origin of these inhomogeneous electronic states is the ionic potential associated with the off-stoichiometry oxygen dopants disordered in the BiO layer [107]. It is very important to clarify whether this is of bulk or surface nature. This remains to be seen. It should be noted here that nanoscale inhomogeneity have proposed in connection with bulk-sensitive NMR measurements for LSCO [109].

Electron Doping

In electron-doped systems such as $Nd_{2-x}Ce_xCuO_4$ (NCCO), the AF phase exists over a much larger doping range when compared with hole doping. This means that the AF correlations with wave vector $\mathbf{Q} = (\pi, \pi)$ remain in the CuO_2 plane of the electron-doped system [110]. Based on the t-t'-t''-J model, one can say that the signs of t' and t'' for the electron doping favor the stability of AF as discussed in the previous section: The sublattice hoppings by t' and t'' stabilize the Néel-type configurations in the spin background. On the electron-doped side of the phase diagram, there are no reports of a superconducting "dome" in bulk samples. Also, there is no evidence from NMR measurements for a spingap in the normal state, this in contrast to the case of hole doping [111]. The presence of pseudogap in electron doping is, however, still controversial, depending on the experiment. The contrasting behavior of hole and electron doped materials will be discussed below.

2.3.3 Optical Conductivity

Carrier doping into the insulating CuO_2 plane induces dramatic changes in the electronic states over a wide energy range. This is different from a standard doped semiconductor where electron correlation effects are not important. For the cuprates, such changes are seen in the doping dependence of the optical conductivity $\sigma(\omega)$. Figure 2.32a shows $\sigma(\omega)$ for LSCO determined from the reflectivity measurements after a Kramers-Kronig transformation [112]. The Mott gap excitation at 2 eV in La_2CuO_4 rapidly loses weight with increasing x, while the weight in the low-energy region (< 1.5 eV) increases. The total weight seems to remain almost constant. This phenomenon has also been observed in an electron-doped material NCCO (Fig. 2.32b) [112, 113], where weight from above the Mott gap is transferred below ~ 1.2 eV. This reconstruction of the electronic states cannot be explained by a rigid-band picture since then only a small decrease of the inter-band feature is expected.

Using strong-coupling perturbation theory (t/U expansion) for the single-band Hubbard model, Eskes et al. [114] found that the spectral weight for the excitation from the lower to upper Hubbard bands, is related to the strength of the spin-spin correlation function along the current direction, $-\sum_i \langle (\boldsymbol{S}_i \cdot \boldsymbol{S}_{i+\boldsymbol{\delta}} - \frac{1}{4} n_i n_{i+\boldsymbol{\delta}}) \rangle$, where $\boldsymbol{\delta}$ is the unit vector in the current direction and $\langle \ldots \rangle$ denotes the expectation value in the ground state. We can

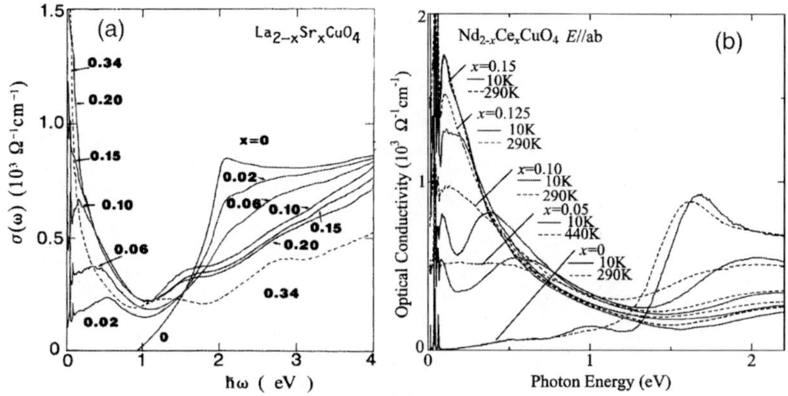

Fig. 2.32. The doping dependence of the optical conductivity for (**a**) hole-doped $La_{2-x}Sr_xCuO_4$ at room temperature (From [112]) and (**b**) electron-doped $Nd_{2-x}Ce_xCuO_4$. (From [113])

understand this very easily. The current transfers an electron to a neighboring site. In order to make the Mott gap excitation, i.e., create a doubly occupied site, first of all there has to be an electron present, and second the electrons have to form a singlet. Both conditions are reflected in the spin-spin correlation function. This relationship explains the rapid disappearance of the Mott gap excitation, since when we introduce one carrier into the 2D AF spin system, two spin-spin bonds are removed in the $\boldsymbol{\delta}$ direction and thus the initial weight W_0 is reduced to $(1-2x)W_0$.

The low-energy spectral weight in Fig. 2.32a seems to consist of two parts: One is the Drude-like components centered at $\hbar\omega = 0$ and the other is the component appearing in the mid-infrared (IR) region (< 0.5 eV). The latter is understood to correspond to the incoherent motion of carriers coupled to the spin fluctuations [115]. This mid-IR component is clearly seen below $x = 0.1$, but as the concentration increases it merges to the Drude component. This implies that all of carriers can move coherently. The two component feature is also seen in NCCO, although the Drude part is narrow (Fig. 2.32b). More interestingly, a notable dip with an energy as large as $\omega_{\rm dip} \sim 0.2$-$0.4$ eV is found below a characteristic temperature $T_{\rm dip}$ for $x < 0.15$ in NCCO. The $\omega_{\rm dip}$ and $T_{\rm dip}$ decrease with doping x, and are related by the approximate relation of $\omega_{\rm dip} \sim 10 k_{\rm B} T_{\rm dip}$. It has be suggested [116] that this dip is related to the fact that AF correlations persist to higher doping in the electron-doped systems. Finite temperature calculations for small clusters of the t-t'-t''-J model demonstrate that a dip emerges in $\sigma(\omega)$ when the AF correlations develop with decreasing temperature.

In order to discuss the spectral-weight transfer upon doping more quantitatively, it is useful to introduce the effective electron number $N_{\rm eff}$ obtained by integrating $\sigma(\omega)$ up to a certain frequency ω,

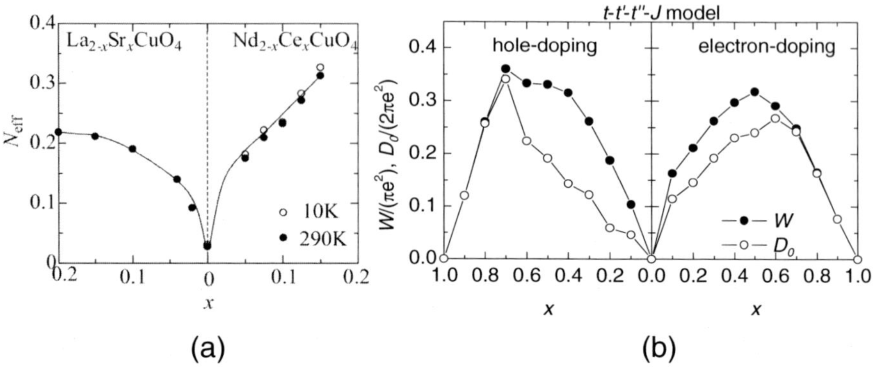

Fig. 2.33. (**a**) The doping dependence of the effective electron number $N_{\mathrm{eff}}(\omega)$ at $\omega = 1.5$ eV for $La_{2-x}Sr_xCuO_4$ (From [112]) and that at $\omega = 1.2$ eV for $Nd_{2-x}Ce_xCuO_4$ (From [113])). (**b**) The doping dependence of the integrated weight W of $\sigma(\omega)$ and the Drude weight D_0 at $T = 0$ for a $\sqrt{20} \times \sqrt{20}$-site lattice of the hole- and electron-doped t-t'-t''-J models. $t'/t = -0.34$, $t''/t = 0.23$, and $J/|t| = 0.4$.

$$N_{\mathrm{eff}}(\omega) = \frac{2m_0}{\pi e^2 N_{\mathrm{Cu}}} \int_0^{\omega} \sigma(\omega')\, \mathrm{d}\omega', \tag{2.83}$$

where m_0 is the free-electron mass, and N_{Cu} is the number of Cu ions a unit volume. Note that (2.83) is based on the simple Drude model (see Appendix A). Figure 2.33a displays N_{eff} for $\omega = 1.5$ eV for LSCO [112] and for $\omega = 1.2$ eV for NCCO [113]. It is found that the spectral-weight transfer from the Mott gap excitation is larger in the electron-doped system than in the hole-doped analog. One of the simplest explanations for this weight difference comes from the difference of the Mott gap (2 eV for La_2CuO_4 and 1.7 eV for Nd_2CuO_4) corresponding to a different magnitude for the effective electron correlations. Actually, it is know that the spectral-weight transfer increases with decreasing the Mott gap values in perovskite titanates [117] and theoretically that the kinetic energy, which corresponds to the integrated weight of $\sigma(\omega)$ (see Appendix A), increases with decreasing the Mott gap (determined by U) [115]. In addition to the gap difference, the different behavior of N_{eff} for hole and electron dopings may be partly reflect the presence of the long-range hoppings t' and t''. To see this, the integrated weight W of $\sigma(\omega)$ is plotted as a function of x in Fig. 2.33b for the hole- and electron-doped t-t'-t''-J models. The value of W near $x = 0$ increases with x more rapidly for electron doping than for hole doping. This reflects the stabilization of the Néel-type spin background by t' and t'' upon electron doping (see Sect. 2.3.1 and Sect. 2.3.2). This in turn enhances the kinetic energy associated with the sublattice hoppings. At the same time, the Drude weight D_0 is also enhanced upon electron doping. This corresponds to the fact that the electron carriers can move more coherently in the high-T_c cuprates.

2.3.4 Single-Particle Spectral Function

In doped semiconductors, a rigid-band like shift of the density of states (DOS) occurs without changing the weight of the DOS. In contrast, the doped-Mott insulators show the spectral-weight transfer from the upper Hubbard band to the lower Hubbard band for hole doping and vise versa for electron doping. A schematic view of the situation is shown for the limit of $U \to \infty$ in Fig. 2.34. Upon hole doping with concentration x, a new state appears just above the Fermi level inside the lower Hubbard band and has weight of $2x$. The weight $2x$ can be easily understood from the fact that there are two low-energy states, with either up or down spin, for every electron removed from the $U \to \infty$ system. The t-J model exhibits this spectral-weight transfer. If double occupation is allowed in the Hubbard model, a further weight transfer is excepted from the upper to lower Hubbard band [114]. The rapid suppression of the weight in the upper Hubbard band has been observed in O1s x-ray absorption and electron-energy loss spectroscopies [118].

The doping and momentum dependences of the single-particle spectral function near the Fermi level have been extensively examined by ARPES experiments. It has been clearly observed that hole carriers doped into the Mott insulators first enter near the $(\pm\pi/2, \pm\pi/2)$ points in the Brillouin zone [119], while electron carriers first appear near $(\pi, 0)$ and $(0, \pi)$ [120]. This difference is to be expected, following our discussion, Sect. 2.3.1, of the motion of a single carrier in the Mott insulator. Such rigid-band-like dopings may be associated with small changes of the magnetic background upon doping. In other words, the carrier-pocket features at the specific momenta may be an indication of strong AF magnetic correlations in the systems. With further doping, the AF

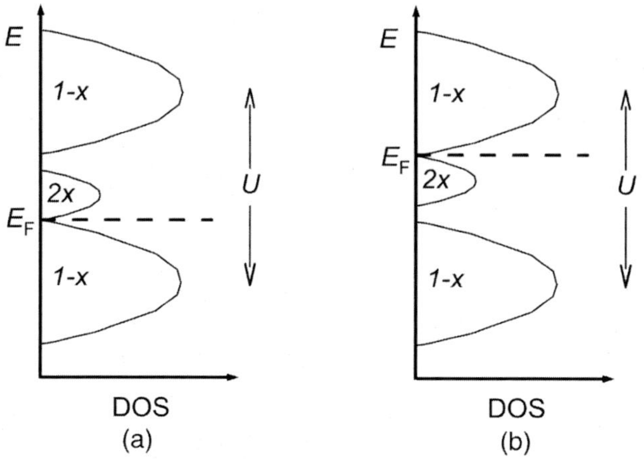

Fig. 2.34. Schematic drawings of the intensities of the electron-addition and -removal spectra for the $U \to \infty$ Hubbard and t-J models. (**a**) Hole doping. (**b**) Electron doping. x is the carrier concentration.

correlations are weakened. At the same time, the Fermi surfaces in the normal state gradually change to eventually be consistent with the band structure calculations based on the local density approximation (LDA). In the underdoped region of hole-doped systems, the pseudogap behavior at around the $(\pi, 0)$ points is observed. Interested readers should consult the recent ARPES review by Damascelli, Hussain, and Shen [64].

2.3.5 Chemical Potential

The doping dependence of the chemical potential μ is very fundamental quantity which gives information about the evolution of the electronic states upon doping. The density derivative of μ is related to the charge susceptibility χ_c,

$$\frac{\partial \mu}{\partial n} = \frac{1}{\chi_c}, \tag{2.84}$$

where n is the electron density and $n = 1 \pm x$ (+ for electron doping and − for hole doping).

The chemical potential shift $\Delta\mu$ has been determined by precise measurements of core-level photoemission spectra for both hole- and electron-doped systems. Unconventional, asymmetric, doping dependences have been reported [121]. Figure 2.35 exhibits the data at $T \sim 80$ K for hole-doped LSCO and $Bi_2Sr_2Ca_{1-x}R_xCu_2O_{8+y}$ (R = Pr, Er) (BSCCO) and for an electron-doped NCCO. In the underdoped LSCO, $\Delta\mu$ is strongly depressed as if the chemical potential is pinned by some levels. This leads to a very large value of χ_c from (2.84), indicating the tendency toward a phase-separated state. This data is consistent with the presence of strong stripe fluctuations

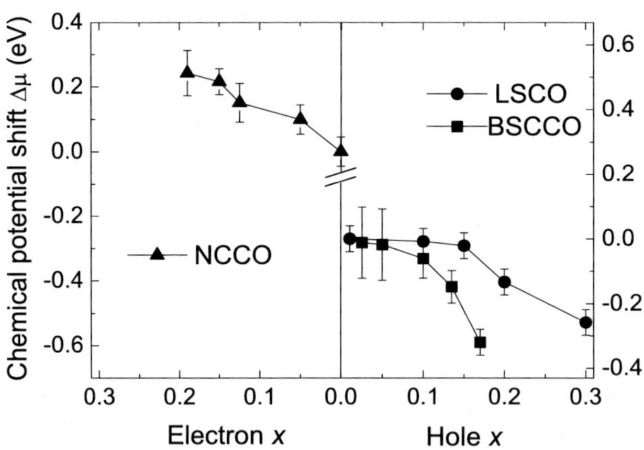

Fig. 2.35. The chemical potential shift $\Delta\mu$ in hole-doped LSCO, BSCCO, and electron-doped NCCO as functions of carrier concentration x. All the data are taken from [121].

or charge inhomogeneity in LSCO. In contrast to LSCO, BSCCO exhibits a large change of $\Delta\mu$ with a quadratic behavior over the whole doping range. This seems to be consistent with a theoretical result for the 2D Hubbard model that predicts $\Delta\mu \propto -x^2$ [122] from the Monte Carlo simulations. Another possibility which might explain the global difference between LSCO and BSCCO is a difference in the values of the long-range hopping terms t' and t''. Exact diagonalization calculations at finite temperatures for the t-t'-t''-J model exhibit [123] this possibility. The magnitude of the hoppings are known to be larger in BSCCO than in LSCO from LDA band calculations and APRES measurements of the Fermi surfaces. Since the tendency toward the phase-separated state becomes weaker with increasing the magnitude of the hoppings, $\Delta\mu$ increases from LSCO to BSCCO.

In NCCO, the chemical potential monotonously increases with electron doping. This in sharp contrast with its hole doped sister system LSCO in the underdoped region. If the suppression of the μ shift in LSCO is attributed to strong stripe fluctuations, the monotonous increase of the chemical potential indicates the absence of such stripe fluctuations in NCCO. Such a monotonous change of $\Delta\mu$ was also reproduced by the t-t'-t''-J model [123].

2.4 Summary

Insulating cuprates are classified as charge-transfer Mott insulators, i.e., the oxygen orbitals have to be explicitly taken into account. The starting Hamiltonian is thus a multi-band Hubbard model composed of Cu $3d_{x^2-y^2}$ and O $2p_\sigma$ orbitals. Due to the formation of the Zhang-Rice singlets induced by carriers introduced into the CuO_2 plane, this Hamiltonian can be reduced to a single-band Hubbard or t-J model. The single-band models with large on-site Coulomb interaction describe very well the electronic and magnetic properties of the cuprates. The fact that the cuprates are low, i.e., one- and two-dimensional lattice systems, provides a unique stage upon which a variety of fundamental solid-state physics scenarios are played out, with high-T_c superconductivity being only one or several possible outcomes. Worthy of special note are the one-dimensional cuprates which exhibit a clear signature spin-charge separation a novel and very fundamental property. It is a wonderful possibility which provides encouragement for fundamental studies of the solid-state that these same one-dimensional insulating cuprates actually have great potential for nonlinear photoelectric devices and this functionality follows from the existence of spin-charge separation.

Within the single-band model, carrier long-range hopping, beyond the nearest-neighbors, plays a crucial role in the electronic states of the high-T_c cuprates. The interplay of strong correlation and longer-range hopping further enriches the physics of the high-T_c cuprates. The presentation in this chapter has been very limited. Interested readers can find further experimental and theoretical information on Mott-insulators, in the next chapter, but also in a number of other books and review articles (see, for example, [8, 64, 99, 105, 106, 124–126]).

Cuprates with ladder structure have also been studied extensively, since they provide a potential bridge between one- and two-dimensions. It might be expected that any tendency towards superconductivity should develop rungs are added. In addition, the ladder cuprates have unique properties such as a spin gap and charge ordering. Much of the novel physic of the ladder cuprates could not be covered in this chapter. Readers interested by this subject might wish consult a review article [127] for more detail.

References

1. J. G. Bednorz, K. A. Muller: Z. Physik **64**, 189 (1986)
2. P. W. Anderson: Sience **235**, 1196 (1987)
3. N. Motoyama, H. Eisaki, S. Uchida: Phys. Rev. Lett. **76**, 3212 (1996)
4. M. Hase, I. Terasaki, K. Uchinokura: Phys. Rev. Lett. **70**, 3651 (1993)
5. M. Uehara, T. Nagata, J. Akimitsu, H. Takahashi, N. Mori, K. Kinoshita: J. Phys. Soc. Jpn. **65**, 2764 (1996)
6. J. Zaanen, G. A. Sawatzky: Can. J. Phys. **65**, 1262 (1987)
7. J. Zaanen, G. A. Sawatzky, J. W. Allen: Phys. Rev. Lett. **55**, 418 (1985)
8. M. Imada, A. Fujimori, Y. Tokura: Rev. Mod. Phys. **70**, 1039 (1998)
9. Y. Ohta, T. Tohyama, S. Maekawa: Phys. Rev. Lett. **66**, 1228 (1991)
10. Y. Ohta, T. Tohyama, S. Maekawa: Phys. Rev. B **643**, 2968 (1991)
11. Y. Mizuno, T. Tohyama, S. Maekawa, T. Osafune, N. Motoyama, H. Eisaki, S. Uchida: Phys. Rev. B **57**, 5326 (1998)
12. J. Callaway: *Quantum Theory of Solid State*, 2nd ed. (Academic Press, San Diego, 1991), p. 381
13. V. J. Emery: Phys. Rev. Lett. **58**, 2794 (1987)
14. A. K. McMahan, R. M. Martin, S. Satpathy: Phys. Rev. B **38**, 6650 (1988)
15. M. S. Hybertsen, M. Schhlüter, N. E. Christensen: Phys. Rev. B **39**, 9028 (1989)
16. H. Eskes, G. A. Sawatzky, L. F. Feiner: Physica C **160**, 424 (1989)
17. P. W. Anderson: Phys. Rev. **115**, 2 (1959) and *Magnetism*, edited by G. T. Rado and H. Suhl (Academic, New York, 1963), Vol. I, p. 25
18. H. Eskes, J. H. Jefferson: Phys. Rev. B **48**, 9788 (1993)
19. H. Suzuura, H. Yasuhara, A. Furusaki, N. Nagaosa, Y. Tokura: Phys. Rev. Lett. **76**, 2579 (1996)
20. J. Lorenzana, G. A. Sawatzky: Phys. Rev. Lett. **74**, 1867 (1995); Phys. Rev. B **52**, 9576 (1995)
21. J. Lorenzana, R. Eder: Phys. Rev. B **55**, R3358 (1997)
22. M. Takigawa, O. A. Starykh, A. W. Sandvik, R. R. P. Singh: Phys. Rev. B **56**, 13681 (1997)
23. W. A. Harrison: *Electronic Structure and the Properties of Solids: The Physics of the Chemical Bond* (W. H. Freeman and Company, San Francisco,1980)
24. P. E. Sulewski, P. A. Fleury, K. B. Lyons, S-W. Cheong, Z. Fisk: Phys. Rev. B **41**, 225 (1990); Y. Tokura, S. Koshihara, T. Arima, H. Takagi, S. Ishibashi, T. Ido, S. Uchida: Phys. Rev. B **41**, 11657 (1990); S. Sugai, M. Sato, T. Kobayashi, J. Akimitsu, T. Ito, H. Takagi, S. Uchida, S. Hosoya, T. Kajitani, T. Fukuda: Phys. Rev. B **42**, 1045 (1990)

25. P. Bourges, H. Casalta, A. S. Ivanov, D. Petitgrand: Phys. Rev. Lett. **79**, 4906 (1997); S. M. Hayden, G. Aeppli, P. Dai, H. A. Mook, T. G. Perring, S-W. Cheong, Z. Fisk , F. Doğan, T. E. Mason: Physica B **241-243**, 765 (1998)
26. Y. Mizuno, T. Tohyama, S. Maekawa: Phys. Rev. B **58**, R14713 (1998)
27. J. C. Slater, G. Koster: Phys. Rev. **94**, 1498 (1954)
28. M. Matsuda, K. Katsumata, R. S. Eccleston, S. Brehmer, H.-J. Mikeska: Phys. Rev. B **62**, 8903 (2000)
29. G. Castilla, S. Chakravarty, V. J. Emery: Phys. Rev. Lett. **75**, 1823 (1995); J. Riera, A. Dorby: **51**, 16098 (1995)
30. G. S. Uhrig: Phys. Rev. Lett. **79**, 163 (1997); W. Brenig: Phys. Rev. B **56**, 14441 (1997); Y. Mizuno, T. Tohyama, S. Maekawa: Phys. Rev. B **60**, 6230 (1999)
31. K. Yoshida: *Theory of Magnetism* (Springer, Berlin, Heidelberg 1998)
32. R. Coldea, S. M. Hayden, G. Aeppli, T. G. Perring, C. D. Frost, T. E. Mason, S.-W. Cheong, Z. Fisk: Phys. Rev. Lett. **86**, 5377 (2001)
33. E. Müller-Hartmann, A. Reischl: Eur. Phys. J. B **28**, 173 (2002)
34. H. J. Schmidt, Y. Kuramoto: Physica C **167**, 263 (1990); Y. Mizuno, T. Tohyama, S. Maekawa: J. Low Temp. Phys. **117**, 389 (1999)
35. M. Roger, J. H. Hetherington, J. D. Delrieu: Rev. Mod. Phys. **55**, 1 (1983)
36. A. Yoshimori, H. Fukuda: J. Phys. Soc. Jpn. **46**, 1663 (1979); K. Yoshida, S. Inagaki: J. Phys. Soc. Jpn. **50**, 3268 (1981)
37. F. C. Zhang, T. M. Rice: Phys. Rev. B **37**, 3759 (1988)
38. J. H. Jefferson, H. Eskes, L. F. Feiner: Phys. Rev. B **45**, 7959 (1992)
39. W. Kohn: Phys. Rev. **133**, A171 (1964)
40. C. Kittel: *Introduction to Solid State Physics* (John Wiley & Sons, Inc., New York 1995)
41. T. Arima, K. Kikuchi, M. Kasuya, S. Koshihara, Y. Tokura, T. Ido, S. Uchida: Phys. Rev. B **44**, 917 (1991)
42. S. Tomonaga: Prog. Theor. Phys. **5**, 544 (1950)
43. J. M. Luttinger: J. Math. Phys. **4**, 1154 (1963)
44. F. D. M. Haldane J. Phys. C: Solid State Phys. **14**, 2585 (1981)
45. J. Sólyom: Adv. Phys. **28**, 201 (1979)
46. J. Voit: Rep. Prog. Phys. **58**, 977 (1995)
47. H. J. Schulz, G. Cuniberti, P. Pieri: 'Fermi Luquid and Luttinger Liquids'. In: *Field Theories for Low-Dimensional Condensed Matter Systems.* ed. by G. Morandi, P. Sodano, A. Tagliacozzo, V. Tognetti (Springer, Berlin, Heidelberg, New York 2000) pp. 9–81
48. H. J. Schulz: Int. J. Mod. Phys. B **5**, 57 (1991)
49. V. J. Emery: 'Theory of the One-Dimensional Electron Gas'. In: *Highly Conducting One-Dimensional Solids.* ed. by J. T. Devreese, R. E. Evrard, V. E. van Doren (Plenum Press, New York 1979) pp. 247-302
50. G. D. Mahan: *Many-Particle Physics*, Second Edition (Plenum Press, New York 1990)
51. A. O. Gogolin, A. A. Nersesyan, A. M. Tsvelik: *Bosonization and Strongly Correlated Systems* (Cambridge University Press, Cambridge 1998)
52. E. H. Lieb and F. Y. Wu: Phys. Rev. Lett. **20**, 1445 (1968)
53. M. Ogata, H. Shiba: Phys. Rev. B **41**, 2326 (1990)
54. M. Ogata, M. U. Luchini, S. Sorella, F. F. Assaad: Phys. Rev. Lett. **66**, 2388 (1991)
55. S. Maekawa, T. Tohyama: Rep. Prog. Phys. **64**, 383 (2001)
56. P. A. Bares, G. Blatter: Phys. Rev. Lett. **64**, 2567 (1990); P. A. Bares, G. Blatter, M. Ogata: Phys. Rev. B **44**, 130 (1991)

57. H. J. Schulz: Phys. Rev. Lett. **64**, 2831 (1990); N. Kawakami, S-K. Yang: Phys. Lett. **148A**, 359 (1990); H. Frahm, V. E. Korepin: Phys. Rev. B **42**, 10553 (1990)
58. Z. Y. Weng, D. N. Sheng, Y-C. Chen, C. S.Ting: Phys. Rev. B **55**, 3894 (1997)
59. H. Suzuura, N. Nagaosa: Phys. Rev. B **56**, 3548 (1997)
60. V. Meden, K. Schönhammer: Phys. Rev. B **46**, 15753 (1992); J. Voit: Phys. Rev. B **47**, 6740 (1993)
61. A. Luther, V. J. Emery: Phys. Rev. Lett. **33**, 589 (1974)
62. J. Voit: J. Phys.: Condens. Matter **8**, L779 (1996); Eur. Phys. J. B **5**, 505 (1998)
63. T. Tohyama, S. Maekawa: J. Phys. Soc. Japan **65**, 1902 (1996)
64. A. Damascelli, Z. Hussain, Z.-X. Shen: Rev. Mod. Phys. **75**, 473 (2003)
65. C. Kim, A. Y. Matsuura, Z.-X. Shen, N. Motoyama, H. Eisaki, S. Uchida, T. Tohyama, S. Maekawa: Phys. Rev. Lett. **77**, 4054 (1996)
66. H. Fujisawa, T. Yokoya, T. Takahashi, S. Miyasaka, M. Nibune, H. Takagi: Phys. Rev. B **59**, 7358 (1999)
67. K. Kobayashi, T. Mizokawa, A. Fujimori, M. Isobe, Y. Ueda, T. Tohyama, S. Maekawa: Phys. Rev. Lett. **82**, 803 (1999)
68. P. A. Lee, N. Nagaosa: Phys. Rev. B **46**, 5621 (1992)
69. W. Marshall: Proc. R. Soc. A **232**, 48 (1955)
70. D. H. Lu, D. L. Feng, N. P. Armitage, K. M. Shen, A. Damascelli, C. Kim, F. Ronning, Z.-X. Shen, D. A. Bonn, R. Liang, W. N. Hardy, A. I. Rykov, S. Tajima: Phys. Rev. Lett. **86**, 4370 (2001)
71. T. Mizokawa, C. Kim, Z.-X. Shen, A. Ino, A. Fujimori, M. Goto, H. Eisaki, S. Uchida, M. Tagami, K. Yoshida, A. I. Rykov, Y. Siohara, K. Tomimoto, S. Tajima: Phys. Rev. B **60**, 12335 (1999); T. Mizokawa, K. Nakada, C. Kim, Z.-X. Shen, T. Yoshida, A. Fujimori, S. Horii, Yuh Yamada, H. Ikuta, U. Mizutani: Phys. Rev. B **65**, 193101 (2002)
72. P. Y. Yu, M. Cardona: *Fundamentals of Semiconductors* (Springer, Berlin, Heidelberg 1996)
73. W. Stephan, K. Penc: Phys. Rev. B **54**, 17269 (1996)
74. E. Jeckelmann, F. Gebhard, F. H. L. Essler: Phys. Rev. Lett. **85**, 3910 (2000)
75. R. Neudert, M. Knupfer, M. S. Golden, J. Fink, W. Stephan, K. Penc, N. Motoyama, H. Eisaki, S. Uchida: Phys. Rev. Lett. **81**, 657 (1998)
76. M. Z. Hasan, P. A. Montano, E. D. Isaacs, Z.-X. Shen, H. Eisaki, S. K. Sinha, Z. Islam, N. Motoyama, S. Uchida: Phys. Rev. Lett. **88**, 177403 (2002)
77. H. Kishida, H. Matsuzaki, H. Okamoto, T. Manabe, M. Yamashita, Y. Taguchi, Y. Tokura: Nature **405**, 929 (2000); H. Kishida, M. Ono, H. Matsuzaki, M. Yamashita, Y. Taguchi, Y. Tokura, H. Okamoto: Synthetic Metals **135-136**, 315 (2003)
78. T. Ogasawara, M. Ashida, N. Motoyama, H. Eisaki, S. Uchida, Y. Tokura, H. Ghosh, A. Shukla, S. Mazumdar, M. Kuwata-Gonokami: Phys. Rev. Lett. **85**, 2204 (2000)
79. H. Kishida, M. Ono, K. Miura, H. Okamoto, M. Izumi, T. Manako, M. Kawasaki, Y. Taguchi, Y. Tokura, T. Tohyama, K. Tsutsui, S. Maekawa: Phys. Rev. Lett. **87**, 177401 (2001)
80. P. N. Butcher, D. Cotter: *The Elements of Nonlinear Optics* (Cambridge University Press, Cambridge, 1990)
81. R. W. Boyd: *Nonlinear Optics* (Academic Press, New York, 1992)
82. Y. Mizuno, K. Tsutsui, T. Tohyama, S. Maekawa: Phys. Rev. B **62**, R4769 (2000)
83. J. des Cloizeaux, J. J. Pearson: Phys. Rev. **128**, 2131 (1962)

84. S. Uchida, H. Takagi, K. Kitazawa, S. Tanaka: Jpn. J. Appl. Phys. **26**, L1 (1987)
85. B. O. Wells, Z.-X. Shen, A. Matsuura, D. M. King, M. A. Kastner, M. Greven, R. J. Birgeneau: Phys. Rev. Lett. **74**, 964 (1995)
86. C. Kim, P. J. White, Z.-X. Shen, T. Tohyama, Y. Shibata, S. Maekawa, B. O. Wells, Y. J. Kim, R. J. Birgeneau, M. A. Kastner: Phys. Rev. Lett. **80**, 4245 (1998)
87. S. LaRosa, I. Vobornik, F. Zwick, H. Berger, M. Grioni, G. Margaritondo, R. J. Kelley, M. Onellion, A. Chubukov: Phys. Rev. B **56**, R525 (1997)
88. S. Schmitt-Rink, C. M. Varma, A. E. Ruckenstein: Phys. Rev. Lett. **60**, 2793 (1988); C. L. Kane, P. A. Lee, N. Read: Phys. Rev. B **39**, 6880 (1989); G. Martinez, P. Horsh: Phys. Rev. B **44**, 317 (1991)
89. T. Tohyama, Y. Shibata, S. Maekawa, Z.-X. Shen, N. Nagaosa, L. L. Miller: J. Phys. Soc. Jpn. **69**, 9 (2000)
90. M. Z. Hasan, E. D. Isaacs, Z.-X. Shen, L. L. Miller, K. Tsutsui, T. Tohyama, S. Maekawa: Science **288**, 1811 (2000)
91. Y. Ando, A. N. Lavrov, S. Komiya, K. Segawa, X. F. Sun: Phys. Rev. Lett. **87**, 017001 (2001)
92. H. Yasuoka, T. Imai, T. Shimizu: *Strong Correlation and Superconductivity*, ed. by H. Fukuyama, S. Maekawa, A. P. Malozemoff (Springer-Verlag, 1989); W. W. Warren, Jr., R. E. Walstedt, G. F. Brennert, R. J. Cava, R. Tycko, R. F. Bell, G. Dabbagh: Phys. Rev. Lett. **62**, 1193 (1989)
93. J. Rossat-Mignod, L. P. Regnault, C. Vettier, P. Bourges, P. Burlet, J. Bossy, J. Y. Henry, G. Lapertot: Physica C **185-189**, 86 (1991)
94. J. W. Loram, K. A. Mirza, J. R. Cooper, W. Y. Liang: Phys. Rev. Lett. **71**, 1740 (1993)
95. C. C. Homes, T. Timusk, R. Liang, D. A. Bonn, W. N. Hardy: Phys. Rev. Lett. **71**, 1645 (1993)
96. A. G. Loeser, Z.-X. Shen, D. S. Dessau, D. S. Marshall, C.-H. Park, P. Fournier, A. Kapitulnik: Science **273**, 325 (1996); H. Ding, T. Yokoya, J. C. Campuzano, T. Takahashi, M. Randeria, M. R. Norman, T. Mochiku, K. Kadowaki, J. Giapinzakis: Nature **382**, 51 (1996)
97. Ch. Renner, B. Revaz, J.-Y. Genoud, K. Kadowaki, Ø. Fischer: Phys. Rev. Lett. **80**, 149 (1998)
98. J. L. Tallon, J. W. Loram: Physica C **349**, 53 (2001)
99. T. Timusk, B. Statt: Rep. Prog. Phys. **62**, 61 (1999)
100. J. M. Tranquada, B. J. Sternlieb, J. D. Axe, Y. Nakamura, S. Uchida: Nature **375**, 561 (1995)
101. S-W. Cheong, G. Aeppli, T. E. Mason, H. Mook, S. M. Hayden, P. C. Canfield, Z. Fisk, K. N. Clausen, J. L. Martinez: Phys. Rev. Lett. **67**, 1791 (1991)
102. X. J. Zhou, P. Bogdanov, S. A. Kellar, T. Noda, H. Eisaki, S. Uchida, Z. Hussain, Z.-X. Shen: Science **286**, 268 (1999)
103. T. Noda, H. Eisaki, S. Uchida: Science **286**, 265 (1999)
104. V. J. Emery, E. Fradkin, S. A. Kivelson, T. C. Lubensky: Phys. Rev. Lett. **85**, 2160 (2000); P. Prelovšek, T. Tohyama, S. Maekawa: Phys. Rev. B **64**, 052512 (2001)
105. S. A. Kivelson, E. Fradkin, V. Oganesyan, I. P. Bindloss, J. M. Tranquada, A. Kapitulnik, C. Howald: Rev. Mod. Phys. **75**, 1201 (2003)
106. M. A. Kastner, R. J. Birgeneau, G. Shirane, Y. Endoh: Rev. Mod. Phys. **70**, 897 (1998)
107. S. H. Pan, J. P. O'Neal, R. L. Badzey, C. Chamon, H. Ding, J. R. Engelbrecht, Z. Wang, H. Eisaki, S. Uchida, A. K. Gupta, K.-W. Ng, E. W. Hudson, K. M. Lang, J. C. Davis: Nature **413**, 282 (2001)

108. K. M. Lang, V. Madhavan, J. E. Hoffman, E. W. Hudson, H. Eisaki, S. Uchida, J. C. Davis: Nature **415**, 412 (2002)
109. P. M. Singer, A. W. Hunt, T. Imai: Phys. Rev. Lett. **88**, 047602 (2002)
110. K. Yamada, K. Kurahashi, T. Uefuji, M. Fujita, S. Park, S.-H. Lee, Y. Endoh: Phys. Rev. Lett. **90**, 137004 (2003)
111. G.-q. Zheng, T. Sato, Y. Kitaoka, M. Fujita, K. Yamada: Phys. Rev. Lett. **90**, 197005 (2003)
112. S. Uchida, T. Ido, H. Takagi, T. Arima, Y. Tokura, S. Tajima: Phys. Rev. B **43**, 7942 (1991)
113. Y. Onose: Ph. D. Thesis, The University of Tokyo, Tokyo 2002: Y. Onose, Y. Taguchi, K. Ishizaka, Y. Tokura: Phys. Rev. Lett. **87**, 217001 (2001); Y. Onose, Y. Taguchi, K. Ishizaka, Y. Tokura: Phys. Rev. B **69**, 024504 (2004)
114. H. Eskes, A. M. Oleś: Phys. Rev. Lett. **73**, 1279 (1994): H. Eskes, A. M. Oleś, M. B. J. Meinders, W. Stephan: Phys. Rev. B **50**, 17980 (1994)
115. E. Dagotto: Rev. Mod. Phys. **66**, 763 (1994)
116. T. Tohyama, S. Maekawa: Phys. Rev. B **64**, 212505 (2001)
117. T. Katsufuji, Y. Okimoto, Y. Tokura: Phys. Rev. Lett. **74**, 3497 (1996)
118. C. T. Chen, F. Sette, Y. Ma, M. S. Hybertsen, E. B. Stechel, W. M. C. Foulkes, M. Schulter, S.-W. Cheong, A. S. Cooper, L. W. Rupp, Jr., B. Batlogg, Y. L. Soo, Z. H. Ming, A. Krol, Y. H. Kao: Phys. Rev. Lett. **66**, 104 (1991); H. Romberg, M. Alexander, N. Nücker, P. Adelmann and J. Fink, Phys. Rev. B **42**, 8768 (1990)
119. T. Yoshida, X. J. Zhou, T. Sasagawa, W. L. Yang, P. V. Bogdanov, A. Lanzara, Z. Hussain, T. Mizokawa, A. Fujimori, H. Eisaki, Z.-X. Shen, T. Kakeshita, S. Uchida: Phys. Rev. Lett. **91**, 027001 (2003)
120. N. P. Armitage, F. Ronning, D. H. Lu, C. Kim, A. Damascelli, K. M. Shen, D. L. Feng, H. Eisaki, Z.-X. Shen, P. K. Mang, N. Kaneko, M. Greven, Y. Onose, Y. Taguchi, Y. Tokura: Phys. Rev. Lett. **88**, 257001 (2002)
121. N. Harima, J. Matsuno, A. Fujimori, Y. Onose, Y. Taguchi, Y. Tokura: Phys. Rev. B **64**, 220507(R) (2001); N. Harima, A. Fujimori, T. Sugaya, I. Terasaki: Phys. Rev. B **67**, 172501 (2003)
122. N. Furukawa, M. Imada: J. Phys. Soc. Jpn. **60** (1991) 3669
123. T. Tohyama, S. Maekawa: Phys. Rev. B **67**, 092509 (2003)
124. P. W. Anderson: *The Theory of Superconductivity in the High-Tc Cuprates* (Princeton University Press, Princeton NJ 1997)
125. J. Orenstein, A. J. Millis: Science **288**, 46 (2000)
126. S. Sachdev: Rev. Mod. Phys. **75**, 913 (2003)
127. E. Dagotto: Rep. Prog. Phys. **62**, 1525 (1999)
128. Very recently, an inelastic neutron scattering measurement for $SrCuO_2$ has been reported. I. A. Zaliznyak, H. Woo, T. G. Perring, C. L. Broholm, C. D. Frost, H. Takagi: cond-mat/0312724

3. Theory of Superconductivity

There are a number of transition metal oxide superconductors listed with their transition temperatures in Table 3.1. However there exist only a single family of transition metal oxide superconductors with high transition temperatures and all of these have as a common element the existence of Cu-O planes. These materials are commonly called "the cuprates" and will be the focus of this chapter. While, no doubt, the other oxides represent interesting case studies in superconductivity they do not have either the same technological promise nor represent the intellectual challenge that do the cuprates. The key experimental facts and a comparison, principally with the results of exact diagonalization calculations, have been described in Chap. 2. Unfortunately, despite a deluge of articles, there is no real consensus on the form of an appropriate theory. The somewhat delicate task here is therefore to explain the *elements* that might make up a *future* widely accepted theory of cuprate superconductivity.

The Bardeen-Cooper-Schrieffer, (BCS) theory of superconductivity [1,2] is almost certainly relevant. This must be thought of as being made of two parts. The most important, first, part is the pairing theory of superconductivity and second is the mechanism for that pairing. The BCS theory was designed to explain superconductivity in conventional metals and most intermetallic compounds, for which it is very widely accepted that phonons provide the necessary attractive interaction to mediate pairing. This chapter is based

Table 3.1. A selection of transition metal oxide superconductors with the superconductive transition temperature T_c.

Material	T_c
NbO	1 K
TiO	2 K
$SrTiO_{3-x}$	0.7 K
$Na_xCoO_2 \cdot yH_2O$	5 K
K_xWO_3	6 K
$LiTi_2O_4$	13 K
$La_{2-x}Sr_xCuO_4$	38 K
$YBa_2Cu_3O_{6+x}$	92 K
$Tl_2Ca_2Ba_2Cu_3O_{10}$	125 K

upon the premise that something akin to the BCS pairing theory applies to the cuprates. As the name suggests, at the heart of the pairing theory is the concept of electron pairs which carry the supercurrent. These pairs can be classified by their spin and orbital quantum numbers (adapted to lattice problems). Most superconductors have s-wave singlet pairs, i.e., the spin and orbital quantum numbers are both zero. Based upon strong experimental evidence [3], it is widely accepted that the cuprates have rather d-wave singlet pairing, correspond to orbital quantum number $\ell = 2$ but still $s = 0$. While the evidence for this assumption, which will be adopted here, is strong it is far from being evident that the phonons are responsible for pairing interaction, although this possibility still has a well respected clan of supporters [4–6]. For this reason, in addition to the BCS pairing theory, the elements of the phonon pairing mechanism will be developed, this in order that the reader will have the background to understand these proposals.

The rest of this chapter is based upon a prevalent prejudice that the proximity of the superconducting and the anti-ferromagnetic phases of these presumed Mott insulators is relevant to cuprate superconductivity, or put succinctly that the pairing interaction is of magnetic origin. It is therefore necessary to develop the theory of highly correlated electrons based upon Hamiltonians such as the Hubbard or $t - J$-models.

Within this subset of magnetic mechanisms there are too many proposals to be adequately covered in an introductory chapter [6,7]. Dealt with are only a few over arching themes. One of these is the resonant valence bond (RVB) proposal of P. W. Anderson [8]. This is a common thread connecting many suggestions. Another is the SO(5) theory of S.-C. Zhang [9] which permits the magnetic-superconductor phase diagram to be pictured as a single entity. Yet another common thread is "thin" superconductivity. Even in the BCS pairing theory, the number of condensed electrons $\sim \Delta/E_F$ is small, where Δ is the gap and E_F the Fermi energy. In the cuprates the number is even smaller and seems [10] to be proportional to the transition temperature T_c. A particularly elegant approach to this gossamer nature of the superconductivity has been presented [11] by R. B. Laughlin, and this will also be described.

The "tools of the trade" for the theory of highly correlated electron systems are many [12,13]. Theory must handle the conceptual problem that, for a lightly doped Mott insulator, there are many more spin degrees of freedom than there are charge carriers. For this reason formalisms based on the physical electron field $c^\dagger_{\boldsymbol{k}\sigma}$ tend to run into difficulties. Within the auxiliary particle method [14], in the simplest case, the site i operator $c^\dagger_{\sigma i} \approx f^\dagger_{\sigma i} b_i$ factors into a spin field $f^\dagger_{\sigma i}$ and a charge field b_i which reflects holes. The Mott insulator is thereby described in terms of the $f^\dagger_{\sigma i}$-particles to which, upon doping, are added a few hole $b^\dagger_i$-particles. This method will be introduced in this chapter and used to transcribe the above mentioned theory into a uniform notation.

Since the prejudice is that the superconductivity evolves from the anti-ferromagnetism upon doping, there is also an introduction to the Holstein-

Primakoff theory [15] for the parent compound along with the adaptation of this to spin polarons, i.e., to the problem [16] of a single hole in the antiferromagnetism parent. The Jordan-Wigner transformation is also introduced since this is perhaps better at describing spin liquids, i.e., the spin state in the presence of a finite density of holes. For a two dimensional system this latter transformation converts the hard core bosons, which correspond to the $S = 1/2$ algebra, to fermions plus flux tubes. The mean-field approximation to this scheme will be explored and related to the properties of flux states. Also in the introductory material is the Bogoliubov theory of a superfluid which is adapted to describe the potential condensation of hard core bosonic holes.

3.1 The BCS Pairing Theory

Whatever the real complications of the electronic band structure, for illustration purposes, it is here assumed that the non-interacting electrons involved in superconductivity are described by,

$$\mathcal{H}_0 = -t \sum_{\langle ij \rangle \sigma} (c^\dagger_{i\sigma} c_{j\sigma} + H.c) + (\epsilon_d - \mu) \sum_{i\sigma} c^\dagger_{i\sigma} c_{i\sigma} = \sum_{\boldsymbol{k}\sigma} (\epsilon_{\boldsymbol{k}} - \mu) c^\dagger_{\boldsymbol{k}\sigma} c_{\boldsymbol{k}\sigma}. \quad (3.1)$$

Here, as is common practice, μ the chemical potential has been included in the particle energies and there is no magnetic field. The energy $\epsilon_{\boldsymbol{k}} = \epsilon_d - 2t\gamma_{\boldsymbol{k}}$, where specializing to a two dimensional square lattice $\gamma_{\boldsymbol{k}} = \cos k_x a + \cos k_y a$, lattice spacing a, appropriate to the cuprates, Chap. 1. There is a band of width $8t$ centered at ϵ_d. To this is added an effective attractive electron-electron interaction to give a Hamiltonian of the form,

$$\mathcal{H} = \sum_{\boldsymbol{k}\sigma} (\epsilon_{\boldsymbol{k}} - \mu) c^\dagger_{\boldsymbol{k}\sigma} c_{\boldsymbol{k}\sigma} - V \sum_{\boldsymbol{qkk}'\sigma\sigma'} c^\dagger_{\boldsymbol{k}'+\boldsymbol{q}\sigma'} c_{\boldsymbol{k}'\sigma'} c^\dagger_{\boldsymbol{k}-\boldsymbol{q}\sigma} c_{\boldsymbol{k}\sigma}. \quad (3.2)$$

Following BCS [1, 2], the superconductive state has a broken symmetry and an off-diagonal long range order parameter is defined by,

$$\Delta = V \sum_{\boldsymbol{k}} \langle c_{-\boldsymbol{k}\downarrow} c_{\boldsymbol{k}\uparrow} \rangle. \quad (3.3)$$

Clearly, since $c_{-\boldsymbol{k}\downarrow} c_{\boldsymbol{k}\uparrow}$ destroys two particles the existence of a finite Δ implies finite off-diagonal matrix elements which connect states with different numbers of particles. This order parameter is factored out of the interaction term, i.e., $\mathcal{H} \to \sum_{\boldsymbol{k}\sigma} (\epsilon_{\boldsymbol{k}} - \mu) c^\dagger_{\boldsymbol{k}\sigma} c_{\boldsymbol{k}\sigma} + V \sum_{\boldsymbol{qk}} c^\dagger_{\boldsymbol{k}-\boldsymbol{q}\uparrow} c^\dagger_{-\boldsymbol{k}+\boldsymbol{q}\downarrow} \langle c_{-\boldsymbol{k}\downarrow} c_{\boldsymbol{k}\uparrow} \rangle \ldots$. Dropping all quantities which are higher than second order in the c operators gives, $\mathcal{H} = \sum_{\boldsymbol{k}} \mathcal{H}_{\boldsymbol{k}}$, where,

$$\mathcal{H}_{\boldsymbol{k}} = (\epsilon_{\boldsymbol{k}} - \mu)(c^\dagger_{\boldsymbol{k}\uparrow} c_{\boldsymbol{k}\uparrow} + c^\dagger_{-\boldsymbol{k}\downarrow} c_{-\boldsymbol{k}\downarrow}) - \Delta c^\dagger_{\boldsymbol{k}\uparrow} c^\dagger_{-\boldsymbol{k}\downarrow} - \Delta^* c_{-\boldsymbol{k}\downarrow} c_{\boldsymbol{k}\uparrow}, \quad (3.4)$$

and where use if made of $\epsilon_{\boldsymbol{k}} = \epsilon_{-\boldsymbol{k}}$. Defining $v = (c^\dagger_{\boldsymbol{k}\uparrow}, c_{-\boldsymbol{k}\downarrow})$, and dropping a constant, this can be written as

$$\mathcal{H}_{\boldsymbol{k}} = vM_k v^\dagger; \quad M_{\boldsymbol{k}} = 2(\epsilon_{\boldsymbol{k}} - \mu)S_z - 2\Delta S_x = \boldsymbol{H}\cdot\boldsymbol{S} \tag{3.5}$$

where e.g., S_z is the standard $S = 1/2$ matrix. This can but put in a diagonal form by a rotation $R(\theta)$ as follows:

$$2E_{\boldsymbol{k}}S_z = R(\theta)M_{\boldsymbol{k}}R^{-1}(\theta), \quad \text{where} \quad \tan\theta = \frac{H_x}{H_z} = \frac{\Delta}{\epsilon_{\boldsymbol{k}} - \mu}. \tag{3.6}$$

Trivially then,

$$E_{\boldsymbol{k}} = \sqrt{(\epsilon_{\boldsymbol{k}} - \mu)^2 + \Delta^2}. \tag{3.7}$$

Now, inserting rotations matrices, $\mathcal{H}_{\boldsymbol{k}} = vR^{-1}RM_{\boldsymbol{k}}R^{-1}Rv^\dagger$ leads to the definitions, $w \equiv (\alpha^\dagger_{\boldsymbol{k}}, \beta_{\boldsymbol{k}}) = vR^{-1}$, or that $w^{\dagger T}$ is

$$\begin{pmatrix} \alpha_{\boldsymbol{k}} \\ \beta^\dagger_{\boldsymbol{k}} \end{pmatrix} = R(\theta)\begin{pmatrix} c_{\boldsymbol{k}\uparrow} \\ c^\dagger_{-\boldsymbol{k}\downarrow} \end{pmatrix}, \quad \text{where} \quad R(\theta) = \begin{pmatrix} \cos\theta/2, & -\sin\theta/2 \\ \sin\theta/2, & \cos\theta/2 \end{pmatrix}, \tag{3.8}$$

for $S = 1/2$ [17]. The original *fermion* commutation rules correspond to $v^\dagger v + v^T v^{\dagger T} = 1$, where, e.g., v^T is the simple transpose of v. (Bosons would have $v^\dagger v - v^T v^{\dagger T} = 2S_z$.) Since R is real it follows that $w^\dagger w + w^T w^\dagger T = 1$, i.e., fermion (but not boson) commutation rules are preserved by R. It now is possible to write the Hamiltonian in diagonal form, i.e.,

$$\mathcal{H}_{\boldsymbol{k}} = E_{\boldsymbol{k}}(\alpha^\dagger_{\boldsymbol{k}}\alpha_{\boldsymbol{k}} + \beta^\dagger_{\boldsymbol{k}}\beta_{\boldsymbol{k}}) \tag{3.9}$$

again dropping a constant. Explicitly,

$$\alpha_{\boldsymbol{k}} = \cos\frac{\theta}{2}c_{\boldsymbol{k}\uparrow} - \sin\frac{\theta}{2}c^\dagger_{-\boldsymbol{k}\downarrow} \quad \text{while} \quad \beta_{\boldsymbol{k}} = \cos\frac{\theta}{2}c_{-\boldsymbol{k}\downarrow} + \sin\frac{\theta}{2}c^\dagger_{\boldsymbol{k}\uparrow}. \tag{3.10}$$

It is usual to define,

$$u_{\boldsymbol{k}} = \cos\frac{\theta}{2} = \frac{1}{\sqrt{2}}[1+\cos\theta]^{1/2} = \frac{1}{\sqrt{2}}\left[1 + \frac{(\epsilon_{\boldsymbol{k}} - \mu)}{E_{\boldsymbol{k}}}\right]^{1/2} \tag{3.11}$$

using the result $\tan\theta = \Delta/(\epsilon_{\boldsymbol{k}} - \mu)$, while

$$v_{\boldsymbol{k}} = \sin\frac{\theta}{2} = \frac{1}{\sqrt{2}}\left[1 - \frac{(\epsilon_{\boldsymbol{k}} - \mu)}{E_{\boldsymbol{k}}}\right]^{1/2}. \tag{3.12}$$

Since the $\alpha^\dagger_{\boldsymbol{k}}$ and $\beta^\dagger_{\boldsymbol{k}}$ correspond to excitations with positive definite energies, it must be that the destruction operators $\alpha_{\boldsymbol{k}}$ and $\beta_{\boldsymbol{k}}$ destroy the appropriate ground state. If $|0\rangle$ is the true vacuum then,

$$|BCS\rangle \propto \prod_{\boldsymbol{k}} \alpha_{\boldsymbol{k}}\beta_{\boldsymbol{k}}|0\rangle \tag{3.13}$$

has the desired property that, e.g., $\alpha_{\boldsymbol{k}}|BCS\rangle = 0$ since any $\alpha_{\boldsymbol{k}}\alpha_{\boldsymbol{k}} = 0$, and at least for $\theta \neq 0$ this state is non-zero. Using $\alpha_{\boldsymbol{k}}\beta_{\boldsymbol{k}}|0\rangle = (v_{\boldsymbol{k}})(u_{\boldsymbol{k}} + v_{\boldsymbol{k}}c^\dagger_{\boldsymbol{k}\uparrow}c^\dagger_{-\boldsymbol{k}\downarrow})|0\rangle$, it is easy to check that

$$|BCS\rangle = \prod_{\boldsymbol{k}}(u_{\boldsymbol{k}} + v_{\boldsymbol{k}}c^\dagger_{\boldsymbol{k}\uparrow}c^\dagger_{-\boldsymbol{k}\downarrow})|0\rangle, \tag{3.14}$$

is the normalized ground state.

In order to close the calculation, it is necessary to calculate, $\Delta = V\sum_{\boldsymbol{k}}\langle c_{-\boldsymbol{k}\downarrow}c_{\boldsymbol{k}\uparrow}\rangle$, and needed are expressions for e.g., $c_{\boldsymbol{k}\uparrow}$. The inverse of (3.8) is,

$$\begin{pmatrix} c_{\boldsymbol{k}\uparrow} \\ c^{\dagger}_{-\boldsymbol{k}\downarrow} \end{pmatrix} = R^{-1}(\theta)\begin{pmatrix} \alpha_{\boldsymbol{k}} \\ \beta^{\dagger}_{\boldsymbol{k}} \end{pmatrix} = R(-\theta)\begin{pmatrix} \alpha_{\boldsymbol{k}} \\ \beta^{\dagger}_{\boldsymbol{k}} \end{pmatrix}, \tag{3.15}$$

which gives,

$$c_{\boldsymbol{k}\uparrow} = u_{\boldsymbol{k}}\alpha_{\boldsymbol{k}} + v_{\boldsymbol{k}}\beta^{\dagger}_{\boldsymbol{k}} \quad \text{and} \quad c^{\dagger}_{-\boldsymbol{k}\downarrow} = u_{\boldsymbol{k}}\beta^{\dagger}_{\boldsymbol{k}} - v_{\boldsymbol{k}}\alpha_{\boldsymbol{k}}, \tag{3.16}$$

so that the diagonal part of $c_{-\boldsymbol{k}\downarrow}c_{\boldsymbol{k}\uparrow} = u_{\boldsymbol{k}}v_{\boldsymbol{k}}(\beta_{\boldsymbol{k}}\beta^{\dagger}_{\boldsymbol{k}} - \alpha^{\dagger}_{\boldsymbol{k}}\alpha_{\boldsymbol{k}}) = u_{\boldsymbol{k}}v_{\boldsymbol{k}}(1 - \beta^{\dagger}_{\boldsymbol{k}}\beta_{\boldsymbol{k}} - \alpha^{\dagger}_{\boldsymbol{k}}\alpha_{\boldsymbol{k}})$.

The α and β operators correspond to good fermions, and so the *thermal* expectation value $\langle\alpha^{\dagger}_{\boldsymbol{k}}\alpha_{\boldsymbol{k}}\rangle = \langle\beta^{\dagger}_{\boldsymbol{k}}\beta_{\boldsymbol{k}}\rangle = f(E_{\boldsymbol{k}})$ where $f(x) = (\mathrm{e}^{\beta x} + 1)^{-1}$; $\beta = 1/k_BT$, is the usual fermion occupation factor. It follows that,

$$\Delta = V\sum_{\boldsymbol{k}} u_{\boldsymbol{k}}v_{\boldsymbol{k}}(1 - 2f(E_{\boldsymbol{k}})) = V\sum_{\boldsymbol{k}} u_{\boldsymbol{k}}v_{\boldsymbol{k}}\tanh(E_{\boldsymbol{k}}/2). \tag{3.17}$$

Now $u_{\boldsymbol{k}}v_{\boldsymbol{k}} = \frac{1}{2}\sin\theta = (\Delta/2E_{\boldsymbol{k}})$ so that, $\Delta = V\sum_{\boldsymbol{k}}(\Delta/2E_{\boldsymbol{k}})\tanh(E_{\boldsymbol{k}}/2)$. The factor Δ cancels to give,

$$1 = V\sum_{\boldsymbol{k}}\frac{1}{2E_{\boldsymbol{k}}}\tanh(\beta E_{\boldsymbol{k}}/2), \tag{3.18}$$

as an implicit equation for Δ.

To summarize: the pairing Hamiltonian has been put in the form, $\mathcal{H} = \mathcal{H} = \sum_{\boldsymbol{k}}\left[E_{\boldsymbol{k}}(\alpha^{\dagger}_{\boldsymbol{k}}\alpha_{\boldsymbol{k}} + \beta^{\dagger}_{\boldsymbol{k}}\beta_{\boldsymbol{k}})\right]$, where the *positive* excitation energies of the system are, $E_{\boldsymbol{k}} = \sqrt{(\epsilon_{\boldsymbol{k}} - \mu)^2 + \Delta^2}$, and there are *two* branches. Logically it must be that the non-interacting Fermi sea is recovered in the limit that the interaction $V \to 0$. In this limit $E_{\boldsymbol{k}} = |\epsilon_{\boldsymbol{k}}|$. That the system can only have positive excitations is of course correct. The negative energies $\epsilon_{\boldsymbol{k}}$ correspond to filled states and when the $V \to 0$ limit of the transformations are examined, it is found that a particle-hole transformation has been made on these negative energy states.

What is described so far is singlet s-wave pairing. The electrons which form pairs and which enter in the definition of the order parameter have opposite spin directions. It was also assumed that there is no momentum dependence to the pairing interaction and therefore no $\boldsymbol{k}$ dependence of the order parameter. The pairing for the model described by (3.2) is s-wave because the attractive interaction V has no structure. Specializing to two dimensions, the simplest way to induce d-wave pairing is to assume instead,

$$V_{\boldsymbol{k}\boldsymbol{k}'} = V\gamma_{d\boldsymbol{k}}\gamma_{d\boldsymbol{k}'} \tag{3.19}$$

where $\gamma_{d\boldsymbol{k}} = (\cos k_xa - \cos k_ya)$. Then the equation for $\Delta_{\boldsymbol{k}}$ becomes $\Delta_{\boldsymbol{k}} = \sum_{\boldsymbol{k}'}V_{\boldsymbol{k}\boldsymbol{k}'}(\Delta_{\boldsymbol{k}'}/2E_{\boldsymbol{k}'})\tanh(E_{\boldsymbol{k}'}/2)$ which has a solution with

$$\Delta_{\boldsymbol{k}} = \Delta\gamma_{d\boldsymbol{k}} \tag{3.20}$$

where Δ has a value determined by the equivalent of (3.18). Inspection of $V_{\boldsymbol{kk}'}$ shows that it is equal to $V(\cos k_x - \cos k_y)^2$, i.e., large and *attractive* when $\boldsymbol{k} = \boldsymbol{k}'$ but is $-V(\cos k_x - \cos k_y)^2$, i.e., *repulsive* when $\boldsymbol{k} = \boldsymbol{k}' + \boldsymbol{\pi}$. Since $V_{\boldsymbol{kk}'}$ reflects electron-electron scattering it is seen that d-wave pairing is favored when forward scattering is strong and attractive in nature. Assuming near neighbor interactions on a square lattice the interaction is of the form $V_{\boldsymbol{kk}'} = V\gamma_{\boldsymbol{k}-\boldsymbol{k}'} = (V/2)(\gamma_{\boldsymbol{k}}\gamma_{\boldsymbol{k}'} + \gamma_{d\boldsymbol{k}}\gamma_{d\boldsymbol{k}'} + \gamma^s_{\boldsymbol{k}}\gamma^s_{\boldsymbol{k}'} + \gamma^s_{d\boldsymbol{k}}\gamma^s_{d\boldsymbol{k}'})$ where the odd parity quantities γ^s have the cosines replaced by sines. Clearly the other factor in the integral for the gap, i.e., $[(\tanh\frac{1}{2}\beta E_{\boldsymbol{k}})/E_{\boldsymbol{k}}]$, is invariant to the operations of the cubic group. In particular it is unchanged under reflection, i.e., $k_x \to -k_x$ and $k_y \to -k_y$ and to a rotation by $\frac{\pi}{2}$, i.e., $k_x \to k_y$ and $k_y \to k_x$. For spin singlet pairing, the order parameter must have even parity, and these symmetries guarantee that the odd parity terms average to zero, as do cross terms between s and d-wave pairing. While the s and d-wave components have the same weight, the difference in the $E_{\boldsymbol{k}}$ implies a d-wave gap *near half-filling.* The real space transform of $\Delta_{\boldsymbol{k}}$ reflects the pair wave function. If a singlet pair of electrons is to have a maximum amplitude on near neighbor sites it is an advantage to have nodes in the directions between those sites. This describes precisely a d-wave pair wave-function.

The density of states is defined by $N(E_{\boldsymbol{k}}) = \mathrm{d}n/\mathrm{d}E_{\boldsymbol{k}} = (\mathrm{d}n/\mathrm{d}\epsilon_{\boldsymbol{k}})(\mathrm{d}E_{\boldsymbol{k}}/\mathrm{d}\epsilon_{\boldsymbol{k}})$ which gives,

$$N(E) = 2N(0)\frac{E}{\sqrt{E^2 - \Delta^2_{\boldsymbol{k}}}}, \tag{3.21}$$

where the factor of two corresponds to the two branches. Essentially this density of states is accessible via tunneling experiments, however a mirror image of the density of states is seen for negative energies corresponding to the tunneling of holes rather than electrons. The theoretical and experimental [18] results are shown in Fig. 3.1.

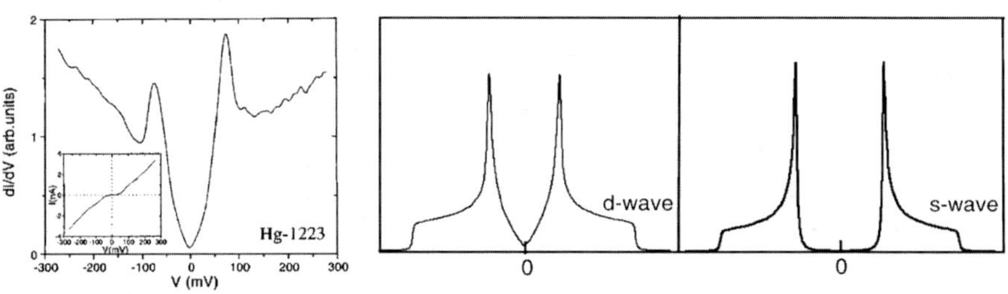

Fig. 3.1. To the left, experimental tunneling data for polycrystalline Hg-1223, with $T_c \approx 135$ K. In the middle is a simulation of the data assuming d-wave pairing and to the right for the s-wave equivalent. For the simulations the units are arbitrary but the scale is chosen to facilitate the comparison with experiment.

3.2 Phonons in Solids

Consider the motion of the atoms of a solid about their equilibrium positions. Assuming that the zero point motion about these positions is not too large, the restoring potential will be, to a good approximation, harmonic for low lying excitations of the lattice. Even at the level of classical mechanics, the normal mode solutions of this problem can be very complicated and these complications are of course present in the quantum mechanical problem. However these are not essential considerations for an understanding of the theory and hence the simplest case of a mono-atomic cubic lattice with near neighbor interactions and the appropriate dimensions will be considered. This contains the essential physics. The Hamiltonian can be written as,

$$\mathcal{H} = \sum_i \frac{{P_i}^2}{2M} + \frac{K}{2} \sum_{\langle ij \rangle} (\boldsymbol{Q}_i - \boldsymbol{Q}_j)^2, \tag{3.22}$$

where $\boldsymbol{P}_i$ is the momentum, $\boldsymbol{Q}_i$ the displacement from equilibrium, and $\boldsymbol{r}_i$ the position of the i-th atom. The advantage of the simplifying assumptions is that motions in the x, y and z directions decouple and are equivalent, thus

$$\mathcal{H} = \mathcal{H}_x + \mathcal{H}_y + \mathcal{H}_z \tag{3.23}$$

where, with e.g., P_{ix} being the x-component of $\boldsymbol{P}_i$, etc.,

$$\mathcal{H}_x = \sum_i \frac{{P_{ix}}^2}{2M} + \frac{K}{2} \sum_{\langle ij \rangle} (Q_{ix} - Q_{jx})^2. \tag{3.24}$$

In the following, the subscript x will be dropped and only $\mathcal{H}_x$ sub-problem will be considered.

The quantized solution to this problem introduces a set of Bose *phonon* creation operators $a_{\boldsymbol{k}}^{\dagger}$ with commutation rules $[a_{\boldsymbol{k}}, a_{\boldsymbol{k}'}^{\dagger}] = \delta_{\boldsymbol{k},\boldsymbol{k}'}$. In terms of these,

$$\mathcal{H} = \sum_{\boldsymbol{k}} \omega_{\boldsymbol{k}} \left(n_{\boldsymbol{k}} + \frac{1}{2} \right) \tag{3.25}$$

where the number operator $n_{\boldsymbol{k}} = a_{\boldsymbol{k}}^{\dagger} a_{\boldsymbol{k}}$ and where $\omega_{\boldsymbol{k}} = \frac{2K}{M}[1 - \gamma_{\boldsymbol{k}}]$. The ground state energy is $E_0 = \frac{1}{2}\sum_{\boldsymbol{k}} \hbar\omega_{\boldsymbol{k}}$ which corresponds to the zero point motion of the atoms, the corresponding state is denoted $|\ >$. The excited states of the system are obtained by operating on this vacuum with the creation operators $a_{\boldsymbol{k}}^{\dagger}$ which create a *phonon* of energy $\hbar\omega_{\boldsymbol{k}}$, i.e., an excited state energy is determined by giving a set of *occupation numbers* $n_{\boldsymbol{k}}$ whence, corresponding to the state,

$$\prod_{\boldsymbol{k}} [(1 + n_{\boldsymbol{k}})!]^{-1/2} \left(a_{\boldsymbol{k}}^{\dagger} \right)^{n_{\boldsymbol{k}}} |\ > \quad \text{with energy} \quad E_0 + \sum_{\boldsymbol{k}} \hbar\omega_{\boldsymbol{k}} n_{\boldsymbol{k}}. \tag{3.26}$$

Each phonon with a wave vector $\boldsymbol{k}$ has an energy $\hbar\omega_{\boldsymbol{k}}$ and any number of phonons can be created with such a single such wave vector. Recall that these results are for $\mathcal{H}_x$ and therefore that each energy level is three times degenerate. If, for a given wave-vector $\boldsymbol{k}$, one of the axes is chosen together the direction of propagation, it is easily appreciated that there is one longitudinal mode with the displacement of the atoms along the direction of propagation along with two transverse modes. More generally these modes are *not* degenerate and mix.

3.3 Phonons as Intermediate Bosons

The interactions between elementary particles are mediated by intermediate bosons. For example [19] the photon mediates the electromagnetic interactions. Here it will shown how the phonons can play a similar role mediating an interaction between electrons, this leading to an attractive interaction, and one possibility for superconductivity in the cuprates.

Consider the effects of atomic displacements on electronic Hamiltonian $\mathcal{H}_0 = -\sum_{\langle ij \rangle \sigma}(t_{ij}c^\dagger_{i\sigma}c_{j\sigma} + H.c) + \sum_{i\sigma}(\epsilon_{id} - \mu)c^\dagger_{i\sigma}c_{i\sigma}$ where to the near neighbor interaction t_{ij} and atomic electronic energy ϵ_{id} have been added site indices. When two atoms come closer together the hopping integral t_{ij} gets bigger and so if $\boldsymbol{Q}_i$ is the atomic displacement of site i then, e.g., for a bond in the x-direction $t_{ij} = t - v(Q_{xi} - Q_{xj})$ which adds an electron-displacement interaction of the form,

$$v2\mathrm{i}c^\dagger_{\boldsymbol{k}+\boldsymbol{q}}c_{\boldsymbol{k}}[\sin(q_x a)Q_{x\boldsymbol{q}} + \sin(q_y a)Q_{y\boldsymbol{q}} + \sin(q_z a)Q_{z\boldsymbol{q}}], \tag{3.27}$$

where here, and the in the rest of this section, the spin label has been dropped for clarity. Because of the factors $\sin q_x a$, $\sin q_y a$ and $\sin q_z a$ the amplitude for small angle electron scattering, i.e., when $\boldsymbol{q} \to 0$, is strongly suppressed. This is the tight binding version of the usual electron phonon interaction found in the standard texts [15]. For small $\boldsymbol{q}$, $[\sin(q_x a)Q_{x\boldsymbol{q}} + \sin(q_y a)Q_{y\boldsymbol{q}} + \sin(q_z a)Q_{z\boldsymbol{q}}]$ is just the Fourier transform of $a\boldsymbol{\nabla} \cdot \boldsymbol{Q}_i$. As mentioned in Sect. 3.1, strong small angle scattering is a requirement for d-wave pairing. The displacement of the electron bearing atom described by $\mathcal{H}_0$ cannot have a first order effect on ϵ_{id} however the movement of the *ligands* to this atom *can* have such an effect. The simplest conceptual possibility is for the ligands to move in and out together, what is usually called a breading mode [20]. Then, either because the electrostatic potential increases or decreases, or because of a covalent correction, it is possible to have $\epsilon_{id} = \epsilon_d + vQ_i$ where here the Q_i is the displacement amplitude for this mode. The Fourier transform of such an interaction gives

$$\mathcal{H}_{ep} = v\frac{1}{\sqrt{N}}\sum_{\boldsymbol{kq}} c^\dagger_{\boldsymbol{k}+\boldsymbol{q}}c_{\boldsymbol{k}}Q_{\boldsymbol{q}}. \tag{3.28}$$

In terms of phonon operators $Q_{\boldsymbol{q}} = (2m\omega_{\boldsymbol{q}})^{-1/2}(a_{\boldsymbol{q}} + a^\dagger_{-\boldsymbol{q}})$, and in the context of d-wave pairing, it is to be observed that,

$$(2m\omega_{\boldsymbol{q}})^{-1/2} \sim \frac{1}{k^{1/2}}, \tag{3.29}$$

i.e., because the factor of k which would come from the gradient operator is absent, this electron-phonon interaction is in fact *singular* for the forward direction. This divergence has to do with the fact that the (acoustic) phonons, like photons, are massless. There is a second issue which must also be considered in the context of enhanced forward scattering. Here the interaction between the atomic level and the lattice displacement was assumed local while in fact this reflects the Coulomb interaction the Fourier transform of which $\sim q^{-2}$ and therefore which also enhances the small q pairing amplitude. This divergence is removed when screening is included. The efficiency by which the cuprates screen the Coulomb interaction is therefore also important to the present considerations. This has been discussed in the literature [5].

Absorbing all $\boldsymbol{q}$ dependences into a matrix element, the electron-phonon interaction becomes,

$$\mathcal{H}_{\mathrm{ep}} = \sum_{\boldsymbol{kq}} M_{\boldsymbol{q}} c^{\dagger}_{\boldsymbol{k}+\boldsymbol{q}} c_{\boldsymbol{k}} (a_{\boldsymbol{q}} + a^{\dagger}_{-\boldsymbol{q}}) \equiv \sum_{\boldsymbol{k}} \mathcal{H}_{\boldsymbol{q}}. \tag{3.30}$$

This expression is more general than the derivation indicated above. The difference in the weighting of forward scattering, the details of the potential, etc., are all hidden in the $\boldsymbol{q}$ dependence of $M_{\boldsymbol{q}}$. In second-order $\mathcal{H}_{\mathrm{ep}}$ leads to a self-energy and also to a coupling between two electrons. The coupling is an indirect interaction through the phonon field. One electron polarizes the lattice; the other electron interacts with the polarization. The total Hamiltonian of electrons, phonons, and their interaction is $\mathcal{H} = \mathcal{H}_0 + \mathcal{H}_{\mathrm{ep}}$ where the non-interacting electrons and phonons are described by $\mathcal{H}_0 = \sum_{\boldsymbol{q}} \hbar\omega_{\boldsymbol{q}} a^{\dagger}_{\boldsymbol{q}} a_{\boldsymbol{q}} + \sum_{\boldsymbol{k}} (\epsilon_{\boldsymbol{k}} - \mu) c^{\dagger}_{\boldsymbol{k}} c_{\boldsymbol{k}}$. A canonical transformation is used to eliminate $\mathcal{H}_{\mathrm{ep}}$ in favor of a new Hamiltonian with only electron-electron interactions. The new,

$$\tilde{\mathcal{H}} \equiv \mathrm{e}^{-S} \mathcal{H} \mathrm{e}^{S} = \mathcal{H} + [\mathcal{H}, S] + \frac{1}{2}[[\mathcal{H}, S], S] + \ldots \tag{3.31}$$

The operator S is chosen so that $[\mathcal{H}_0, S] + \mathcal{H}_{\mathrm{ep}} = 0$ which causes the second term to eliminate the direct electron-phonon interaction $\mathcal{H}_{\mathrm{ep}}$, whence $\tilde{\mathcal{H}} \approx \mathcal{H}_0 - \frac{1}{2}[\mathcal{H}_{\mathrm{ep}}, S]$. It is easy to show, $\langle n|S|m\rangle = \langle n|\mathcal{H}_{\mathrm{ep}}|m\rangle/(E_m - E_n)$. An expression for S can be written without needing to specify the matrix elements involved, i.e.,

$$S = M_{\boldsymbol{q}} \sum_{\boldsymbol{k}} \frac{a^{\dagger}_{\boldsymbol{q}} c^{\dagger}_{\boldsymbol{k}-\boldsymbol{q}} c_{\boldsymbol{k}}}{\epsilon_{\boldsymbol{k}} - \epsilon_{\boldsymbol{k}-\boldsymbol{q}} - \hbar\omega_{\boldsymbol{q}}} + M_{\boldsymbol{q}} \sum_{\boldsymbol{k}} \frac{a_{-\boldsymbol{q}} c^{\dagger}_{\boldsymbol{k}-\boldsymbol{q}} c_{\boldsymbol{k}}}{\epsilon_{\boldsymbol{k}} - \epsilon_{\boldsymbol{k}-\boldsymbol{q}} + \hbar\omega_{\boldsymbol{q}}}. \tag{3.32}$$

Then evaluating $\mathcal{H}_{ee} = -\frac{1}{2}[\mathcal{H}_{\mathrm{ep}}, S]$ ignoring self-energy terms, i.e, corrections to the bare parameters of $\mathcal{H}_0$,

$$\mathcal{H}_{\mathrm{ee}} = \sum_{\boldsymbol{q}} {M_{\boldsymbol{q}}}^2 \sum_{\boldsymbol{kk'}} \frac{\hbar\omega_{\boldsymbol{q}}}{(\epsilon_{\boldsymbol{k}} - \epsilon_{\boldsymbol{k}-\boldsymbol{q}})^2 - (\hbar\omega_{\boldsymbol{q}})^2} c^{\dagger}_{\boldsymbol{k'}+\boldsymbol{q}} c_{\boldsymbol{k'}} c^{\dagger}_{\boldsymbol{k}-\boldsymbol{q}} c_{\boldsymbol{k}}. \tag{3.33}$$

This represents an attractive, i.e., negative interaction if $|\epsilon_{\boldsymbol{k}} - \epsilon_{\boldsymbol{k}-\boldsymbol{q}}| < \hbar\omega_{\boldsymbol{q}}$, for those phonon modes with appreciable matrix elements M_q. Since the energy scale for phonons is the Debye energy ω_D this criterion is usual expressed as $|\epsilon_{\boldsymbol{k}} - \epsilon_{\boldsymbol{k}-\boldsymbol{q}}| < \hbar\omega_D$.

Of course, this is only part of the interaction between the electrons, it remains the case that there is also the (screened) Coulomb interaction which is repulsive. Under suitable circumstances the attractive interaction outweighs the screened Coulomb interaction for electrons near the Fermi surface, i.e., for electrons such that $|\epsilon_{\boldsymbol{k}} - \mu| < \omega_D$ the effective attractive interaction is of the form,

$$\mathcal{H}_{\text{ee}} = V \sum_{\boldsymbol{q}\boldsymbol{k}\boldsymbol{k}'} c^\dagger_{\boldsymbol{k}'+\boldsymbol{q}} c_{\boldsymbol{k}'} c^\dagger_{\boldsymbol{k}-\boldsymbol{q}} c_{\boldsymbol{k}}, \tag{3.34}$$

where $V \sim {M_q}^2/\omega_D$. For excitation energies which are large compared with $\hbar\omega_D$ the interaction becomes repulsive. To account for this fact the attractive interaction reflected by $\mathcal{H}_{\text{ee}}$ will be cut-off for energies greater than this. This will introduce $\hbar\omega_D$ as an important *characteristic energy scale* in the phonon theory of superconductivity which follows. In particular, with these considerations in mind the equation for the superconductive order parameter and T_{c} becomes,

$$1 = V \sum_{|\epsilon_{\boldsymbol{k}} - \mu| < \omega_D} \frac{1}{2E_{\boldsymbol{k}}} \tanh\left(\frac{\beta E_{\boldsymbol{k}}}{2}\right), \tag{3.35}$$

i.e., only electrons within ω_D of the Fermi surface are assumed involved in the superconductivity.

For not too small energies, such that $|\epsilon_{\boldsymbol{k}} - \mu| > \Delta_{\boldsymbol{k}}$ it follows that $E_{\boldsymbol{k}} \approx |\epsilon_{\boldsymbol{k}} - \mu|$ and the above sum approximates to a logarithmic integral, i.e.,

$$V \sum_{|\epsilon_{\boldsymbol{k}} - \mu| < \omega_D} \frac{1}{2E_{\boldsymbol{k}}} \tanh\left(\frac{\beta E_{\boldsymbol{k}}}{2}\right) \approx \rho V \int_{E_R}^{E_V} \frac{1}{x} \mathrm{d}x = \rho V \ln \frac{E_V}{E_R} \tag{3.36}$$

where the ultraviolet cut-off $E_V \approx \omega_D$ while the infrared equivalent $E_R \approx [\Delta_{\boldsymbol{k}}]_{\max}$, i.e., the maximum value of $\Delta_{\boldsymbol{k}}$, *or* $k_B T$ whichever is the larger. The condition that this integral be unity then gives T_{c} *and* $[\Delta_{\boldsymbol{k}}]_{\max}$ *both* to be,

$$\sim \hbar\omega_D \mathrm{e}^{-\frac{1}{\rho V}}. \tag{3.37}$$

The difference in the efficiency of these two *infrared cut-offs* in turn gives the ratio of T_{c} to $[\Delta_{\boldsymbol{k}}]_{\max}$. For, e.g., a classical s-wave $\boldsymbol{k}$ independent gap $2\Delta \approx 3.5 k_b T_{\text{c}}$ [1, 2]. These are results for the weak coupling theory when $g = \rho V < 1$. Even if $g \sim 1$ the transition temperature is limited by the Debye energy $\hbar\omega_D$ which is typically a few hundred degrees. The issues which determine the maximum T_{c} which can be obtained by this mechanism are complicated. The reader is referred to some recent literature [4, 5, 20].

3.4 Theory of the Anti-ferromagnetic Parent Compounds

The superconductive state arises from the doping of an Mott insulator, i.e., an anti-ferromagnet. In fact, the SO(5) theory presented in Sect. 3.18 makes the bold suggestion that the order parameter for anti-ferromagnetism and superconductivity belong to the same space. Superconductivity simply appears due to a chemical potential induced rotation in this larger space. Clearly, it is important to understand the description of these parent Mott anti-ferromagnets.

This problem of a two dimensional anti-ferromagnet has been studied by the full arsenal of theoretical tools, i.e., exact diagonalization [21] of small systems, quantum Monte Carlo [22], variational wave functions [23] etc., etc., with the conclusion that one of the simplest descriptions of the ground state based upon the Holstein-Primakoff transformation [15] is adequate for most purposes. Following a theorem of Mermin and Wagner [24], there cannot be long range anti-ferromagnetic order at finite temperature in a two dimensional Heisenberg magnet, however with the addition of even very weak interplane interactions a finite transition temperature is entirely possible. Given that there is long range anti-ferromagnetic order, for the Heisenberg model which is symmetric in spin space, there remains a continuous symmetry and, after a general theorem, there should exist two Goldstone bosons, i.e., excitations which have zero energy as $\boldsymbol{q} \to 0$. Unlike the anomalous case of a ferromagnetic which has quadratic dispersion relationship, here the energy $\hbar\omega_{\boldsymbol{q}} = vq$ for small $\boldsymbol{q}$ where the spin wave velocity $v \sim J_a/\hbar$, and where a is the lattice constant.

Because there remains a continuous symmetry, it is technically necessary to add a vanishing symmetry breaking field h_A in order to define the direction of the staggered magnetization, or Néel vector, $\boldsymbol{N}$. The Heisenberg model with such a field has a Hamiltonian,

$$H = J\sum_{\langle i\,j\rangle} \boldsymbol{S}_i \cdot \boldsymbol{S}_j - h_A {\sum_i}^a S^a_{i\,z} + h_A {\sum_i}^b S^b_{i\,z} \quad h_A \equiv 2\mu_0 H_A \tag{3.38}$$

where, as usual, the notation $\langle i\,j\rangle$ indicates that the sum is over near neighbors only and where the superscripts ^{a} and ^{b} indicate that the relative sums run over *different* alternating sub-lattices, i.e., for the sub-lattice ^{a} the moments tend to align along the positive z-axis and the inverse for the ^{b} sub-lattice. The *natural* axis of quantization alternates between the a and b-lattices. Take the axis of quantization for the $a(b)$-lattice to correspond to the positive(negative) z-axis, then Holstein-Primakoff representation is,

$$S^a_{i\,z} = S - a_i^\dagger a_i \quad \text{while} \quad S^b_{i\,z} = -S + b_i^\dagger b_i, \tag{3.39}$$

where, e.g., $[a_i, a_j^\dagger] = \delta_{ij}$. The state in the absence of the a or b- bosons is the vacuum state of the system. Here the vacuum, referred to as the Néel state, has all the spins "up"("down") on the $a(b)$-lattice. The raising operator on the a-lattice is written as

$$S^+_{a\,i} = (2S)^{1/2}(1 - a_i^\dagger a_i)^{1/2} a_i, \tag{3.40}$$

so that $S^-_{a\,i} = (2S)^{1/2} a_i^\dagger (1 - a_i^\dagger a_i)^{1/2}$ but:

$$S^+_{b\,i} = (2S)^{1/2} b_i^\dagger (1 - b_i^\dagger b_i)^{1/2}, \tag{3.41}$$

i.e., the representations of say S^+ and S^- are interchanged as one passes from one sub-lattice to another. If $c_{\boldsymbol{k}}$ is the spin wave variable (or Fourier transform) associated with the a-sub-lattice and $d_{\boldsymbol{k}}$ corresponds to the b-lattice then, in terms of spin wave variables

$$H = -2NzJS^2 - 2Nh_A S + H_0 + H_1 \tag{3.42}$$

where z is the number of near-neighbors and N is the number of lattice sites. The constants correspond to the classical ground state energy while H_1 reflects the interaction between the a and b-bosons. The axis of quantization have been chosen to correspond to the classical solution, i.e., to a choice of directions for the spins which leads to the lowest energy if the spins are treated as classical vectors. In general, an incorrect choice for the axis of quantization results in spin waves with negative energies.

The bi-linear term is:

$$H_0 = \frac{JzS}{2} \sum_{\boldsymbol{k}} \left[\gamma_{\boldsymbol{k}} \left(a_{\boldsymbol{k}}^\dagger b_{\boldsymbol{k}}^\dagger + a_{\boldsymbol{k}} b_{\boldsymbol{k}}\right) + \left(a_{\boldsymbol{k}}^\dagger a_{\boldsymbol{k}} + b_{\boldsymbol{k}}^\dagger b_{\boldsymbol{k}}\right)\right]$$
$$+ h_A \left(a_{\boldsymbol{k}}^\dagger a_{\boldsymbol{k}} + b_{\boldsymbol{k}}^\dagger b_{\boldsymbol{k}}\right) \equiv \sum_{\boldsymbol{k}} H_{\boldsymbol{k}}. \tag{3.43}$$

The Hamiltonian $H_{\boldsymbol{k}}$ can be diagonalized by a Bogoliubov transformation. This transformation for *fermions* was already introduced in connection with the BCS theory of superconductivity. Here for bosons

$$\alpha_{\boldsymbol{k}} = u_{\boldsymbol{k}} a_{\boldsymbol{k}} - v_{\boldsymbol{k}} b_{\boldsymbol{k}}^\dagger; \quad \alpha_{\boldsymbol{k}}^\dagger = u_{\boldsymbol{k}} a_{\boldsymbol{k}}^\dagger - v_{\boldsymbol{k}} b_{\boldsymbol{k}} \tag{3.44}$$

and

$$\beta_{\boldsymbol{k}} = u_{\boldsymbol{k}} b_{\boldsymbol{k}} - v_{\boldsymbol{k}} a_{\boldsymbol{k}}^\dagger; \quad \beta_{\boldsymbol{k}}^\dagger = u_{\boldsymbol{k}} b_{\boldsymbol{k}}^\dagger - v_{\boldsymbol{k}} a_{\boldsymbol{k}}. \tag{3.45}$$

The new operators satisfy Boson commutation rules $[\alpha_{\boldsymbol{k}} \alpha_{\boldsymbol{k}}^\dagger] = 1$; $[\beta_{\boldsymbol{k}} \beta_{\boldsymbol{k}}^\dagger] = 1$; $[\alpha_{\boldsymbol{k}} \beta_{\boldsymbol{k}}^\dagger] = 0$; etc., *if* ${u_{\boldsymbol{k}}}^2 - {v_{\boldsymbol{k}}}^2 = 1$ (taking a phase convention that $v_{\boldsymbol{k}}$ and $u_{\boldsymbol{k}}$ are real). This follows since, e.g., $[\alpha_{\boldsymbol{k}}, \alpha_{\boldsymbol{k}}^\dagger] = {u_{\boldsymbol{k}}}^2 [a_{\boldsymbol{k}}, a_{\boldsymbol{k}}^\dagger] + {v_{\boldsymbol{k}}}^2 [b_{\boldsymbol{k}}^\dagger, b_{\boldsymbol{k}}] = {u_{\boldsymbol{k}}}^2 - {v_{\boldsymbol{k}}}^2$ which must therefore be equal to unity. A difference as compared with fermions is the negative sign in ${u_{\boldsymbol{k}}}^2 - {v_{\boldsymbol{k}}}^2 = 1$.

The coefficients $u_{\boldsymbol{k}}$ and $v_{\boldsymbol{k}}$ are chosen such that $H_{\boldsymbol{k}} = \hbar\omega_{\boldsymbol{k}} (\alpha_{\boldsymbol{k}}^\dagger \alpha_{\boldsymbol{k}} + \beta_{\boldsymbol{k}}^\dagger \beta_{\boldsymbol{k}})$. With this, e.g., ${v_{\boldsymbol{k}}}^2 = \frac{1}{2}\left[\frac{1}{\sqrt{1-\gamma_{\boldsymbol{k}}}} - 1\right]$ and

$$H = -\frac{1}{2} NzJS(S+1) - Nh_A \left(S + \frac{1}{2}\right) + \sum_{\pm\boldsymbol{k}} \hbar\omega_{\boldsymbol{k}} \left(\hat{n}_{\boldsymbol{k}}^{\pm} + \frac{1}{2}\right), \tag{3.46}$$

where $\alpha_{\boldsymbol{k}}^{\dagger}\alpha_{\boldsymbol{k}} = \hat{n}_{\boldsymbol{k}}^{+}$ and $\beta_{\boldsymbol{k}}^{\dagger}\beta_{\boldsymbol{k}} = \hat{n}_{\boldsymbol{k}}^{-}$ can be identified as number operators and the spin wave energies are

$$\hbar\omega_{\boldsymbol{k}} = \sqrt{(JzS + h_A)^2 - (JzS)^2(\gamma_{\boldsymbol{k}}/2)^2}. \tag{3.47}$$

There are two degenerate spin wave modes for each $\boldsymbol{k}$ value however there are only $N/2$ values of $\boldsymbol{k}$ since the Fourier transforms are defined relative to one sub-lattice. If we take the limit that the symmetry breaking field $h_A \to 0$ and use the small wavelength limit approximation to $\gamma_{\boldsymbol{k}}$ with $z = 4$ and $S = 1/2$ then

$$\hbar\omega_{\boldsymbol{k}} = \left(\sqrt{2}Ja\right) k, \tag{3.48}$$

i.e., the initial dispersion is linear. Physically since the Cu in the cuprates corresponds to $S = 1/2$ the anisotropy field $h_A = 0$. What *is* possible is a small difference between the exchange constants for the different spin components. This has a similar effect of producing a gap in the excitation spectrum. The neutron experiments [25] on YBCO indicate the absence of such a gap. These experiments, in fact, exhibits two branches. These are attributed to an acoustic and optical mode and arise because YBCO is a bi-layer material, i.e., one which has two closely spaced Cu-O layers.

It is to be observed that even in such a simple theory the nature of the *vacuum* has been modified due to quantum fluctuations. This occurs because the new $\alpha_{\boldsymbol{k}}^{\dagger}$ and $\beta_{\boldsymbol{k}}^{\dagger}$ operators are a mixture of creation and destruction operators. The vacuum is no longer destroyed by the $a_{\boldsymbol{k}}$ and $b_{\boldsymbol{k}}$ but rather by $\alpha_{\boldsymbol{k}}$ and $\beta_{\boldsymbol{k}}$. As will be discussed in more detail in the next section, such a state is a coherent mixture of states which contain a different number of excited $a_{\boldsymbol{k}}^{\dagger}$ and $b_{\boldsymbol{k}}^{\dagger}$ spin waves. Relative to the classical ground state, these account for quantum fluctuations. The easiest such correction to calculate is the total energy with no spin waves (including $\mathrm{O}(S^2)$ and $\mathrm{O}(S)$)

$$E_0 = -\frac{1}{2}NJzS(S+1) + \sum_{\boldsymbol{k}} \hbar\omega_{\boldsymbol{k}} \equiv -\frac{1}{2}NJzS'^2. \tag{3.49}$$

This has a clear interpretation. The first term has the classical value S^2 of the magnetization squared replaced by $\hat{S}^2 = S(S+1)$ while the second term is just the zero point motion associated with a *two* sets of spin waves viewed as a set of harmonic oscillators, i.e., is similar to the corresponding term associated with sound waves (phonons). The left hand side defines an effective classical spin value

$$S' = (S + f); \quad \text{where} \quad f = \frac{1}{N}\sum_{\boldsymbol{k}} \left(1 - \frac{2\hbar\omega_{\boldsymbol{k}}}{zJS}\right), \tag{3.50}$$

which occurs not only in the energy but also in the spin wave velocity. In this context it accounts for a Hartree term reflecting the factors of, e.g., $(1 - a_i^{\dagger}a_i)^{1/2}$ in the definitions (3.40) and (3.41) which tends to reduce S' and a Fock correction from the interactions in $\mathcal{H}_1$ which has the opposite effect.

For $S = 1/2$ of interest here $(S+f) = S(1+0.158)$, i.e., implies an 16% increase in the spin wave velocity and an appreciable lowering of the ground state energy. With this correction the spin wave spectrum for YBCO gives a value [25] of $J \approx 100\text{meV}$ for the parent anti-ferromagnet.

Upon performing the sum on $\boldsymbol{k}$ in (3.49) the result corresponds to a ground state energy of $-0.335J$ per bond. The best estimates of this quantity by numerical and other methods [21–23] agree with this essentially within the error bar of the method. *However* it is notoriously the case that bad wave functions give good ground state energies. A direct test of the wave function is the magnetic moment per site. The *total* a-sublattice magnetization is defined to be

$$\mathcal{S}_z^a = \sum_i^a S_{i\,z}^a = NS - \sum_{\boldsymbol{k}} a_{\boldsymbol{k}}^\dagger a_{\boldsymbol{k}} . \tag{3.51}$$

What is desired is the expectation value of this sublattice magnetization evaluated with the ground state wave function. This is re-written using the transformed operators:

$$\mathcal{S}_z^a = NS - \sum_{\boldsymbol{k}} \left(u_{\boldsymbol{k}}{}^2 \alpha_{\boldsymbol{k}}^\dagger \alpha_{\boldsymbol{k}} + v_{\boldsymbol{k}}{}^2 \beta_{\boldsymbol{k}} \beta_{\boldsymbol{k}}^\dagger + \text{off-diagonal terms} \right) . \tag{3.52}$$

While the first operator is $\hat{n}_{\boldsymbol{k}}^+$ and is zero for the ground state, $\beta_{\boldsymbol{k}} \beta_{\boldsymbol{k}}^\dagger = 1 - \hat{n}_{\boldsymbol{k}}^-$ and is unity. It follows that

$$\langle \mathcal{S}_z^a \rangle = NS - \sum_{\boldsymbol{k}} v_{\boldsymbol{k}}{}^2 . \tag{3.53}$$

Doing the integral in two dimensions gives

$$S_{\text{eff}} = \frac{\langle \mathcal{S}_z^a \rangle}{N} = (S - 0.197) , \tag{3.54}$$

i.e., with $S = 1/2 = 0.500$, the expectation value of the individual spin $S_{\text{eff}} = 0.303$. It is remarkable that this agrees within a few percent with the best numerical estimates [21–23]. Physically S_{eff} is the z-component of $\boldsymbol{S}$ which is reduced by fluctuations while S' is its effective length which is *also* reduced when compared $S + \frac{1}{2}$ or $\sqrt{S(S+1)}$ but *is* increased relative to S.

These results for the ground state energy and aligned magnetic moment can be compared with those obtained by simple perturbation theory [26] on the Néel vacuum where the spin-flip, i.e., transverse matrix elements are considered to be the perturbation. The perturbation expansion for the energy is $(J/8)(-2 - \frac{2}{3} + 0.0037 - 0.0063 - 0.003 - 0.001)$ per bond while that for the moment is $\frac{1}{2} - \frac{1}{9} - \frac{4}{225} - 0.009 - 0.007 - 0.004$ both of which appear to initially converge rapidly. However a detailed analysis of these expansions suggest that the Heisenberg model is at the limit of convergence, i.e., the series would converge if the transverse matrix elements are smaller than the longitudinal part [26]. In this case there is a gap in the excitation spectrum.

Technically there is a potentially serious problem with the spin wave approach since it is the interactions contained in H_1 which constrains the system to not mix into the ground state wave function unphysical states which have two or more of the spin wave deviation bosons per site. The spin waves are $S = 1$ objects and hence the correction ~ 0.197 is a measure of the average number of such deviations. That this is a relatively small number compared to unity lends support to the results obtained with the present method.

Of course the fact that Néel vector, $\boldsymbol{N}$, here points in the z-direction was a matter of choice. Even in a quasi-two-dimensional system there is a SO(3) symmetry which put simply means that $\boldsymbol{N}$ can be made to point in any direction. (It should however be observed that there are always small anisotropy terms which break this symmetry, i.e., there is in reality a some preferred direction for $\boldsymbol{N}$. However, for a $S = 1/2$ system the energies are very small and will be ignored in the present discussion.)

It is not straightforward to calculate the thermodynamic transition temperature T_N for Néel ordering within this formalism, however an elementary mean-field theory gives $\langle \mathcal{S}_z \rangle = \frac{1}{2} \tanh[zJ \langle \mathcal{S}_z \rangle / 2k_B T]$ which demonstrates that $T_N \sim zJ/2$. It is left as an exercise for the reader to show that for low temperatures compared with T_N, spin waves have an internal energy in d-dimensions $U \sim T^{d+1}$ while the specific heat $C \sim T^d$.

The superconductive cuprates have a finite doping of holes and in fact holes can be added relatively easily to the above formalism. When it acts on the Néel state, which is the vacuum for this method, the physical electron destruction operator $c_{i\uparrow}$ only has a finite result if it acts on the a-sublattice. In this case the result is a hole at site i. Similarly $c_{i\downarrow}$ acting on the b-sublattice also produces a hole with, in both cases, no other effect. It is possible to define a new fermion operator $h_i^\dagger$ with no spin label such that $h_i^\dagger = c_{i\uparrow}$ for a-sublattice and $h_i^\dagger = c_{i\downarrow}$ otherwise. Now, e.g., $c_{i\downarrow}$ for the a-sublattice can be written as $c_{i\downarrow} \equiv c_{i\uparrow} a_i = h_i^\dagger a_i$ for $S = 1/2$. It follows, in the spin wave approximation, that the $t-J$-model, $H = -t \sum_{\langle ij \rangle \sigma} (\hat{c}_{\sigma i}^\dagger \hat{c}_{\sigma j} + \text{h.c.}) - J \boldsymbol{S}_i \cdot \boldsymbol{S}_j$; $\hat{c}_{\sigma i}^\dagger = c_{\sigma i}^\dagger (1 - c_{-\sigma i}^\dagger c_{-\sigma i})$, becomes

$$H = \mathcal{H}_0 - t \sum_{ij} [h_i^\dagger h_j b_j + h_i^\dagger a_i h_j + \text{h.c.}], \tag{3.55}$$

and it is seen that a hole cannot hop without the emission or absorption of a spin wave. For the problem of a single hole in an anti-ferromagnet, this has the effect of reducing the hole bandwidth from $\sim zt$ to $\sim zJ$ in agreement with exact diagonalization results [28]. Equation (3.55) is of the same form as is encountered in the theory of polarons [27]. It has been used as the basis for theories of magnetic polarons in the context of the cuprates. In particular a certain self-consistent Born approximation has been shown [16,28] to reproduce quite well the results of exact diagonalization studies on small systems [28]. The same approximation was introduced some time ago in connection with standard polarons [27].

However, the formalism is subject to a constraint which is violated by the approximations. It is required that the sum of the hole $h_i^\dagger$ and spin particle $a_i^\dagger$ number be less than or equal to unity, i.e.,

$$h_i^\dagger h_i + a_i^\dagger a_i \leq 1. \tag{3.56}$$

While the similar constraint on the spin wave boson number was, at least in principle, enforced by the interaction $\mathcal{H}_I$ this is not the case here. Such constraints are a common feature in the theory of high correlated electrons systems as will be seen in the following sections. Here this should be viewed as being an effective interaction between the charge and spin carriers, making the spin wave frequencies, and interactions, depend upon the presence of holes. The degree to which this effects the self-localization of holes to form magnetic polarons is an unsolved problem.

3.5 The Jordan–Wigner Transformation and Flux Tubes

It is possible to reproduce the $S = 1/2$ algebra with a set of fermions using the Jordan-Wigner transformation [12, 13]. This spin algebra corresponds to that of a set of hard core bosons and this transformation is a fashion by which to replace such bosons by fermions which is well adapted to one dimensional systems. This is the most elementary illustration of the notional equation: **hard core boson = fermion plus flux tube**. In this section along with explaining the meaning of this equality, it will be shown how this method can be used to solve exactly for the energy of the $S = 1/2$ XY-model in one spatial dimension and how this transformation can be adapted to two or higher dimensions by introducing a "string gauge" to perform the transformation. A more elaborate formulation based upon the general auxiliary particle method will be presented in Sect. 3.13.

Given that $\boldsymbol{S}_n$ is the spin operator for the nth site of a one dimensional system then it is trivial, for $S = 1/2$, that S_n^+ obey on-site Fermi commutation rules and that,

$$f_n^\dagger = \mathrm{e}^{\mathrm{i}\pi \sum_{m=1}^{n-1} \hat{Q}_m} S_n^+ \tag{3.57}$$

obeys such rules even when the site indices are different. Here $\hat{Q}_n = f_n^\dagger f_n$, i.e., is the number of particles, or up spins, at site n. The quantity $\mathrm{e}^{\mathrm{i}\pi \sum_{m=1}^{n-1} \hat{Q}_m}$ includes a factor of -1 for every occupied site which has a lower site index than the site of interest. It is a simple exercise to verify that this suffices to change hard core bosons into fermions. It is then straightforward to show that the one dimensional Heisenberg Hamiltonian $\mathcal{H} = -J \sum_n \boldsymbol{S}_n \cdot \boldsymbol{S}_{n+1}$ becomes,

$$\mathcal{H} = \frac{J}{2} \sum_n (f_n^\dagger f_{n+1} + \text{h.c.}) + J \sum_n f_n^\dagger f_n f_{n+1}^\dagger f_{n+1} + J\hat{N} \tag{3.58}$$

where, with $\hat{N} = \sum_n f_n^\dagger f_n$, the last term reflects an effective chemical potential $\mu = -J$. Because of the longitudinal part of the exchange interaction, this contains an interaction term. The Hamiltonian for the XY-model $\mathcal{H} = -(J/2)\sum_n [S_n^+ S_{n+1}^- + \text{h.c.}]$ is simpler and reduces to

$$\mathcal{H} = \frac{J}{2}\sum_n (f_n^\dagger f_{n+1} + \text{h.c.}) \tag{3.59}$$

which is just a one dimensional tight binding model for fermions. Ignoring potential problems with periodic boundary conditions, this can be solved by the usual Fourier transformation methods, i.e.,

$$\mathcal{H} = J\sum_k (\cos ka)\, f_k^\dagger f_k . \tag{3.60}$$

In the absence of an external field there is no chemical potential in the model and hence there is a half filled band with a ground state energy $E_0 = -(N/\pi)J$ or an energy per bond of $-0.32J$ which might be compared to $-0.25J$ for a simple Néel state and $-0.44J$ for the one dimensional Heisenberg model. Notice that since J is positive, the particles occupy k-space between $\pi/2a$ and $3\pi/2a$ or the equivalent. There is evidently no long range order. While this approach is simple as concerns the energies of the model, it is highly non-trivial to determine the correlation functions. In, e.g., the transverse spin correlation function the factors of $\mathrm{e}^{\mathrm{i}\pi\sum_{m=1}^{n-1}\hat{Q}_m}$ do *not* cancel as they do in $\mathcal{H}$.

A magnetic system which can be accurately described by a Fermi sea is often called a "spin liquid". It is speculated, when the anti-ferromagnetic order is dissolved by the addition of holes, the cuprates correspond to such a spin liquid. The theories of Sects. 3.4 and 3.15 suggest that the adjacent *anti-ferromagnetic* phase has boson spin waves which lie in an order induced, spin liquid gap. The branches meet near $\boldsymbol{k} = \{\pi/2, \pi/2\}$.

A finite two of higher dimensional lattice Hamiltonian can always be mapped to a one dimensional model albeit one with long range interactions. One mapping is via the zig-zag path shown in Fig. 3.2a. Now the factors of $\mathrm{e}^{\mathrm{i}\pi\sum_{m=1}^{n-1}\hat{Q}_m}$ cannot be eliminated from $\mathcal{H}$, however it is an observation which is easily checked that the changes of sign which these factors entail can be evaluated by a graphical method illustrated in Fig. 3.2b. To each particle is attached two lines which leave the system without crossing the zig-zag path. The sign of any hopping matrix element is $(-1)^L$ where L is the number of lines crossed in passing along any path which connects the two sites in question. This prescription is nothing more of less than the definition of phase factors which arise from attaching a fictitious unit flux tube to each particle, Fig. 3.3, using a rather special gauge for the magnetic field.

A flux tube corresponds to a region in space in which there is a finite magnetic field in the limit that the size of this region is reduced to zero maintaining the total flux through the cross-section fixed. The magnetic flux

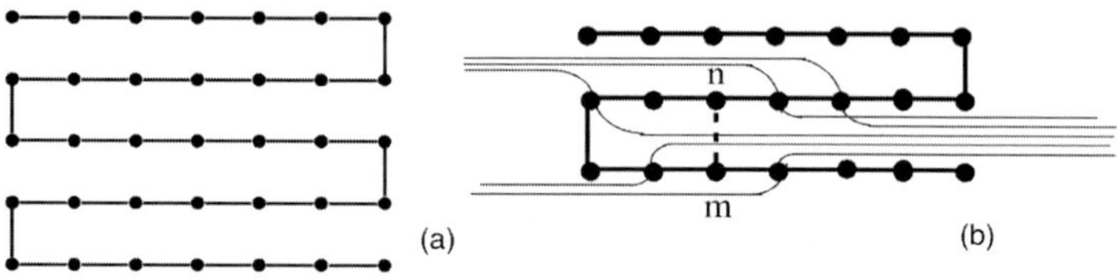

Fig. 3.2. (**a**) Shows one possible fashion by which to define a one dimensional path on a two dimensional square lattice. (**b**) Illustrates the convention by which it is possible to evaluate the Jordan-Wigner factor $\mathrm{e}^{\mathrm{i}\pi\sum_{m=1}^{n-1}\hat{Q}_m}$ using a graphical method. Each particle has two lines attached which leave in opposite directions without crossing the zig-zag path. This factor is equal to $(-1)^L$ where L is the number of lines crossed in hopping between sites. The dashed line shows the hopping path between sites n and m. The factor $\mathrm{e}^{\mathrm{i}\pi\sum_{m=1}^{n-1}\hat{Q}_m} = (-1)^{L=2} = 1$, since two lines are crossed.

is, $\Phi = \int \boldsymbol{B}\cdot \mathrm{d}\boldsymbol{S}$ where $\boldsymbol{B}$ is the magnetic field and $d\boldsymbol{S}$ is an element of surface. With the two dimensional XY-model in mind, a model might be,

$$\mathcal{H} = -\sum_{\langle ij\rangle} t_{ij} f_i^\dagger f_j + \text{h.c.}; \quad \text{where} \quad t_{ij} = \frac{J}{2}\mathrm{e}^{\mathrm{i}\int_{\boldsymbol{r}_i}^{\boldsymbol{r}_j}\boldsymbol{a}\cdot\mathrm{d}\boldsymbol{r}}, \tag{3.61}$$

where $\boldsymbol{r}_i \equiv (n_x\hat{x} + n_y\hat{y})a$ involves the integer coordinates n_x, n_y of the site i and the integral is along the hopping path. The vector potential $\boldsymbol{a}$ reflects the fictitious field. For a single particle with an attached flux Φ, Stokes theorem implies $\oint \boldsymbol{a}\cdot\mathrm{d}\boldsymbol{r} = \Phi$. For a unit of flux $\oint \boldsymbol{a}\cdot\mathrm{d}\boldsymbol{r} = 2\pi$.

The special Jordan-Wigner or *string* gauge described above simply corresponds to the identification of $\int_{\boldsymbol{r}_i}^{\boldsymbol{r}_j}\boldsymbol{a}\cdot\mathrm{d}\boldsymbol{r} = \pi L$. With the above definition of L it is trivial that $\oint \boldsymbol{a}\cdot\mathrm{d}\boldsymbol{r} = 2\pi$ since in passing around a single particle there are precisely two lines which are cut independent of the path, i.e., the Jordan-Wigner transformation can be considered as converting the hard core

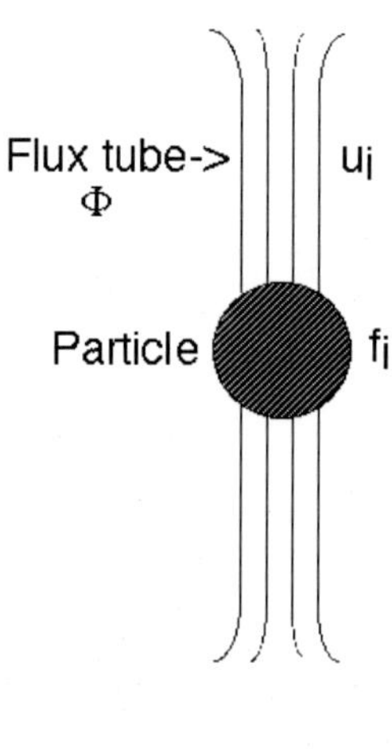

Fig. 3.3. A fermion is converted into a hard core boson by attaching a unit flux tube. When one particle is interchanged with another the tube adds a phase of π.

bosons into fermions by attaching a unit flux tube to the latter. This rather particular gauge can be related to a more standard expression for $\boldsymbol{a}$ by a suitable gauge transformation. Perhaps the simplest expression would be

$$\boldsymbol{a} = \hat{\phi}\frac{1}{r}; \qquad \hat{\phi} = \boldsymbol{r} \times \hat{\mathbf{z}} \tag{3.62}$$

where r is the distance from the particle and $\hat{\phi}$ is a unit vector defined by the Euler angle ϕ, with the z axis out of the plane. This is designed to trivially have $\oint \boldsymbol{a} \cdot \mathrm{d}\boldsymbol{r} = 2\pi$, however now there is a phase change for every hopping matrix element t_{ij}. In Fig. 3.4 is illustrated how, by systematic gauge transformations, the string gauge can be converted to that described by (3.62). Here the gauge transformation involve $f_i^\dagger \to \mathrm{e}^{\mathrm{i}\alpha} f_i^\dagger$ for all of the operators associated with the blocks of sites as shown in the figure.

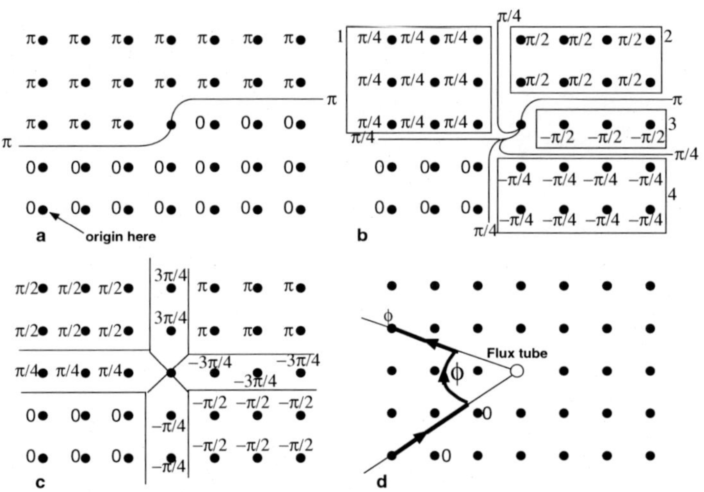

Fig. 3.4. (**a**) This is the string gauge. Taking a convention that the site at the bottom left hand corner has phase zero, the figure shows the phases $\int_0^i \boldsymbol{a} \cdot \mathrm{d}\boldsymbol{r}$ in the string gauge when there is a flux tube on the central site. There is a π phase change at the string. (**b**) Shows the phase due to $\int_0^i \boldsymbol{a} \cdot \mathrm{d}\boldsymbol{r}$ after a gauge transformation $f_i^\dagger \to \mathrm{e}^{\mathrm{i}\alpha} f_i^\dagger$ for all the sites in a given block. For the blocks 1-4 the phases $\alpha = 3\pi/4, \pi/2, \pi/2, \pi/4$. In effect, one of the π strings has been split into four $\pi/4$ strings, as shown. (**c**) Both π strings are split into four $\pi/4$ strings in a more symmetric way. (**d**) The phase $\int_0^i \boldsymbol{a} \cdot \mathrm{d}\boldsymbol{r}$ calculated with $\boldsymbol{a} = \hat{\phi}/r$ is path independent. Using the indicated path, $\int_0^i \boldsymbol{a} \cdot \mathrm{d}\boldsymbol{r} = \phi$ the angle between the two straight lines. In (**c**) near and next-near-neighbors have a phase which agrees with (**d**). A systematic application of an infinite number of gauge transformations converts the string gauge (**a**) to that defined by (**d**) for all sites.

When other than a unit flux tube is attached to the fermions, this constitutes a representation of an *anyon* [29] which are particles which have statistics between those of fermions and bosons and which can only exist in two dimensions. The idea that the elementary excitations in HTSC are anyons has been actively pursued. It is known that in the ground state, semions, i.e., particles with one half a flux tube, condense in pairs and it has been suggested that T_c might be associated with such a condensation. This proposition is consistent with the fact that true magnetic flux is quantized in units which correspond to 2e. Such ideas have been largely abandoned since, in the relevant materials, there is no real evidence of persistent currents which would accompany semions.

This notwithstanding, here the interest is in representing hard core bosons as fermions and this is a special case. It *is* possible to extended the formalism beyond two dimensions and in particular to multi-layer systems [30]. It is also possible to define flux tube operators

$$u_i^\dagger = \mathrm{e}^{\mathrm{i}\int_0^{\boldsymbol{r}_i} \boldsymbol{a}\cdot\mathrm{d}\boldsymbol{r}} \tag{3.63}$$

where the lower limit of integration is some arbitrary origin. For these operators to be uniquely defined, independent of the integration path, implies that only unit flux tubes are permitted. With this, different paths which pass around a different number of particles have integrals $\int_0^{\boldsymbol{r}_i} \boldsymbol{a}\cdot\mathrm{d}\boldsymbol{r}$ which differ by multiples of 2π and thereby have the same value for u_i. Clearly when evaluated with the string gauge these u_i correspond to the exponential factors in the beginning definition of the Jordan-Wigner transformation, i.e., $f_i^\dagger = u_i^\dagger S_i^+$ and otherwise differ only by a gauge transformation.

3.6 Coherent States, Grassman Variables and Flux Tubes

The flux tube operators u_n have been introduced as a means of converting spin operator S_n^+, which obey the algebra of hard core bosons, into a Fermi field, i.e., $f_n^\dagger = u_n^\dagger S_n^+$. Grassman variables play a related role and arise in the theory of fermionic coherent states. In connection with HTCS, it is often suggested that the doped holes are (hard core) bosons which undergo Bose-Einstein condensation. Such condensates might be associated with bosonic coherent states. This section is dedicated to a brief review of the relevant mathematical concepts.

Bosonic coherent states are commonplace in quantum mechanics. Consider a simple harmonic oscillator $H = \omega a^\dagger a$, for which in suitable units the displacement $x = a^\dagger + a$. Adding an constant external force, this becomes $H_f = \omega a^\dagger a + f(a^\dagger + a)$. This problem is diagonalized by defining $b^\dagger = a^\dagger + c$ where c is a real number. One has that $\omega b^\dagger b = \omega a^\dagger a + \omega c(a^\dagger + a) + \omega c^2$ so that with $c = f/\omega$: $H_f = \omega b^\dagger b - f^2/\omega$, i.e., the old problem is equivalent to the new one with a shift in energy and a displacement of the center of oscillation by c. Reflecting this displacement, the definition of the vacuum is changed. The original vacuum state is such that $a|\rangle = 0$ whereas now it must be that $b|c\rangle = (a+c)|c\rangle = 0$, i.e., $a|c\rangle = c|c\rangle$ and $|c\rangle$ is an *eigenstate* of a (but *not* $a^\dagger$). It is easy to verify that $|c\rangle = \mathrm{e}^{-|c|^2}\mathrm{e}^{ca^\dagger}|\rangle$, here generalizing to the possibility that c is complex.

Next consider a material Bose system described by a set of creation operators $a_{\boldsymbol{k}}^\dagger$ for which all the N_0 particles are condensed at $\boldsymbol{k} = 0$ in the ground state. If this ground state is considered as the vacuum state $|{N_0}^{1/2}\rangle$ for a set of Bose excitations $b_{\boldsymbol{k}}^\dagger$ it should be that $b_{\boldsymbol{k}}|{N_0}^{1/2}\rangle = 0$. Given that the $\langle {N_0}^{1/2}|a_{\boldsymbol{k}=0}^\dagger a_{\boldsymbol{k}=0}|{N_0}^{1/2}\rangle = N_0$, it must be that $a_{\boldsymbol{k}=0}^\dagger = b_{\boldsymbol{k}=0}^\dagger + \mathrm{e}^{\mathrm{i}\phi}{N_0}^{1/2}$ and that $|{N_0}^{1/2}\rangle = \mathrm{e}^{-N_0}\mathrm{e}^{{N_0}^{1/2}a_{\boldsymbol{k}=0}^\dagger}|\rangle$. This construction of the ground state leaves a well defined bosonic $b_{\boldsymbol{k}=0}^\dagger$ however it is the case that $|{N_0}^{1/2}\rangle$ does not have a fixed number of particles but is rather is a coherent combination of states with different particle number such that the expectation value of the number operator is N_0. In this sense most states which represent condensates are coherent states, e.g., the BCS ground state (3.14) is a coherent superposition of states with different fermion number.

Actually it is important to distinguish between material bosons such as ^{4}He which, in the absence of a condensate, have a finite negative chemical potential μ and non-material bosons such as photon, phonons or spin waves, which are not conserved, and for which the chemical potential μ is always zero. As will be seen in Sect. (3.8), for material bosons the substitution $a_{\boldsymbol{k}=0}^\dagger = b_{\boldsymbol{k}=0}^\dagger + \mathrm{e}^{\mathrm{i}\phi}{N_0}^{1/2}$ results in many terms which do not conserve particle number. That these be zero must be equivalent to the requirement that $\mu = 0$, i.e., reduce to Schrödinger's equation for the *condensed bosons*. There are two additional important related observations to be made. First a state with N condensed bosons is proportional to $\mathrm{e}^{\mathrm{i}N\phi}$ so that the condensate number operator $\hat{N} = -\mathrm{i}(\partial/\partial\phi)$ and second if $\langle\hat{O}\rangle$ is the expectation value of some physical operator $\hat{O}$ then $\int(\mathrm{d}\phi/2\pi)\langle\hat{O}\rangle$ reduces to a sum over number diagonal states. The distribution of the number of bosons N is strongly peaked at N_0.

In order to define *fermion* coherent states it is usual to introduce Grassman variables. Given a set of fermion creation operators $c_n^\dagger$ it is desired to define a state such that $c_n|\{\xi_n\}\rangle = \xi_n|\{\xi_n\}\rangle$ while $c_{n'}|\{\xi_n\}\rangle = \xi_{n'}|\{\xi_n\}\rangle$, and this implies that $c_{n'}c_n|\{\xi_n\}\rangle = \xi_{n'}\xi_n|\{\xi_n\}\rangle$. However $c_nc_{n'}|\{\xi_n\}\rangle = \xi_n\xi_{n'}|\{\xi_n\}\rangle$ while $c_nc_{n'}|\{\xi_n\}\rangle = -c_nc_{n'}|\{\xi_n\}\rangle$ and so it must be that $\xi_n\xi_{n'} = -\xi_{n'}\xi_n$, i.e, the quantities ξ_n and $\xi_{n'}$ are *not* plain c-numbers but are anticommuting Grassman variables which obey

$$\{\xi_n, \xi_{n'}\} = \{\xi_n^\dagger, \xi_{n'}\} = 0 \tag{3.64}$$

for all n and n'. Since they are by nature operators, to accommodate Grassman variables requires that the fermion Hilbert space be extended.

It is perhaps therefore surprising that a set of anti-commuting fields $f_n^\dagger = u_n^\dagger S_n^+$ can be defined via the Jordan-Wigner transformation *without* extending the Hilbert space. However, as is, this transformation has an undesirable property, namely the fermion field $f_n^\dagger$ does not commute with its own bosonic equivalent S_m^+, $n \neq m$. Correcting this defect *does* involve extending the Hilbert space. The extended space accommodates an extra false spin, $\boldsymbol{\sigma}$, degree of freedom per site, where $\boldsymbol{\sigma}$ is the vector of $s = 1/2$ Pauli matrices. The operators $u_n^\dagger = \sigma_{nx} \mathrm{e}^{[\mathrm{i} \int_0^{r_n} \boldsymbol{a} \cdot \mathrm{d}\boldsymbol{r}]}$ involve this false spin, via σ_{nx}, as does the definition of the source of flux, i.e., in (3.57), $\hat{Q}_n = (1/2)(\sigma_{nz} + 1)$. It remains the case that the u_n and $u_n^\dagger$ are unitary, i.e., $u_n^\dagger u_n = \sigma_{nx} \mathrm{e}^{[\mathrm{i} \int_0^{r_n i} \boldsymbol{a} \cdot \mathrm{d}\boldsymbol{r}]} \mathrm{e}^{[-\mathrm{i} \int_0^{r_n} \boldsymbol{a} \cdot \mathrm{d}\boldsymbol{r}]} \sigma_{nx} = 1 = u_n u_n^\dagger$ since ${\sigma_{nx}}^2 = 1$ but, now the $u_n^\dagger$ and $u_m^\dagger$ or u_m anti-commute when $n \neq m$ and it is the case that $f_n^\dagger$ and S_m^+ commute. These $u_n^\dagger$ are an example of a Clifford algebra. The state of the false spin has no physical consequence, it simply keeps track of the fermion number. In a theory in which this is conserved the expectation value of any physical quantity must be diagonal in the false spin space. The choice of the vacuum for this space reflects a gauge symmetry. To be specific, this is taken to have all the false spins down. Formally, superconductivity breaks this gauge symmetry and the order parameter Δ will be off-diagonal in both the real and false space, however again physical quantities involve $\Delta\Delta^\dagger$ which *is* diagonal. The Jordan-Wigner transformation which results from the newly defined u_i differs only in having the desired commutation rules and the technical issues discussed in this paragraph can be largely ignored in what follows.

3.7 Mean-Field Approximations and Flux States

The example of the one dimensional XY-model illustrates that under certain conditions, in particular in the absence of long range anti-ferromagnetic order, the elementary excitations of a spin system can map to those of a spinless Fermi sea, i.e., to a spin liquid. The original idea behind the RVB state of Anderson [8] was that a two dimensional anti-ferromagnet might be such a liquid with the addition feature of having a BCS like condensation of the spins. In fact, the two dimensional anti-ferromagnetic Heisenberg model in two dimensions is now widely believed to have an ordered ground state and with the addition of weak three dimensional interactions has a finite ordering temperature T_N. However T_N is rapidly driven to zero by hole doping and so it remains a real possibility that a *doped* two dimensional anti-ferromagnetic Heisenberg model has a spin liquid ground state. The simplest fashion by which to arrive at such a RVB class of ground state is to use the Jordan-Wigner formulation, to treat the interaction term in the Hartree-Fock approximation and to make the mean-field approximation of spreading the flux of the individual particles so that it makes a uniform field of one-half a flux quantum per square. In the absence of order the Hartree terms are an unimportant constant and the Fock term renormalizes J. In connection with the anyon problem, this mean-field approximation has been compared to the results [31] of exact diagonalization on small systems with the conclusion that the representation of hard core bosons as fermions plus flux tubes gives a large overlap between the mean-field and exact ground states. While there seems to be no formal proof that this approximation provides an upper bound for the ground state energy, this is the case numerically [31].

The resulting mean-field problem of spinless fermions in a uniform field on a square lattice has been much studied. (Actually added to every other square is a whole flux quantum Φ_0. This results in an equivalent problem but reduces the average field to zero.) It has been shown [32] that the ground state of the current problem of a half-flux-quantum per square at half-filling has a lower ground state energy than in the absence of the magnetic field. In the absence of a lattice the field would lead to Landau levels of which the lowest level would be exactly filled. The effect of the lattice is to broaden these levels to form bands. There is single state per site and so if each Landau level is completely occupied at half-filling it is necessarily the case that there are only two such levels and the system has particle-hole symmetry. In the simplest gauge $\boldsymbol{a} = n_x\hat{y}/2a$ where n_x is defined by $\boldsymbol{r}_i = a(n_x\hat{x} + n_y\hat{y})$, n_x and n_y integers. With this $J_{ij} \equiv J u_i u_j^\dagger = J\mathrm{e}^{\mathrm{i}\pi n_x}$ for bonds in the y-direction while $J_{ij} = J$ for x-directed bonds. The result has chains with a hopping between chains which adds a momentum π to the x-component of the momentum. The resulting determinant is

$$\begin{vmatrix} J\cos k_y a - E_{\boldsymbol{k}}, & J\cos k_x a \\ J\cos k_x a, & -J\cos k_y a - E_{\boldsymbol{k}} \end{vmatrix} = 0 \tag{3.65}$$

and leads to energies

$$E_{\boldsymbol{k}} = \frac{J}{\sqrt{2}}\sqrt{{\gamma_{\boldsymbol{k}}}^2 + {\gamma_{d\boldsymbol{k}}}^2}. \tag{3.66}$$

The resulting density of states, shown in the central panel of Fig. 3.1, has two bands corresponding to Landau levels which touch at zero energy. As compared with the field free dispersion $E_{\boldsymbol{k}} = J\gamma_{\boldsymbol{k}}$ there is a d-wave gap at the Fermi surface while the band width narrows from $4J$ to $4J/\sqrt{2}$. However, the effect of the gap is greater than that of the band narrowing, and it is the case that the total energy of the flux state, i.e., $-0.240J$ is lower that $-0.204J$ the energy without a field [32]. Both energies are per bond. This is rather disappointing since the Néel state gives $-0.250J$. However including a Fock correction this becomes and $\approx -0.26J$. Anti-ferromagnetic ordering improves this to a respectable $\approx -0.32J$ within a percent or so of the exact result, see Sects. 3.4 and 3.15.

In order to make a comparison with the RVB theory it is better to use a gauge which respects the symmetry of the lattice when

$$J_{ij} = \frac{J}{\sqrt{2}}(1 \pm \mathrm{i}\mathrm{e}^{\mathrm{i}\boldsymbol{\pi}\cdot\boldsymbol{r}_i}) \tag{3.67}$$

where the plus and minus signs correspond to horizontal and vertical bonds. In momentum space the part $\pm(\mathrm{i}J/\sqrt{2})\mathrm{e}^{\mathrm{i}\boldsymbol{\pi}\cdot\boldsymbol{r}_i}$ is reflected by a pairing term

$$\mathrm{i}\sum_{\boldsymbol{k}} \Delta_{\boldsymbol{k}}(f^{\dagger}_{\boldsymbol{k}} f_{\boldsymbol{k}+\boldsymbol{\pi}} - \mathrm{h.c.}) \tag{3.68}$$

which causes particle-hole pairing with a pair momentum of $\boldsymbol{\pi}$. The "order parameter" for this pairing

$$\Delta_{\boldsymbol{k}} = \frac{J}{\sqrt{2}}(\cos k_x a - \cos k_y a) = \frac{J}{\sqrt{2}}\gamma_{d\boldsymbol{k}} \tag{3.69}$$

is temperature independent and reflects the formation of Landau levels. Thus such a theory involves ever present *pre-formed pairs* of particles. The only characteristic temperature T^* is the band width $\sim J$. This T^* will be modified upon doping, see Sect. 3.17. In common with the RVB theory, Sect. 3.11, the order parameter has d-wave symmetry with nodes along the diagonal directions in momentum space. There is no Fermi surface except at the points $\boldsymbol{k} = [(\pm\boldsymbol{\pi}/2), (\pm\boldsymbol{\pi}/2)]$.

3.8 Bogoliubov Theory for a Bose Superfluid

While free bosons undergo Bose-Einstein condensation they are not a superfluid unless there are also interactions. This has been much discussed [15] in the context of liquid ^{4}He and one of the earliest theories for interacting bosons was that of Bogoliubov [15] which will be briefly described in this section. This is relevant to the study of HTSC materials since in a number of models it is envisaged that the doped holes are bosons and these become a superfluid through Bose-Einstein condensation. Invariably the bosonic holes are hard core in nature and therefore interact via this exclusion effect if not in other ways. The hard core constraint implies that the wave function must have a curvature with an associated wave vector $k_\ell \sim \hbar/\ell$ where ℓ is the inter-particle distance. This might be replaced by an effective interaction. The exercise shows that such an interaction has the effect of linearizing the initial dispersion relationship corresponding to the fact that the lowest lying excitations should be sound waves and that this implies that in two dimensions there is a pseudo-gap in the charge spectrum. The size of this gap is such that it reflects a transfer of spectral weight which is proportional to the condensate density. Finally it is shown that even at $T = 0$, due to scattering, there are about as many bosons above, as in, the condensate. It might be expected that similar effects occur in the charge sector for well constructed models of the cuprate HTSC.

The model comprises the kinetic energy plus a local *repulsive* interaction $v\sum_i n_i n_i$; $v > 0$ so that

$$\mathcal{H} = \sum_{\boldsymbol{k}}(\frac{(\hbar k)^2}{2m} - \mu)a^{\dagger}_{\boldsymbol{k}} a_{\boldsymbol{k}} + \frac{v}{2N}\sum_{\boldsymbol{q}\boldsymbol{k}\boldsymbol{k}'} a^{\dagger}_{\boldsymbol{k}-\boldsymbol{q}} a^{\dagger}_{\boldsymbol{k}'+\boldsymbol{q}} a_{\boldsymbol{k}'} a_{\boldsymbol{k}} \tag{3.70}$$

where the factor of two accounts for double counting. It is assumed that there is a condensate in the $\boldsymbol{k} = 0$ state. The ground state is therefore a Bose coherent state and $a^\dagger_{\boldsymbol{k}=0} = b^\dagger_{\boldsymbol{k}=0} + c$. If the condensate amplitude c is considered to be a complex parameter, then the minimal principle insists that the ground state expectation value of

$$\frac{\partial \mathcal{H}}{\partial c} = 0 \quad \text{and} \quad \frac{\partial \mathcal{H}}{\partial c^*} = 0 \tag{3.71}$$

and a little through about the structure of the resulting equation shows that, e.g.,

$$\frac{\partial \mathcal{H}}{\partial c} = [\mathcal{H}, b^\dagger_{\boldsymbol{k}=0}] \tag{3.72}$$

i.e., the above minimal conditions are just equivalent to the requirement that $i\hbar\partial_t b^\dagger_{\boldsymbol{k}=0} = [\mathcal{H}, b^\dagger_{\boldsymbol{k}=0}] = 0$. Which physically says that the $\boldsymbol{k} = 0$ level must have zero energy. Ignoring interactions this reduces to $\mu = vn_0$ where $n_0 = N_0/N = |c|^2/N$ is the condensate fraction. There is a shift in the chemical potential due to interactions.

Collecting the terms bi-linear in the field operators, dropping a constant, leads to $\mathcal{H} = \sum_{\boldsymbol{k}} \mathcal{H}_{\boldsymbol{k}}$,

$$\mathcal{H}_{\boldsymbol{k}} = \left[(\xi_{\boldsymbol{k}} + vn_0)(b^\dagger_{\boldsymbol{k}} b_{\boldsymbol{k}} + b^\dagger_{-\boldsymbol{k}} b_{-\boldsymbol{k}}) + vn_0(b^\dagger_{\boldsymbol{k}} b^\dagger_{-\boldsymbol{k}} + b_{-\boldsymbol{k}} b_{\boldsymbol{k}})\right], \tag{3.73}$$

where $\xi_{\boldsymbol{k}} = \frac{(\hbar k)^2}{2m} - \mu'$ and where $\mu' = \mu - vn_0$ is the shifted chemical potential. This is mathematically equivalent to the problem of anti-ferromagnetic spin waves. It is trivial that $\mathcal{H}_{\boldsymbol{k}} = vMv^\dagger$ where $v = (b^\dagger_{\boldsymbol{k}}, b_{-\boldsymbol{k}})$,

$$M = \begin{pmatrix} \xi_{\boldsymbol{k}} + vn_0 & vn_0 \\ vn_0 & \xi_{\boldsymbol{k}} + vn_0 \end{pmatrix}. \tag{3.74}$$

The Bose commutation rules are conserved by a pure Lorentz boost. Using such a boost to diagonalize $\mathcal{H}_{\boldsymbol{k}}$, results in the eigenenergy,

$$\mathcal{E}_{\boldsymbol{k}} = \sqrt{\xi_{\boldsymbol{k}}\,(\xi_{\boldsymbol{k}} + 2vn_0)}, \tag{3.75}$$

and coefficient

$$v_{\boldsymbol{k}}{}^2 = \frac{1}{2}\left(\left(\frac{\xi_{\boldsymbol{k}} + 2vn_0}{\xi_{\boldsymbol{k}}}\right)^{1/2} - 1\right). \tag{3.76}$$

Hard core repulsions introduce the length scale $\ell \sim 1/n_0{}^{1/2}$ so $vn_0 \sim \hbar^2/2m\ell^2 \sim \hbar^2 n_0/2m$. The initial dispersion $\mathcal{E}_{\boldsymbol{k}} = \hbar(n_0/2m)^{1/2}k - \mu'$ is linear and only becomes quadratic with $\mathcal{E}_{\boldsymbol{k}} = \frac{(\hbar k)^2}{2m} - \mu'$ when $k > n_0{}^{1/2}$. The resulting two dimensional density of states is illustrated in Fig. 3.5. It is initially linear and then becomes a constant when $k > n_0{}^{1/2}$, i.e., there is a pseudo gap in the particle spectral density. The number of missing states is $\sim n_0$ and the result can be interpreted in terms of a transfer of spectral weight from low energies into the condensate. It is also the case that not all of the

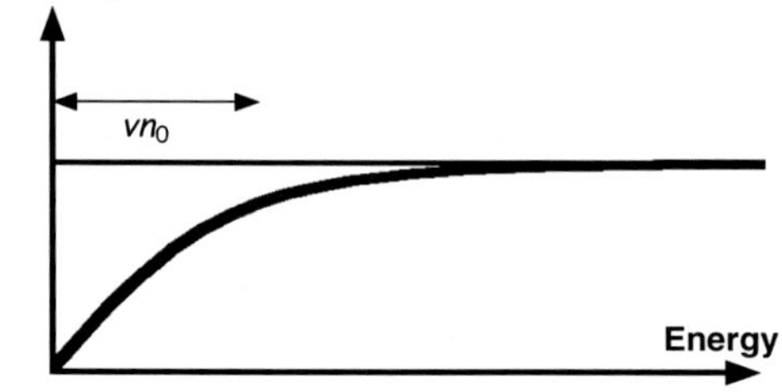

Fig. 3.5. Compared with the constant two dimensional density of states, horizontal line, there is a pseudo-gap in the density of states which has a width vn_0. For hard core bosons the condensate accounts for the missing states.

bosons are condensed even at zero temperature, in fact the occupation of the $\boldsymbol{k}$ state is $v_{\boldsymbol{k}}{}^2$ which implies that, for this model, there are roughly as many non-condensed as there are condensed bosons at zero temperature.

Bose-Einstein condensation occurs when the thermal wave length λ_T, given by $(\hbar^2/2m\lambda_T{}^2) \approx k_B T$ is approximately equal to the inter-particle distance. In two dimensions this implies $k_B T_0 \approx \hbar^2 n_0/2m$ and for hard core bosons implies that $k_B T_0 \approx vn_0$.

The observations made in this section must also be true for any adequate theory for the condensation of hard core bosonic holes in the context of the cuprates.

3.9 Auxiliary Particle Methods

The example of the Heisenberg Hamiltonian, which corresponds to a half-filled Hubbard model illustrates well that a physical system can exhibit the spin degrees of freedom of its component electrons in a situation where the low lying charge degrees of freedom are completely suppressed. Methods for handling these spin degrees of freedom alone have been described in previous sections. A formalism [14] for dealing with lightly doped systems is more recent. This auxiliary particle scheme, often called the slave boson method, effectively separates the spin and degrees of freedom of the individual electrons, i.e., an electron is split into two distinct spin and charge fields. As discussed in Chap. 2, in one dimension it is known that there is spin charge separation with spin and charge excitations having quite different velocities. The auxiliary particle method was first devised [14] for magnetic impurities problems where spin-charge separation also occurs but is reflected by the fact that the relaxation times, rather than the velocities, are vastly different for the spin and charge degrees of freedom.

At a more mathematical level, the usefulness of these methods is that the largest U-term of the Hubbard model appears as a particle energy and *not* as an interaction. The cost is the existence of a *constraint* that the total number of these new particles be exactly unity. (The same method can be used [33] to deal with more complicated sub-units than a single lattice site but this will not be developed here.) This method will be used in later sections in order to have a unity of approach. The basic methods are developed in this section.

Consider a single site in the Hubbard model. There are only four possible states for this site, $|0\rangle$ corresponding to the absence of an electron, $|\uparrow\rangle \equiv c^\dagger_{\uparrow i}|0\rangle$, and $|\downarrow\rangle \equiv c^\dagger_{\downarrow i}|0\rangle$ reflecting the two possible spin values for a site with a single electron, and finally $|\uparrow\downarrow\rangle \equiv c^\dagger_{\uparrow i}c^\dagger_{\downarrow i}|0\rangle$. Clearly there is a certain ambiguity in the definition of $|\uparrow\downarrow\rangle$ this might be either $c^\dagger_{\uparrow i}c^\dagger_{\downarrow i}|0\rangle$ or $c^\dagger_{\downarrow i}c^\dagger_{\uparrow i}|0\rangle = -c^\dagger_{\uparrow i}c^\dagger_{\downarrow i}|0\rangle$. (Actually a number of arbitrary phase factors might be included here.) In the so called "slave boson" scheme four auxiliary operators: $b^\dagger_i$, $f^\dagger_{\sigma i}$ and $b^\dagger_{\uparrow\downarrow i}$ are defined where the b-particles are bosons while the f-particles are fermions. In defining these operators it is important to realize that the auxiliary particle vacuum state $|\rangle$ is *unphysical* and quite distinct from $|0\rangle$. A sufficient set of identifications is:

$$b^\dagger_i|\rangle \equiv |0\rangle \quad f^\dagger_{\sigma i}|\rangle \equiv |\sigma\rangle = c^\dagger_{\sigma i}|0\rangle \quad \text{and} \quad b^\dagger_{\uparrow\downarrow i}|\rangle \equiv |\uparrow\downarrow\rangle = c^\dagger_{\uparrow i}c^\dagger_{\downarrow i}|0\rangle, \tag{3.77}$$

where it might be observed that the bosons are associated with states with an even number of electrons while the states with an odd number are reflected by fermions. (While logical, this convention might be inverted with fermions being associated with even electron states and in fact a continuum of anyons schemes is possible.) *Only* the states with a single auxiliary particle have physical equivalents so not only is $|\rangle$ unphysical but also all states with more than a single auxiliary particle have no physical counterpart. These unphysical states are excluded by enforcing the *constraint*:

$$Q_i = \sum_\sigma f^\dagger_{\sigma i}f_{\sigma i} + b^\dagger_i b_i + b^\dagger_{\uparrow\downarrow i}b_{\uparrow\downarrow i} = 1. \tag{3.78}$$

Nearly all approximation schemes found in the literature *do not* satisfy this constraint. Typically it is only maintained on average. The principal such approximation will be presented in what follows along with an approach which *does* satisfy the constraint and which is of the same class of theory.

The physical electron operator,

$$c^\dagger_{\sigma i} = f^\dagger_{\sigma i}b_i + \sigma b^\dagger_{\uparrow\downarrow i}f_{-\sigma i} \tag{3.79}$$

where the factor of σ in the second term reflects the difference in sign between $c^\dagger_{\downarrow i}c^\dagger_{\uparrow i}|0\rangle$ and $c^\dagger_{\uparrow i}c^\dagger_{\downarrow i}|0\rangle$. It is an attractive feature of the method that $c^\dagger_{\sigma i}$ is separated into two parts, e.g., $\hat{c}^\dagger_{\sigma i} = c^\dagger_{\sigma i}(1 - n_{-\sigma i})$ which does not create doubly occupied states and which occurs in the definition of the $t-J$-model introduced in Chap. 2 is equal to $f^\dagger_{\sigma i}b_i$.

Clearly, since the original $\mathcal{H}$ is written in terms of the conduction electron operators, $c^\dagger_{\sigma i}$ or $c^\dagger_{\sigma i}$, an auxiliary particle mapping might be obtained, e.g., by substituting (3.79) in $\mathcal{H}$, however, this is certainly not the simplest possibility.

In fact there are an infinite number of ways in which to write $\mathcal{H}$, or for that matter any other operator.

However an operator can always be written in a form which is bi-linear in the operators associated with a given site. This form is unique to within certain phase degrees of freedom. Consider an arbitrary operator $\hat{O}$ with a matrix representation M which corresponds to the row vector $r = (|\uparrow\downarrow\rangle, |\uparrow\rangle, |\downarrow\rangle, |0\rangle)$ i.e, such that $\hat{O} = rMr^\dagger$. The bi-linear auxiliary particle version of this operator

$$\hat{O} = RMR^\dagger \quad \text{where} \quad R = \left(b^\dagger_{\uparrow\downarrow i}, f^\dagger_{\uparrow i}, f^\dagger_{\downarrow i}, b^\dagger_i\right). \tag{3.80}$$

Given the mapping described above, by construction, the matrix elements in the auxiliary space are the same as in the original space. Multiplication rules, e.g., $\hat{O}_1\hat{O}_2 = RM_1M_2R^\dagger$, and hence commutation rules are satisfied. To prove this it is noted that in the reduction $RM_1R^\dagger RM_2R^\dagger = RM_1M_2R^\dagger$, the matrix $R^\dagger R = 1$ by virtue of the constraint. A physical state to the right of the product $RM_1R^\dagger RM_2R^\dagger$ has a single auxiliary particle which, in finite matrix elements, is destroyed by the operators in the $R^\dagger$ to the right. It follows that the middle $R^\dagger R$ acts on the auxiliary particle vacuum. The off-diagonal elements of $R^\dagger R$ comprise products, e.g., $b_i f^\dagger_{\downarrow i}$ which are zero since if $f^\dagger_{\downarrow i}$ gives a finite result b_i destroys the resulting state. Because auxiliary particle vacuum lies to the right, the first operator in a diagonal elements, e.g., $f_{\downarrow i} f^\dagger_{\downarrow i}$ acts with a finite result, while the second operators restores the beginning vacuum so the result is $f_{\downarrow i} f^\dagger_{\downarrow i} = 1$. This auxiliary particle contraction of a creation-destruction product occurring in the middle of a particle string is most useful. Applying (3.80), e.g., to $c^\dagger_{\sigma i}$ results in (3.79).

The bi-linear form of the Hubbard model, $H = -t\sum_{\langle ij\rangle\sigma}(c^\dagger_{\sigma i}c_{\sigma j} + \text{h.c.}) + U\sum_i \hat{n}_{i\uparrow}\hat{n}_{i\downarrow}$, see Chapters 1-2, is,

$$\begin{aligned}
\mathcal{H} &= \mathcal{H}_0 + \mathcal{H}' \\
\mathcal{H}_0 &= -t\sum_{\langle ij\rangle\sigma}(b^\dagger_i b_j - b^\dagger_{\uparrow\downarrow i}b_{\uparrow\downarrow j})f^\dagger_{\sigma i}f_{\sigma j} + U\sum_i b^\dagger_{\uparrow\downarrow i}b_{\uparrow\downarrow i} \\
\mathcal{H}' &= -t\sum_{\langle ij\rangle}\left[b^\dagger_i b^\dagger_{\uparrow\downarrow j}f_{i\uparrow}f_{j\downarrow} + b^\dagger_j b^\dagger_{\uparrow\downarrow i}f_{\uparrow i}f_{\downarrow i} + \text{h.c.}\right]
\end{aligned} \tag{3.81}$$

and, as promised, the largest energy U appears on the diagonal of the Hamiltonian matrix. In addition to the constraint, the cost of this transformation is that the hopping term, which was bi-linear in the c-operators, is now a four field interaction. Here $\mathcal{H}_0$ contains a part involving the $b^\dagger_{\uparrow\downarrow i}$ which reflects the upper Hubbard band, Fig. 3.6 and which, because of the diagonal energy $U\sum_i b^\dagger_{\uparrow\downarrow i}b_{\uparrow\downarrow i}$, is shifted up by an energy U as compared with the lower such band and which involves the hole operators $b^\dagger_i$. The interactions contained in $\mathcal{H}'$ are inter-Hubbard-band couplings which lead to an anti-ferromagnetic interaction.

The *physical* chemical potential is a Lagrange multiplier which controls the average number of physical electrons. This usual average constraint is

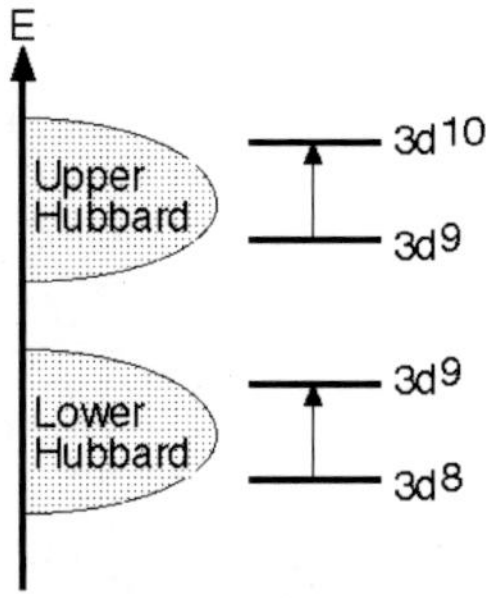

Fig. 3.6. For highly correlated electrons, when U is large, there is an upper and lower Hubbard band. In the context of Cu^9 the lower and upper bands correspond to transitions between the electronic *configurations* $3d^8 \to 3d^9$ and $3d^9 \to 3d^{10}$ respectively.

enforced by adding $-\mu\hat{N}$ to the Hamiltonian where in terms of auxiliary particles $\hat{N} = \sum_i \left[2b^\dagger_{i\uparrow\downarrow}b_{i\uparrow\downarrow} + \sum_\sigma f^\dagger_{i\sigma}f_{i\sigma}\right]$. Using the auxiliary particle constraint, there is a more particle-hole symmetric version namely,

$$\hat{N} = \mathcal{N} - \sum_i \left[b^\dagger_i b_i - b^\dagger_{\uparrow\downarrow i}b_{\uparrow\downarrow i}\right] \tag{3.82}$$

where $\mathcal{N}$ is the number of *sites* in the crystal. Thus the thermodynamic Hamiltonian $\mathcal{H} - \mu\hat{N}$ becomes $\mathcal{H} - \mu\mathcal{N} + \mu\sum_i \left[b^\dagger_i b_i - b^\dagger_{\uparrow\downarrow i}b_{\uparrow\downarrow i}\right]$, i.e., the effective chemical potential can be made zero if μ is accounted for as a diagonal energy in *only* the "charge particles" $b^\dagger_i$ and $b^\dagger_{\uparrow\downarrow i}$. In this picture $b^\dagger_i$ corresponds to a *holon* with charge $+e$ while $b^\dagger_{\uparrow\downarrow i}$ reflects *doublon* with the opposite charge each relative to the singly occupied states which are considered to be neutral.

Evidently, the operator corresponding to the current is also of considerable importance in the study of superconductivity. The charge at a given site is $q_i = -e\sum_\sigma c^\dagger_{\sigma i}c_{\sigma i} = -e\sum_\sigma f^\dagger_{\sigma i}f_{\sigma i} - 2eb^\dagger_{\uparrow\downarrow i}b_{\uparrow\downarrow i} = -e + e(b^\dagger_i b_i - b^\dagger_{\uparrow\downarrow i}b_{\uparrow\downarrow i})$, where the last equality uses the constraint. It is observed that the charge can be associated with the Bose particles and it might be anticipated that only the bosons are involved in the current operator. This is *not* the case.

For a bond $\langle ij\rangle$ the current flowing down the bond in the direction $i \to j$ is defined to be j_{ij}. The current continuity equation for site i is then,

$$\frac{\partial q_i}{\partial t} + \sum_j j_{ij} = 0. \tag{3.83}$$

In the usual way, using the Heisenberg equation of motion, this leads to

$$j_{ij} = -\mathrm{i}et\sum_\sigma (c^\dagger_{\sigma j}c_{\sigma i} - \text{h.c.}); \qquad c^\dagger_{\sigma i} = f^\dagger_{\sigma i}b_i + \sigma b^\dagger_{\uparrow\downarrow i}f_{\sigma i}, \tag{3.84}$$

which is just the usual definition for a tight binding model except that the physical conduction electron operators must be written in terms of the auxiliary particles. Clearly the spin particles $f^\dagger_{\sigma i}$ cannot be eliminated from this equation. Because there is always a single particle at a given site a hopping event corresponds to the *interchange* of a Bose and a Fermi particle. To be more specific consider the individual particle currents. A relationship between these is obtained by differentiating the constraint, assuming that there are no $b^\dagger_{\uparrow\downarrow i}$-particles i.e.,

$$\frac{dQ_i}{dt} = \sum_\sigma \frac{dn_{\sigma i}}{dt} + \frac{dn_i}{dt} = 0, \quad \text{where} \quad n_{\sigma i} = f^\dagger_{\sigma i} f_{\sigma i}, \; n_i = b^\dagger_i b_i. \tag{3.85}$$

Using the continuity equations, it is deduced that,

$$j^h_{ij} = -j^s_{ij}, \tag{3.86}$$

where, j^h_{ij} and j^h_{ij} are the number currents for the charge and spin particles respectively. *The spin and charge number currents are always equal and opposite.* This implies a Bose $b^\dagger_i$-particle superfluid is *not* a sufficient condition for superconductivity, a suitable pairing of the $f^\dagger_{\sigma i}$-particles is also necessary in order that the system as a whole be a superconductor.

In many mean-field theories based upon this auxiliary particle formalism, there is spin charge separation, and the number currents in the two sectors are defined independently and have their own resistivity. The relationship $j^h_{ij} = -j^s_{ij}$ implies that the two sectors are effectively in *series* so that $\rho = \rho_s + \rho_h$ where ρ_s and ρ_h are the resistivity of the spin and charge degrees of freedom.

3.10 Magnetic Exchange Interactions via Intermediate Bosons

Using the canonical transformation method of Sect. 3.3 and starting with the Hubbard model, the $b^\dagger_{\uparrow\downarrow i}$-boson can be eliminated in favor of a $-J\boldsymbol{S}_i \cdot \boldsymbol{S}_j$ interaction between the $f^\dagger_{\sigma i}$-particles and this constitutes a derivation of the $t-J$-model from the Hubbard model for large enough U. The idea is to eliminate the inter-Hubbard-band coupling term,

$$\mathcal{H}_P = -t \sum_{\langle ij\rangle\sigma} \sigma(b^\dagger_{\uparrow\downarrow i} f_{-\sigma i} b^\dagger_j f_{\sigma j} + b^\dagger_i f_{\sigma i} b^\dagger_{\uparrow\downarrow j} f_{-\sigma j} + \text{h.c.}). \tag{3.87}$$

Since U is considered large, in requiring that $[S, \mathcal{H}_0] = \mathcal{H}_P$, the energies are calculated ignoring t, i.e.,

$$S = -t \sum_{\langle ij\rangle\sigma} \sigma(\frac{b^\dagger_{\uparrow\downarrow i} f_{-\sigma i} b^\dagger_j f_{\sigma j}}{(U-\mu)+(\mu)} + \frac{b^\dagger_i f_{\sigma i} b^\dagger_{\uparrow\downarrow j} f_{-\sigma j}}{(U-\mu)+(\mu)} - \text{h.c.}), \tag{3.88}$$

where from (3.81), ignoring t but accounting for the chemical potential, the f-particles have zero energy while $b^\dagger_{\uparrow\downarrow i}$ has energy $U-\mu$ and $b^\dagger_i$ energy $+\mu$. A typical term in $\mathcal{H}_1 = \frac{1}{2}[S, \mathcal{H}_P]$ is of the form,

$$\frac{t^2}{U}[f^\dagger_{-\sigma i} b_{\uparrow\downarrow i} f^\dagger_{\sigma j} b_j b^\dagger_{\uparrow\downarrow i} f_{\sigma i} b^\dagger_j f_{-\sigma j}] \tag{3.89}$$

and involves for site i, the string of operators, $f^\dagger_{-\sigma i} b_{\uparrow\downarrow i} b^\dagger_{\uparrow\downarrow i} f_{\sigma i} = f^\dagger_{-\sigma i} f_{\sigma i}$. This simplification reflects the fact that $b_{\uparrow\downarrow i} b^\dagger_{\uparrow\downarrow i} = 1$ when $f^\dagger_{-\sigma i} b_{\uparrow\downarrow i} b^\dagger_{\uparrow\downarrow i} f_{\sigma i}$ is finite, and follows since, when $f_{\sigma i}$ acts on a constraint satisfying state, the result is the unphysical vacuum. With a similar simplification for site j the exemplified term becomes,

$$\frac{t^2}{U}[f^\dagger_{-\sigma i} f_{\sigma i} f^\dagger_{\sigma j} f_{-\sigma j},] \tag{3.90}$$

which exchanges opposite spin f-particles. The net result for the spin interaction for the sites i and j is,

$$\mathcal{H}_1 = J\left(\boldsymbol{S}_i \cdot \boldsymbol{S}_j - \frac{1}{4} n_i n_j\right); \quad \text{with} \quad J = \frac{4t^2}{U} \tag{3.91}$$

where, e.g., $S_i^+ = f^\dagger_{\uparrow i} f_{\downarrow i}$ involves only the f-particles and where $n_i = \sum_\sigma f^\dagger_{\sigma i} f_{\sigma i}$. This interaction can only occur if the near neighbor spins are of opposite directions, since otherwise the intermediate state ↑↓ is not possible. The appearance of the term $(J/4)n_i n_j$ reflects this fact. (A further discussion of the value of J in the $t - J$-model in relation to Zhang-Rice singlets [34] will be found in Chap. 2.) In addition, are two site terms which exchange the Bose particles and which are usually not important since there are no $b^\dagger_{\uparrow\downarrow i}$-particles. What is usually also ignored are three site terms which involve,

$$\frac{t^2}{U}[f^\dagger_{-\sigma i} b_{\uparrow\downarrow i} f^\dagger_{\sigma k} b_k b^\dagger_{\uparrow\downarrow i} f_{\sigma i} b^\dagger_j f_{-\sigma j}] = \frac{t^2}{U}[f^\dagger_{-\sigma i} f^\dagger_{\sigma k} b_k f_{\sigma i} b^\dagger_j f_{-\sigma j}], \tag{3.92}$$

which are often called the "pair hopping terms" since the net effect is to take a hole from site k to site j while at the same time moving an opposite spin pair which had up spin σ and site i and $-\sigma$ at j to finish with σ and site k and $-\sigma$ at i. It is reasoned that this term which involves hole hopping is small compared with the principle such terms, given that $J < t$.

The *effective* model for the lower Hubbard band when written in terms of auxiliary particles is,

$$\mathcal{H} = -t\sum_{\langle ij\rangle\sigma} f^\dagger_{\sigma i} b_i b^\dagger_j f_{\sigma j} + J\left(\boldsymbol{S}_i \cdot \boldsymbol{S}_j - \frac{1}{4} n_i n_j\right); \quad \text{with } J = \frac{4t^2}{U}. \tag{3.93}$$

This is just the $t - J$ model, $\mathcal{H} = -t\sum_{\langle ij\rangle\sigma}(\hat{c}^\dagger_{\sigma i}\hat{c}_{\sigma j} + \text{h.c.}) - J\boldsymbol{S}_i \cdot \boldsymbol{S}_j$; $\hat{c}^\dagger_{\sigma i} = c^\dagger_{\sigma i}(1 - c^\dagger_{-\sigma i} c_{-\sigma i})$ but where kinetic-energy term is written in terms of the b and f-particles, and as described above, the operator $\boldsymbol{S}_i \cdot \boldsymbol{S}_j$ is expressed using f rather than c-operators.

3.11 The (Mean-Field) RVB Slave Boson Theory

When the constraint $Q_i = 1$ is relaxed to an average requirement $\langle Q_i\rangle = 1$, the $f_{\sigma i}$-particles become independent. This approximation scheme was used to develop, very early on [35, 36], the RVB approach to cuprate superconductors and has been used extensively in the literature. Usually mean-field theories are justified by a minimum principle, i.e., the ground state given by a mean-field theory is usually an upper bound to the real such energy. Unfortunately here this type of principle cannot be applied since relaxing the constraint in conjunction with the mean-field approximation introduces unphysical states into the physical manifold and invalidates this minimal principle. (What *can* and has be done is to construct a mean-field solution, perform a Gutzwiller projection and then use this to evaluate a bound on the ground state energy, see Sect. 3.12.)

The mean-field solutions are of the form of product wave functions comprising *three* parts. There is are two Fermi seas, one each for the two $f_{\sigma i}$-particle spin directions. The third part describes the charge boson $b_i^\dagger$. Thus for the spin part, the Bose operators are replaced by their expectations values:

$$\mathcal{H} = -\tilde{\delta} t \sum_{\langle ij\rangle\sigma} f_{\sigma i}^\dagger f_{\sigma j} + J\left(\boldsymbol{S}_i \cdot \boldsymbol{S}_j - \frac{1}{4} n_i n_j\right) - \mu \hat{N}, \tag{3.94}$$

with $\tilde{\delta} = \langle b_i^\dagger b_j \rangle$. This effective $\mathcal{H}$ only involves spin f-particles. Assuming, for the ground state all the Bose particles condense at $\boldsymbol{q} = 0$, $\tilde{\delta} = x$ the concentration of holes. (This to be confirmed self-consistently below.) As expected on physical grounds, there is no kinetic energy derived from the t-term in the absence of doped holes. Since the total number of f-particles is identical to the total number of physical particles (given no occupancy of the upper Hubbard band) the choice has been made to account for the true chemical potential in this spin sector. There is a second effective chemical potential μ_h which must be introduced in the mean-field equation for the holes and which reflects the Lagrange multiplier for the constraint $\langle Q_i \rangle = 1$, more on this below.

Consider first the undoped case when x and therefore $\tilde{\delta}$ are zero and try to find the lowest energy pairing state, i.e., the BCS like state which maximizes in the negative sense $J(\boldsymbol{S}_i \cdot \boldsymbol{S}_j - \frac{1}{4} n_i n_j)$, written in terms of the f-operators. For this situation $\frac{1}{4} n_i n_j$ is an constant since there is a f-particle at each site. Searching only in the space of particle-particle pairing order parameters the lowest energy state has $s + \mathrm{i}d$-pairing. Within this mean-field approach [37] the energy of this state is $J\langle \boldsymbol{S}_i \cdot \boldsymbol{S}_j \rangle = -0.344J$, *per bond*, and for a square lattice. This is actually *lower* that the best estimates [38] of the ground state energy for the true anti-ferromagnetic ground state which has an energy per bond of $(-0.328 \pm 0.003)J$ and illustrates the above mentioned problem with the unreliable nature mean-field energies. The result of minimizing the ground state energy is $\langle f_{\uparrow j} f_{\downarrow i} \rangle \approx 2(0.479)(1 \pm \mathrm{i})$ where, with a certain phase convention, the sign is positive for vertical and negative for horizontal bonds. Taking a Fourier transform, the order parameter $\Delta_{\boldsymbol{k}} = (0.479J)(\gamma_{\boldsymbol{k}} + \mathrm{i}\gamma_{d\boldsymbol{k}})$, i.e., with s *and* d-wave components. This leads to a dispersion relationship,

$$E_{\boldsymbol{k}} = 0.479J\left[(\cos k_x a)^2 + (\cos k_y a)^2\right]^{1/2} . \tag{3.95}$$

There are a number of other mean-field factorizations of $J\boldsymbol{S}_i \cdot \boldsymbol{S}_j$ which, for the undoped case, lead to the same energy and which are dubbed with a number of names. This degeneracy is lifted when doping is added. The Fourier transform of $-\tilde{\delta}t\sum_{\langle ij\rangle\sigma} f^{\dagger}_{\sigma i}f_{\sigma j}$ is real and proportional to $\gamma_{\boldsymbol{k}}$, i.e., has s-wave symmetry. To maximize the binding energy the s-wave component of $J\boldsymbol{S}_i\cdot\boldsymbol{S}_j$ should add to this and also be real. This corresponds to a particle-hole mean-field amplitude $\chi_{ij} = \langle f^{\dagger}_{\uparrow i}f_{\downarrow j}\rangle$ and a pair amplitude $\langle f_{\uparrow j}f_{\downarrow i}\rangle = \pm 2\Delta$. Where again, by convention and corresponding to d-wave pairing, the sign of $\langle f_{\uparrow j}f_{\downarrow i}\rangle$ is taken positive for vertical bonds and negative for the horizontal equivalents. This reduces the Hamiltonian to,

$$\mathcal{H} = -\sum_{\langle ij\rangle\sigma}\left[\left(\tilde{\delta}t + \frac{1}{2}\chi_{ij}\right) f^{\dagger}_{\sigma i}f_{\sigma j}\right] + \sum_{\langle ij\rangle}(\pm)\Delta(f^{\dagger}_{\uparrow i}f^{\dagger}_{\downarrow j} + f_{\downarrow i}f_{\uparrow j}) \tag{3.96}$$

where the order parameters χ_{ij} and Δ are to be determined by minimizing the ground state energy. The pairing must be d-wave if this is to give the same energy as the $s + id$ state when the doping is zero. What is described by this Hamiltonian is a set of chargeless $S = 1/2$ *Fermi* excitations which undergo BCS condensation. These excitations are often called *spinons*. They contrast with the $S = 1$ *Bose* spin waves which are the low lying (Goldstone) excitations of a anti-ferromagnet, Sect. 3.4. This BCS like condensate associated with the spin degrees of freedom is called the "resonant valence bond", or RVB state [8].

The relative sign in $(\tilde{\delta}t + \frac{1}{2}\chi_{ij})$ needs some attention. The relevant part of the exchange interaction is, $(J/2)(S^+_i S^-_j + S^-_i S^+_j) = (J/2)(f^{\dagger}_{\uparrow i}f_{\downarrow i}f^{\dagger}_{\downarrow j}f_{\uparrow j} + \text{h.c.})$ which in mean-field theory becomes,

$$\begin{aligned}&\frac{J}{2}(\langle f_{\downarrow i}f^{\dagger}_{\downarrow j}\rangle f^{\dagger}_{\uparrow i}f_{\uparrow j} + \langle f^{\dagger}_{\uparrow i}f_{\uparrow j}\rangle f_{\downarrow i}f^{\dagger}_{\downarrow j} + \text{h.c.})\\ &\qquad = -\frac{J}{2}\chi_{ij}(f^{\dagger}_{\uparrow i}f_{\uparrow j} + f^{\dagger}_{\downarrow j}f_{\downarrow i} + \text{h.c.})\end{aligned} \tag{3.97}$$

where the minus sign comes from changing the order of the creation and destruction operators, e.g., $\chi_{ij} = \langle f^{\dagger}_{\downarrow j}f_{\downarrow i}\rangle = -\langle f_{\downarrow i}f^{\dagger}_{\downarrow j}\rangle$. Experimentally this sign is of some importance. Clearly $\chi_{ij} \leq 1/2$ (a value of 0.2-0.3 is realistic) and with accepted values of $t \sim 500$meV and $J \sim 120$meV one has $(\tilde{\delta}t + \frac{1}{2}\chi_{ij}J) \sim 100 - 130$meV. The part of the band which is below the Fermi surface has a dispersion which is about four times this value and is of the correct order of magnitude to explain the dispersive bands seen near the Fermi surface in ARPES [40]. Also, following BCS theory, the fact that the bandwidth $\propto (\tilde{\delta}t + \frac{1}{2}\chi_{ij}J)$ associated with these spinons broadens with doping implies that the transition temperature to the RVB state decreases strongly with $\tilde{\delta}$, i.e., with the doping x. This defines a line, decreasing with increasing x, above which superconductivity will not occur. Finally, again to the extent that this effective Hamiltonian describes the ARPES bands, the fact that this band broadens with increasing doping conforms with the general prejudices about the expected behavior of bands near a Mott-Hubbard metal-insulator transition.

The effective Hamiltonian for the charge sector results from using the RVB wave-function to evaluate the expectation values of the spin operators. Since the b-operators only occur in the t-term, the result is simply,

$$\mathcal{H} = -t \sum_{\langle ij \rangle \sigma} b_i^\dagger b_j \langle f_{\sigma i}^\dagger f_{\sigma j} \rangle \tag{3.98}$$

i.e.,

$$\mathcal{H} = -2t \sum_{\langle ij \rangle} \chi_{ij} b_i^\dagger b_j . \tag{3.99}$$

This describes a set of free bosons, usually referred to as *holons*, which have a low energy quadratic dispersion and therefore a Bose-Einstein condensation temperature $T_0 \propto x$ for a two dimensional plane. Condensation does indeed occur at $q = 0$ as was assumed above in connection with the spin sector. The straight line corresponding to this condensation provides a second line, increasing with x above which superconductivity will not occur. Notice the energy scale $\sim 4t$ for the holon dispersion is *not* strongly doping dependent.

The phase diagram [36, 41, 42] for this slave boson theory is illustrated in Fig. 3.7, and admitting an anti-ferromagnetic region for small x captures the essential fact that there is a limited region over which there is high temperature superconductivity with a maximum T_c for some modest value of x. Actually the realistic estimate for the T_0 is much too high. It is argued [43] that strong fluctuations strongly suppress T_0. It should be observed that in the underdoped region and descending in temperature, the spinon RVB transition occurs first this followed by the condensation of holons whereas in the overdoped region the inverse if true. Since most of the entropy is associated with the spins, it might be expected that RVB transition would be most evident in specific heat measurements. Since there is only clear evidence for a single true phase transition it is argued that the transition which occurs at the higher transition becomes a cross-over. This notwithstanding, there should

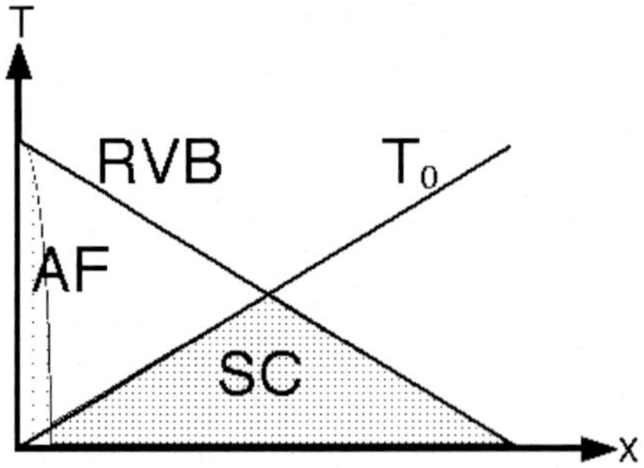

Fig. 3.7. The two lines in this phase diagram correspond to Bose-Einstein condensation of the holons, labeled T_0, and that due to the BCS like RVB transition. The anti-ferromagnetic (AF) phase for small concentrations in not predicted by the theory. The RVB transition moves to lower temperature with increasing x since the spinon band broadens. There is a finite region under the T_0 and RVB lines where superconductivity (SC) is predicted.

always be a sharp BCS like anomaly at T_c in the overdoped region. In fact, in underdoped systems it is claimed that the a large amount of entropy [44] is indeed lost along a line associated with the "pseudo-gap" however, while e.g., YBaCuO shows a well developed BCS-like specific heat feature, this is not universal and only near optimal doping.

On the face of it, this theory makes a remarkable prediction. The Bose holon charge carriers only have a charge $+e$ and it would be expected that the flux quantum for the superconducting state would be twice as big as for the usual BCS superconductors. This early prediction was soon proved wrong [45] and has lead proponents of theories derived from these roots to insist that there is an attractive interaction between holons which cause them to condense in pairs.

Many physical quantities can be calculated once the physical electron Green's function,

$$\mathcal{G}_{\sigma ij}(t) = \langle T c^{\dagger}_{\sigma i}(t) c_{\sigma j}(0) \rangle, \tag{3.100}$$

is know. Since the physical electrons have been decomposed into a spinon and holon, evaluating this quantity involves a convolution of the spinon and holon Green's functions and, in principle at least, vertex corrections must be accounted for. These complications only occur for a Landau Fermi liquid when higher order Green's functions, such as those for the conductivity or dynamical susceptibility, are calculated [12]. Consider the contributions from the lower Hubbard band. With this limitation $c^{\dagger}_{\sigma i} \approx f^{\dagger}_{\sigma i} b_i$ and since there is, in general, a holon condensate, $b_i = \langle b_i \rangle + b^r_i$ where $c = \langle b_i \rangle^2$ is the concentration of condensed bosons and where b^r_i reflects those holons which are not condensed. Immediately that part of $\mathcal{G}_{ij}(t)$ which involves the holon condensate, i.e., the entire Green's function at zero temperature, is,

$$\mathcal{G}^c_{\sigma ij}(t) = c \langle T f^{\dagger}_{\sigma i}(t) f_{\sigma j}(0) \rangle, \tag{3.101}$$

and requires only a detailed knowledge of the *spinon* Green's function. At least when there is a condensate, this should correspond to the dominant contribution near the Fermi surface and validates the earlier discussion of the ARPES spectra. The part of $\mathcal{G}^r_{\sigma ij}(t)$ which involves b^r_i requires a mathematically difficult convolution to be made. The low energy density of states for the holons is a constant while that for the spinons, given that there is a d-wave gap, is linear so that their convolution must, for low energies, go as the energy squared. In addition a convolution gives a very broad relatively weakly dispersive background. The coherent part $\mathcal{G}^c_{\sigma ij}(t)$ on the other hand mimics the spinon density of states which is initially linear and has well defined quasi-particle peaks. In a quantitative manner this phenomenology is also reflected in ARPES experiments [40, 46]. In the underdoped region, where the holon T_0 signals the onset of superconductivity, the ARPES bands are indeed broad and poorly defined as compared with those in the overdoped regime where there is a holon condensate *above* T_c and where there are, in fact, much better defined quasi-particles. The superconductivity is also "thin" [11] in the sense that the quasi-particle peaks have a strength which is $\sim c \ll 1$, i.e., much less than for Landau liquid based superconductors where poles have strength ~ 1. It *is* indeed observed in ARPES [40], for the underdoped region, that, in particular, the area under the superconductive peaks is roughly proportional to the superconducting fraction measured by muon rotation and to T_c, which from above is proportional to c in this regime.

3.12 The Gutzwiller Projection and a U(1) Symmetry

The weakness of the mean-field methods presented above is that the wave functions of the theory do not obey the constraint $Q_i = 1$. This constraint is intimately related to the non-double-occupancy constraint which must be imposed upon the $t-J$-model. An alternative approach [39] which embodies the same physical ideas but which does obey the latter constraint is to construct BCS of other wave functions and then apply the Gutzwiller projection which eliminates the doubly occupied states. The result is considered as as a variational ansatz for the fully constrained problem. Then, e.g., the magnitude of the BCS and/or anti-ferromagnetic order parameters are varied in order to minimize the energy. Within this approach there is an important U(1) symmetry. A projection implies a mapping from a larger to a smaller Hilbert space. For the Gutzwiller projection, *in the absence of holes*, the operator $c^\dagger_{\uparrow i}$ creates an up spin at site i, *but* the fact that $c_{\downarrow i}$ destroys a down spin implies also an up spin at this site since the two spin directions are the only two possibilities. More generally the combination $\alpha c^\dagger_{\uparrow i} + \beta c_{\downarrow i}$, with $|\alpha|^2 + |\beta|^2 = 1$ has the same property and defines the U(1) symmetry associated with the Gutzwiller projection. (The U(1) classification reflects the fact that the SU(2) rotation before projection leads only to a local phase change in the physical target space.) While $c^\dagger_{\uparrow i}$ and $c_{\downarrow i}$ are equivalent, the phase of the resulting states can change from site to site. Consider the down spin ferromagnetic ground state. With an anti-ferromagnetic exchange the ground state with S_z one greater is obtained by acting with $\sum_i \mathrm{e}^{\mathrm{i}\boldsymbol{\pi}\cdot\boldsymbol{r}_i} S_i^+$. Clearly if this is to result, after a projection, from the action of $c^\dagger_{\boldsymbol{k}\uparrow}$, then reflecting the factor of $\mathrm{e}^{\mathrm{i}\boldsymbol{\pi}\cdot\boldsymbol{r}_i}$ in the spin operator, the equivalent down spin operator must be $c_{-\boldsymbol{k}+\pi\downarrow}$, i.e., for an up spin which is created with $\boldsymbol{k}$ the equivalent down spin is destroyed with $-\boldsymbol{k}+\boldsymbol{\pi}$. It can then be appreciated that several order parameters are in fact equivalent. In particular the BCS-RVB order parameter $\langle c^\dagger_{\boldsymbol{k}\uparrow} c^\dagger_{-\boldsymbol{k}\downarrow}\rangle$ which reflects particle-particle pairing is equivalent to $\langle c^\dagger_{\boldsymbol{k}\uparrow} c_{\boldsymbol{k}+\pi\uparrow}\rangle$ corresponding to particle-hole pairing. As was be seen in Sect. 3.7, this latter pairing occurs in the theory of *flux states* when there is one-half of a flux quantum per square for a two dimensional a square lattice. The U(1) symmetry implies that these two apparently different pairing schemes are in fact the same at half-filling. Since this symmetry is lost when holes are present this equivalence is also lost and the BCS-RVB has lower energy since the chemical potential is pinned at the center of the gap.

3.13 Auxiliary Particles and the Introduction of Flux Tubes

There is a striking difference between the Jordan-Wigner and auxiliary particles methods of representing the system without holes. In the former scheme there are only as many particles as there are up spins whereas with the latter approximation the total number of particles is equal to the number of sites. The auxiliary particle scheme evidently greatly over-specifies the problem since the spin state is known once the position of all the up spins is given. The only role played by the down spin fermion operators is to conspire to change the sign of matrix elements in just the correct fashion that the up spin fermions mimic hard core bosons, i.e., they play the same role of the flux tube operators u_i.

While the flux tube operators are not essential to the auxiliary particle methods their introduction permits a certain simplification. The picture of the spin problem is then the same as within the Jordan-Wigner formulation and this permits the method to be extended to the doped regime in a simple fashion. (The Jordan-Wigner formulation *can* be directly extended [30] but this will not be pursued here.)

Using the definition of the flux operators given in Sect. 3.9 the combination,

$$(f^\dagger_{\downarrow i}u_i); \tag{3.102}$$

is a hard core Bose and hence products of these *commute* for different sites. It follows that the order of these down spin operators is immaterial in a basis set written as

$$\Big[\prod_i f^\dagger_{\uparrow i}\Big]\Big[\prod_i (f^\dagger_{\downarrow i}u_i)\Big]\Big|\Big\rangle. \tag{3.103}$$

On the other hand the order of the $f^\dagger_{\uparrow i}$ *does* matter and these are considered as the equivalent of the particles in the Jordan-Wigner formulation.

Since the u_i are unitary operators, i.e., $u^\dagger_i u_i = u_i u^\dagger_i = 1$, the spin-flip part of the exchange can be written as,

$$\frac{J}{2}[f^\dagger_{\uparrow i}u_i(u^\dagger_i f_{\downarrow i})(f^\dagger_{\downarrow j}u_j)u^\dagger_j f_{\uparrow j} + \text{h.c.}]. \tag{3.104}$$

Because, e.g., $(f^\dagger_{\downarrow j}u_j)$ is a (hard core) bosonic field it can be commuted to left without any changes of sign. If the matrix element is finite then there must be a $(u^\dagger_j f_{\downarrow j})$ in the basis vector to the left and hence it is possible to form the combination $(u^\dagger_j f_{\downarrow j})(f^\dagger_{\downarrow j}u_j) = u^\dagger_j u_j = 1$ where the fact that $f_{\downarrow j}f^\dagger_{\downarrow j} = 1$

for finite matrix elements which satisfy the constraint has been used, see Sect. 3.9. Thus, for the finite matrix elements of this interaction, the factors of $(u_i^\dagger f_{\downarrow i})$ and $(f_{\downarrow j}^\dagger u_j)$ with the corresponding factors in the basis vectors reduce to factors unity, i.e., these play no role other than to ensure that the final state obeys the constraint if the initial state did. If the constraint is enforced by other means they play no role at all and this part of the interaction might be contracted to,

$$\frac{J}{2}\left[f_{\uparrow i}^\dagger u_i u_j^\dagger f_{\uparrow j} + \text{h.c.}\right], \tag{3.105}$$

which is identical to the direct Jordan-Wigner representation, except that the particle operators have an up spin label. The combination, e.g., $f_{\uparrow i}^\dagger u_i$ is again a hard core boson. Operators contracted in this way are understood to act on a similarly contracted basis vectors, i.e., $[\prod_i f_{\uparrow i}^\dagger][\prod_i (f_{\downarrow i}^\dagger u_i)]|\rangle$ contracts to $[\prod_i f_{\uparrow i}^\dagger]|\rangle$. It is implied that the vacuum has all the spins down.

Following this philosophy the contracted Hamiltonian is, to within a constant,

$$\mathcal{H}_J = \frac{J}{2}\sum_{\langle ij\rangle}\left[(f_{\uparrow i}^\dagger u_i u_j^\dagger f_{\uparrow j} + \text{h.c.}) + \hat{n}_i\hat{n}_j\right] - J\sum_i \hat{n}_i, \tag{3.106}$$

where, $\hat{n}_i = f_{\uparrow i}^\dagger f_{\uparrow i}$. (Flux tube operators can be bracketed into $\hat{n}_i = f_{\uparrow i}^\dagger u_i u_i^\dagger f_{\uparrow i}$ if desired.) This Hamiltonian is the same as (3.58) deduced *directly* using a Jordan-Wigner transformation.

By adding flux tube operators the $f_{\downarrow i}^\dagger$ operators can be contracted out of the hopping term. Within the $t-J$ model the down spin physical electron hopping term is $-t\sum_{\langle ij\rangle} f_{\downarrow i}^\dagger b_i b_j^\dagger f_{\downarrow j} = -t\sum_{\langle ij\rangle}(f_{\downarrow i}^\dagger u_i)u_i^\dagger b_i b_j^\dagger u_j(u_j^\dagger f_{\downarrow j})$ which contracts to $-t\sum_{\langle ij\rangle} u_i^\dagger b_i b_j^\dagger u_j$, i.e. the hole operators acquire a flux tube to become, e.g., $u_i^\dagger b_i$, i.e., effective fermions. The full $t-J$ model contracted in this way is

$$\begin{aligned}\mathcal{H} &= \mathcal{H}_t + \mathcal{H}_J\\ \mathcal{H}_t &= -t\sum_{\langle ij\rangle}\left[f_{\uparrow i}^\dagger b_i b_j^\dagger f_{\uparrow j} + u_i^\dagger b_i b_j^\dagger u_j\right] + \text{h.c.}\end{aligned} \tag{3.107}$$

In analogy to the quantum Hall effect, the flux tube operators play a non-trivial role and this ultimately reflects the fact that the total spin operator $\boldsymbol{S}$ commutes with $\mathcal{H}$. This is of critical importance, since when the flux tubes are accounted for, in the absence of symmetry breaking anti-ferromagnetism, *the spin excitations of the system are $S = 1/2$ spinons.* Take $|G\rangle$ to be the lowest energy half-filled Fermi sea. This is clearly non-degenerate with $S_z = 0$ and must have a total spin $S = 0$. The lowest energy state $f_{\uparrow \boldsymbol{k}}|G\rangle$ has $S_z = -1$ and is not degenerate with any state with larger negative value of S_z. It follows that $f_{\uparrow \boldsymbol{k}}|G\rangle$, in fact, has the quantum numbers $S = 1$ and $S_z = -1$. Such a state can always be raised twice with a finite result. The state $(S^+)^2 f_{\uparrow \boldsymbol{k}}|G\rangle$ must be identified with $f^\dagger_{\uparrow \boldsymbol{k}+\boldsymbol{\pi}}|G\rangle$ which has the same energy and $S = +1$. However there is no such simple identification for the state $S^+ f_{\uparrow \boldsymbol{k}}|G\rangle$ which has quantum numbers $S = 1$ and $S_z = 0$ except that it is a *longitudinal* excitation of the ground state. The relevant operator is easily determined: $S^+ f_{\uparrow \boldsymbol{k}}|G\rangle = [S^+, f_{\uparrow \boldsymbol{k}}]|G\rangle + f_{\uparrow \boldsymbol{k}} S^+|G\rangle$, but $S^+|G\rangle = 0$ so the longitudinal operator is $[S^+, f_{\uparrow \boldsymbol{k}}] = \frac{1}{\sqrt{N}} \sum_i e^{i\boldsymbol{k}\cdot\boldsymbol{r}_i} u_i (f^\dagger_{\uparrow i} f_{\uparrow i} - f_{\uparrow i} f^\dagger_{\uparrow i})$. The appropriate fermionic projection operators are defined by

$$p_{\boldsymbol{k}} = \frac{1}{\sqrt{N}} \sum_i \mathrm{e}^{\mathrm{i}\boldsymbol{k}\cdot\boldsymbol{r}_i} u_i f^\dagger_{\uparrow i} f_{\uparrow i}; \quad t_{\boldsymbol{k}} = \frac{1}{\sqrt{N}} \sum_i \mathrm{e}^{\mathrm{i}\boldsymbol{k}\cdot\boldsymbol{r}_i} u_i f_{\uparrow i} f^\dagger_{\uparrow i} \tag{3.108}$$

where the presence of the u_i in the Fourier transform is all important. Since S^+ is a generator of SU(2), the new operator $(p_{\boldsymbol{k}} - t_{\boldsymbol{k}}) = [S^+, f_{\uparrow \boldsymbol{k}}]$ simply arises from a group operation on $f_{\uparrow \boldsymbol{k}}$. The states $p_{\boldsymbol{k}}|G\rangle$ and $t_{\boldsymbol{k}}|G\rangle$ are related via time reversal, $\mathcal{T}$, and reflection, R, symmetry. Since $\mathcal{T}$ causes $\uparrow \to \downarrow$, $p_{-\boldsymbol{k}}|G\rangle = \mathcal{T} t_{\boldsymbol{k}}|G\rangle$ so that with a suitable definition $p_{\boldsymbol{k}}|G\rangle = R\mathcal{T} t_{\boldsymbol{k}}|G\rangle$ and since both $\mathcal{T}$ and R commute with $\mathcal{H}$ its follows that both $p_{\boldsymbol{k}}|G\rangle$ and $t_{\boldsymbol{k}}|G\rangle$ are eigenstates. With these definitions, the triplet state is then $(p_{\boldsymbol{k}} - t_{\boldsymbol{k}})|G\rangle$. The sum $p_{\boldsymbol{k}} + t_{\boldsymbol{k}} = \frac{1}{\sqrt{N}} \sum_i \mathrm{e}^{\mathrm{i}\boldsymbol{k}\cdot\boldsymbol{r}_i} u_i$ is also fermionic and, since it does not involve the f-operators, commutes with a bosonic objects such as $\boldsymbol{S}$. Since the ground state is a singlet $\boldsymbol{S}^2|G\rangle = \hbar^2 S(S+1)|G\rangle = 0$ and since $[(p_{\boldsymbol{k}} + t_{\boldsymbol{k}}), \boldsymbol{S}] = 0$ it follows that $\boldsymbol{S}^2(p_{\boldsymbol{k}} + t_{\boldsymbol{k}})|G\rangle = 0$, i.e., the combination $(p_{\boldsymbol{k}} + t_{\boldsymbol{k}})|G\rangle$ is a spin *singlet*. Since they have different quantum numbers for the total spin, it follows that $(p_{\boldsymbol{k}} - t_{\boldsymbol{k}})|G\rangle$ and $(p_{\boldsymbol{k}} + t_{\boldsymbol{k}})|G\rangle$ are orthogonal, which in turn implies that $p_{\boldsymbol{k}}|G\rangle$ and $t_{\boldsymbol{k}}|G\rangle$ are also orthogonal. The normalization is also easily found. Assume that $f_{\uparrow \boldsymbol{k}}|G\rangle$ is normalized. The matrix elements of the raising operator S^+ for $S = 1$ are $\sqrt{2}$ so that

$$|\boldsymbol{k}, t\rangle = \frac{1}{\sqrt{2}} [p_{\boldsymbol{k}} - t_{\boldsymbol{k}}]|G\rangle \tag{3.109}$$

is the normalized triplet state, and, since the $p_{\boldsymbol{k}}|G\rangle$ and $t_{\boldsymbol{k}}|G\rangle$ are orthogonal, they are also normalized. Similarly,

$$|\boldsymbol{k}, s\rangle = \frac{1}{\sqrt{2}} [p_{\boldsymbol{k}} + t_{\boldsymbol{k}}]|G\rangle \tag{3.110}$$

is the normalized singlet excitation.

The existence of the singlet state $|\boldsymbol{k}, s\rangle$ implies directly that the basic spin excitations are $S = 1/2$ *spinons*. It is evident that the operators $\boldsymbol{S}$, which in the absence of charge carriers, are the only way to couple to the spin sector, imply a selection rule, $\Delta S_z = 0, \pm 1, \pm 2, \ldots$, so that spinons are always involved in "particle-hole" type excitations. For the lowest lying excitations discussed above, the possible excitations correspond to the product space $S_{1/2} \otimes S_{1/2} = S_1 \oplus S_0$. The fact that there is degenerate S_1 triplet and singlet S_0 therefore reflects the fact that the underlying excitations are indeed $S = 1/2$ *spinons*. (See below for more detail.) The present can be thought of a more abstract version of the origin demonstration [47] of the existence of spinons (without using the name) and which was based upon the Bethe ansatz solution of the one dimensional Heisenberg model.

The excitation created by $u_{\boldsymbol{k}} \equiv p_{\boldsymbol{k}} + t_{\boldsymbol{k}} = \frac{1}{\sqrt{N}} \sum_i e^{i\boldsymbol{k}\cdot\boldsymbol{r}_i} u_i$ is the singlet equivalent of the triplet excitation created by $f_{\uparrow\boldsymbol{k}} = \frac{1}{\sqrt{N}} \sum_i e^{i\boldsymbol{k}\cdot\boldsymbol{r}_i} f_{\uparrow i}$. Comparing these expressions, it is manifest that u_i has exactly the same effect at does $f_{\uparrow i}$ *except* it does not change the value of S_z. Making a parallel with the two dimensional electron gas, it might be expected that the flux tube operator u_i would create a *hole* at site i, *however* the present system has particle-hole symmetry and hence a *particle* is created with equal weight. This leads to a physical picture in which there is a singlet formed between the spin(on) at site i and a net spin one half of the vacuum, i.e., the rest of the system. With this picture, e.g., $p_i = u_i f_{\uparrow i}^\dagger f_{\uparrow i}$ projects out that part of the singlet with has the spin up at site i. This emphasizes the off-diagonal parts of u_i. The diagonal part is just a phase factor determined by the state of the system and is the reason u_i was introduced in the Jordan-Wigner formulation.

It is easy to appreciate how, say, $S_i^+ = f_{\uparrow i}^\dagger u_i$ creates particle-hole spinon excitations. In the following, an explicit u_i, *in a final result*, implies the diagonal parts of the operator while, e.g., $p_i = u_i f_{\uparrow i}^\dagger f_{\uparrow i}$ involves only the off-diagonal part. With this, the Fourier transform of u_i is not well defined. Dropping this diagonal part $S_i^+ = f_{\uparrow i}^\dagger d_i$, where, while a local relationship is indicated, it is intended that this is only valid when Fourier transformed. After such a transform $S_{\boldsymbol{q}}^+ = (1/N) \sum_{\boldsymbol{k}} f_{\uparrow \boldsymbol{q}-\boldsymbol{k}}^\dagger d_{\boldsymbol{k}}$. The final interpretation of the $f_{\uparrow i}$ particles is similar to that of u_i and by inference p_i and t_i. The triplet is made of a $S = 1/2$ spinon at site i with again a vacuum $S = 1/2$ so that in the pair, e.g., $f_{\uparrow \boldsymbol{q}-\boldsymbol{k}}^\dagger d_{\boldsymbol{k}}$ the vacuum spins cancel and this is viewed as a $S = 1/2$ spinon particle-hole pair.

The operators p_i and $f_{\uparrow i}$ are not independent and it is useful for later use to express the off-diagonal parts of p_i in terms of $f_{\uparrow i}$. By definition $p_{\boldsymbol{k}} - t_{\boldsymbol{k}} = [S^+, f_{\uparrow \boldsymbol{k}}]$ while $p_{\boldsymbol{k}} + t_{\boldsymbol{k}} = u_{\boldsymbol{k}}$ so that $p_{\boldsymbol{k}} = (1/2)\{[S^+, f_{\uparrow \boldsymbol{k}}^\dagger] + u_{\boldsymbol{k}}\}$. Thus, in the triplet excitation sector, $S = 1$, $S_z = 1$, where the singlet, $u_{\boldsymbol{k}}$, and $f_{\uparrow \boldsymbol{k}} S^+$ operators annihilate all states, it is possible to write $p_i = u_i f_{\uparrow i}^\dagger f_{\uparrow i} + (1/2) S^+ f_{\uparrow i}$ with, again, the understanding that only the *diagonal* elements of the displayed u_i are to be accounted for.

3.14 Spin Pairing

In the search for a constraint satisfying RVB type solution to the spin problem there is an immediate difficulty. The RVB order parameter $\langle f_{\uparrow i} f_{\downarrow j} \rangle$ implies a BCS type coherent state which does *not* obey the constraint. However as discussed in Sect. 3.12, corresponding to a U(1) symmetry [39], when the constraint *is* satisfied in the absence of holes, destroying a down spin particle must be accompanied by the creation of an up spin. The hole with the same energy as a particle with momentum $-\boldsymbol{k}$ has momentum $\boldsymbol{k}+\boldsymbol{\pi}$ and order parameter is therefore, $\mathrm{e}^{\mathrm{i}\boldsymbol{\pi}\cdot\boldsymbol{r}_i}\langle f^\dagger_{\uparrow i} f_{\uparrow j}\rangle$ where $\boldsymbol{r}_i$ is the integer coordinate of site i. This is precisely an amplitude which occurs in the mean-field theory, i.e., in connection with flux states. In a gauge which is symmetric with respect to the x and y directions, the average field of 1/2 a flux quantum per site corresponds to, see Sect. 3.7,

$$u_i u_j^\dagger = \frac{1}{\sqrt{2}}[1 \pm \mathrm{i}\mathrm{e}^{\mathrm{i}\boldsymbol{\pi}\cdot\boldsymbol{r}_i}], \tag{3.111}$$

where the plus sign is for horizontal bonds, and where the coefficient reflects the unitary nature of the operator $u_i u_j^\dagger$. The "order parameter" is immediately,

$$\Delta_0 = \frac{J}{\sqrt{2}}. \tag{3.112}$$

The existence of this amplitude is associated with the formation of two Landau levels which are broadened into bands through the interaction with the lattice. There can be no thermodynamic phase transition associated with the "condensation" of the particles in the lowest such level since the level is macroscopically occupied for *all* temperatures.

The effective mean-field Hamiltonian for the flux state ground state is,

$$\mathcal{H} = \sum_{\boldsymbol{k}} [\epsilon_{\boldsymbol{k}} f^\dagger_{\boldsymbol{k}} f_{\boldsymbol{k}} + \mathrm{i}(\Delta_{\boldsymbol{k}} f^\dagger_{\boldsymbol{k}+\boldsymbol{\pi}} f_{\boldsymbol{k}} - \mathrm{h.c.})] \tag{3.113}$$

where $\epsilon_{\boldsymbol{k}} = \frac{J}{\sqrt{2}}\gamma_{\boldsymbol{k}}$ and $\Delta_{\boldsymbol{k}} = \Delta_0 \gamma^d_{\boldsymbol{k}} = \frac{J}{2\sqrt{2}}\gamma^d_{\boldsymbol{k}}$, see Sect. 3.7. Also from Sect. 3.7 is the dispersion relationship

$$E_{\boldsymbol{k}} = \sqrt{\frac{J^2}{2}{\gamma_{\boldsymbol{k}}}^2 + {\Delta_{\boldsymbol{k}}}^2} \tag{3.114}$$

i.e., of BCS form but with a well specified value of $\Delta_{\boldsymbol{k}}$, i.e., without a "gap equation" for this quantity.

It might be thought that such a particle-hole pairing cannot lead to superconductivity. It will be shown in Sect. 3.13 that a finite physical electron pair amplitude $\langle c_{-\boldsymbol{k}\downarrow} c_{\boldsymbol{k}\uparrow}\rangle$ results from this flux state order parameter $\Delta_{\boldsymbol{k}}$ in the presence of condensed holes. There then exist general arguments due to S. Weinberg [50] that the existence of this physical off-diagonal order suffices to explain the hallmark of superconductivity namely the Meissner effect, etc.

3.15 Fermionic Excitations in an Anti-ferromagnet

Anti-ferromagnetism has been described in Sect. 3.4 using the Holstein-Primakoff transformation. This gives quite satisfactory description of the undoped parent material along with its low lying bosonic, i.e., spin wave excitations.

In Sect. 3.5 has been sketched an alternative *fermionic* description of a $S = 1/2$ magnet based upon the Jordan-Wigner transformation. A mean-field approximation within this method leads to a flux state which has a certain pairing order parameter which is finite for all temperatures. It must be that this flux-state can undergo a spontaneously broken symmetry towards an anti-ferromagnetic ground state. In later sections this fermionic approach will be used as the basis for the doped superconducting cuprates. The predictions of such a theory might therefore be compared with the fermionic, i.e., ARPES [40] quasi-electron excitations of lightly doped anti-ferromagnets.

Up until here the longitudinal part of the exchange interaction has been largely ignored since, in the absence of a broken magnetic symmetry its mean field expectation value, i.e., the Hartree term, along with the chemical potential term give zero. When such a symmetry *is* broken it is precisely this term which is responsible. Assuming anti-ferromagnetic ordering along the z-direction,

$$\mathcal{H}_{\mathrm{AF}} \equiv \sum_{\langle ij \rangle} J S_{zi} S_{zj} \to \sum_i 2 J m S_{zi} \mathrm{e}^{\mathrm{i}\boldsymbol{\pi}\cdot\boldsymbol{r}_i} \tag{3.115}$$

where $-1 < m < 1$ is the anti-ferromagnetic order parameter $m = 2\mathrm{e}^{-\mathrm{i}\boldsymbol{\pi}\cdot\boldsymbol{r}_i} \langle S_{zi} \rangle$ and a two dimensional square lattice is assumed. In terms of the f-particles,

$$S_{zi} = \left(f^{\dagger}_{\uparrow i} f_{\uparrow i} - \frac{1}{2} \right) \tag{3.116}$$

and so, taking a Fourier transform,

$$\mathcal{H}_{\mathrm{AF}} = 2Jm \sum_{\boldsymbol{k}} f^{\dagger}_{\uparrow \boldsymbol{k}} f_{\uparrow \boldsymbol{k}+\boldsymbol{\pi}} \tag{3.117}$$

which causes scattering across the Fermi surface. It is easy to include $\mathcal{H}_{\mathrm{AF}}$ in the calculation of the energies for the flux state problem, with the result,

$$E_{\boldsymbol{k}} = \sqrt{\frac{J^2}{2} {\gamma_{\boldsymbol{k}}}^2 + {\Delta_{\boldsymbol{k}}}^2 + (2Jm)^2}. \tag{3.118}$$

The effect of a finite m is to open a gap over the entire Fermi surface and it might be noticed that there is no fermionic mode which comes to zero at $\boldsymbol{q} = \boldsymbol{\pi}$. If the dispersion terms $\frac{J}{\sqrt{2}} \gamma_{\boldsymbol{k}}$ and $\Delta_{\boldsymbol{k}}$ are dropped $E_{\boldsymbol{k}}$ reduces to the simplest mean-field theory result $+2Jm$ for a spin-flip. The dispersive terms reflect the effects of quantum fluctuations. In order to obtain the ground state energy it is necessary to calculate both Fock corrections and m self-consistently. Such an approach gives an energy per bond $\approx -0.32J$ which is comparable with the best results for the energy but $m \approx 0.8$ indicating the "wave function" does not account well enough for quantum fluctuations. The

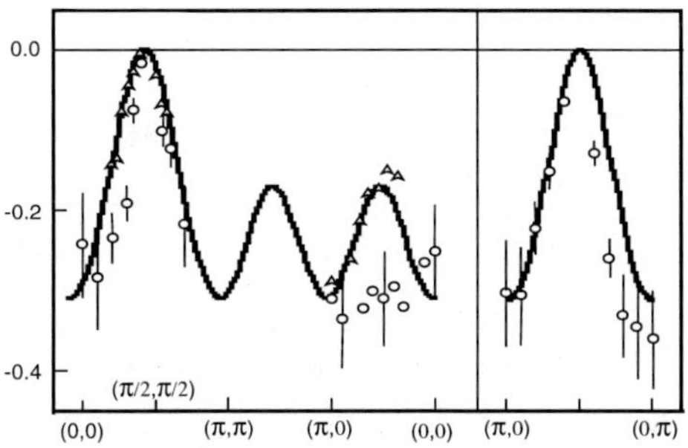

Fig. 3.8. A comparison of (3.118) with the experimental ARPES data (open circles, Wells et al., and open triangles, LaRosa et al., [47]) for undoped anti-ferromagnetic $Sr_2CuO_2Cl_2$. The exchange was set to be 100meV and the fit corresponds to a, theoretically rather small, but experimentally realistic, value of $m = 0.35$.

evident defect is the absence of a, collective gapless $\boldsymbol{q} = \boldsymbol{\pi}$ *bosonic*, Goldstone mode.

These bosonic excitations can be exhibited by considering the equation of motion for the bosonic object $f^\dagger_{\uparrow i} u_i$, and this reproduces the results similar to those of the Holstein-Primakoff theory. It is suggested that the appropriate approach is to use the results of the Holstein-Primakoff theory to calculate the averages which appear in (3.118) and that the fermionic excitations should be considered as being *additional*, with the bosons lying in the fermionic gap. As will be shown in detail in Sect. 3.16, with coherent hole doping $E_{\boldsymbol{k}}$ becomes the dispersion relationship for the low lying physical quasi-electrons. It is interesting to note that, Fig. 3.8, this theory agrees well with the ARPES results [48] for single hole excitations in the anti-ferromagnet. This contrasts with the explanation to be found Chap. 2 for this same spectrum but is similar in concept to a proposal of Laughlin [49] that ARPES sees the spinon spectrum.

Without doing any detailed calculation, it is evident, upon extending this approach to finite temperatures, that there is a finite temperature $T_N \sim J$ above which $m = 0$. Evidently, *it is the case* that there should be no second transition temperature associated with the flux state since the Néel temperature can be considered as *the* condensation temperature of the spin hard core bosons.

3.16 SU(3) Approach to Hole Coherent States

It was shown in Sect. 3.6 that a macroscopic condensation of bosons in the ground state should be viewed a renormalization of the ground state which involves a *coherent state*. If such a coherent state is to reflect hole condensation for the present problem it is evident that it *cannot* be defined in the same fashion since for a finite ground state hole condensate concentration n_0 there is a finite probability of finding two bosons as the same site and this is forbidden by the constraint. Even worse, with such a coherent state as the vacuum, creating the large density of spin f-particles even more strongly violates the constraint. It is necessary to construct coherent states which are consistent with the constraint and which permit a hole condensate [51]. In fact, the constraint defines a three dimensional sphere in the bosonic space of the operators $b_i^\dagger$, $f_{\uparrow i}^\dagger u_i$ and $f_{\downarrow i}^\dagger u_i$, i.e., if $v_{i0} = (b_i^\dagger, f_{\uparrow i}^\dagger u_i, f_{\downarrow i}^\dagger u_i)$, then $Q_i = v_{i0} v_{i0}^\dagger = 1$. The constraint is then invariant to the action of unitary operators, R_i:

$$v_{i0} v_{i0}^\dagger = v_{i0} R_i^\dagger R_i v_{i0}^\dagger = v_i v_i^\dagger = 1 \tag{3.119}$$

where $v_i = R_i v_{i0}$ and $R_i^\dagger R_i = R_i R_i^\dagger = 1$. Assuming v_i reduces to v_{i0} for small angles, the local rotations R_i belong to SU(3). Such rotations mix the spin and charge degrees of freedom and hence permit a (bosonic) hole condensate to be sub-summed into the spin (fermion) degrees of freedom. The simplest case is when $R_i = R_j = R$. Phase separation implies a non-uniform R_i.

The, $n(n-1)/2 = 10$, SO(5) rotations of magnetic and superconductive order parameters are included within this SU(3) theory with only $n^2 - 1 = 8$ generators since the SO(5) operators are defined on two sites. The O(3) rotations of the magnetic order parameter are effected by the usual SU(2) rotations of the $f_{\sigma i}^\dagger$ operators. As later developments will show, the U(1) symmetry of the superconductive order parameter is reflected by the similar symmetry of the hole Bose condensation amplitude. Clearly SU(2)×U(1) is included in SU(3). This will be taken up again in later subsections.

The immediate idea is to begin with some spin only candidate ground state and find a rotation R which, for a given chemical potential, minimizes the ground state energy. Here only a single possibility will be really considered namely the constraint satisfying flux state ground state introduced in Sect. 3.7, and which is fully equivalent to a Gutzwiller projected RVB state. Clearly the beginning spin state must have an order parameter consistent with experiments and indeed these flux states have a *d*-wave gap structure. Given a choice of the spin ground state there remains an infinite number of different rotations R. When the spin operators are rotated towards the hole operators, holes are added to the system producing a hole coherent state, i.e., a state which is a superposition of states with different hole numbers. The description of the flux state developed in Sects. (3.13) and (3.14) has $f_{\uparrow i}^\dagger$ fermions, along with $f_{\downarrow i}^\dagger u_i$ and $b_i^\dagger$ bosons, and a general rotation R will introduce *both* fermion and bosonic holes as these are sub-summed into $f_{\uparrow i}^\dagger$ or $f_{\downarrow i}^\dagger u_i$. The co-existence of these two types of holes is what motivated the introduction of the extended Hilbert space in connection with the flux tubes as described at the end of Sect. (3.6). In the absence of a broken symmetry, the up and down spin directions should be equivalent and this implies that the each spin direction rotates towards the hole axis by an equal amount, i.e., the rotation is that illustrated in Fig. 3.9.

For a small SU(3) rotation, the spin operators can be expressed as

$$R f_{\sigma i}^\dagger R^{-1} \equiv \tilde{f}_{\sigma i}^\dagger = \left(1 - \frac{\psi_i^{*\sigma} \psi_i^\sigma}{2}\right) f_{\sigma i}^\dagger + \psi_i^\sigma b_i^\dagger u_i^\dagger, \tag{3.120}$$

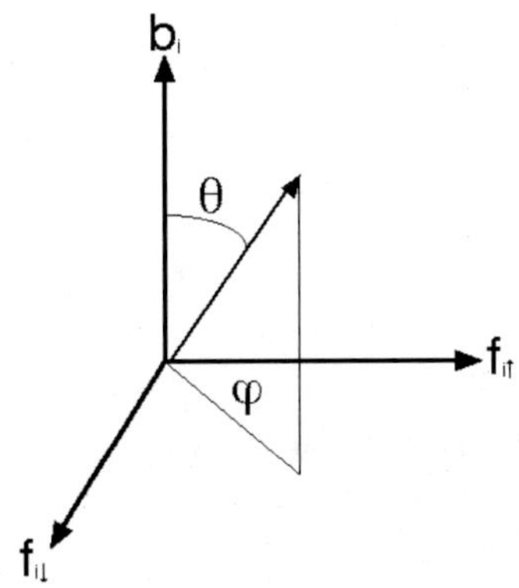

Fig. 3.9. Holes are coherently added by a rotation in the space of the auxiliary particle operators for each site. The angle θ is proportional to the condensate wave function, assumed small. It will turn out that the down spin condensate wave function alternates while the up spin equivalent does not. This implies that the angle $\phi = \pm\pi/4$ changing sign from site to site. More generally, corresponding to SU(3), the rotation angles are not real, this corresponding to a complex hole wave function.

which coherently adds a hole with the *spin dependent* wave function ψ_i^σ. The rotation of the hole operator is specified by,

$$Rb_i^\dagger R^{-1} \equiv \tilde{b}_i^\dagger = \left(1 - \frac{\psi_i^{\downarrow *}\psi_i^\downarrow}{2} - \frac{\psi_i^{*\uparrow}\psi_i^\uparrow}{2}\right) b_i^\dagger - \psi_i^\uparrow f_{\uparrow i}^\dagger u_i - \psi_i^\downarrow f_{\downarrow i}^\dagger u_i . \quad (3.121)$$

By construction, the $\tilde{f}_{\sigma i}^\dagger$ and $\tilde{b}_i^\dagger$ obey the appropriate commutation rules.

The SU(3) rotation adds holes to the system *even* when there are none of the new $\tilde{b}_i^\dagger$ particles are present. The concentration of such holes is determined by evaluating $x = \langle b_i^\dagger b_i\rangle$ which involves the *original* hole operators. One has,

$$\begin{aligned} b_i^\dagger b_i &\approx (1 - \psi_i^{\downarrow *}\psi_i^\downarrow - \psi_i^{*\uparrow}\psi_i^\uparrow)\tilde{b}_i^\dagger \tilde{b}_i \\ &+\psi_i^{*\uparrow}\psi_i^\uparrow \tilde{f}_{\uparrow i}^\dagger \tilde{f}_{\uparrow i} + \psi_i^{\downarrow *}\psi_i^\downarrow \tilde{f}_{\downarrow i}^\dagger \tilde{f}_{\downarrow i} + \psi_i^{*\downarrow}\psi_i^\uparrow \tilde{f}_{\uparrow i}^\dagger \tilde{f}_{\downarrow i} + \psi_i^{\uparrow *}\psi_i^\downarrow \tilde{f}_{\downarrow i}^\dagger \tilde{f}_{\uparrow i}. \end{aligned} \quad (3.122)$$

In the absence of $\tilde{b}_i^\dagger$ particles this reduces to $n_i^0 = b_i^\dagger b_i \approx \psi_i^{*\uparrow}\psi_i^\uparrow \tilde{n}_i^\uparrow + \psi_i^{\downarrow *}\psi_i^\downarrow \tilde{n}_i^\downarrow$. Assuming a non-polarized spin system,

$$x = \frac{1}{2}(|\psi_i^\uparrow|^2 + |\psi_i^\downarrow|^2) = |\psi_i^\uparrow|^2 = |\psi_i^\downarrow|^2, \quad (3.123)$$

since then $|\psi_i^\uparrow| = |\psi_i^\downarrow|$.

The *original* physical electron operators can be expressed in terms of these new operators, e.g., $\hat{c}_{\sigma i}^\dagger = f_{\sigma i}^\dagger b_i$ is equal to,

$$\begin{aligned} &\left[\left(1 - \frac{\psi_i^{\sigma *}\psi_i^\sigma}{2}\right)\tilde{f}_{\sigma i}^\dagger - \psi_i^\sigma \tilde{b}_i^\dagger u_i\right] \\ &\quad\times\left[\left(1 - \frac{\psi_i^{\sigma *}\psi_i^\sigma}{2} - \frac{\psi_i^{*-\sigma}\psi_i^{-\sigma}}{2}\right)\tilde{b}_i + \psi_i^{*-\sigma}\tilde{f}_{-\sigma i}u_i^\dagger + \psi_i^{\sigma *}\tilde{f}_{\sigma i}u_i^\dagger\right] \end{aligned} \quad (3.124)$$

Keeping only the leading order terms,

$$\begin{aligned} \hat{c}_{\sigma i}^\dagger \approx (1 - \psi_i^{\sigma *}\psi_i^\sigma - \tfrac{\psi_i^{*-\sigma}\psi_i^{-\sigma}}{2})\tilde{f}_{\sigma i}^\dagger \tilde{b}_i + (\psi_i^{*-\sigma}\tilde{f}_{\sigma i}^\dagger \tilde{f}_{-\sigma i}u_i^\dagger + \psi_i^{\sigma *}\tilde{f}_{\sigma i}^\dagger \tilde{f}_{\sigma i}u_i^\dagger) \\ -\psi_i^\sigma \tilde{b}_i^\dagger u_i \tilde{b}_i - \psi_i^\sigma \tilde{b}_i^\dagger(\psi_i^{*-\sigma}\tilde{f}_{-\sigma i} + \psi_i^{*\sigma}\tilde{f}_{\sigma i}). \end{aligned} \quad (3.125)$$

Examining the terms in order:

(i) The first term $\tilde{f}^{\dagger}_{\sigma i}\tilde{b}_i$ is of the original form except that it is the tilde operators which are involved and there is a prefactor, $(1-\psi_i^{\sigma *}\psi_i^{\sigma} - \frac{\psi_i^{*-\sigma}\psi_i^{-\sigma}}{2}) \approx (1-3x/4)$. The hopping term thereby picks up a factor $(1-3x/4)^2 \approx (1-3x/2)$ which implies a direct band *narrowing* for overdoped cuprates.
(ii) The combination $\tilde{f}^{\dagger}_{\uparrow i}(\tilde{f}_{\downarrow i}u_i^{\dagger})$ causes spin flips. When, following Sect. 3.13, the bosonic combinations $(\tilde{f}_{\downarrow i}u_i^{\dagger})$ are contracted out, there is, e.g., a part $\psi_i^{*-\sigma}\tilde{f}^{\dagger}_{\uparrow i}$ of $\hat{c}^{\dagger}_{\uparrow i}$ which corresponds to the destruction of a condensate hole with the production of a spin particle. As with the coherent excitations in the mean-field-slave boson approach, the physical electron creation operator for momentum $\hbar\boldsymbol{k}$ creates a spin particle with the same momentum. Also, as discussed in detail the next section, it is implied that the hopping term contains a transverse part of the exchange interaction proportional to xt.
(iii) Next is $\psi_i^{\sigma *}\tilde{f}^{\dagger}_{\sigma i}\tilde{f}_{\sigma i}u_i^{\dagger}$. This reflects the existence of non-condensed fermionic holes, even for the ground state. Involved is a convolution of the effective boson $\tilde{f}_{\sigma i}u_i^{\dagger}$ with the fermion $\tilde{f}^{\dagger}_{\sigma i}$ and this causes $\hat{c}^{\dagger}_{\uparrow i}$ to generate incoherent excitations. Via the hopping term the incoherent and coherent parts of the spectrum interact and the results have been compared, with success, with the ARPES spectrum of the cuprates [51]. Also via the hopping term (iii) produces a longitudinal part of the exchange proportional to xt. Both (ii) and (iii) come from the fact that the hole is rotated towards the spin particles.
(iv) In contrast the combination $\tilde{b}_i^{\dagger}\tilde{b}_i$ appears because the spin particles rotate towards the hole. In the hopping term this leads to an interaction between the new hole particles. This is not of particular importance to the present discussion since the next term has a bigger effect.
(v) In fact the last $O(|\psi_i^{\sigma}|^2)$ part leads to a $b_i^{\dagger}b_j^{\dagger} + b_i b_j$ interaction $\sim xt$ when the $(\tilde{f}_{\downarrow i}u_i^{\dagger})$ are contracted out of the Hamiltonian. This is identical to the interaction which occurs in the Bogoliubov theory of an interacting superfluid of Sect. 3.8, and as in that theory the factor of x reflects the condensate amplitude squared. While doping with these new particles will not be considered explicitly, it is this interaction which produces a pseudogap in the hole spectrum and which governs T_{c} as described in Sect. 3.8.

The condensate wave-functions $\psi_i^{\uparrow *}$, $\psi_i^{\uparrow}$, $\psi_i^{*\downarrow}$ and $\psi_i^{\downarrow}$ are determined by the minimal conditions, e.g.,

$$\frac{\partial}{\partial \psi_i^{\uparrow *}} E_0(\psi_i^{\uparrow *}, \psi_i^{\uparrow}, \psi_i^{*\downarrow}, \psi_i^{\downarrow}) = 0 \tag{3.126}$$

which yields a time independent, non-linear, Schrödinger's equation for the $\psi_i^{\uparrow}$. How the minimization of the ground state energy determines the ground state energy is taken up again in the next section.

3.17 The Effective Exchange for Coherent Doping

It has been shown that the *inter-band* kinetic energy t-term in the Hubbard model leads to an effective *anti-ferromagnetic* exchange interaction $J = 4t^2/U$. Evidently in the infinite U limit this interaction vanishes. However it is known [12] that under the same conditions, for the Anderson model of magnetic impurities, there is a second process which comes from the virtual doping on the impurity of holes from above the Fermi surface of the conduction band. In the present notation the exchange would be $\sim 4t^2/\mu$ and is also anti-ferromagnetic. (Here $\epsilon_d \to \mu$ is the distance of the magnetic level below the Fermi surface.) This result can be interpreted as being $\sim 2xt$ since $x \sim 2t/\mu$ is the probability of finding an hole on the magnetic site. On the other hand there is a theorem [52] due to Nagaosa which states, for a finite size $t - J$ model with $J = 0$ and a single doped hole the ground state is ferromagnetic, i.e., doping holes produces a ferromagnetic interaction. The difference between the two situations is the hole wave function. For the present problem, when this is determined by minimizing the total ground state energy, the result is a *ferromagnetic* interaction.

Consider the terms bi-linear in the spin particles which arose in $\mathcal{H}_t + \mathcal{H}_J$ of Sect. 3.13. These are

$$\sum_{\langle ij\rangle}\left[\frac{J}{2}f^{\dagger}_{\uparrow i}u_i u^{\dagger}_j f_{\uparrow j} - t f^{\dagger}_{\uparrow i}b_i f_{\uparrow j}b^{\dagger}_j\right] + \text{h.c.} \tag{3.127}$$

Naïvely it might be imagined that the bosons condense into the $k = 0$ state so that $b^{\dagger}_i b_j = x$. Although not accurate, see below, this, with (3.111) for $u_i u^{\dagger}_j$, clearly implies a ferromagnetic contribution $\sim xt$ to the exchange. This might be compared with (3.96) obtained with the same assumption and for which the $\tilde{\delta t}$ term has the opposite sign! The difference stems from mean-field approximation used to decouple the up and down spin fermions in Sect. 3.11 and which strongly violates the constraint.

A more accurate description of the hole condensate is based upon the the spin symmetric rotation of Fig. 3.9, and (3.121) which implies the original $b^{\dagger}_i \approx \psi^{\downarrow}_i f^{\dagger}_{\downarrow i}u_i + \psi^{\uparrow}_i f^{\dagger}_{\uparrow i}u_i$ where the tildes have been omitted on the right hand side for clarity and where it is assumed that there are no $\tilde{b}^{\dagger}_i$-particles. It should first be appreciated that $b^{\dagger}_i$ is replaced by an operator and *not* a complex number. A more useful expression is for the physical field $f^{\dagger}_{\uparrow i}b_i = \psi^{\downarrow}_i f^{\dagger}_{\uparrow i} + \psi^{\uparrow}_i p_i + \psi^{\uparrow}_i f^{\dagger}_{\uparrow i}f_{\uparrow i}u^{\dagger}_i + \psi^{\downarrow}_i f^{\dagger}_{\uparrow i}p_i u^{\dagger}_i$ here adapted to the, Sect. 3.13, Jordan-Wigner formulation. It is seen that sub-summing holes into the down spins, *via* $\psi^{\downarrow}_i$, implies a up spin physical operator $f^{\dagger}_{\uparrow i}b^{\dagger}_i$ creates a simple up spin $f^{\dagger}_{\uparrow i}$ excitation. This term alone reproduces the *naïve* result $-t\psi^{\downarrow}_i\psi^{*\downarrow}_j f^{\dagger}_{\uparrow i}f_{\uparrow i}$ given that $|\psi^{\downarrow}_i| = |\psi^{*\downarrow}_j| = x^{1/2}$. However, in addition, when holes are similarly rotated into the up spins the same operator causes a flux tube to be added to an up spin site and hence creates a longitudinal

p_i excitation through the off-diagonal matrix elements of the $u_i^\dagger$. Following the convention introduced near the end of Sect. 3.13, the last two terms are proportional to the diagonal elements of $u_i^\dagger$ and are necessary if the resulting interaction term is to have longitudinal parts, as it should in a rotationally invariant theory. The p_i have equal triplet and singlet character and cannot be ignored in the derivation of the effective Hamiltonian for the triplet $f_{\uparrow i}^\dagger$ excitations. Explicitly, account must be taken of a term $-t\psi_i^\uparrow p_i \psi_j^{*\uparrow} p_j^\dagger$ which, following Sect. 3.13, when projected unto the triplet sector becomes $-(t/4)\psi_i^\uparrow S^+ f_{\uparrow i}\psi_j^{*\uparrow} f_{\uparrow i}^\dagger S^- = +(t/2)\psi_i^\uparrow \psi_j^{*\uparrow} f_{\uparrow i}^\dagger f_{\uparrow i}$ commuting the operators and using the fact that the matrix elements of $S^\pm = \sqrt{2}$ for the triplet manifold. This cancels one-half of above naïve result. The physics behind this cancelation is simple: The interaction $S_i^+ S_j^-$ transfers an $S=1$ spin deviation from sites $j \to i$, while $f_{\uparrow i}^\dagger b_i^\dagger b_i^\dagger f_{\uparrow i}$ transfers only an $S = 1/2$ electron, i.e., is only half as efficient at transferring angular momentum per hopping event. The naïve result would implies equal efficiency.

Ignoring interactions which have no diagonal parts (3.127) is,

$$\sum_{\langle ij\rangle} \left[\frac{J}{2} u_i u_j^\dagger - \frac{t}{2}\psi_i^{*\downarrow}\psi_j^\downarrow \right] f_{\uparrow i}^\dagger f_{\uparrow j} + \frac{t}{2}\psi_j^\uparrow \psi_i^{*\uparrow} f_{\uparrow i}^\dagger f_{\uparrow i} u_i^\dagger u_j f_{\uparrow j}^\dagger f_{\uparrow j} + \text{h.c.} \quad (3.128)$$

which now contains not only a transverse part, but also a longitudinal interaction between parallel up spins. Since the SU(3) rotation causes up spins to be admixed into the holes, it is also necessary to account for the bi-linear part of the charge term: $+t\sum_{\langle ij\rangle} b_i^\dagger u_i u_j^\dagger b_j + \text{h.c.}$, which becomes

$$+\frac{t}{2}\sum_{\langle ij\rangle}[\psi_j^{*\uparrow}\psi_i^\uparrow f_{\uparrow i}^\dagger (u_i)^2 (u_j^\dagger)^2 f_{\uparrow j} - \psi_i^\downarrow \psi_j^{*\downarrow} f_{\uparrow i} f_{\uparrow i}^\dagger u_i u_j^\dagger f_{\uparrow j} f_{\uparrow j}^\dagger] + \text{h.c.} \quad (3.129)$$

and again generates both a second transverse and longitudinal contribution.

The last step is to determine the hole condensate wave functions ψ_i^σ by minimizing the ground state energy, i.e., using (3.126). It is assumed that this is equivalent to minimizing the longitudinal parts since these have diagonal matrix elements in the real space basis set. This assumption will examined in more detail in Sect. 3.19. This is accomplished by having, e.g., $\psi_i^\uparrow u_i = 1$ however this equates a function to (the diagonal part of) an operator, and this is only possible when the state on which the combination acts is clearly specified. In fact, both the wave functions $\psi_i^\uparrow$ and flux tube operators u_i which appear above are proxies for similar quantities which appear in $R|H\rangle$. Rotated *fermion* up spins have both of these appearing together in the ket $R|H\rangle$, and $\psi_i^\uparrow u_i = 1$ is appropriate, while for rotated *boson* down spins the $\psi_i^\downarrow$ in the ket must cancel against a $u_i^\dagger$ in the bra and so $\psi_i^\downarrow u_i^\dagger = 1$ is required. In the rotated operators it is always the combination $\psi_i^\sigma u_i^\dagger$ which appears and the equivalent statement is that this combinations must act to the right or left as the spin is up or down. Thus the "normal ordering" of the flux tube operators depends on the spin. The correct ordering occurs if the $\hat{c}_{\uparrow i}$ ($\hat{c}_{\downarrow i}$) are (anti-) normal ordered as they are in the above. With this

$$\psi_i^{\downarrow} = \mathrm{e}^{\mathrm{i}\phi}(x/2)^{1/2}u_i \quad \text{and} \quad \psi_i^{\uparrow} = \mathrm{e}^{\mathrm{i}\phi}(x/2)^{1/2}\mathrm{e}^{\mathrm{i}\boldsymbol{\pi}\cdot\boldsymbol{r}_i}u_i^{\dagger}. \tag{3.130}$$

In collecting together the hole induced *longitudinal* interactions it is to be observed that there is a contribution equal to the displayed terms which arises from the Hermitian conjugate part. The result is

$$-2xt(S_{iz}S_{jz} + (1/4)\hat{n}_i\hat{n}_j). \tag{3.131}$$

The sum of the transverse terms is

$$-(x/2)t(u_iu_j^{\dagger} + u_i^{\dagger}u_j)f_{\uparrow i}^{\dagger}f_{\uparrow j}. \tag{3.132}$$

Both the exchange of a up and down spin with a hole lead to the transport of spin *but* in opposite directions and this physics is represented by the manifestly real combination $(1/2)(u_iu_j^{\dagger}+u_i^{\dagger}u_j)$ of flux tube operators. For the flux state with $u_iu_j^{\dagger} = \frac{1}{\sqrt{2}}[1 \pm \mathrm{i}\mathrm{e}^{\mathrm{i}\boldsymbol{\pi}\cdot\boldsymbol{r}_i}]$ this reduces to a factor of $1/\sqrt{2}$. Physically the counter rotating up and down spin currents cannot couple to the charge current and visa versa. The net hole induced transverse, i.e., hopping, term is therefore $-(xt/\sqrt{2})f_{\uparrow i}^{\dagger}f_{\uparrow j}$. For the longitudinal terms, which lead, e.g., to potential magnetic ordering it is simply the case that the effective exchange is

$$J_{\text{eff}} = J - 2xt \tag{3.133}$$

however there is no transverse contribution to the flux state order parameter, so that while, e.g., in (3.111), $[J/\sqrt{2}]\gamma_{\boldsymbol{k}} \Rightarrow [J_{\text{eff}}/\sqrt{2}]\gamma_{\boldsymbol{k}}$ it remains the case that $\Delta_0 = J/\sqrt{2}$.

There is a somewhat hidden quantum critical point for x_c for which J_{eff} passes through zero, i.e., when $x_c = J/2t \sim 0.16$, or approximately optimal doping for the cuprate HTSC materials. At this point the longitudinal term which induces anti-ferromagnetic correlations goes to zero while the fluctuations effects represented by a finite Δ_0 do not. The presence of pairing somewhat masks the associated Fermi surface change, since the Fermi edge in momentum space is not defined (except near $\boldsymbol{k} = \pm\boldsymbol{\pi}/2, \pm\boldsymbol{\pi}/2$).

The fact that near to optimal doping the dispersion term $[J_{\text{eff}}/\sqrt{2}]\gamma_{\boldsymbol{k}}$ for the spin-particles goes to zero implies that the coherent part of the physical Green's function discussed at the end of Sect. 3.13 cannot possibly correspond to the dispersive band near the Fermi surface as seen in ARPES. As will be shown immediately below, superconductivity, signaled by a finite BCS order parameter, corresponds to the existence of condensed holes and this condensate part of Green's function is to be identified with the coherence peaks which appear only in the superconductive state. The above mentioned dispersive feature must be identified with the second particle-hole part (iii) also mentioned at the end of Sect. 3.13.

When there is un-equal doping of the two spin directions the factor involved in hopping from site i to j or the inverse becomes $(|\psi^{\uparrow}|^2u_i^{\dagger}u_j + |\psi^{\downarrow}|^2u_j^{\dagger}u_i)$, where, e.g., $|\psi^{\downarrow}|^2$ is the uniform magnitude of $|\psi_i^{\downarrow}|^2$. Thus if say only the down spin direction is doped this becomes, e.g., $u_i^{\dagger}u_j$ and the total

$J \to (J - 2xt)$, pairing is absent, and x_c becomes a simple quantum critical point with a divergent density of states. Some evidence consistent with such a quantum critical point exists when superconductivity is suppressed by a large field [53].

Finally for this section, having determined the condensate wave function it is easy to evaluate the BCS amplitude $\langle \hat{c}_{-\boldsymbol{k}\downarrow}\hat{c}_{\boldsymbol{k}\uparrow}\rangle$. In real space what is needed is the ground state expectation value of $\langle \hat{c}_{\downarrow i}\hat{c}_{\uparrow j}\rangle$, i.e.,, $\langle b_i^\dagger f_{\downarrow i} b_j^\dagger f_{\uparrow j}\rangle$. Using the inverse of the expressions for the rotated operators, (3.120) and (3.121), this contains,

$$\langle \psi_i^\uparrow f_{\uparrow i}^\dagger u_i f_{\downarrow i}\psi_j^\downarrow f_{\downarrow j}^\dagger u_j f_{\uparrow j}\rangle = \langle \psi_i^\uparrow f_{\uparrow i}^\dagger (u_i)^2 (u_i^\dagger f_{\downarrow i})\psi_j^\downarrow (f_{\downarrow j}^\dagger u_j) f_{\uparrow j}\rangle . \tag{3.134}$$

In using $\psi_i^\uparrow = x^{1/2}\mathrm{e}^{\mathrm{i}\phi}\mathrm{e}^{\mathrm{i}\boldsymbol{\pi}\cdot\boldsymbol{r}_i}u_i^\dagger$ and $\psi_i^\downarrow = x^{1/2}\mathrm{e}^{\mathrm{i}\phi}u_i$ it is important to note that $\psi_j^\downarrow$ eliminates a u_j because of the normal ordering convention for down spin sites. The result is $\langle \hat{c}_{\downarrow i}\hat{c}_{\uparrow j}\rangle = x\mathrm{e}^{2\mathrm{i}\phi}\mathrm{e}^{\mathrm{i}\boldsymbol{\pi}\cdot\boldsymbol{r}_i}\langle f_{\uparrow i}^\dagger u_i u_j^\dagger f_{\uparrow j}\rangle$. For the flux state $u_i u_j^\dagger = \frac{1}{\sqrt{2}}(1 \pm \mathrm{e}^{\mathrm{i}\boldsymbol{\pi}\cdot\boldsymbol{r}_i})$ and it follows that,

$$\langle \hat{c}_{-\boldsymbol{k}\downarrow}\hat{c}_{\boldsymbol{k}\uparrow}\rangle \propto x\mathrm{e}^{2\mathrm{i}\phi}\langle f_{\uparrow\boldsymbol{k}}^\dagger f_{\uparrow\boldsymbol{k}+\boldsymbol{\pi}}\rangle , \tag{3.135}$$

i.e., this Fourier transform just picks out the "order parameter" for the flux state of Sect. 3.7. The conclusion to be drawn is that a spin flux state plus a coherent hole condensate is equivalent to a BCS physical electron pair, with all the associated consequences [50]. This extends the result that RVB-BCS and flux state pairing are equivalent at half-filling to states which coherent hole doping. Since the holes are sub-summed into the spin degrees of freedom it is evident that results valid half-filling are extended into the doped region of the phase diagram.

The order parameter for the condensed holes is $\mathrm{e}^{\mathrm{i}\phi}x^{1/2}$ which has a U(1) symmetry corresponding to the fact that the global phase of this hole wave function is arbitrary. This can be decomposed into two real order parameters $n_1 = \cos\phi$ and $n_5 = \sin\phi$ corresponding to the real and imaginary parts of $\mathrm{e}^{\mathrm{i}\phi}x^{1/2}$ and defined so that the two dimensional vector (n_1, n_5) has unit length. (The odd notation has to do with SO(5) theory introduced in the next section .)

3.18 SO(5) Theory

It has been noted that the symmetry group for the Néel vector $\boldsymbol{N}$ is SO(3) while that associated with the BCS order parameter is U(1). It follows that the symmetry group which accommodates both order parameters is SO(3) $\times$ U(1). The combined order parameter has five components. With some convention, the two components n_1 and n_5 are those associated with the U(1) symmetry and superconductivity. The remaining n_2, n_3 and n_4 are the three components of $\boldsymbol{N}$. The minimal group which can accommodate the sub-symmetry SO(3) $\times$ U(1) is SO(5). A SO(5) rotation not only turns magnetism into superconductivity but also adds doping to the Mott insulator. It is implicit in the theory that the hole concentration is proportional to the magnitude of the superconductive component of the order parameter. Thus if superconductivity is turned off, by a SO(5) rotation, the resulting state is an insulator. It seems to be indeed the case that applying a large magnetic field induces anti-ferromagnetic correlations and that the resulting non-superconductive state is an insulator.

Following S.-C. Zhang [9], it is observed that the *total* spin operators $\boldsymbol{S}$ are the generators of rotations of $\boldsymbol{N}$, i.e., form the Lie algebra for SO(3). It is charge operator Q which performs rotations in the superconductive space and therefore is the unique element of the Lie algebra for the U(1). These operators must commute with H. In the five dimensional SO(5) space there are many more additional rotation operators which rotate $\boldsymbol{N}$ towards the superconducting state. These new π-operators are all important. If they commute with H then there is an exact SO(5) symmetry. This can only be the case for some special value of the chemical potential μ, taken to be zero, since to within a constant $-\mu\hat{N} = -\mu Q$ and the operators of the Lie algebra do not commute, see below. The predictive powers of the theory remain good if the weaker condition $[H, \pi] = \omega\pi$ is satisfied and implies that the π-operators reflect collective modes of the model. Physically Demler and Zhang [54] associate this mode with the neutron resonance [55] which has an energy $\sim$ 40meV at optimal doping. This energy is a measure of the difference in energy between the anti-ferromagnetic and superconductive states and such a small energy π mode implies that they are indeed close in energy.

Depending upon in which direction the new five dimensional order parameter points, the system is either superconducting or anti-ferromagnetic or a mixture of the two. Experimentally the anti-ferromagnetic state is stable in the absence of doping but when chemical potential μ is reduced superconductivity becomes the preferred phase. There is therefore some critical μ_c which brings the two phases to the same energy. It should be imagined that, the chemical potential plays the same role as the magnetic field in the spin-flop of an anti-ferromagnetic. In this case the smaller SO(3) symmetry is broken by the existence of an easy axis energy $-A{S_z}^2$ which aligns the Néel vector, $\boldsymbol{N}$, along the S_z axis. The applied field corresponds to hS_z, and when $h_c = \sqrt{JA}$, $\boldsymbol{N}$ flops into the $x-y$-plane. A finite value of A implies that the Goldstone bosons, i.e., the spin-waves, have finite energy even in the limit $\boldsymbol{q} \to 0$. The spin-flop transition is signaled by the fact that one of these collective modes reaches zero energy for h_c. Within the SO(5) theory, the "external field" $-\mu\hat{N}$ causes a superspin-flop in which $\boldsymbol{n}$ flops from the anti-ferromagnetic SO(3) sphere to the U(1) superconductive plane.

In an n dimensional space a rotation turns the $\boldsymbol{n}$-vector in some plane. There are therefore as many independent rotations as there are planes and the number is $n(n-1)/2$. There are therefore ten independent rotation operators for SO(5). In addition to these ten operators there are the five operators of which order parameters $\boldsymbol{n}$ is the expectation value. As is the case for the SO(3) anti-ferromagnetic problem, the rotation operators are independent from those which define $\boldsymbol{n}$. In the scheme envisaged by Zhang [9] there are therefore 15 independent operators. The present development makes use of the auxiliary particles methods developed in earlier sections, this in order to permit the reader to have an overview of the different approaches in a common notation.

The three operators corresponding to the magnetic order parameter can be written as,

$$n_2 = \sum_{\alpha,\alpha' i} \mathrm{e}^{\mathrm{i}\boldsymbol{\pi}\cdot\boldsymbol{r}_i} f_{\alpha i}^{\dagger} \sigma_{\alpha\alpha'}^{x} f_{\alpha' i} \tag{3.136}$$

and where n_3 and n_4 are given by the same expression with $\sigma_{\alpha\alpha'}^{x} \to \sigma_{\alpha\alpha'}^{y}$ and $\sigma_{\alpha\alpha'}^{x} \to \sigma_{\alpha\alpha'}^{z}$, respectively. In connection with superconductivity Zhang [9] defines the equivalent of,

$$\Delta^{\dagger} = \frac{1}{4} \sum_{\langle ij \rangle} (\pm) b_j^{\dagger} f_{\uparrow j} b_i^{\dagger} f_{\downarrow i} \tag{3.137}$$

where this order parameter has d-symmetry corresponding to, e.g., the sign being positive for vertical bonds and negative for the horizontal equivalents. Then,

$$n_1 = (\Delta^{\dagger} + \Delta) \quad \text{and} \quad n_5 = -\mathrm{i}(\Delta^{\dagger} + \Delta), \tag{3.138}$$

which have spin singlet symmetry. In this notation the spin rotation operators are,

$$S_{\gamma} = \sum_{\alpha,\alpha' i} f_{\alpha i}^{\dagger} \sigma_{\alpha\alpha'}^{\gamma} f_{\alpha' i} \tag{3.139}$$

while the less familiar operator which causes $n_1 \leftrightarrow n_5$ is

$$Q = \sum_{i} b_i^{\dagger} b_i \tag{3.140}$$

which is the operator for the total charge relative to half-filling. It is easy to check that $[\Delta^{\dagger}, Q] = \Delta^{\dagger}$ but $[\Delta, Q] = -\Delta$.

These make up 4 rotation operators and so there are 6 new such operators which are required to complete SO(5). Such operators can be readily found by requiring that it rotate n_2, n_3 and n_4 into n_1 and n_5. Adapting Zhang [9], these are given in terms of,

$$\pi_{\alpha}^{\dagger} = \sum_{\langle ij \rangle \alpha,\alpha'} (\pm) \mathrm{e}^{\mathrm{i}\boldsymbol{\pi}\cdot\boldsymbol{r}_i} b_j^{\dagger} f_{\alpha j} (\sigma^{\alpha} \sigma^{y})_{\alpha\alpha'} b_i^{\dagger} f_{\alpha' i}. \tag{3.141}$$

The SO(5) Lie algebra then corresponds to the ten generators such that,

$$[L_{ab}, L_{cd}] = \mathrm{i}(\delta_{ac} L_{bd} - \delta_{ad} L_{bc} - \delta_{bc} L_{ad} + \delta_{bd} L_{ac}) \tag{3.142}$$

where the L_{ab} correspond to the matrix,

$$\begin{pmatrix} 0 & & & & \\ \pi_x^\dagger + \pi_x^\dagger & 0, & & & \\ \pi_y^\dagger + \pi_y^\dagger & -S_z & 0 & & \\ \pi_z^\dagger + \pi_z^\dagger & S_y & -S_x & 0 & \\ Q & -\mathrm{i}(\pi_x^\dagger - \pi_x^\dagger) & -\mathrm{i}(\pi_y^\dagger - \pi_y^\dagger) & -\mathrm{i}(\pi_z^\dagger - \pi_z^\dagger) & 0 \end{pmatrix}, \tag{3.143}$$

where the value of the matrix elements on the upper right triangle are determined by antisymmetry, $L_{ab} = -L_{ba}$. Explicitly evaluation of these commutation rules in the continuum limit shows that the expressions for the L_{ab} obey this SO(5) algebra. This limit involves replacing the average $\langle(\gamma_{\boldsymbol{k}})^2\rangle = 1$ where $\gamma_{\boldsymbol{k}}$ is defined in Sect. 3.1.

An infinitesimal rotation $d\theta$ of say the component i towards j corresponds to $R = \mathrm{e}^{\mathrm{i}d\theta L_{ij}}$, i.e., for the component $Rn_iR^{-1} \approx n_i + \mathrm{i}d\theta[R, n_i] = n_i + d\theta n_j$ which implies in general that the L_{ab} obey,

$$[L_{ab}, n_c] = -\mathrm{i}\delta_{bc}n_a + \mathrm{i}\delta_{ac}n_b \tag{3.144}$$

again a relationship which is explicitly valid in the continuum limit.

The SO(5) symmetry is only exact if,

$$[\mathcal{H}, L_{ab}] = 0 \tag{3.145}$$

for all L_{ab}. Using the explicit definitions above and the auxiliary particle version of say the Hubbard $\mathcal{H}$, it is easy to see that $[\mathcal{H}, \pi_\alpha^\dagger] \neq 0$. *However* Demler and Zhang [54] show within a certain approximation scheme,

$$[\mathcal{H}, \pi_\alpha^\dagger] = \omega\pi_\alpha^\dagger. \tag{3.146}$$

When this equation is valid it expresses the fact that $\pi_\alpha^\dagger$ is an operator which creates a well defined collective excitation. Numerical work by Meixner and Hanke is cited [56] as showing that correlation function for the π-operators has indeed a sharp collective peak. This supports the the claim that the operator $\pi_\alpha^\dagger$ while not being exactly *the* operator for this mode contains it with a large weight. A weaker requirement is then that this collective mode equation is satisfied if the Hilbert space is sufficiently reduced. The minimum such space would be spanned by the mean-field ground states of $\mathcal{H}$. In this case the π-mode would be identified as involving transitions in which the condensed particle in the anti-ferromagnetic state, i.e., f-particle-holes-pairs with momentum π are excited to, or from, d-wave singlet superconductive pairs. Given that such a mode exists implies that the SO(5) Casimir operator,

$$C = \sum_{a<b} {L_{ab}}^2 \tag{3.147}$$

does commute with $\mathcal{H}$ at least in this restricted space. This permits the ground and excited states to be classified by the eigenvalue of C. Since the ground state of the anti-ferromagnetic problem has $C = 0$ it follows, in a suitably restricted Hilbert space, the effective Hamiltonian in the absence of doping is the definite positive, $\frac{1}{2\chi}\sum_{a<b} L_{ab}{}^2$, to which might be added gradient terms which would reflect the slow variation of the order parameters. That $(1/\chi) \neq 0$ follows from the fact that χ is the spin susceptibility when SO(5) is exact. When this symmetry is *not* exact $\mathcal{H} = \sum_{a<b} \frac{1}{2\chi_{ab}} L_{ab}{}^2$, but since O(3)×U(1) is a good symmetry one has that $\chi_{ab} = \chi_s$ when a and b involves only the spin labels $2, 3, 4$, is χ_c, the charge susceptibility, when these are $1, 5$ and otherwise is a new χ_π. The chemical potential term $-\mu\hat{N} = -2\mu L_{15}$ so that the effective Hamiltonian is,

$$\mathcal{H} = \sum_{a<b} \frac{1}{2\chi_{ab}} L_{ab}{}^2 - 2\mu L_{15} + V(\boldsymbol{n}) + \text{gradient terms}, \tag{3.148}$$

where $V(\boldsymbol{n})$ represents terms which explicitly break the SO(5) symmetry. The first $L_{ab}{}^2$ term reflects the kinetic energy for bosonic excitations associated with the order parameter. The equation of motion for $\boldsymbol{n}$ is equivalent to the Lagrangian,

$$\mathcal{L} = \frac{1}{2} \sum_{a<b} \chi_{ab} \left(n_a (i\partial_t n_b - \mathrm{i} \sum_c B_{ac} n_c) \right)^2 + V(\boldsymbol{n}) + \text{gradient terms}, \tag{3.149}$$

where the only finite effective field is $B_{15} = -B_{51} = 2\mu$. In the usual BCS theory, a finite chemical potential implies that the order parameter has a time dependent phase $\mathrm{e}^{2\mathrm{i}\mu t}$ which in the present language implies that the vector $(n_1, n_5) = n_{15}(\cos 2\mu t, \sin 2\mu t)$ will process. Given this time dependence the kinetic energy produces an energy shift $-(\chi_c/2)(2\mu)^2 n_{15}{}^2$. Because the π-operators cause a coupling between these superspin components and the spin degrees of freedom, the corresponding time dependent motion is more complicated, *however* the same phase factor $\mathrm{e}^{2\mathrm{i}\mu t}$ occurs everywhere and it is not hard to appreciate that the equivalent energy shift is,

$$V_\mu = -\frac{(2\mu)^2}{2}(n_1{}^2 + n_5{}^2)\left(\chi_c(n_1{}^2 + n_5{}^2) + \chi_\pi(n_2{}^2 + n_3{}^2 + n_4{}^2)\right) \tag{3.150}$$

which adds to the explicit symmetry breaking term $V(\boldsymbol{n})$.

The effective potential $V_{\text{eff}} = V(\boldsymbol{n}) + V_\mu(\boldsymbol{n})$ determines the equilibrium direction of the superspin $\boldsymbol{n}$ and hence the phase diagram. The mean-field theory for an anti-ferromagnet, without quantum fluctuations, schematically corresponds to a free-energy function of the form,

$$F = a'(T - T_{\text{MF}})|\boldsymbol{n}|^2 + b|\boldsymbol{n}|^4 \tag{3.151}$$

where T_{MF} is the mean-field transition temperature. Given exact SO(5) symmetry the only generalization needed is that $\boldsymbol{n}$ is now the superspin vector rather than just the Néel vector. Well below T_{MF}, the magnitude $|\boldsymbol{n}| = \sqrt{2a'T_{\text{MF}}/b}$ is fixed and it is envisaged that $\boldsymbol{n}$ is redefined so that it is of unit length. If the deviations from symmetry have the $|\boldsymbol{n}|^2$ terms as the origin then the deviation from SO(5) is reflected by,

$$V(\boldsymbol{n}) = -\frac{g}{2}(n_2{}^2 + n_3{}^2 + n_4{}^2). \tag{3.152}$$

Since, when $x = 0$, the anti-ferromagnetic phase is stable, it must be that $g > 0$. (There must also be some weak part of the potential which causes the Néel vector (n_2, n_3, n_4) to assume some preferred direction in real space.) In contrast $V_\mu(\boldsymbol{n})$ has a maximum for the anti-ferromagnetic direction and a minimum when the order parameter for that phase is a minimum. In fact if $\chi_c = \chi_\pi$, then for a critical value,

$$\mu_c = \frac{1}{2}\sqrt{\frac{g}{\chi}}, \tag{3.153}$$

$V_{\text{eff}} = -(g/2)|\boldsymbol{n}|^2$ and there is SO(5) symmetry. For $\mu < \mu_c$ the system is anti-ferromagnetic with no holes present while for $\mu > \mu_c$ the system is superconducting and has a finite hole density. This density has a first order phase jump at $\mu = \mu_c$. If considered as a function of the hole doping x and if the SO(5) symmetry really is exact at $\mu = \mu_c$ then $\boldsymbol{n}$ would undergo a free rotation at constant μ with the coexistence of superconductivity and magnetism. In Sect. 3.12 is was shown that the existence of magnetism implies a finite gap in the quasi-particle spectrum and so this free rotation would have a definite experimental signature for the underdoped region. When the SO(5) symmetry is not exact at $\mu = \mu_c$, then the system will phase separate into hole rich and hole poor regions, see Fig. 3.10.

3.19 SO(5) and SU(3) Theories

As S.-C. Zhang [9] points out, the superconducting state of the SO(5) approach as described in the previous section, is not equivalent to that derived

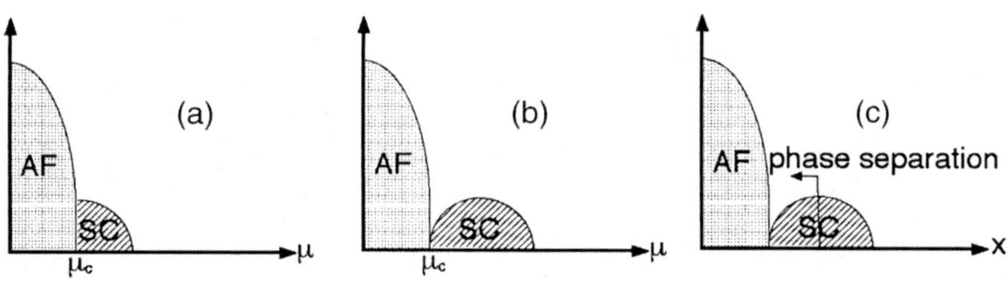

Fig. 3.10. Shown are the two most likely phase diagrams within the SO(5) theory. The superspin flop occurs for a critical value μ_c. For case (**a**) the transition temperature remains finite at the flop, while in (**b**) the transition temperature is driven to zero due to fluctuations. These are plotted as a function of μ the chemical potential. When plotted as a function of concentration, x, and for (**a**) and in the underdoped region there is either phase separation with hole rich and hole poor regions as indicated in (**c**) *or* the superspin undergoes a free rotation as a function of x in this regime with a co-existence of superconductive and magnetic order parameters. Experimentally this would be reflected by a gap in the electronic density of states for underdoped materials. When this diagram is applied to the flux state, RVB, theory of the next section (**a**) and (**c**) always apply, i.e., a free rotation never occurs.

from Anderson's RVB construction and by extension to constraint satisfying flux-state approach presented in the previous few sections and which is of the same class. However these statements are based upon the identification of the π rotation operators of the theory and, in fact, when these are redefined the current particular flavor of the RVB theory, presented in Sects. 3.7, 3.13–3.16 of this chapter does map onto the SO(5) scheme contained in the SU(3) theory. There are pre-existing (spin) pairs which reflect the existence Landau levels and the flux state order parameter at all temperatures. Superconductivity occurs when these spin pairs are given charge coherence with the appearance of a hole condensate. The U(1) symmetry must now be associated with this hole condensation and no longer with the physical BCS order parameter. The commutator $[H, \pi] = \omega\pi$, gains an extra significance since $[H, \pi] = 0$ is shown to imply that the lowest hole bosonic level lies at the chemical potential. The value of ω in the anti-ferromagnetic state is therefore the energy of the lower Hubbard band relative to the chemical potential.

Starting from the spin flux state, the anti-ferromagnetic state is obtained by performing a SU(2) Bogoliubov rotation of the spin $f^\dagger_{\uparrow i}$-particles involving momenta $\boldsymbol{k}$ and $\boldsymbol{k}+\boldsymbol{\pi}$, while the superconductive state is obtained by a SU(3) rotating or the $\boldsymbol{k}$ spin towards the hole $b^\dagger_i$-particle. The U(1) invariance corresponds to the overall phase of the hole wave functions ψ^σ_i. Unlike the BCS order parameter in the previous section the hole condensate wave function is local. Specifically,

$$\tilde{f}^\dagger_{\uparrow i} = \left(1 - \frac{1}{2}(m^2 - \psi^{*\uparrow}_i \psi^\uparrow_i)\right) f^\dagger_{\uparrow i} + m\mathrm{e}^{\mathrm{i}\boldsymbol{\pi}\cdot\boldsymbol{r}_i} f^\dagger_{\uparrow i} + \psi^\uparrow_i b^\dagger_i u_i \tag{3.154}$$

where m is the expectation value of the n_4,

$$n_4 = \sum_i \mathrm{e}^{\mathrm{i}\boldsymbol{\pi}\cdot\boldsymbol{r}_i} f^\dagger_{\uparrow i} f_{\uparrow i} \quad \text{while} \quad \Delta^\dagger = \sum_i (\psi^\uparrow_i / x^{1/2}) b^\dagger_i u_i f_{\uparrow i} \tag{3.155}$$

reflects superconductivity. With the new, simpler, definition of $\Delta^\dagger$ comes a new definition of, e.g.,

$$\pi^\dagger_z = \sum_i \pi^\dagger_{zi} = \sum_i \mathrm{e}^{-\mathrm{i}\boldsymbol{\pi}\cdot\boldsymbol{r}_i} (\psi^\uparrow_i / x^{1/2}) b^\dagger_i u_i f_{\uparrow i}, \tag{3.156}$$

where the wave-function $\psi^\uparrow_j = \mathrm{e}^{\mathrm{i}\boldsymbol{\pi}\cdot\boldsymbol{r}_i} x^{1/2} \mathrm{e}^{\mathrm{i}\phi} u_j$ while $\psi^\downarrow_j = x^{1/2}\mathrm{e}^{\mathrm{i}\phi} u^\dagger_j$. This π mode has a strength x in neutron scattering a $\boldsymbol{k} = \boldsymbol{\pi}$, since given $S_{z\boldsymbol{\pi}} = \sum_i \mathrm{e}^{-\mathrm{i}\boldsymbol{\pi}\cdot\boldsymbol{r}_i}(f^\dagger_{\uparrow i} f_{\uparrow i} - 1/2)$ and using (3.120) for the rotated f-operators, this contains $x^{1/2}(\pi^\dagger_z + \pi_z)$. It is evident that, in general, $\pi^\dagger_z$ does *not* commute exactly with $\mathcal{H}$. Explicitly, such a commutator

$$\sum_i [\mathcal{H}, \pi^\dagger_{zi}] = \sum_i \mathrm{e}^{-\mathrm{i}\boldsymbol{\pi}\cdot\boldsymbol{r}_i} (\psi^\uparrow_i / x^{1/2})([\mathcal{H}, b^\dagger_i] u_i f_{\uparrow i} + b^\dagger_i [\mathcal{H}, u_i f_{\uparrow i}]). \tag{3.157}$$

As was discussed in Sect. 3.16, the term proportional to $[\mathcal{H}, b^\dagger_i] f_{\uparrow i}$ substitutes $f_{\uparrow i}$ into $\mathcal{H}$ at every location where b_i occurs, while the term involving

$[\mathcal{H}, u_i f_{\uparrow i}]$ replaces $f_{\uparrow i}^\dagger$ with $b_i^\dagger$. The approach assumes that there are no $b_i^\dagger$ particles and so the latter commutator can be dropped. If follows that the SO(5) symmetry is good if,

$$[\mathcal{H}, b_i^\dagger] = -\mathrm{i}\hbar \frac{\partial b_i^\dagger}{\partial t} = 0. \tag{3.158}$$

Schrödinger's equation for $\psi_i^\dagger$ is then $[\mathcal{H}, b_i^\dagger]|G\rangle = -\mathrm{i}\hbar \frac{\partial b_i^\dagger}{\partial t}|G\rangle$. The physical requirement for SO(5) symmetry is therefore that the energy associated with the Bose-Einstein condensation lies at the chemical potential. That this be the case directly implies that the coefficient of x in the ground state expectation value of $\mathcal{H}$ be zero. From (3.131) and (3.132), this leads to

$$\mu = t[1 - m^2 + \sqrt{2}(\langle f_{\uparrow i}^\dagger f_{\uparrow j}\rangle + \text{c.c.})]. \tag{3.159}$$

For the anti-ferromagnet, see Sect. 3.15, both m^2 and $\sqrt{2}(\langle f_{\uparrow i}^\dagger f_{\uparrow j}\rangle + \text{c.c.}) \sim 1/3$. This implies a small and positive value of the chemical potential, μ_i, which then locates the top of the lower Hubbard band for the insulator. For optimal doping, $x = x_c$, the value of J_{eff} as defined by (3.133) is zero and m^2 and $\langle f_{\uparrow i}^\dagger f_{\uparrow j}\rangle \sim 0$. As a result μ *increases* upon doping the anti-ferromagnetic insulator, to have a value μ_c on superconducting plane, and implies that the system is *unstable* to doping. Phase separation will occur as described, Fig. 3.10, in Sect. 3.15. Since the anti-ferromagnetic and superconducting states have the same energy when $x = x_c$, the lowest Bose level is well above the chemical potential in the anti-ferromagnet.

Specifically the Fermi surface of the superconductor is at $\mu_c - \mu_i$ above the lower Hubbard band, i.e., doping adds states *inside* the insulator gap as seems to be indicated by a number of experiments, see e.g., [40]. From (3.159), as an estimate, $\mu_c - \mu_i \sim 0.2t$. In a tunneling experiment $\mu_c - \mu_i$ is the height of tunneling barrier for hole tunneling. It is a remarkable fact, [57], that holes cannot tunnel efficiently through even a single layer of the anti-ferromagnetic phase consistent with a large value of $\mu_c - \mu_i$.

Additional insight into the nature of the phase diagram can be obtained using symmetry considerations. Since the holes are sub-summed into the spin degrees of freedom, the SU(3) $\to$ SU(2) group associated with the spin space. It is customary to think of the expectation value of $J\boldsymbol{S}_i \cdot \boldsymbol{S}_j$ as $JS^2\langle\cos\theta\rangle$ where θ is the angle between the spin at sites i and j and the angular bracket implies an average over quantum fluctuations. Mathematically this can be can be obtained by starting with the ferromagnetic state with $J\langle \boldsymbol{S}_i \cdot \boldsymbol{S}_j\rangle = JS^2$ and rotating the axis of quantization of, say, the spin at site j. The fact that there are *two* f-operators in the definition of $\boldsymbol{S}_j$ implies a $\cos\theta$ despite the fact that the SU(2) rotations, see Sect. 3.1, involve the half angle $\theta/2$. For this same ferromagnetic state and a small hole concentration $-t\langle (f_{\uparrow i}^\dagger b_i b_j^\dagger f_{\uparrow j} + \text{h.c.})\rangle = -2xt$, but now there is only *one* f-operator so $-t\langle (f_{\uparrow i}^\dagger b_i b_j^\dagger f_{\uparrow j} + \text{h.c.})\rangle = -2xt\langle\cos\theta/2\rangle$. The Holstein-Primakoff theory of Sect. (3.4) suggests that correct ground state is given when fluctuations are ignored, even for $S =$

1/2. The spin state is then determined by minimizing $(J/4)\cos\theta - 2xt\cos\theta$. If it is assumed that $\theta_i = \boldsymbol{Q}\cdot\boldsymbol{r}_i$, $\theta = \theta_j - \theta_i$, then $\boldsymbol{Q} = \boldsymbol{\pi}$ only for $x = 0$ and otherwise shifts towards zero. A ferromagnetic state results for $x_c = J/2t$ which is the same critical concentration as obtained earlier. It is easy to check that, in fact, the phase separated state has lower energy which, if the Coulomb repulsion is accounted for, is suggestive of stripes. Spin polarons with self trapped holes and holes localized on anti-ferromagnetic Bloch walls are also possible. The different transformation properties of the t and J terms under SU(2) is also reflected by the fact that the flux state order parameter Δ_0 remains unaffected by doping. Intrinsically the "fluctuations" associated Δ_0, i.e., superconductivity, dominate near this critical point at $x = x_c$.

In Sect. 3.16 it was shown that the SU(3) hole rotation introduces a hole pairing term

$$\approx 2xt\sum_{\langle ij\rangle}(b_i^\dagger b_j^\dagger + b_i b_j)$$

and which is exactly of the form studied in connection with Bogoliubov theory of a superfluid, Sect. 3.8. This adds a further μ shift for the superconductor but more importantly induces a pseudo-gap, Fig. 3.5, in the charge spectrum and sets the scale for the hole condensation temperature, and hence T_c. This is large $2xt \sim 1000$K, *however* this current approach assumes that all the holes can be condensed resulting in a condensed faction $c \sim x/2$, but in a phase separated state $c < x/2$. Experimentally the condensed faction c varies from HTSC material to material. There is an accumulation of evidence in favor of the relationship $k_B T \propto c$.

3.20 Gossamer Superconductivity

Laughlin [11] has suggested the idea of Gossamer or "thin" superconductivity. The Gutzwiller projection, eliminates all doubly occupied states and renders a half filled BCS state an insulator. Laughlin has observed that *if* the Gutzwiller projector had an inverse one could in fact deduce the Hamiltonian for which the projected eigenstates are eigenvalues and that this might be considered *the* model for the cuprate problem. Instead of considering the limiting projection Laughlin considers an operator which simply reduces the weight of the doubly occupied states. By varying the parameter in this operator it is possible to construct a Hamiltonian which adiabatically connects the standard BCS Hamiltonian ($\mathcal{H} = \sum_{\boldsymbol{k}}\mathcal{H}_{\boldsymbol{k}}$ with $\mathcal{H}_{\boldsymbol{k}}$ is given by (3.4)) with one for a highly correlated system.

The Laughlin projection operator is,

$$\mathcal{P}_\alpha = \prod_i^N z_0{}^{(n_{j\uparrow}+n_{j\downarrow}/2)}(1-\alpha_0 n_{j\uparrow}n_{j\downarrow}) \tag{3.160}$$

and the state on which it acts is simply the BCS ground state,

$$|\Phi\rangle = \prod_{\boldsymbol{k}} (u_{\boldsymbol{k}} + v_{\boldsymbol{k}} c^{\dagger}_{\boldsymbol{k}\uparrow} c^{\dagger}_{-\boldsymbol{k}\downarrow})|0\rangle \tag{3.161}$$

where $|0\rangle$ is the true vacuum without any particles. The state,

$$|\tilde{\Phi}\rangle \equiv \mathcal{P}_{\alpha}|\Phi\rangle \tag{3.162}$$

evidently reduces to the original $|\Phi\rangle$ when $\alpha_0 = 0$, given that z_0 is chosen to normalize the state, while when $\alpha_0 \to 1$ this becomes a Gutzwiller projection which gives zero weight to doubly occupied sites. It is important that $\mathcal{P}_{\alpha}$ has an inverse,

$$\mathcal{P}_{\alpha}^{-1} = \prod_{i}^{N} {z_0}^{-(n_{j\uparrow}+n_{j\downarrow}/2)}(1 + \beta_0 n_{j\uparrow} n_{j\downarrow}), \tag{3.163}$$

where $\beta_0 = \alpha_0/(1-\alpha_0)$. Whatever the value of this parameter, x the doping relative to half-filling is given by,

$$\frac{1}{N}\sum_{\boldsymbol{k}} {v_{\boldsymbol{k}}}^2 = 1 - \frac{1}{N}\sum_{\boldsymbol{k}} {u_{\boldsymbol{k}}}^2 = \frac{1-x}{2}. \tag{3.164}$$

Taking results from Sect. 3.1, the BCS problem in diagonal form is,

$$\mathcal{H} = \sum_{\boldsymbol{k}} E_{\boldsymbol{k}} (\alpha^{\dagger}_{\boldsymbol{k}} \alpha_{\boldsymbol{k}} + \beta^{\dagger}_{\boldsymbol{k}} \beta_{\boldsymbol{k}}) \tag{3.165}$$

after dropping a constant, and where,

$$\alpha_{\boldsymbol{k}} = u_{\boldsymbol{k}} c_{\boldsymbol{k}\uparrow} - v_{\boldsymbol{k}} c^{\dagger}_{-\boldsymbol{k}\downarrow} \quad \text{and} \quad \beta_{\boldsymbol{k}} = u_{\boldsymbol{k}} c_{-\boldsymbol{k}\downarrow} + v_{\boldsymbol{k}} c^{\dagger}_{\boldsymbol{k}\uparrow}. \tag{3.166}$$

When $\mathcal{H}$ is written this way the ground state energy $E_0 = 0$ since both $\alpha_{\boldsymbol{k}}$ and $\beta_{\boldsymbol{k}}$ destroy the ground state. Using these definitions $\mathcal{P}_{\alpha}\mathcal{H}|\Phi\rangle = \mathcal{P}_{\alpha}\mathcal{H}\mathcal{P}_{\alpha}^{-1}\mathcal{P}_{\alpha}|\Phi\rangle$ so that $\mathcal{P}_{\alpha}\mathcal{H}\mathcal{P}_{\alpha}^{-1}$ is a Hamiltonian for which $\mathcal{P}_{\alpha}|\Phi\rangle$ is an eigenstate. Then since $\mathcal{P}_{\alpha}\mathcal{H}|\Phi\rangle = E_0\mathcal{P}_{\alpha}|\Phi\rangle = 0$ the eigenvalue is again zero. Then,

$$\mathcal{P}_{\alpha}\mathcal{H}\mathcal{P}_{\alpha}^{-1} = \mathcal{H}_{\alpha} = \sum_{\boldsymbol{k}} E_{\boldsymbol{k}} (\tilde{\alpha}^{\dagger}_{\boldsymbol{k}} \tilde{\alpha}_{\boldsymbol{k}} + \tilde{\beta}^{\dagger}_{\boldsymbol{k}} \tilde{\beta}_{\boldsymbol{k}}), \tag{3.167}$$

where, e.g., $\tilde{\beta}_{\boldsymbol{k}} = \mathcal{P}_{\alpha}\beta_{\boldsymbol{k}}\mathcal{P}_{\alpha}^{-1}$. In fact $\mathcal{P}_{\alpha}|\Phi\rangle$ is necessarily a ground state, since for any state $|\chi\rangle$,

$$\langle\chi|\mathcal{H}_{\alpha}|\chi\rangle = \sum_{\boldsymbol{k}} E_{\boldsymbol{k}} (|\langle\alpha_{\boldsymbol{k}}\chi|\alpha_{\boldsymbol{k}}\chi\rangle|^2 + |\langle\beta_{\boldsymbol{k}}\chi|\beta_{\boldsymbol{k}}\chi\rangle|^2 \geq 0 \tag{3.168}$$

and hence the equality corresponds to the lowest energy possible. Thus $\mathcal{H}_{\alpha}$ can be deformed from the BCS form, for $\alpha_0 = 0$, to an unknown Hamiltonian for a strongly correlated system when $\alpha_0 = 1$!

Equivalent steps can be used to show that, e.g., $\tilde{\alpha}^{\dagger}_{\boldsymbol{k}}|\tilde{\Phi}\rangle$ is an eigenstate with energy $E_{\boldsymbol{k}}$. However it is *very* important to observe that, since the operator $\mathcal{P}_{\alpha}$ is not Hermitian, it this the case that, $\tilde{\alpha}^{\dagger}_{\boldsymbol{k}} = \mathcal{P}_{\alpha}\alpha^{\dagger}_{\boldsymbol{k}}\mathcal{P}_{\alpha}^{-1} \neq (\tilde{\alpha}^{\dagger}_{\boldsymbol{k}})^{\dagger} =$

$(\mathcal{P}_\alpha \alpha_{\boldsymbol{k}}^\dagger \mathcal{P}_\alpha^{-1})^\dagger$, i.e., as observed by Laughlin [11] $\tilde{\alpha}_{\boldsymbol{k}}^\dagger$ does not properly anti-commute with its Hermitian conjugate. It is also a fact that $\mathcal{H}_\alpha$ is not Hermitian as explicitly expressions give below will confirm.

It is easy to write down an expression for the projected operators, e.g.,

$$\tilde{c}_{i\sigma} = {z_0}^{-1/2}(1 + \beta_0 n_{i-\sigma})c_{i\sigma} = {z_0}^{-1/2}\left(b_i^\dagger f_{i\sigma} + \frac{1}{1-\alpha_0}\sigma f_{i-\sigma}^\dagger b_{i\uparrow\downarrow}^\dagger\right) \tag{3.169}$$

or

$$\tilde{c}_{i\sigma}^\dagger = {z_0}^{1/2}(1 - \alpha_0 n_{i-\sigma})c_{i\sigma}^\dagger = {z_0}^{1/2}(b_i^\dagger f_{i\sigma} + (1-\alpha_0)\sigma f_{i-\sigma}^\dagger b_{i\uparrow\downarrow}^\dagger), \tag{3.170}$$

where the auxiliary particle expressions are given to the right.

The special case of half filling contains the essential physics. With this simplification particle-hole symmetry dictates,

$$\tilde{c}_{i\sigma} = ({z_0}^{-1/2} b_i^\dagger f_{i\sigma} + \sigma {z_0}^{1/2} f_{i-\sigma}^\dagger b_{i\uparrow\downarrow}) \tag{3.171}$$

and,

$$\tilde{c}_{i\sigma}^\dagger = ({z_0}^{1/2} f_{i\sigma}^\dagger b_i + \sigma {z_0}^{-1/2} b_{i\uparrow\downarrow}^\dagger f_{i-\sigma}), \tag{3.172}$$

where $1/2{z_0}^2$ is the concentration of holes and/or doubly occupied sites, see below.

The original the BCS ground state is $|\Phi\rangle = \prod_{\boldsymbol{k}} \alpha_{\boldsymbol{k}} \beta_{\boldsymbol{k}} |0\rangle$ and inserting $\mathcal{P}_\alpha^{-1}\mathcal{P}_\alpha = 1$, it follows that,

$$\mathcal{P}_\alpha|\Phi\rangle = |\tilde{\Phi}\rangle = \prod_{\boldsymbol{k}} \tilde{\alpha}_{\boldsymbol{k}} \tilde{\beta}_{\boldsymbol{k}} |0\rangle \tag{3.173}$$

however $\tilde{\beta}_{\boldsymbol{k}} = u_{\boldsymbol{k}} \tilde{c}_{-\boldsymbol{k}\downarrow} + v_{\boldsymbol{k}} \tilde{c}_{\boldsymbol{k}\uparrow}^\dagger$ acting upon $|0\rangle$ reduces to $v_{\boldsymbol{k}} \tilde{c}_{\boldsymbol{k}\uparrow}^\dagger$ and therefore with a different normalization, $|\tilde{\Phi}\rangle = \prod_{\boldsymbol{k}} \tilde{\alpha}_{\boldsymbol{k}} |S_z = +N/2\rangle$ where $|S_z = +N/2\rangle$ is the up spin ferromagnetic state. The $\tilde{\alpha}_{\boldsymbol{k}}$ are to be though of as creating the "particles" on this ferromagnetic vacuum. There are $\mathcal{N}$ = number of sites operators in the product on $\tilde{\alpha}_{\boldsymbol{k}}$. The BCS state does not have a fixed number of electrons. Because of the factors of z_0, and its inverse, the states which have a total number of electrons N which differs from half filling, i.e., $N = \mathcal{N}/2$, have their weight reduced by ${z_0}^{-1/2}$, as compared to the usual BCS state, for each excess hole or doubly occupied site. For a *fixed* N the state which has the highest weight has a *minimum* of such hole or doubly occupied sites. Hole and doubly occupied sites must then be added in pairs and a state with one such pair has its weight decreased by ${z_0}^{-1}$ while for n such pairs the weight is reduced by ${z_0}^{-n}$, for large z_0. Consider, say, states with an excess of electrons. The weight of a state with a hole at a given site is determined by the weight $1/\sqrt{2}$ for a state on which $c_{i\downarrow}$ acts times ${z_0}^{-1}$ in order that this produce a hole rather than a site which has a single electron. The probability of finding such a hole is therefore $\delta = (1/2){z_0}^{-2}$ and this must be the average hole doping at half filling.

The beginning Hamiltonian might be taken to be,

$$\mathcal{H} = -\sum_{\langle ij\rangle}[t\sum_{\sigma}(c^{\dagger}_{i\sigma}c_{j\sigma} + \text{h.c.}) + \mathrm{i}\Delta_0(\pm 1)(c^{\dagger}_{i\uparrow}c^{\dagger}_{j\downarrow} + \text{h.c.})] \tag{3.174}$$

and where, as usual, for *d*-wave pairing the sign of the order parameter is negative for vertical bonds and positive for the horizontal equivalents. This differs from $\mathcal{H} = \sum_{\boldsymbol{k}} E_{\boldsymbol{k}}(\alpha^{\dagger}_{\boldsymbol{k}}\alpha_{\boldsymbol{k}} + \beta^{\dagger}_{\boldsymbol{k}}\beta_{\boldsymbol{k}})$ only by an additive constant.

In order to again simplify the physics involved consider first the case when $\Delta_0 = 0$. Then $\mathcal{H}_\alpha$ reduces to the kinetic energy,

$$\tilde{T} = -t\sum_{\langle ij\rangle}\sum_{\sigma}(\tilde{c}^{\dagger}_{i\sigma}\tilde{c}_{j\sigma} + \tilde{c}^{\dagger}_{j\sigma}\tilde{c}_{i\sigma}) = \tilde{T}_\ell + \tilde{T}_u + \tilde{T}_i \tag{3.175}$$

where for the lower Hubbard band $\tilde{T}_\ell = -t\sum_{\langle ij\rangle}\sum_{\sigma}(f^{\dagger}_{i\sigma}b_i b^{\dagger}_j f_{j\sigma} + \text{h.c.})$ while $\tilde{T}_u = -t\sum_{\langle ij\rangle}\sum_{\sigma}(f^{\dagger}_{j\sigma}b_{j\uparrow\downarrow}b^{\dagger}_{i\uparrow\downarrow}f_{i\sigma} + \text{h.c.})$ corresponds to the upper band. Notice intra-band matrix elements are t and that these terms are Hermitian. In contrast the inter-band,

$$\begin{aligned}\tilde{T}_i = &-\frac{t}{z_0}\sum_{\langle ij\rangle}\sum_{\sigma}\sigma(b^{\dagger}_j f_{j\sigma}b^{\dagger}_{i\uparrow\downarrow}f_{i-\sigma} + b^{\dagger}_i f_{i\sigma}b^{\dagger}_{j\uparrow\downarrow}f_{j-\sigma})\\ &-z_0 t\sum_{\langle ij\rangle}\sum_{\sigma}\sigma(f^{\dagger}_{i\sigma}b_i f^{\dagger}_{j-\sigma}b_{j\uparrow\downarrow} + f^{\dagger}_{j\sigma}b_j f^{\dagger}_{i-\sigma}b_{i\uparrow\downarrow})\end{aligned} \tag{3.176}$$

is manifestly non-Hermitian. Destroying an adjacent doubly occupied plus hole pair gives two sites with single occupation has the large matrix element $-tz_0$ while the Hermitian conjugate matrix element for the inverse process is the much smaller $+t/z_0$.

A simple excitation at site i is created by $\tilde{c}^{\dagger}_{i\sigma} = ({z_0}^{1/2}f^{\dagger}_{i\sigma}b_i + \sigma {z_0}^{-1/2}b^{\dagger}_{i\uparrow\downarrow}f_{i-\sigma})$. If a local excitation is defined by, $|i\sigma\rangle = \tilde{c}^{\dagger}_{i\sigma}|\tilde{\Phi}\rangle$, it is a remarkable fact that $\tilde{T}$ hops the excitation to near neighbors with a uniform matrix element of $-t$. The factors of z_0 in $\tilde{T}$ exactly compensate for the similar factors which weight different states in the state vectors. This should be viewed as being the consequence of the construction, reflecting the fact that $\tilde{c}^{\dagger}_{\boldsymbol{k}\sigma}|\tilde{\Phi}\rangle$ corresponds to a well defined excitation. The normalization of $|i\sigma\rangle$ and therefore this excited state is easily determined. In $|\tilde{\Phi}\rangle$, the state which has an single $-\sigma$ electron has weight $1/\sqrt{2}$ while that with a hole has the much smaller weight $1/z_0\sqrt{2}$. The action of $\tilde{c}^{\dagger}_{i\sigma}$ is to project out these two states and leads to a doubly occupied state and a single occupied site and *both* final states have weight $1/{z_0}^{1/2}\sqrt{2}$. It follows that $|\boldsymbol{k}\rangle = {z_0}^{1/2}\tilde{c}^{\dagger}_{\boldsymbol{k}\sigma}|\tilde{\Phi}\rangle$ is normalized for large z_0. The action of the physical $c^{\dagger}_{i\sigma}$ on $|\tilde{\Phi}\rangle$ is somewhat different. It projects out the part of the state vector which has a single $-\sigma$ electron to give a doubly occupied final state with a weight factor $1/\sqrt{2}$ while that part which starts with a hole ends with a single σ electron and weight $1/\sqrt{2}z_0$. In $\tilde{c}^{\dagger}_{\boldsymbol{k}\sigma}|\tilde{\Phi}\rangle$ the doubly occupied state has the usual weight $1/z_0\sqrt{2}$

while a site with spin σ has weight $1/\sqrt{2}$. The net overlap photoemission matrix element,

$$\langle \boldsymbol{k}|c^{\dagger}_{\boldsymbol{k}\sigma}|\tilde{\Phi}\rangle = {z_0}^{-1/2}, \tag{3.177}$$

i.e., is reduced by a factor of ${z_0}^{-1/2}$.

Turning to the superconducting state, it is a direct consequence of the fact that, in the projected BCS state, the relative weight of states which differ by two holes have weights which differ by $1/z_0$ that for the order parameter,

$$\langle c^{\dagger}_{\boldsymbol{k}\uparrow}c^{\dagger}_{-\boldsymbol{k}\downarrow}\rangle = \frac{1}{z_0}\langle c^{\dagger}_{\boldsymbol{k}\uparrow}c^{\dagger}_{-\boldsymbol{k}\downarrow}\rangle_{\mathrm{BCS}}. \tag{3.178}$$

When compared with the $t-J$-model result it is observed that $\frac{1}{z_0}$ is *twice* the number or holes. The extra factor of two *correctly* reflects the fact that here there are two processes one each for the lower and upper Hubbard band. These results reproduce in a different notation those of Laughlin [11].

Given that z_0 is large, the above calculations illustrate that most of the spectral weight is missing in the states with the small energy $E_{\boldsymbol{k}}$. The remaining weight is shifted by an effective effective Hubbard U to higher energies. This weight corresponds to the upper and lower Hubbard bands. A good estimate of U for very large z_0 can be obtained by considering in more detail the action of $c^{\dagger}_{i\sigma}$ on the ground state. This creates an excess electron which must go into the upper Hubbard band. As described explicitly above, the absence of the factors involving z_0 in $c^{\dagger}_{i\downarrow}$ implies that the new state has different weights as compared with those for the ground state. The largest matrix element of the $\tilde{T}$ involves $-tz_0 f^{\dagger}_{j\uparrow}b_j f^{\dagger}_{i\downarrow}b_{i\uparrow\downarrow}$ and taking the diagonal matrix element gives,

$$tz_0\langle\Phi|b^{\dagger}_i f_{i\downarrow} f^{\dagger}_{j\uparrow}b_j f^{\dagger}_{i\downarrow}b_{i\uparrow\downarrow}b^{\dagger}_{i\uparrow\downarrow}f_{i\uparrow}|\Phi\rangle = tz_0\langle\Phi|f^{\dagger}_{j\uparrow}b_j b^{\dagger}_i f_{i\uparrow}|\Phi\rangle, \tag{3.179}$$

which is z_0 times a piece of the kinetic energy $\tilde{T}$. It is then observed that each part of $\tilde{T}$ makes an equal contribution to $\langle\Phi|\tilde{T}|\Phi\rangle$ and therefore the piece of interest gives $(1/8)\langle\Phi|\tilde{T}|\Phi\rangle = (1/4)\sum_{\boldsymbol{k}}\epsilon_{\boldsymbol{k}}{v_{\boldsymbol{k}}}^2$. Lastly since this contribution corresponds to one of four neighbors for a square lattice, the result is,

$$U = z_0\sum_{\boldsymbol{k}}\epsilon_{\boldsymbol{k}}{v_{\boldsymbol{k}}}^2 \sim 4tz_0, \tag{3.180}$$

which is similar to the estimate of Laughlin [11].

A small doping does not alter the situation greatly, except that now $E_{\boldsymbol{k}} = \sqrt{(\epsilon_{\boldsymbol{k}}-\mu)^2 + {\Delta_0}^2}$, i.e., the weak spectral feature at the Fermi surface slowly moves to lower energy. A superconducting density of states with a low weight lying in the gap has been shown to be a feature of hole doping in Sects. 3.11 and 3.16.

3.21 Summary

Phonons are almost certainly responsible for superconductivity in most materials. If this is so for the high T_c cuprates then the electron-phonon coupling must be strong and that the pairing be d-wave implies strong attractive forward scattering. The strength of this scattering depends crucially on both the phonon modes involved and on the electrostatic screening length. These issues have been addressed in Sects. 3.1–3.3.

If on the other hand the superconductivity is associated with the proximity of the anti-ferromagnetic phase then the theoretical description of a lightly doped Mott insulator is needed. The theory of the parent anti-ferromagnet is described in Sect. 3.4 using the Holstein-Primakoff transformation. The Jordan-Wigner transformation described in Sect. 3.5 is an alternative description of the $S = 1/2$ spin algebra which is well adapted to spin liquids, exemplified by the one dimensional XY model. The hard core bosons of this algebra are converted into fermions by this transformation since this transformation adds a unit flux tube to the bosons. Section 3.6 contains a mathematical interlude with explains coherent states, Grassman variables and flux tubes. In Sect. 3.7 is introduced a mean-field theory in which the flux tubes are replaced by the equivalent uniform field. A connection is made with the theory of flux states and Landau levels, the description of which implies pre-existing particle-hole pairs at all temperatures. As a preliminary to the introduction of holes, in Sect. 3.8 is described the old theory of Bogoliubov for a Bose superfluid. With a suitable interaction this reproduces the charge gap which is to be expected in the condensation of hard core bosons and relates the transition temperature to this gap size and hence the condensed fraction. The auxiliary particle method is a useful way of representing the spin and charge degrees of freedom of highly correlated systems. This method is presented in some detail in Sect. 3.9 and, as and exercise, in Sect. 3.10 this method is used to make a canonical transformation of the Hubbard model and which reduces it to the $t - J$-model.

The conceptual principles, the RVB spin pairing state and Bose-Einstein condensation of the holes, are perhaps captured by the mean-field slave boson theory described in Sect. 3.11, however this has the serious mathematical defect that it does not satisfy the constraint $Q_i = 1$. One approach which does not have this defect is to use, e.g., the RVB wave function deduced from the mean-field theory and use this as a variational wave function for the fully constrained problem. Sections 3.13-3.17 provide a mathematical framework in which these issues of spin pairing and hole condensation can be addressed within a formalism which satisfies this constraint. An important outcome, for this particular flavor of the RVB theory, is the equivalence of the RVB and "flux" spin states when coherent doping of a condensate is induced by an SU(3) rotation in the auxiliary particle space. This flux state pairing reflects the filling of lowest Landau level and, unlike RVB pairing in the mean-field theory, there is no thermodynamic transition for spin pairing.

Superconductivity occurs when the holes condense. Also the energy scales become modified. Whereas, in the mean-field approach, the hopping matrix element for spin particles is $\sim J+2xt$, with the constraint satisfying approach this becomes $J-2xt$. This has important consequences for the interpretation of experiment.

The SO(5) approach is present in Sect. 3.17. This is a bosonic theory which unifies the BCS superconductive, Δ, and anti-ferromagnetic order, $\boldsymbol{N}$, parameters into a single superspin $\boldsymbol{n}$. If this is to be an appropriate symmetry group, then members of the Lie algebra, π_α, which rotate $\boldsymbol{n}$ from magnetism to superconductivity are needed. If $[\mathcal{H}, \pi_\alpha] = 0$ then the symmetry is exact, however when the weaker condition $[\mathcal{H}, \pi_\alpha] = \omega\pi_\alpha$ is satisfied, the bosonic collective modes can be deduced from symmetry considerations. It is envisaged that the cuprates are close to SO(5) symmetry. If this symmetry is exact at a value μ_c which makes the magnetic and superconductive states degenerate then the underdoped region would correspond to a free rotation with coexisting anti-ferromagnetic and superconductive order parameters. Otherwise there is phase separation with hole rich and hole poor regions. In large measure, the phenomenology reflects an implicit assumption that holes are *only* added by turning on Δ.

Again for the constraint satisfying flavor of the RVB theory presented here, the order parameter in the superconductive state reflects hole condensation and this changes the definitions behind the SO(5) theory as described in Sect. 3.19. In this approach, which *does* have fermionic degrees of freedom, the energy of the π mode reflects the distance of the chemical potential from the lower Hubbard band and SO(5) symmetry is never exact. As a consequence phase separation is predicted for the underdoped region, the doping of holes puts states into the insulator gap, and there is a barrier for the tunneling of holes in the fully anti-ferromagnetic state.

The gossamer superconductors described in Sect. 3.20 also have fermionic states in the insulator gap. Being related to the Hubbard model there are compensating doping of holes in the lower, and electrons in the upper, Hubbard band and this leads to a finite density of states mid-gap even for the insulator. While this is an excellent toy model by which to introduce the concept of thin superconductivity its physical interpretation is complicated by the fact that $\mathcal{H}$ is not Hermitian.

This chapter is far from complete. This is necessarily the case since there is not yet *a* theory of the cuprate superconductors, but also since the aim is to give the reader an unified introduction to some general principles rather than develop detailed, but controversial, predictions for one theory.

References

1. J. Bardeen, L. N. Copper, J. R. Schrieffer: Phys. Rev. **108**, 1175 (1957)
2. See, e.g., M. Tinkham: *Introduction to Superconductivity*, McGraw-Hill, New York (1975)
3. See review: C. C. Tsuei, J. R. Kirtley: Rev. Mod. Phys. **72**, 969 (2000)
4. See, e.g., A. A. Abrikosov: cond-mat/9912394 (1999)
5. See also e.g., J. Bouvier, J. Bok: Physica C**217-225**, 217 (1997); J. I. Gersten, M. Weger: Phys. Rev. B**65**, 214530 (2002)
6. See references in, D. K. Morr, D. Pines: Phys. Rev. B**61**, R6483 (2000)
7. V. J. Emery, S. Kivelson: Nature **374**, 434 (1995)
8. P. W. Anderson: Science **235**, 1196 (1987)
9. S.-C. Zhang: Science **275**, 1089 (1997)
10. S. Uemura: Phys. Rev. Lett. **62**, 2317 (1989)
11. R. B. Laughlin: cond-mat/0209269 (2002); see also F. C. Zhang: cond-mat/0209272 (2002)
12. G. D. Mahan: *Many-particle physics*, Plenum Press, New York, (1981)
13. J. W. Negele, H. Orland: *Quantum Many-Particle Systems*, Addison-Wesley, Reading (1987)
14. S. E. Barnes: J. Phys. F**6**, 115; J. Phys. F6 1376 (1976); pg. 301, Proc. XIX Congress Ampere, North Holland (1977); J. Phys. F**7** 2637 (1977); J. de Physique, **39** C5-828 (1978); Advances in Physics *30* 801-938 (1981); J. Phys. C.M., **6**, 719, (1994)
15. See e.g., C. Kittel: *Quantum theory of solids*, Wiley, New York (1963)
16. S. Schmitt-Rink, C. M. Varma, A. E. Ruckenstein: Phys. Rev. Lett. **60**, 2793 (1988)
17. J.J. Sakurai: *Modern quantum mechanics*, Benjamin, Menlo Park, (1985)
18. J. Y. T. Wei, C. C. Tsuei, P. J. M. van Bentum, Q. Xiong, C. W. Chu, M. K. Wu: Phys. Rev. B**57** 3650 (1998)
19. See, e.g., A. Messiah: *Quantum mechanics*, North Holland, Amsterdam (1961)
20. See discussion in: Z.-X. Shen, A. Lanzara S., Ishihara, N. Nagaosa: Phil. Mag. **82**, 1349 (2002)
21. E. Dagotto, A. Moreo: Phys. Rev. B**38**, 5087 (1988)
22. M. Gross, E. Sánches-Velasco, E. Siggia: Phys. Rev. B**39**, 2484; Phys. Rev. B**40**, 11328 (1989)
23. S. Liang, B. Doucot, P. W. Anderson: Phys. Rev. Lett. **61**, 365 (1988)
24. N. D. Mermin, H. Wagner: Phys. Rev. Lett. **22**, 1133 (1966)
25. D. Reznik, P. Bourges, H.F. Fong, L.P. Regnault, J. Bossy, C. Vettier, D.L. Milius, I.A. Aksay, B. Keimer: Phys. Rev. B53 R14741 (1996)
26. R. R. P. Singh: Phys. Rev. B**39**, 9760 (1989)
27. J. Appel: in *Solid State Physics*, **21**, ed. F. Seitz, D. Turnbull, H. Ehrenreich (Academic, New York, 1968)
28. G. Martinez, P. Horsch: Phys. Rev. B**44** 317 (1991)
29. See e.g., A. Khare: *Fractional statistics and quantum field theory*, World Scientific, Singapore (1997)
30. S. E. Barnes, S. Maekawa: J. Phys. Cond. Matter **14**, L19-L28 (2002)
31. See: G. S. Canright, S. M. Girvin, A. Brass: Phys. Rev. Lett. **63**, 2291 (1989)
32. See: Y. Hasegawa, P. Lederer, T. M. Rice, P. B. Wiegmann: Phys. Rev. Lett. **63**, 907 (1989)
33. See: S. E. Barnes: J. de Physique, **39** C5-828 (1978); D. Vacaru, S. E. Barnes: J. Phys. C.M. **6**, 719, 1994
34. F. C. Zhang, T. M. Rice: Phys. Rev. B**37**, 3759 (1988)
35. Y. Isawa, S. Maekawa, H. Ebisawa: Physica **148**B, 391 (1987)

36. G. Baskaran, Z. Zou, P. W. Anderson: Solid State Commun. **63**, 973 (1987)
37. G. Kotliar: Phys. Rev. **37**, 3664 (1988))
38. See: F. C. Zhang, C. Gros, T. M. Rice, H. Shiba: Supercond. Sci. & Tech. **1**, 36 (1988); T. K. Lee, S. Feng: Phys. Rev. B**38**, 11 809 (1988)
39. F. C. Zhang, C. Gros, T. M. Rice, H. Shiba: Supercond. Sci. & Tech. **1**, 36 (1988)
40. A. Damascelli, Z.-X. Shen, Z. Hussain: Rev. Mod. Phys. **75**, 473 (2003)
41. Y. Suzumura, Y. Hasegawa, H. Fukuyama: J. Phys. Soc. Jpn. **57**, 401 (1988)
42. G. Kotliar, J. Liu: Phys. Rev. B**38**, 5142 (1988)
43. See, e.g., P. A. Lee: Physica C**317**, 194 (1999); X. G. Wen, P. A. Lee: Phys. Rev. Lett. **76**, 503 (1996)
44. See, e.g., recent survey: J. L. Tallon, J. W. Loram , C. Panagopoulos: cond-mat/0211048 (2002)
45. C. E. Gough, M. S. Colclough, E. M. Forgan, R. G. Jordan, M. Keene, C. M. Muirhead, A. I. M. Rae, N. Thomas, J. S. Abell, S. Sutton: Nature (London) **326**, 855 (1987)
46. T. Tohyama, S. Maekawa: Supercond. Sci. Technol. **13**, R17 (2000)
47. L. D. Faddeev, L.A. Takhtajan: Phys. Lett. **85**A 375 (1981); J. Sov. Math. **24**, 241 (1984)
48. B. O. Wells, Z. -X. Shen, A. Matsuura, D. M. King, M. A. Kastner, M. Greven, R. J. Birgeneau: Phys. Rev. Lett. **74**, 964 (1995); S. LaRosa, I. Vobornik, F. Zwick, H. Berger, M. Grioni, G. Margaritondo: Phys. Rev. B**56**, R525 (1997)
49. R. B. Laughlin: Phys. Rev. Lett. **79**, 1726 (1997)
50. S. Weinberg: *The Quantum Theory of Fields*, Cambridge University Press (August 1996)
51. S. E. Barnes, S. Maekawa: Phys. Rev. B**67**, 224513 (2003)
52. Y. Nagaoka: Sol. State Commun. **3**, 409 (1965); Phys. Rev. **147** 392 (1966)
53. See, F. F. Balakirev, J. B. Betts, A. Miglioro, S. Ono, Y. Ando, S. Boebinger: Nature **424**, 912 (2003); H. J. Kang, P. Dai, J. W. Lynn, M. Matsuura, J. R. Thompson, S.-C. Zhang, D. N. Argyriou, Y. Onose, Y. Tokura: Nature **423**, 522 (2003) and references therein
54. E. Demler, S.-C. Zhang: Phys. Rev. Lett. **75**, 4126 (1995)
55. See, H. F. Fong, P. Bourges, Y. Sidis, L. P. Regnault, A. Ivanov, G. D. Gu, N. Koshizuka, B. Keimer: Nature **398**, 588 (1999) and references therein
56. S. Meixner, W. Hanke, E. Demler, S.-C. Zhang: Phys. Rev. Lett. **79**, 4902 (1997)
57. I. Bozovic, G. Logvenov, M. A. J. Verhoeven, P. Caputo, E. Goldobin, T. H. Geballe: Nature **422**, 873 (2003)

4. Manganites

4.1 Manganite with Cubic Perovskite Structure

4.1.1 Introduction

Following the discovery of the high Tc superconducting cuprates [1], the study of a variety of transition-metal (TM) oxides was revived. Modern theoretical and experimental methods have been used to study both materials which had been fabricated in earlier times and new and/or doped materials. The most attractive such compounds are the manganites $R_{1-x}A_x\mathrm{MnO}_3$ [2,3] (R: a trivalent cation and A: a divalent cation) with pseudo-cubic perovskite structure and related materials. The crystal structures of the manganite family are shown in Fig. 4.1. There is only a single MnO_2 layer in a unit cell of $R_{1-x}A_{1+x}\mathrm{MnO}_4$, while $R_{2-2x}A_{1+2x}\mathrm{Mn}_2\mathrm{O}_7$ contains a double layer. The pseudo-cubic $R_{1-x}A_x\mathrm{MnO}_3$ reflects the limit of an infinite number of layers.

The colossal magnetoresistance (CMR) [4–7] is the best known phenomenon observed in manganites. There is a "colossal" reduction of the electrical resistivity upon applying a magnetic field. This exotic phenomenon has promoted both fundamental and applied physics condensed matter research and resulted in new magnetic devices. Fundamental studies of the perovskite manganites were started in the 1950's by Jonker and van Santen who have found a correlation between ferromagnetic ordering and metallic conductivity [2, 3]. This new phenomenon has been explained by the double exchange interaction proposed by Zener [8], and developed by Anderson and Hasegawa [9], and de Gennes [10]. In this model there is an effective inter-site ferromagnetic spin interaction which results from the combined effects of the intra-atomic Hund coupling and electron transfer. A modest negative magnetoresistance (MR) of the manganites was discovered in the 1970's [11]. The revival of theoretical and experimental interest in the manganites dates to the 1990's, leading to (1) the discovery of "colossal" MR ratios $[R(0\mathrm{T}) - R(7\mathrm{T})]/R(7\mathrm{T}) \sim 10^{11}$ in $\mathrm{Pr}_{0.7}\mathrm{Ca}_{0.3}\mathrm{MnO}_3$ [12], (2) to the recognition that the CMR effect is *not* explained within the simple double exchange interaction scenario [13], and (3) to various other novel exotic phenomena and which have stimulated the development of number of new physical concepts.

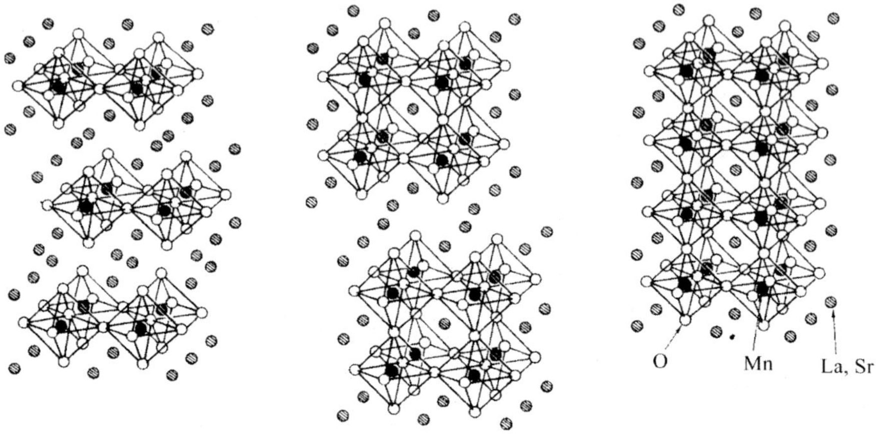

Fig. 4.1. Crystal structures of manganites $R_{1-x}A_{1+x}\mathrm{MnO}_4$ (left), $R_{2-2x}A_{1+2x}\mathrm{Mn}_2\mathrm{O}_7$ (center), and $R_{1-x}A_x\mathrm{MnO}_3$ (right).

The manganites are now recognized to be correlated electron systems, similar in this respect to the high T_c cuprates. When coupled to the lattice degree of freedom, exhibit a variety of phenomena associated with spin, charge and orbital degrees of these electrons. It is instructive to compare the electronic structure in CMR manganites with that of high Tc cuprates. The electron configuration of the ion Cu^{2+} in La_2CuO_4 is $3d^9$ and it is located at the center of a CuO_6 octahedron with a tetragonal distortion. One hole occupies the $3d_{x^2-y^2}$ orbital. The Mn^{3+} ion in $LaMnO_3$ has four $3d$ electrons and sits in a MnO_6 octahedron. These $3d$ electrons form the high spin state with three electrons in the t_{2g} orbitals with the remaining electron occupying the doubly degenerate e_g orbitals, i.e. the configuration is $t_{2g}^3 e_g^1$ (see Fig. 4.2). As a result this ion has an *orbital degree of freedom* resulting from the e_g orbital degeneracy. In general the orbital state of such a Mn^{3+} ion at site i is a linear combination of the $3d_{3z^2-r^2}$ and $3d_{x^2-y^2}$ states, i.e., can be written as

$$|\theta_i, \phi_i\rangle = \cos\left(\frac{\theta_i}{2}\right)|d_{3z^2-r^2}\rangle + \mathrm{e}^{\mathrm{i}\phi_i}\sin\left(\frac{\theta_i}{2}\right)|d_{x^2-y^2}\rangle, \tag{4.1}$$

where θ_i and ϕ_i are angles in the orbital space. (If $\phi_i = 0$ the wave function is real.) This state is an eigen-state of a pseudo $S = 1/2$ spin operator defined by

$$\boldsymbol{T}_i = \frac{1}{2}\sum_{\sigma\gamma\gamma'} d^\dagger_{i\gamma\sigma}\boldsymbol{\sigma}_{\gamma\gamma'}d_{i\gamma'\sigma}, \tag{4.2}$$

where $d_{i\gamma\sigma}$ is the annihilation operator of an e_g electron with spin $\sigma(=\uparrow,\downarrow)$ and orbital $\gamma(= 3z^2 - r^2, x^2 - y^2)$ at site i, and where the $\boldsymbol{\sigma}$ are the Pauli matrices. The eigen-states of the operator T_{iz} with the eigen-values of $1/2$ and $-1/2$ are $|d_{3z^2-r^2}\rangle$ and $|d_{x^2-y^2}\rangle$, respectively.

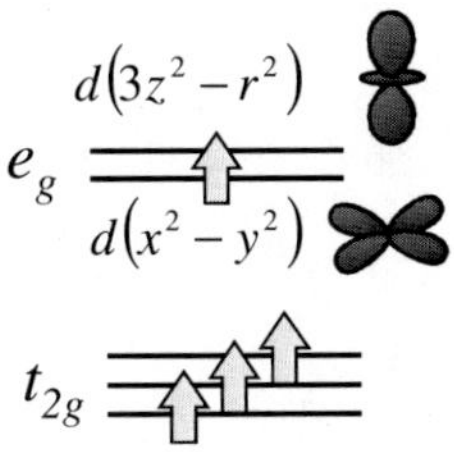

Fig. 4.2. A schematic energy levels of a Mn^{3+} ion.

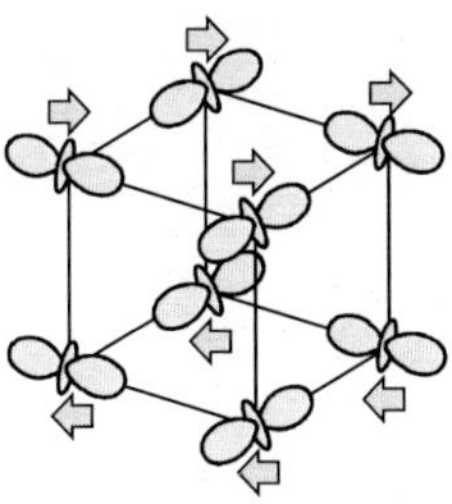

Fig. 4.3. The orbital ordered state and the A-type AFM ordered state in $LaMnO_3$.

In a solid, the orbital degree of freedom often exhibits a symmetry breaking long range *orbital* order resulting from inter-site orbital interactions. A well known example is $LaMnO_3$, where, below 780K, two different kinds of orbitals are alternately aligned in the *ab* plane with orbital stacking along the *c* axis (see Fig. 4.3) [14,15]. The wave functions of these orbitals are supposed to be the $3d_{3x^2-r^2}$- and $3d_{3y^2-r^2}$-types which correspond to $\theta = 4\pi/3$ and $8\pi/3$ with $\phi = 0$ in (4.1), respectively. This orbital order dominates the A-type antiferromagnetic (AFM) anisotropic spin order in which the spins are ferromagnetically aligned in the *ab* plane but antiferromagnet ordered along the *c* axis. Another type of the orbital order is observed in $Pr_{0.5}Ca_{0.5}MnO_3$ for which the nominal valence of a Mn ion is 3.5+. Below a common ordering temperature, there is charge order, i.e. spatially the Mn have two different valencies, in addition, for one of these charge sublattices there is an orbital order [16], and, finally there is a CE-type AFM order [17]. There are CMR effects in the vicinity of the charge and orbital ordering transition implying that the orbital degree of freedom plays a crucial roles in CMR, and other, dramatic phenomena in the manganites.

In this section, we review the CMR manganites with an emphasis on the orbital degree of freedom and its coupling to the spin, charge and lattice degrees of freedom. In addition several up-to-date theories and experiments relevant to electronic and crystal structures are introduced. This is an very active area of research and new materials, experimental data and theoretical concepts arise on a daily basis. In addition to introducing the well developed facts, this chapter aims as clarifying this bewildering situation. Also see other superb reviews and books for manganites and related topics [18–23].

4.1.2 Electronic Hamiltonian and Exchange Interaction

Let us introduce an electronic model Hamiltonian for manganites with pseudo-cubic perovskite structure. As mentioned previously, the nominal electronic configuration of an isolated Mn^{3+} is the $3d^4$ state with $210(={}_{10}C_4)$ degeneracy. This is lifted by the electron-electron interactions at a single ion and forms the LS multiplets. The lowest multiplets are determined by the Hund rule to be a 25 fold degenerate ^5D state. This state is further split by the crystalline field due to the surrounding negatively charged O ions. In the cubic crystal symmetry, the lowest multiplet is denoted by the Γ_3 state corresponding to the high spin $t_{2g}^3 e_g^1$ configuration. A number of the degeneracy is $10(=5\times 2)$ where 5 is ascribed to the $S=2$ state and the remaining double degeneracy comes from the orbital degree of freedom.

Consider the ionic model where the electron hybridization is neglected. In general, for the transition-metal oxides, there are two candidate for the lowest electronic excitation; (1) the $d-d$ excitation where one electron is removed from a TM ion and added to another TM ion. Explicitly, the process is given by $|d^n d^n\rangle \rightarrow |d^{n+1} d^{n-1}\rangle$. (2) the $p-d$ (charge transfer) excitation where one O $2p$ electron is excited to a TM ion, denoted by $|d^n\rangle \rightarrow |d^{n+1}\underline{L}\rangle$ where $\underline{L}$ represents the O $2p$ state occupied by a hole. These energies are represented by the Coulomb interaction energy at the same site, U and the charge transfer energy Δ, respectively. In addition to the defined U and Δ, it is convenient to introduce the "effective" Coulomb interaction $U_{\rm eff}$ and charge transfer energy $\Delta_{\rm eff}$ [24]. These are the excitation energies between the lowest multiplet of each configuration. $U_{\rm eff}$ and $\Delta_{\rm eff}$ determine most of the physical properties in the ground state and low energy excited states. As mention later, in more detail, magnitudes of these energies in manganites are estimated as (U, Δ) ~(8eV, 4.5eV) and $(U_{\rm eff}, \Delta_{\rm eff})$ ~(7eV, 2eV). This implies that the system is classified as a charge-transfer type insulator where the lowest excitation is determined by the $p-d$ excitation. Beyond the ionic model, there is a strong hybridization between the Mn $3d$ and O $2p$ orbitals. In order to describe the low energy electronic structure in manganites, up to the order of several eV, one of the most generic electronic model is the extended $p-d$ model where the five Mn $3d$ and three O $2p$ orbitals are introduced at each site and the electron hopping and electron-electron interactions are introduced. Instead of such a standard way, here we introduce a more simple electronic Hamiltonian for manganites where the O $2p$ electron degrees of freedom are integrated out, and the e_g and t_{2g} electrons at Mn sites are treated as mobile electrons and localized spins, respectively. Some interaction parameters between them are introduced phenomenologically. One of the most advantages in this model is in reduction of the electronic degrees of freedom, which makes possible theoretical analysis where the electron correlation effects are taken into account progressively.

Because of the large overlap implied by the σ bond between the O $2p_\sigma$ and doubly degenerate Mn e_g orbitals, the e_g-electrons are considered to mobile. In contrast, the t_{2g}-electrons are assumed to be localized with a Hund's rule spin $S = 3/2$, because (1) the covalent π bond between the t_{2g} and O $2p$ orbitals is very much weaker than σ bond described above, and (2) the energy splitting, $10Dq$, between the e_g and t_{2g} orbitals, is smaller than the Hund coupling J_H between e_g and t_{2g} spins, and as a consequence the low spin state of the $3d^4$ electron configuration can be neglected. With these considerations, for a simple cubic crystal, the following model Hamiltonian is a good starting point:

$$\mathcal{H} = \mathcal{H}_{e_g} + \mathcal{H}_{t_{2g}} + \mathcal{H}_{e_g-t_{2g}}. \tag{4.3}$$

The first term $\mathcal{H}_{e_g}$ is

$$\mathcal{H}_{e_g} = \varepsilon_d \sum_{i\sigma\gamma} d^\dagger_{i\gamma\sigma} d_{i\gamma\sigma} + \sum_{\langle ij\rangle\sigma\gamma\gamma'} \left(t^{\gamma\gamma'}_{ij} d^\dagger_{i\gamma\sigma} d_{j\gamma'\sigma} + \text{h.c.} \right) + \mathcal{H}_{el-el}, \tag{4.4}$$

where the electron-electron interaction term:

$$\mathcal{H}_{el-el} = \sum_{i\gamma_1\gamma_2\gamma_3\gamma_4\sigma_1\sigma_2} V_{\gamma_1\gamma_2\gamma_3\gamma_4} d^\dagger_{i\gamma_1\sigma_1} d^\dagger_{i\gamma_2\sigma_2} d_{i\gamma_3\sigma_2} d_{i\gamma_4\sigma_1}. \tag{4.5}$$

Here $d^\dagger_{i\gamma\sigma}$ is the creation operator for an e_g electron at site i with spin $\sigma(=\uparrow,\downarrow)$ and orbital label $\gamma(=a,b)$ where a and b imply the $3d_{3z^2-r^2}$ and $3d_{x^2-y^2}$, respectively. The energy ε_d is that of the e_g orbitals, and the two e_g orbitals are assumed to be degenerate. The matrix element $t^{\gamma\gamma'}_{ij}$ governs the electron transfer between γ orbital at site i and γ' at the NN site j. This reflects a second-order process involving t_{pd} the electron transfer matrix element between the Mn $3d$ and O $2p$ orbitals, as is discussed in Chap. 1. Being based on the Slater-Koster parameters, $t^{\gamma\gamma'}_{ij}$ is factorized as $t^{\gamma\gamma'}_{ij} = \alpha(\gamma)\alpha(\gamma')t_0$ where t_0 is the transfer integral between the NN $3d_{3z^2-r^2}$ orbitals along the z direction. Explicitly

$$t^{\gamma\gamma'}_{ij} = t_0 \begin{pmatrix} -\frac{1}{4}, & \frac{\sqrt{3}}{4} \\ \frac{\sqrt{3}}{4}, & -\frac{3}{4} \end{pmatrix}, \quad t_0 \begin{pmatrix} -\frac{1}{4}, & -\frac{\sqrt{3}}{4} \\ -\frac{\sqrt{3}}{4}, & -\frac{3}{4} \end{pmatrix}, \quad t_0 \begin{pmatrix} -1, & 0 \\ 0, & 0 \end{pmatrix}, \tag{4.6}$$

where the position of the j-th ion is defined by $\boldsymbol{r}_j = \boldsymbol{r}_i \pm a\hat{\boldsymbol{x}}$, $\boldsymbol{r}_j = \boldsymbol{r}_i \pm a\hat{\boldsymbol{y}}$ and $\boldsymbol{r}_j = \boldsymbol{r}_i \pm a\hat{\boldsymbol{z}}$, respectively, with the lattice constant a, and the unit vector $\hat{\boldsymbol{x}}$ ($\hat{\boldsymbol{y}}, \hat{\boldsymbol{z}}$) along the x (y, z) axis. The Coulomb interaction matrix elements in (4.5) are

$$V_{\gamma_1\gamma_2\gamma_3\gamma_4} = \int \mathrm{d}\boldsymbol{r}_1 d\boldsymbol{r}_2 \phi_{\gamma_1}(\boldsymbol{r}_1)^* \phi_{\gamma_2}(\boldsymbol{r}_2)^* v(|\boldsymbol{r}_1 - \boldsymbol{r}_2|) \phi_{\gamma_3}(\boldsymbol{r}_2) \phi_{\gamma_4}(\boldsymbol{r}_1), \tag{4.7}$$

where $\phi_\gamma(\boldsymbol{r})$ are the $3d_\gamma$ wave functions and the Coulomb interaction is $v(|\boldsymbol{r}|)$. The Hamiltonian for the AFM superexchange interaction between NN t_{2g} spins is

$$\mathcal{H}_{t_{2g}} = J_{AF} \sum_{\langle ij\rangle} \boldsymbol{S}_{ti} \cdot \boldsymbol{S}_{tj}, \tag{4.8}$$

where $\boldsymbol{S}_{ti}$ is the spin operator for the t_{2g} spins at site i with $S = 3/2$ and $J_{AF} > 0$. This interaction is caused by the virtual exchange of electrons between the t_{2g} orbitals at Mn ions and the $2p_\pi$ orbitals at NN O ions. The e_g and t_{2g} spins at a Mn ion interact ferromagnetically through the Hund coupling:

$$\mathcal{H}_{e_g-t_{2g}} = J_H \sum_i \boldsymbol{S}_{ti} \cdot \boldsymbol{S}_i, \tag{4.9}$$

where $J_H < 0$, and the spin operator for e_g electron is defined by

$$\boldsymbol{S}_i = \frac{1}{2} \sum_{\sigma_1\sigma_2\gamma} d^\dagger_{i\sigma_1\gamma} \boldsymbol{\sigma}_{\sigma_1\sigma_2} d_{i\sigma_2\gamma}. \tag{4.10}$$

Here, J_{AF} and J_H are introduced phenomenologically. The interactions between electrons with orbital degrees of freedom and lattice distortions are considered later. One may imagine that the relativistic spin-orbit coupling (LS coupling) lifts the orbital degeneracy. However, this interaction is irrelevant in the doubly degenerate e_g orbitals as discussed, in more detail, in Sect. 5.4 .

Focus on the electron-electron interaction term $\mathcal{H}_{el-el}$ in (4.5). The matrix elements $V_{\gamma_1\gamma_2\gamma_3\gamma_4}$ when, at least, two of the four orbital suffices γ_l ($l = 1 \sim 4$) are equal, have special meaning. These are

(1) when $\gamma_1 = \gamma_2 = \gamma_3 = \gamma_4$: $\sum_\gamma V_{\gamma\gamma\gamma\gamma} n_{\gamma\uparrow} n_{\gamma\downarrow} \equiv U \sum_\gamma n_{\gamma\uparrow} n_{\gamma\downarrow}$ the intra-orbital Coulomb interaction.
(2) for $\gamma_1 = \gamma_4 \neq \gamma_2 = \gamma_3$: $\frac{1}{2} \sum_{\gamma\neq\gamma'} V_{\gamma\gamma'\gamma'\gamma} n_\gamma n_{\gamma'} \equiv U' n_a n_b$ the direct Coulomb interaction between electrons in different orbitals.
(3) then with $\gamma_1 = \gamma_3 \neq \gamma_2 = \gamma_4$: $-\frac{1}{2} \sum_{\gamma\neq\gamma'\sigma\sigma'} V_{\gamma\gamma'\gamma\gamma'} d^\dagger_{\gamma\sigma} d_{\gamma\sigma'} d^\dagger_{\gamma'\sigma'} d_{\gamma'\sigma} \equiv J \sum_{\sigma\sigma'} d^\dagger_{a\sigma} d^\dagger_{b\sigma'} d_{a\sigma'} d_{b\sigma}$, the exchange interaction.
(4) and finally when $\gamma_1 = \gamma_2 \neq \gamma_3 = \gamma_4$:
$\sum_{\gamma\neq\gamma'} V_{\gamma\gamma\gamma'\gamma'} d^\dagger_{\gamma\uparrow} d^\dagger_{\gamma\downarrow} d_{\gamma'\downarrow} d_{\gamma'\uparrow} \equiv I \sum_{\gamma\neq\gamma'} d^\dagger_{\gamma\uparrow} d^\dagger_{\gamma\downarrow} d_{\gamma'\downarrow} d_{\gamma'\uparrow}$ and is called the pair hopping interaction.

The electron-electron interaction is then

$$\begin{aligned}\mathcal{H}_{el-el} = &\, U \sum_{i\gamma} n_{i\gamma\uparrow} n_{i\gamma\downarrow} + U' \sum_i n_{ia} n_{ib} \\ &+ J \sum_{i\sigma\sigma'} d^\dagger_{ia\sigma} d^\dagger_{ib\sigma'} d_{ia\sigma'} d_{ib\sigma} + I \sum_{i,\gamma\neq\gamma'} d^\dagger_{i\gamma\uparrow} d^\dagger_{i\gamma\downarrow} d_{i\gamma'\downarrow} d_{i\gamma'\uparrow}. \end{aligned} \tag{4.11}$$

In the limit of an isolated ion the parameters in (4.11) are related to the Racah parameters A, B and C via

$$U = A + 4B + 3C, \tag{4.12}$$
$$U' = A - 4B + C, \tag{4.13}$$
$$J = I = 4B + C, \tag{4.14}$$

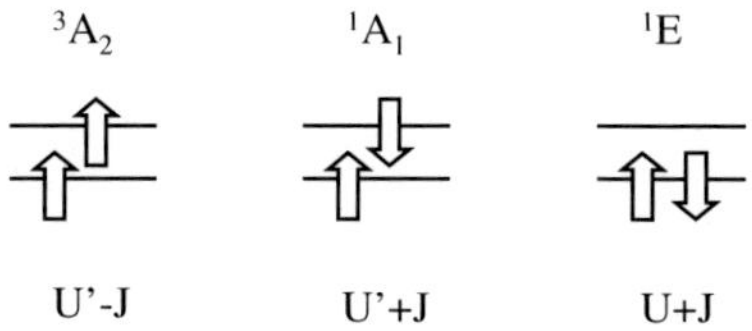

Fig. 4.4. Electron configurations of the e_g^2 states.

and it follows that $U = U' + 2I$. In turn the Racah coefficients are given in terms of Slater parameters by

$$A = F^0 - \frac{49}{441}F^4, \tag{4.15}$$

$$B = \frac{1}{49}F^2 - \frac{5}{441}F^4, \tag{4.16}$$

$$C = \frac{35}{441}F^4, \tag{4.17}$$

where the F^n $(n = 0, 2, 4)$ are defined by

$$F^n = \int_0^\infty \mathrm{d}r_1 \mathrm{d}r_2 r_1^2 r_2^2 R_{32}(r_1)^2 R_{32}(r_2)^2 \frac{r_<^n}{r_>^{n+1}}, \tag{4.18}$$

involving the radial part of the atomic wave function $R_{32}(r)$ of the $3d$ orbital, and where

$$(r_>, r_<) = \begin{cases} (|\boldsymbol{r}_1|, |\boldsymbol{r}_2|) & \text{for} \quad |\boldsymbol{r}_1| > |\boldsymbol{r}_2| \\ (|\boldsymbol{r}_2|, |\boldsymbol{r}_1|) & \text{for} \quad |\boldsymbol{r}_1| < |\boldsymbol{r}_2| \end{cases}. \tag{4.19}$$

In order to clarify the effects of the electron-electron interactions between the e_g electrons, consider the e_g^2 multiplet. The $6(={}_4C_2)$ fold degeneracy is lifted by the electron-electron interactions. The states ${}^3A_2$, ${}^1A_1$ and ${}^1E$ have energies $U' - J$, $U' + J$ and $U + J$. Schematic pictures of these electron configurations are shown in Fig. 4.4. The spin triplet ${}^3A_2$ is the ground state of the e_g^2 configuration reflecting the effects of the Hund coupling energy J. This is the so-called Hund rule. In the case where the pair hopping term in (4.11) is neglected, for simplicity, the eigen states are well understood in the classical sense; the spin triplet states with the different orbitals

$$\psi_T = \left\{ |a\uparrow, b\uparrow\rangle, \quad |a\downarrow, b\downarrow\rangle, \quad \frac{1}{\sqrt{2}}(|a\uparrow, b\downarrow\rangle + |a\downarrow, b\uparrow\rangle) \right\}, \tag{4.20}$$

with energy $E_S = U' - J$, and the spin singlet state with the different orbitals

$$\psi_S = \frac{1}{\sqrt{2}}\left(|a\uparrow, b\downarrow\rangle - |a\downarrow, b\uparrow\rangle\right), \tag{4.21}$$

with energy $E_T = U' + J$, and the double occupied state in the single orbital

$$\psi_D = \{|a\uparrow, a\downarrow\rangle, \quad |b\uparrow, b\downarrow\rangle\}, \tag{4.22}$$

with energy $E_D = U$. Here a or b are orbital labels.

The energy parameters in manganites introduced above are estimated from several experiments. Through the analysis of photoemission spectroscopy (PES) and x-ray absorption spectroscopy (XAS) experiments, Saitoh et al. obtain U =7.8eV, the charge transfer energy Δ =4.5eV and $t_{pd\sigma} = -1.8$eV for $LaMnO_3$ with the assumptions $B = 0.12$eV and $C = 0.55$eV, and U =7.5eV, Δ =2.0eV and $t_{pd\sigma} = -1.5$eV for $CaMnO_3$ with $B = 0.132$eV and C =0.61eV [25]. These energy parameters are also estimated by another PES, inverse PES and XAS experiments [26,27]. The Hund coupling between e_g and t_{2g} spins is estimated from the optical spectra as $3J_H/2 \sim 2 - 3.5$eV in $La_{1-x}Sr_xMnO_3$ [28, 29]. The AFM interaction $J_{AF} \sim$ 1meV between t_{2g} spins is obtained using the mean field approximation from the Néel temperature T_N =120K of $CaMnO_3$. The on-site Coulomb interactions U and U' provide the largest energy scale in this model Hamiltonian. That the Hund coupling $3J_H/2$ is larger than the band width W follows from the almost perfect polarization of the ferromagnetic, metallic, manganites, which also implies that U and U' are much larger than W.

The study of the low energy electronic states is facilitated by a derivation of the effective model Hamiltonian through a reduction of the Hilbert space. In a standard way, such effective Hamiltonian is derived from the generalized Hubbard Hamiltonian where the e_g and t_{2g} electrons are treated in an equal footing at a Mn site. The Hilbert space is restricted as the lowest multiplet for the d^4 electron configuration, i.e. the ^5D high-spin state at each Mn site is permitted. The second-order perturbation with respect to the electron hopping between the nearest-neighboring (NN) Mn sites leads to the effective model. In this procedure, all possible intermediate states of the perturbation, consisting of the multiplets for the $3d^3$ and $3d^5$ electron configurations, are taken into account. Such a derivation of the effective Hamiltonian was performed in [30] and [31]. Instead of the complete manner, here we introduce a more simple version of the effective model Hamiltonian based on the Hamiltonian in (4.3). This is of physically perspective and is more easily applicable to the hole doped and undoped systems.

We assume, to derive the effective Hamiltonian, that the e_g electron occupancy at each Mn site is one or less. The deviations to double occupancy are eliminated by a canonical transformation which introduces a new effective interaction. The matrix elements of this interaction are

$$(\widetilde{\mathcal{H}})_{nm} = (\mathcal{H}_0)_{nm} - \frac{1}{2}\sum_l (\mathcal{H}')_{nl}(\mathcal{H}')_{lm}\left(\frac{1}{E_l - E_m} + \frac{1}{E_l - E_n}\right), \quad (4.23)$$

where E_n and n are the eigen-energy and the index of the eigen-state of $\mathcal{H}_0$, respectively and the result is

$$\mathcal{H}_{e_g} = \mathcal{H}_t + \mathcal{H}_J, \quad (4.24)$$

where a term which is proportional to ε_d has been eliminated by a shift in the energy. Here $\mathcal{H}_t$ and $\mathcal{H}_J$ are equivalent to the t and J terms of the $t-J$ model. Explicitly

$$\mathcal{H}_t = \sum_{\langle ij\rangle\sigma\gamma\gamma'} t_{ij}^{\gamma\gamma'} \left(\widetilde{d}_{i\gamma\sigma}^{\dagger}\widetilde{d}_{j\gamma'\sigma} + \text{h.c.}\right), \tag{4.25}$$

where

$$\widetilde{d}_{i\gamma\sigma} = d_{i\gamma\sigma}\left(1 - n_{i\gamma\bar{\sigma}}\right)\left(1 - n_{i\bar{\gamma}\sigma}\right)\left(1 - n_{i\bar{\gamma}\bar{\sigma}}\right), \tag{4.26}$$

and where $n_{i\gamma\sigma} = d_{i\gamma\sigma}^{\dagger}d_{i\gamma\sigma}$, is the annihilation operator which projects onto the singly occupied state. Both γ and σ take only two values and $\bar{\sigma}$ ($\bar{\gamma}$) indicates the conjugate value to σ (γ). The so-called three-site term is neglected. The interaction $\mathcal{H}_J$ has three parts which involve ψ_T, ψ_S and ψ_D from (4.20), (4.21), and (4.22) as intermediate states, i.e.,

$$\mathcal{H}_J = \mathcal{H}_T + \mathcal{H}_S + \mathcal{H}_D, \tag{4.27}$$

which are explicitly

$$\begin{aligned}\mathcal{H}_T = &-\frac{2}{E_T}\sum_{\langle ij\rangle}\left(\frac{3}{4}n_in_j + \boldsymbol{S}_i\cdot\boldsymbol{S}_j\right)\\ &\times\Bigg[\left(t^{aa2}+t^{bb2}\right)\left(\frac{1}{4}n_in_j - T_{iz}T_{jz}\right) - t^{aa}t^{bb}\left(T_{i+}T_{j-}+T_{i-}T_{j+}\right)\\ &+\left(t^{ab2}+t^{ba2}\right)\left(\frac{1}{4}n_in_j + T_{iz}T_{jz}\right) - t^{ab}t^{ba}\left(T_{i+}T_{j+}+T_{i-}T_{j-}\right)\\ &-\left(t^{ab}t^{aa}-t^{ba}t^{bb}\right)\left(T_{iz}T_{j+}+T_{iz}T_{j-}\right)\\ &-\left(t^{ba}t^{aa}-t^{ab}t^{bb}\right)\left(T_{i+}T_{jz}+T_{i-}T_{jz}\right)\Bigg],\end{aligned} \tag{4.28}$$

$$\begin{aligned}\mathcal{H}_S = &-\frac{2}{E_S}\sum_{\langle ij\rangle}\left(\frac{1}{4}n_in_j - \boldsymbol{S}_i\cdot\boldsymbol{S}_j\right)\\ &\times\Bigg[(t^{aa2}+t^{bb2})\left(\frac{1}{4}n_in_j - T_{iz}T_{jz}\right) + t^{aa}t^{bb}(T_{i+}T_{j-}+T_{i-}T_{j+})\\ &+(t^{ab2}+t^{ba2})\left(\frac{1}{4}n_in_j + T_{iz}T_{jz}\right) + t^{ab}t^{ba}(T_{i+}T_{j+}+T_{i-}T_{j-})\\ &+(t^{ab}t^{aa}+t^{ba}t^{bb})T_{jx} + (t^{ba}t^{aa}+t^{ab}t^{bb})T_{ix}\Bigg],\end{aligned} \tag{4.29}$$

and

$$\begin{aligned}\mathcal{H}_D = &-\frac{4}{E_D}\sum_{\langle ij\rangle}\left(\frac{1}{4}n_in_j - \boldsymbol{S}_i\cdot\boldsymbol{S}_j\right)\\ &\times\Big[t^{aa2}n_{ia}n_{ja} + t^{bb2}n_{ib}n_{jb} + t^{ab2}n_{ia}n_{jb} + t^{ba2}n_{ib}n_{ja}\\ &+t^{aa}t^{ab}n_{ia}T_{jx} + t^{aa}t^{ba}T_{ix}n_{ja} + t^{bb}t^{ab}T_{ix}n_{jb} + t^{bb}t^{ba}n_{ib}T_{jx}\Big].\end{aligned} \tag{4.30}$$

Here $\boldsymbol{T}_i$ is the pseudo-spin operator for the orbital degree of freedom defined in (4.2), and the indices i and j in $t_{ij}^{\gamma\gamma'}$ are omitted. Given that (1) the transfer integrals satisfy the relation in (4.6), and (2) $E_S = E_D$, reduces $\mathcal{H}_J$ to the compact form:

$$\begin{aligned}\mathcal{H}_J = &- 2J_1 \sum_{\langle ij\rangle} \left(\frac{3}{4}n_i n_j + \boldsymbol{S}_i \cdot \boldsymbol{S}_j\right)\left(\frac{1}{4}n_i n_j - \tau_i^l \tau_j^l\right) \\ &- 2J_2 \sum_{\langle ij\rangle} \left(\frac{1}{4}n_i n_j - \boldsymbol{S}_i \cdot \boldsymbol{S}_j\right)\left(\frac{3}{4}n_i n_j + \tau_i^l \tau_j^l + \tau_i^l + \tau_j^l\right), \qquad (4.31)\end{aligned}$$

where

$$\tau_i^l = \cos\left(\frac{2\pi n_l}{3}\right) T_{iz} - \sin\left(\frac{2\pi n_l}{3}\right) T_{ix}, \qquad (4.32)$$

with $(n_x, n_y, n_z) = (1, 2, 3)$. Here, the suffix l in (4.31) indicates the direction of a bond between site i and site j, the up pseudo-spin state for the operator τ_i^l corresponds to an occupied $d_{3l^2-r^2}$. The exchange interactions $J_1 = t_0^2/E_T$ and $J_2 = t_0^2/E_D = t_0^2/E_S$. Taking into account the effects of t_{2g} spins in the mean-field sense, the energy of the intermediate states are re-interpreted as $E_D = U' - J - 2J_H$ and $E_S = U - 2J_H$. We note that each term in $\mathcal{H}_J$ is a product of spin and orbital parts. This is the case because the second-order electron hopping perturbational processes change both the spin and orbital states simultaneously at the i and j sites. It is worth noting the following characteristics of the orbital part: (1) the Hamiltonian does not preserve a continuous symmetry, such as SU(2), *in the orbital space*, (2) even the total z-component of the pseudo-spin, $\sum_i T_{iz}$, is not a conserved quantity, (3) terms linear in the τ_i^l arise. These features are attributed to the off-diagonal, $\gamma \neq \gamma'$, part of the transfer integral $t_{ij}^{\gamma\gamma'}$, (4) the y component of the pseudo-spin operator T_y which indicates the magnetic octupole does not appear in $\mathcal{H}_J$.

In order to clarify roles of the spin and orbital interactions described by this Hamiltonian, we further assume a diagonal $t_{ij}^{\gamma\gamma'} = \delta_{\gamma\gamma'} t$. The Hamiltonian (4.27) then simplifies to

$$\begin{aligned}\mathcal{H}_J = &- 2J_1 \sum_{\langle ij\rangle} \left(\frac{3}{4}n_i n_j + \boldsymbol{S}_i \cdot \boldsymbol{S}_j\right)\left(\frac{1}{4}n_i n_j - \boldsymbol{T}_i \cdot \boldsymbol{T}_j\right) \\ &- 2J_2 \sum_{\langle ij\rangle} \left(\frac{1}{4}n_i n_j - \boldsymbol{S}_i \cdot \boldsymbol{S}_j\right)\left(\frac{3}{4}n_i n_j + \boldsymbol{T}_i \cdot \boldsymbol{T}_j\right). \qquad (4.33)\end{aligned}$$

Now the orbital parts involve only the scalar product of the pseudo-spins. Then, since $U' - J$ is smaller than U and $U' + J$, the first term dominants, favoring staggered orbital and the ferromagnetic (FM) spin order. This FM interaction originates from the Hund coupling J at the same site. (The second term in (4.33) favors uniform orbital order, and as expected in the superexchange interaction in the non-degenerate orbital systems the spins prefer to order anti-ferromagnetically.)

Historically, in 1953, Van Vleck [32] considered the exchange interaction in the case where the orbital degeneracy is present. Goodenough [33] and Kanamori [34] studied the relation between the superexchange interaction and the symmetry of the electron orbitals, and the explicit form of the exchange interactions was derived by Roth [35]. The spin-orbital model Hamiltonian where the orbital degree of freedom is explicitly taken into account was first derived by Kugel and Khomskii [36], and studied many other authors [37–39]. Following the discovery of the CMR effects, this kind of Hamiltonian was recently reinvestigated: The Hamiltonian (4.24) is derived in [40] and is applied to the theoretical analysis in hole doped and undoped manganites. The model Hamiltonians, where the full multiplets at a single Mn site are considered, are derived in [30] and [31].

Apart from the Mott insulating state by doping of holes, another type of ferromagnetic interaction is derived by the Hamiltonian $\mathcal{H}_t + \mathcal{H}_J$, that is, the double exchange interaction [8, 9, 41, 42]. This originates from the intra-atomic Hund coupling J_H between e_g and t_{2g} spins and electron transfer. Consider a pair of NN Mn^{3+} and Mn^{4+}, neglecting the orbital degeneracy for simplicity. At the Mn^{3+} site, there is one e_g electron and a localized t_{2g} spin with $S = 3/2$, and the eigen states have a total spin quantum number $S_{\text{tot}} = S + 1/2$ or $S - 1/2$ and energies $-J_H S$ or $J_H(S+1)$. Because the Hund coupling is much larger than the inter-site electron transfer integral, only the ground state with $S_{\text{tot}} = S + 1/2$ is involved. The Mn^{3+} has no e_g electron but does have a localized t_{2g}, $S = 3/2$, local moment. The local axis of quantization is taken to be along the direction of the local moment at both sites and expectation value for the electron transfer term is evaluated assuming that these directions are canted relative to each other by an angle θ, see Fig. 4.5. The up spin wave function defined at site i is a linear combination of the up and down states defined at site j, i.e.,

$$|\uparrow_i\rangle = \cos\left(\frac{\theta}{2}\right)|\uparrow_j\rangle + \sin\left(\frac{\theta}{2}\right)|\downarrow_j\rangle. \tag{4.34}$$

Define, e.g., $|d_i\rangle$ to be the state where the e_g orbital at site i is occupied by an electron, then, the required matrix elements involving $|d_i\rangle|\uparrow_i\rangle$ and $|d_j\rangle|\uparrow_j\rangle$ are given by

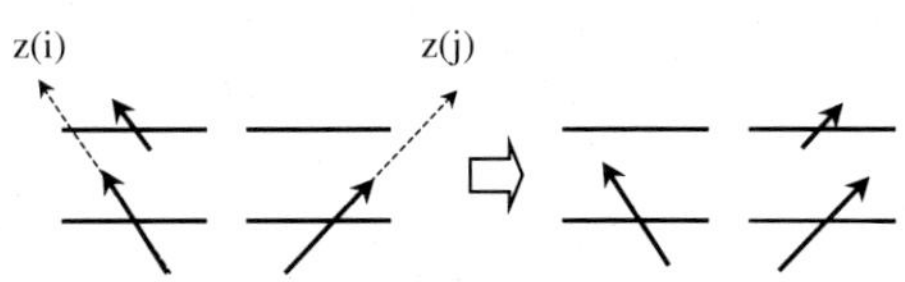

Fig. 4.5. Electron hopping process and the local spin coordinates in the double exchange interaction. The long and short arrows indicate the t_{2g} and e_g spins, respectively. Dotted arrows are the z-axis in the local spin coordinates.

$$E = \begin{pmatrix} -J_H S \, , & b\cos\left(\dfrac{\theta}{2}\right) \\ b\cos\left(\dfrac{\theta}{2}\right) \, , & -J_H S \end{pmatrix}, \tag{4.35}$$

with $b = \langle d_i | \mathcal{H}_t | d_j \rangle$. Diagonalizing this matrix, the lowest energy is trivially given by

$$E_0 = -J_H S - b\cos\left(\frac{\theta}{2}\right), \tag{4.36}$$

which is minimal for $\theta = 0$. Thus the combination effects of the Hund coupling and electron transfer introduces a FM *double exchange interaction* between Mn spins. It is noted that the energy gain due to this double exchange interaction is proportional to $\cos(\theta/2)$, in contrast to $\cos\theta$ for the superexchange interaction.

4.1.3 Jahn-Teller Effect and Cooperative Jahn-Teller Effect

It is widely recognized that, in addition to the electron-electron interaction, the electron-lattice interaction plays key roles on the electronic states in the CMR manganites. More specifically it is important to account for the Jahn-Teller effect. An ion such as Mn^{3+}, with an orbitally degenerate, ground state is usually "Jahn-Teller active", i.e., susceptible to the Jahn-Teller effect. Consider an isolated $Mn^{3+}O_6^{2-}$ octahedron. The potential for the e_g electron at a position $\boldsymbol{r}$ arising from the surrounding O ions at $\boldsymbol{R}_i$ ($i = 1 \sim 6$) is expanded with respect to the normal coordinates of the O vibrations. These are denoted by Q_α where α is an index classifying the normal coordinates. This is given by

$$V(\boldsymbol{r}, \boldsymbol{R}) = V(\boldsymbol{r}, \boldsymbol{R}_0) + \sum_\alpha \frac{\partial V(\boldsymbol{r}, \boldsymbol{R})}{\partial Q_\alpha} Q_\alpha \bigg|_{\boldsymbol{R}=\boldsymbol{R}_0} + O(Q_\alpha^2), \tag{4.37}$$

where $\boldsymbol{R}$ is a symbolic notation denoting the coordinates of all O ions $\boldsymbol{R}_i$ ($i = 1, 6$), and $\boldsymbol{R}_0$ is the equilibrium value of $\boldsymbol{R}$. In the absence of an orbital degeneracy the matrix elements of the second term vanishes. However when there exists a non-Kramers degeneracy, i.e., one which is not a consequence of time reversal symmetry, the second term in (4.37) is usually finite. When the octahedron distorts this costs a lattice energy initially proportional to Q_α^2 whereas the energy gain form (4.37) is linear, and as a result the initial

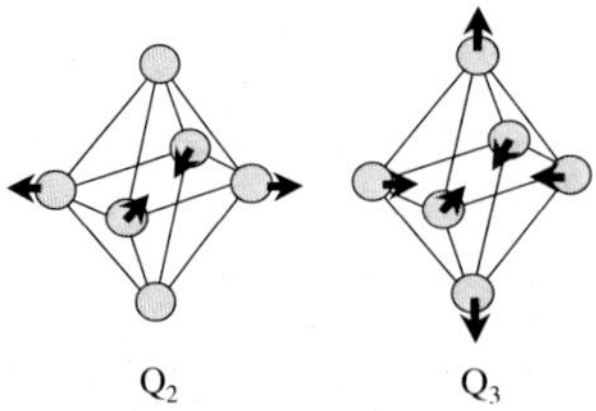

Fig. 4.6. The normal coordinates of the O ion distortions in a MnO_6 octahedra.

structure is unstable. This is the so-called Jahn-Teller effect [43]. Specializing to the doubly degenerate e_g orbitals of current interest, the second term in (4.37) is finite for the following normal coordinates of the MnO_6 octahedra (Fig. 4.6):

$$Q_2 = \frac{1}{2}\left[q_x(\mathrm{d}\hat{x}) - q_y(\mathrm{d}\hat{y}) - q_x(-\mathrm{d}\hat{x}) + q_y(-\mathrm{d}\hat{y})\right], \tag{4.38}$$

$$Q_3 = \frac{1}{12}\big[2q_z(\mathrm{d}\hat{z}) - 2q_z(-\mathrm{d}\hat{z}) \\ - q_x(\mathrm{d}\hat{x}) - q_y(\mathrm{d}\hat{y}) + q_x(-\mathrm{d}\hat{x}) + q_y(-\mathrm{d}\hat{y})\big]. \tag{4.39}$$

where $\boldsymbol{q}(\boldsymbol{R})$ is the displacement vector of the O ions at the position $\boldsymbol{R}$ in a octahedron, and d is the Mn-O distance. The linear interaction between the e_g electrons and these modes is

$$\mathcal{H}_{ele-lat} = g(Q_2 T_x - Q_3 T_z), \tag{4.40}$$

where T_z and T_x are the pseudo-spin operators introduced in (4.2), and

$$g = \int \mathrm{d}\boldsymbol{r} \phi_{\gamma'}(\boldsymbol{r})^* \frac{\partial V(\boldsymbol{r},\boldsymbol{R})}{\partial Q_\alpha} \phi_\gamma(\boldsymbol{r}). \tag{4.41}$$

Diagonalizing $\mathcal{H}_{ele-lat}$ including the harmonic lattice potential, treating Q_2 and Q_3 as classical quantities, gives and effective potential energy

$$U(\boldsymbol{R})_\pm = \frac{1}{2}K(Q_2^2 + Q_3^2) \pm g\sqrt{Q_2^2 + Q_3^2}, \tag{4.42}$$

where K is the spring constant for both modes. The coordinates Q_α define a two dimensional space and defining an amplitude ρ and phase θ in this space

$$Q_2 = \rho \sin\theta, \tag{4.43}$$

$$Q_3 = \rho \cos\theta, \tag{4.44}$$

and the adiabatic potential

$$U(\boldsymbol{R})_\pm = \frac{1}{2}K\rho^2 \pm g\rho = \frac{1}{2}K\left(\rho \pm \frac{g}{K}\right)^2 - \frac{1}{2}\frac{g^2}{K}, \tag{4.45}$$

depends only on the positive semi-definite ρ. The lowest energy corresponds to $U_-(\boldsymbol{R})$ at $\rho = g/K$ independent of θ. This rotational symmetry broken by the anharmonic lattice potential

$$H = B(Q_3^3 - 3Q_2^2 Q_3), \tag{4.46}$$

and the higher order Jahn-Teller interaction

$$H = C\left[(Q_3^2 - Q_2^2)T_z - 2Q_2Q_3T_x\right]. \tag{4.47}$$

In this way the degeneracy of the electronic state is lifted by $H_{ele-lat}$ and the lowest eigen-state is given by

$$|\theta\rangle = \cos\left(\frac{\theta}{2}\right)|d_{3z^2-r^2}\rangle + \sin\left(\frac{\theta}{2}\right)|d_{x^2-y^2}\rangle. \tag{4.48}$$

Except for the absence of the phase factor $e^{i\phi}$, this is the same as (4.1) and implies that the Jahn-Teller interaction stabilizes a *real* orbital wave function. Although the double degeneracy of the e_g orbital is lifted by this interaction, the ground state wave function contains the parameter θ and is therefore not uniquely determined by the linear Jahn-Teller interaction.

Notice when $\theta \to \theta + 2\pi$ the wave function changes sign, i.e. these wave functions are double valued in the θ plane. Here we have an example of a Berry phase [44,45]. Such phases arise when a system with the fast quantum variable, e.g. electron, couples with the slow classical variable, e.g. the lattice. When the fast variable moves adiabatically through a path in the parameter space, the wave function acquires a phase in addition to the conventional kinetic phase. This additional phase is the Berry phase γ_B which is related to the solid angle of the trajectory of the cyclic motion around a topological singular point. In the present case, the degenerate point $\rho = 0$ in the $Q_2 - Q_3$ plane is singular. The Berry phase associated with a adiabatic motion around this point is given by $\gamma_B = \Omega/2 = (2\pi)/2$ where $\Omega(= 2\pi)$ is the solid angle in the two dimensional parameter space, and the factor 1/2 comes from the magnitude of the pseudo-spin operator $\boldsymbol{T}$. Although the electronic wave function is double valued with respect to the parameter θ, the total wave function should be single valued, i.e. the wave function of the nucleus is also double valued with respect to θ. This has an important consequence: the phase mode of the ionic motion, which corresponds to a rotation along the minimum path of the adiabatic potential, is characterized by a half integer angular momentum. The consequences of this are actually observed in Na_3 molecules by the resonant two-photon ionization spectrum [46].

The orbital-lattice interaction in a solid is qualitatively different from that in a single molecule since the O distortions of neighboring octahedra are no longer independent with each other. The energy reduction due to the lifting of the orbital degeneracy drives a structural phase transition referred to as a "cooperative Jahn-Teller effect". [47,48]. The first fundamental treatment of the cooperative phenomena has been done by Kanamori [49]. As a specific example consider the electron lattice interactions for the perovskites $R\mathrm{MnO}_3$ with a number of Mn ions N and a volume V. The displacements of the O ions contain two components (1) the normal coordinate vibrations of the solid, i.e. phonons, and (2) the uniform lattice strains, which is not represented by the phonon at the Γ point. The Hamiltonian for the system where the e_g electrons coupled with such O displacements is given by

$$\mathcal{H} = \mathcal{H}_{el-ph} + \mathcal{H}_{ph} + \mathcal{H}_{el-st} + \mathcal{H}_{st}. \tag{4.49}$$

The first term

$$\mathcal{H}_{el-ph} = \sum_{\boldsymbol{k}\xi l} \sqrt{\hbar\omega_{\boldsymbol{k}\xi}}\, g_{\boldsymbol{k}\xi l} T_{kl} q_{\boldsymbol{k}\xi}, \tag{4.50}$$

describes the coupling of the e_g electron with the Jahn-Teller phonons and originates from the Jahn-Teller coupling;

$$\mathcal{H}_{JT} = g \sum_i (Q_{2i} T_{ix} - Q_{3i} T_{iz}), \tag{4.51}$$

where Q_{i2} and Q_{i3} are the Q_2- and Q_3-modes of the O_6 octahedron around the i-th Mn site. In order to obtain $\mathcal{H}_{el-ph}$ from (4.51) the Q-modes are expressed in terms of the lattice vibration $q_{\boldsymbol{k}\xi}$ normal coordinates with a phonon mode ξ and momentum $\boldsymbol{k}$. The $\omega_{\boldsymbol{k}\xi}$ are the phonon frequencies, and T_{kl} $(l = x, z)$ is the Fourier transform of the pseudo-spin operator for orbital. The resulting coupling constant $g_{\boldsymbol{k}\xi l}$ is proportional to g (4.51). The phonon Hamiltonian

$$\mathcal{H}_{ph} = \sum_{\boldsymbol{k},\xi} \frac{\hbar \omega_{\boldsymbol{k}\xi}}{2} (p^*_{\boldsymbol{k}\xi} p_{\boldsymbol{k}\xi} + q^*_{\boldsymbol{k}\xi} q_{\boldsymbol{k}\xi}), \tag{4.52}$$

is written in terms of $q_{\boldsymbol{k}\xi}$ and the canonical conjugate momenta $p_{\boldsymbol{k}\xi}$. Then

$$\mathcal{H}_{el-st} = g_0 \sqrt{\frac{c_0}{N}} (u_x T_{\boldsymbol{k}=0x} + u_z T_{\boldsymbol{k}=0z}), \tag{4.53}$$

reflects the coupling of the electronic orbitals to the uniform bulk strains

$$u_z = \frac{1}{\sqrt{6}} (2e_{zz} - e_{xx} - e_{yy}), \tag{4.54}$$

and

$$u_x = \frac{1}{\sqrt{2}} (e_{xx} - e_{yy}), \tag{4.55}$$

which are expressed in terms of the elastic strain $e_{\alpha\beta}$ $(\alpha, \beta = x, y, z)$ and where c_0 is the elastic constant. The last term of (4.49)

$$\mathcal{H}_{st} = \frac{c_0}{2} (u_x^2 + u_z^2), \tag{4.56}$$

reflects the elastic energy. The linear coupling term between the orbitals and lattice is eliminated by the canonical transformation

$$\widetilde{q}_{\boldsymbol{k}\xi} = q_{\boldsymbol{k}\xi} - \frac{1}{\sqrt{\hbar \omega_{\boldsymbol{k}}}} \sum_l g_{\boldsymbol{k}\xi l} T_{\boldsymbol{k}l}. \tag{4.57}$$

In terms of these new coordinates and their canonical conjugate momentum $\widetilde{p}_{\boldsymbol{k}\xi}$ along with the τ_i^l defined by (4.32), one finds a Hamiltonian where the orbital and phonon degrees of freedom are decoupled as a matter of form. When we assume that the O distortions are restricted to along the Mn-O bond, and the lattice potential energy originates from the NN Mn-O bond, we have a simple form

$$\mathcal{H}_{el-ph} + \mathcal{H}_{ph} = \frac{g^2}{K} \frac{4}{9} \sum_{\langle ij \rangle} \tau_i^l \tau_j^l + \sum_{\boldsymbol{k}\xi} \frac{\hbar \omega_{\boldsymbol{k}}}{2} (\widetilde{p}^*_{\boldsymbol{k}\xi} \widetilde{p}_{\boldsymbol{k}\xi} + \widetilde{q}^*_{\boldsymbol{k}\xi} \widetilde{q}_{\boldsymbol{k}\xi}), \tag{4.58}$$

where it is assumed that $\widetilde{q}_{\boldsymbol{k}\xi}$ commute with the Hamiltonian, and the self-interactions are neglected. The first term in (4.58) is an effective interaction between orbitals mediated by the virtual exchange of phonons, termed J_{ph}.

It is worth noting that this is of a similar form to the interaction due to the virtual exchange of electrons for strong electron-electron interactions, termed J_{ele}, in (4.31). Both interactions are positive, i.e. g^2/K and $t^2/(U'-I)$, and tend to cooperatively stabilize the staggered orbital order. Thus, one may have a question which interactions are responsible for orbital ordering when it occurs. We introduce an analysis where the two different interactions are estimated in $LaMnO_3$ quantitatively by using (4.31) and (4.58) [50]. The orbital order temperature T_{OO}(=780K) is represented by a function of J_{ele}, J_{ph} and the orbital order parameters $\langle T_z \rangle$ and $\langle T_x \rangle$ in the mean field approximation. Independently, J_{ele} is determined by the Néel temperature of the A-type AFM order (T_N=140K) along with the spin stiffness constant of the spin waves. The best fitting of the exchange interactions are given for the parameter values $J_1 \sim$ 75-85meV, $J_2 \sim$25-40 meV and $g^2/K \sim$10meV [50] which indicates that the electronic contribution to the orbital interaction in $LaMnO_3$ is larger than the phononic contribution.

4.1.4 Phase Diagram and Orbital Order

The CMR manganites are remarkable for the variety and complexity of their electronic and structural phases. Transitions between these phases occur upon, e.g., changing temperature/carrier concentration and/or applying external fields. The CMR phenomena itself arises from a dramatic change of the electronic state of the system which occurs in the vicinity of the phase boundary when a magnetic field is applied. In this subsection, we summarize the phase diagram of manganite $R_{1-x}A_x\mathrm{MnO}_3$ with pseudo-cubic perovskite structure from the view point of the electronic spin-charge-orbital degrees of freedom. It is well known that the electronic and structural phases of this compound can be varied by changing the trivalent cation R and the divalent cation A. The MnO_2 plane is stretched or compressed due differences in the ionic radii. This size effects are characterized by the tolerance factor:

$$t = \frac{r_{R(A)} + r_{\mathrm{O}}}{\sqrt{2}(r_{\mathrm{Mn}} + r_{\mathrm{O}})}, \tag{4.59}$$

where $r_{R(A)}$, r_{O} and r_{Mn} are the ionic radii for $R(A)$, O and Mn ions, respectively. The MnO_2 plane is stretched or compressed as t is greater of less that unity. Almost all manganites with pseudo-cubic structure, have a value of t which is less than one. It is known that the system tends to be insulating (metallic) in the case of small (large) tolerance factors arising from a small (large) ionic radius of R and/or A ions [51]. This has an effect on the electron transfer between NN Mn ions through the deviations of the NN Mn-O-Mn bond angle from 180°. That is, a large bending of the Mn-O bond from 180° implies the small overlap of the Mn $3d$ and O $2p$ wave functions.

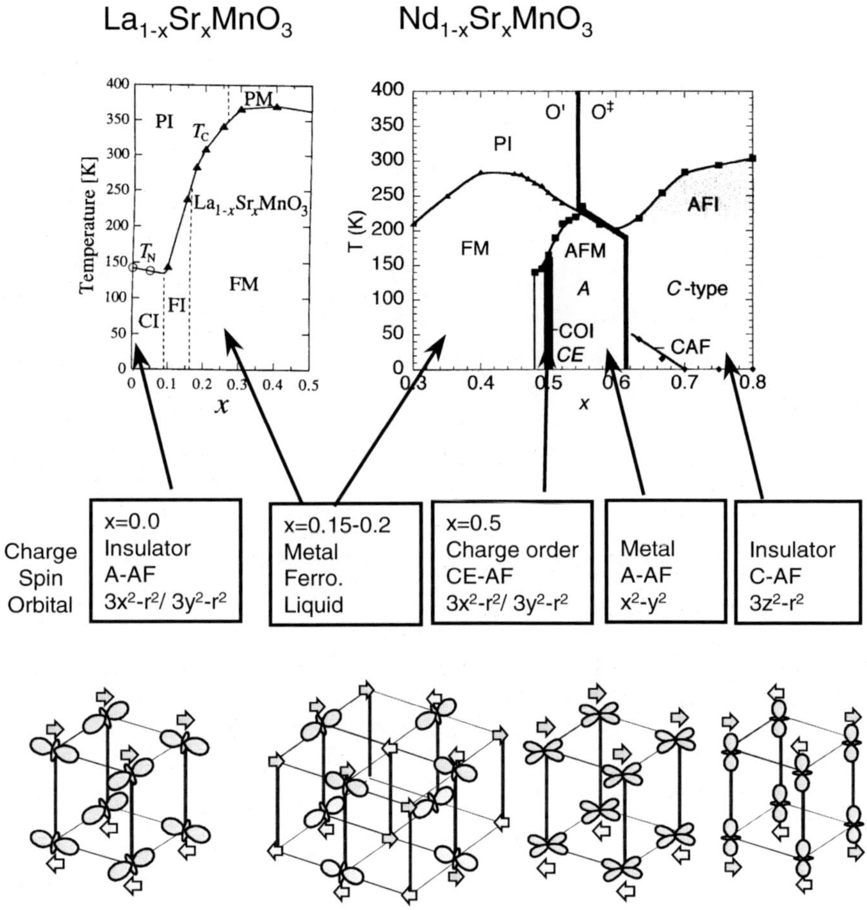

Fig. 4.7. Phase diagram of $La_{1-x}Sr_xMnO_3$ (left), and $Nd_{1-x}Sr_xMnO_3$ (right), and schematic pictures of the orbital ordered states [52, 53].

Figure. 4.7 shows, as a function of the hole concentration x, the schematic electronic phase diagram for manganites with a larger tolerance factor, i.e. $La_{1-x}Sr_xMnO_3$ and $Nd_{1-x}Sr_xMnO_3$, [52,53]. This diagram might be formally classified into the following regions of x:

(I) $x = 0$ (RMnO$_3$): the compound is a Mott insulator due to the strong electron-electron interaction between the Mn 3d electrons. The staggered orbital order associated with the Jahn-Teller type lattice distortion appears below 780K in $LaMnO_3$ where $3d_{3x^2-r^2}$- and $3d_{3y^2-r^2}$-type orbitals are aligned alternately in the ab plane and uniformly along the c axis. A-type AFM order, where Mn spins align ferromagnetically in the ab plane and antiferromagnetically along the c axis, appears

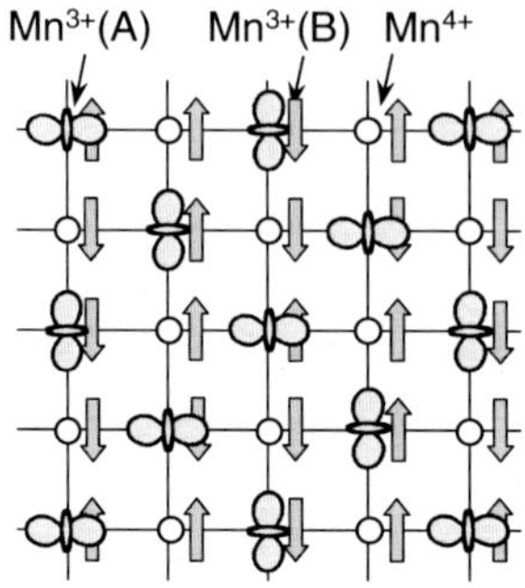

Fig. 4.8. Charge and orbital ordered state in the *ab* plane of $R_{0.5}A_{0.5}MnO_3$. Open circles indicate Mn^{4+} ions.

below 140K. The dispersion relation of the spin waves in this phase is anisotropic where the ratio of the exchange interactions is about $|J_{ab}|/|J_c| = 1.4$. The anisotropies in both the sign and magnitude of the exchange interaction are attributed to orbital order; the stronger ferromagnetic J_{ab} originates from the staggered type of the e_g orbitals in the *ab* plane and the weaker antiferromagnetic J_c caused by J_2 in (4.31) and J_{AF} in (4.8).

(II) $x \sim 0.1$: Upon doping holes into RMnO$_3$, the AFM component of the spin order is reduced. Spins align ferromagnetically, although the electric resistivity exhibits insulating behavior. A variety of electric, magnetic and structural properties are observed experimentally and predicted theoretically, e.g. the spin canting, spiral and flux orders, electronic phase separation associated with different magnetic, structural and orbital states, magnetic field induced structural phase transitions, and so on. Some of them are discussed in more detail in Sect. 4.5. It seems that some of them are directly related to origin of the CMR. However, theory and experiments in this region are still up-to-date and further progress of researches are required.

(III) $0.125 < x < 0.5$: the FM metallic phase is realized at low temperatures. The magnetic moment is almost perfect polarized. This implies that the Hund coupling is larger than the band width. There is no experimental evidence of the long range charge and/or orbital order. It is proposed theoretically that the orbital state is a liquid and/or disordered. These ideas are introduced in Sect. 4.1.5 in more detail. In the low hole doped region, the electric resistivity shows insulating behavior above the Curie temperature and the CMR effect is observed in the vicinity of this temperature.

(IV) $x \sim 0.5$: A charge ordered phase appears in $Nd_{0.5}Sr_{0.5}MnO_3$. The stability of this phase however strongly depends on a kind of cations R and A, e.g. it exists for $0.3 < x < 0.7$ for $Pr_{1-x}Ca_xMnO_3$, but there is no charge ordered phase in $La_{1-x}Sr_xMnO_3$. Two kinds of Mn ions with different charge valence are aligned alternately in the *ab* plane and uniformly along the *c* axis. The unit cell of the charge order is $\sqrt{2}a \times \sqrt{2}b \times c$ where a, b and c are the lattice constants of the cubic cell. At the same temperature of the ordering, the $3d_{3x^2-r^2}$- and $3d_{3y^2-r^2}$-type orbitals has alternately ordered in the ab plane with a $2\sqrt{2}a \times \sqrt{2}b \times c$ unit cell. Below the charge and orbital ordering temperature, CE-type AF order with a $2\sqrt{2}a \times 2\sqrt{2}b \times 2c$ unit cell appears (see Fig. 4.8).

(V) $0.5 < x < 0.7$: there is A-type AFM order. However, this A-type AFM metallic state is much different from that realized in the insulating $LaMnO_3$. In the vicinity of the Néel temperature, the crystal lattice is compressed along the c axis, and uniform $3d_{x^2-y^2}$ orbital order may be realized below T_N. In $Nd_{1-x}Sr_xMnO_3$ with $x \sim 0.6$, with decreasing temperature, the ferromagnetic phase appears at $T_C \sim 280$K and is replaced by an A-type AF order at $T_N \sim 140$K.

(VI) $0.7 < x < 1$: there is C-type AFM order, where spins align ferromagnetically along a certain direction and antiferromagnetically in the plane perpendicular to these one dimensional rods. Since the MnO_6 octahedron are elongated along this FM rod, the $3d_{3z^2-r^2}$ orbital may align along this direction.

(VII) $x = 1$ ($AMnO_3$): There are no Mn e_g electrons and the system becomes a Mott insulator. The AFM interaction J_{AF} between the localized t_{2g} spins leads to G-type (NaCl-type) AFM order.

Some parts of the electronic phase diagram are well reproduced by theoretical calculations. The correlation between a variety of spin and orbital orders in some hole concentration region have been examined by Goodenough [33]. De Gennes calculated the mean-field phase diagram at finite temperature (T) and hole concentration (x) [10]. Here we briefly introduce the calculation method; the theory is based on the Hamiltonian shown in (4.3) with (4.8), (4.9), (4.25) where both the orbital degree of freedom and the Coulomb interaction between the e_g electrons are neglected. The spin operators $\boldsymbol{S}_i$ is treated as the classical vectors characterized by the polar angle θ_i and the azimuthal angle ϕ_i. A thermal distribution of $\boldsymbol{S}_i$ is determined by the distribution function:

$$w_i(\boldsymbol{S}_i) = \frac{1}{\nu_i} \exp\left(\boldsymbol{\lambda}_i \cdot \frac{\boldsymbol{S}_i}{|\boldsymbol{S}_i|}\right), \tag{4.60}$$

where $\boldsymbol{\lambda}_i$ is the mean field acting on $\boldsymbol{S}_i$ and

$$\nu_i = \int_0^{\pi} \mathrm{d}\theta_i \int_0^{2\pi} \mathrm{d}\phi_i \sin\theta_i \exp(|\boldsymbol{\lambda}_i| \cos\theta_i), \tag{4.61}$$

ν_i is the normalization factor. As a function of $\boldsymbol{S}_i$ the expectation value of a local operator O_i,

$$\langle O_i \rangle = \int_0^{\pi} \mathrm{d}\theta_i \int_0^{2\pi} \mathrm{d}\phi_i \sin\theta_i \exp(|\boldsymbol{\lambda}_i| \cos\theta_i) O_i(\theta_i, \phi_i). \tag{4.62}$$

The free energy is given by a summation of the expectation value of the Hamiltonian and an entropy term:

$$\mathcal{F} = \langle \mathcal{H} \rangle_{st} - TN\mathcal{S}, \tag{4.63}$$

where N is the number of Mn ions and where and the entropy is defined by

$$\mathcal{S} = -\langle \ln w(\boldsymbol{S}) \rangle. \tag{4.64}$$

The order parameter is identified as

$$M = \left\langle \frac{\boldsymbol{S} \cdot \boldsymbol{\lambda}}{|\boldsymbol{S}||\boldsymbol{\lambda}|} \right\rangle. \tag{4.65}$$

The mean field solutions are obtained by minimizing $\mathcal{F}$ with respect to the angles and magnitudes of $\boldsymbol{\lambda}_i$. Through the calculation of the phase diagram, a spin canting phase intermediate between AFM and FM phases is predicted in the small hole doping region.

After the discovery of CMR, the spin-charge-orbital and structural phase diagram has been theoretically re-investigated from the several points of view [54–56]. The spin and orbital phase diagram presented in Fig. 4.9 is based on the Hamiltonian (4.3) with (4.8), (4.9), (4.25) and (4.27) [57]. The mean-field theory proposed by de Gennes is modified to be applied to the orbital degenerate system with electron correlation [57]. Starting from $x = 0$, with increasing x the spin structure changes from A-type AFM $\rightarrow$ FM $\rightarrow$ A-type AFM $\rightarrow$ G-type AFM. The orbital states also change; staggered and uniform orbital orders are stabilized in the regions of $x \leq 0.25$ and $x \geq 0.25$, respectively. These spin and orbital ordered states in the regions of low and high hole concentration are dominated by $\mathcal{H}_J$ and $\mathcal{H}_t$, i.e., the superexchange and double exchange interactions, respectively. The ordering temperatures, T_C and T_N for the spin, and T_{OO} for the orbitals, for $Pr_{1-x}Sr_xMnO_3$ and $Nd_{1-x}Sr_xMnO_3$ [52, 53] are qualitatively reproduced in Fig. 4.9 except for a narrow region of charge ordering in $Nd_{0.5}Sr_{0.5}MnO_3$.

4.1.5 Orbital Liquid State

The existence of long-range orbital order has been well established in the low hole-concentration region of manganites with the pseudo-cubic perovskite, i.e., $La_{1-x}Sr_xMnO_3$ ($x < 0.1$), where the system is a Mott insulator (a charge-transfer insulator) with A-type AFM order. As x increases the system shows metallic conduction below the FM transition temperature T_C. In this phase ($x > 0.2$), several anomalous features are observed in experiments.

(1) The optical conductivity shows a broad "Drude like" component up to around 1eV which grows with decreasing temperature [28]. The integrated oscillator strength associated with this broad band is defined by

$$\widetilde{N}_{\text{eff}} = \frac{2m}{\pi e^2 N} \int_0^{\omega_c} \widetilde{\sigma}(\omega) \mathrm{d}\omega, \tag{4.66}$$

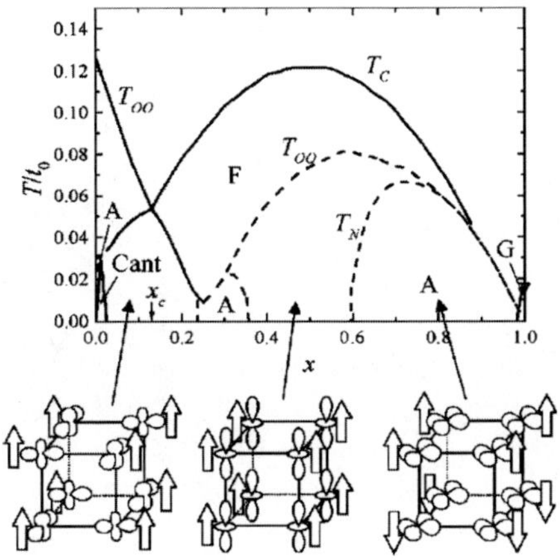

Fig. 4.9. The spin and orbital phase diagram as a function of hole concentration (x) and temperature (T). The temperatures T_C, T_N and T_{OO} are for the ferromagnetic Curie temperature, Néel temperature and orbital ordering temperature, respectively. The labels F, A and G indicate ferromagnetic, A-type antiferromagnetic and G-type antiferromagnetic phases, respectively. The solid (broken) lines are for the second (first) order phase transition. [57].

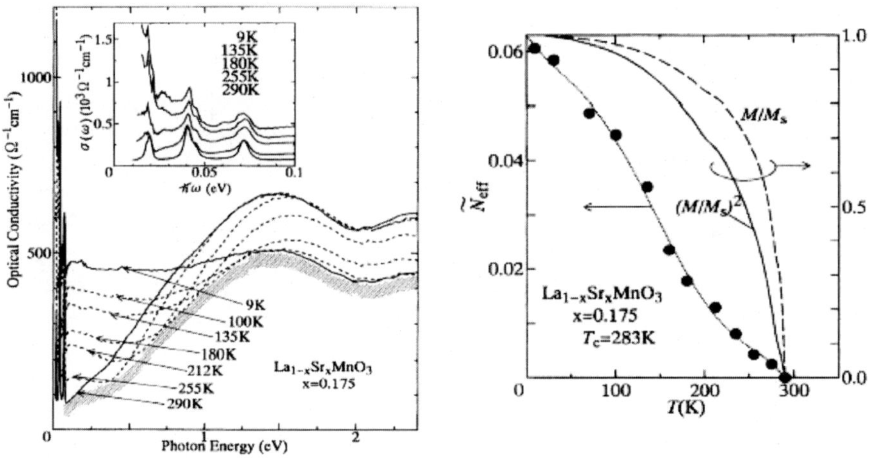

Fig. 4.10. (right): The optical conductivity spectra of for $La_{0.825}Sr_{0.175}MnO_3$. The inset shows a magnification of the far-infrared part. (left): Temperature dependence of the effective number of electrons ($\widetilde{N}_{\rm eff}$). The dashed and solid lines show the temperature dependence of the normalized FM magnetization and its square [28].

where $\widetilde{\sigma}(\omega)$ is the optical conductivity obtained after subtracting a temperature independent part, and ω_c is a cut off energy. This strength $\widetilde{N}_{\rm eff}$ continues to change down to very low temperature despite the fact that the magnetic moment has already saturated, as shown in Fig. 4.10. This suggests that other non-spin degrees of freedom still remain active.

(2) The photoemission spectra show a small discontinuity at the Fermi energy E_F followed by a gap [26,58]. On the other hand, the specific heat is not enhanced with $\gamma \sim 5\text{mJ/K}^2\text{mol}$.

(3) The anisotropy of the conduction and the spin excitations, which is expected with the orbital ordering, are *not* observed, even at low temperatures [59].

Because of the strong on-site electron-electron interactions and the Jahn-Teller coupling, mean-field theory implies that the orbital degree of freedom is almost perfectly polarized and ordered. However, the stability of the orbital ordered states strongly depends on type of the magnetic ordering [54]; the energy barriers between the stable orbital ordered states in the A-type AFM phase are much higher than that in the FM phase. That is, in the anisotropic magnetic order, spin and orbital orders stabilize with each other. On the other hand, large orbital fluctuations are expected in the FM phase. The above experimental and theoretical results suggest, in the FM state, orbital ordering is prevented, and some kind of disordered or liquid orbital state may be realized. This has been called the orbital liquid state [60].

Apart from the orbital system, consider the spin system without randomness. Some class of antiferromagnets are classified as "spin disordered" state or "spin liquid/fluid" state. One of the experimental definition characterizing this class of materials is that the actual magnetic-transition temperature T_C is significantly reduced from that expected from the Weiss constant deduced from the high temperature magnetic susceptibility, corresponding to the mean-field transition temperature. Between these two temperatures, the long-range magnetic ordering is prevented from some reasons, and the system is recognized to be a liquid/fluid state. One of the sources of this liquid state is the geometrical frustration. A well known example is the antiferromangetic Ising model with NN exchange interaction on the triangle lattice which does not show a long-range order at finite temperature. The Kagome and pyrochlore lattices are another examples for the geometrical frustrated systems. Alternative reason of the spin liquid is the low dimensionality. In the spin one-half Heisenberg antiferromagnetic chain, the Neel state is destroyed by quantum fluctuations. The thermal fluctuation also realizes a spin liquid state in two dimensional Heisenberg model at finite temperature. As for the orbital system of the present interest, the two degenerate orbital degree of freedom is described by the pseudo-spin with a magnitude 1/2 where the large quantum fluctuation is expected. On the other hand, the perovskite crystal lattice is essentially the three dimensional network where the frustrated geometry is not seen in this lattice itself. Here we introduce some possible scenario for the orbital liquid state expected in the FM manganites.

Start from the generalized Hubbard Hamiltonian with orbital degeneracy given in (4.4), (4.9) and (4.11). The fact that the on-site electron-electron interactions described by $\mathcal{H}_{el-el}$ are larger than the band width in the manganites implies that not only doubly electron occupied sites can be excluded but also states which have large energies due to the Hund's rule energy J_H. Use is made of the auxiliary-particle method, i.e., the physical electron is written as

$$\widetilde{d}_{i\gamma\sigma} = h_i^\dagger b_{i\gamma}^t b_{i\sigma}^s, \tag{4.67}$$

where the operators h_i, $b_{i\gamma}^t$, and $b_{i\sigma}^s$ correspond to a holon (fermion), a pseudo-spinon (boson), and a spinon (boson), respectively. This is a generalization of the methods described in Chat. 3 to the orbital degenerate case. The spin and orbital configuration $b_{i\gamma}^{\dagger t} b_{i\sigma}^{\dagger s}|\rangle$ in the absence of a hole contains two bosons and has the quantum numbers γ and σ. Here $|\rangle$ is the vacuum in the auxiliary particle space and has no physical counterpart. In order to create the configuration with a single hole, i.e., $h_i^\dagger|\rangle$ from $b_{i\gamma}^{\dagger t} b_{i\sigma}^{\dagger s}|\rangle$ it is therefore necessary to use the operator $h_i^\dagger b_{i\gamma}^t b_{i\sigma}^s$. Strictly speaking the auxiliary particle constraint is $h_i^\dagger h_i + \sum_{\gamma\sigma} b_{i\gamma}^{\dagger t} b_{i\sigma}^{\dagger s} b_{i\sigma}^s b_{i\gamma}^t = 1$, however below a mean-field approximation is to be used. It is useful to define rotations in the spin and orbital spaces described by the unitary matrices U^s and U^t , i.e., new spin states are defined as $\widetilde{|\uparrow_i\rangle} = z_{i\uparrow}^s|\uparrow_i\rangle - z_{i\downarrow}^{s*}|\downarrow_i\rangle$ and $\widetilde{|\downarrow_i\rangle} = z_{i\downarrow}^s|\uparrow_i\rangle + z_{i\uparrow}^{s*}|\downarrow_i\rangle$, as introduced in (4.34), with similar definitions for the orbital states by introducing $z_{i\uparrow}^t$ and $z_{i\downarrow}^t$. Here $z_{i\uparrow}^u$ and $z_{i\downarrow}^u$ are the (1,1) and (2,1) components of the matrices U_i^u, respectively, with $u = (s,t)$. Within the mean-field approximation an arbitrary spin and orbital ground state can be obtained with suitable rotations U_i^s and U_i^t from the state which has the spin $\uparrow$ and the orbital state a. In this approximation, (4.67) is written as

$$\widetilde{d}_{i\gamma\sigma} = U_{i\gamma\gamma'}^t U_{i\sigma\sigma'}^s \bar{d}_{i\sigma'\gamma'}, \tag{4.68}$$

with

$$\bar{d}_{i\sigma\gamma} = \delta_{\sigma\uparrow}\delta_{\gamma a} h_i^\dagger. \tag{4.69}$$

The interpretation of (4.68) is that the original fermion operator $d_{i\sigma\gamma}$ is transformed into the new operator $\bar{d}_{i\sigma\gamma}$ through the rotations of the axes of quantization of the real and pseudo spins described by U_i^s and U_i^t.

It is reasonable to adopt this generalized slave fermion method to examine the possibility of the orbital liquid state since the slave fermion method is known to provide a suitable description of the ordered state and since a doped hole in such an ordered state is clearly identifiable as a spin-less and orbital-less fermion. Also it is clear that the fluctuations in the spin and orbital degrees of freedom relative to some ordered state should be bosons.

In the mean-field approximation for the holons, the transfer term in the Hamiltonian, (4.4), simplifies to

$$\mathcal{H}_t = \sum_{\langle ij\rangle} h_i \widetilde{t}_{ij} h_j^\dagger, \tag{4.70}$$

where the effective transfer for the spin- and orbital-less fermions is given by

$$\widetilde{t}_{ij} = \sum_{\sigma} z_{i\sigma}^{s*}\, z_{j\sigma}^s \sum_{\gamma,\gamma'} z_{i\gamma}^{t*}\, t_{ij}^{\gamma\gamma'}\, z_{j\gamma'}^t\ . \tag{4.71}$$

It is then relevant to examine the orbital fluctuations about the mean-field solution. In the FM phase, in the rotated basis, there is a $b_{i\uparrow}^s$ spin boson at each site. Given that there is a concentration x of holes and that in the absence of a hole there is a spin $b^s_{i\uparrow}$ particle at each site, the constraint reduces to $\sum_\gamma \langle b^{\dagger t}_{i\gamma} b^t_{i\gamma}\rangle = 1 - x \approx 1$ since $x \ll 1$. Following the usual mean-field approach, this constraint is enforced *only on average* by introducing an orbital sector chemical potential $-\lambda$. Then, the energy dispersions for the orbital pseudo-spins for each $\boldsymbol{k}$ are given by

$$\epsilon_\pm(\boldsymbol{k}) = \lambda - xt_0\Big[-(c_x + c_y + c_z) \tag{4.72}$$

$$\pm \sqrt{(c_x^2 + c_y^2 + c_z^2 - c_xc_y - c_yc_z - c_zc_x)}\Big], \tag{4.73}$$

with $c_x = \cos ak_x$ etc. The minimum of this energy corresponds to a flat dispersion along the axis $(\pi/a, \pi/a, k_z)$ and the other two equivalent directions. In the eigen-states for the flat dispersion, $3d_{x^2-y^2}$ ($3d_{y^2-z^2}$, $3d_{z^2-x^2}$) orbitals are almost aligned in the two dimensional layers. The electron in a $3d_{x^2-y^2}$ orbital can not hop in the direction perpendicular to the xy plane because of symmetry. Thus, even in the three dimensional crystal lattice, the orbital fluctuation has a two dimensional character due to their anisotropic electron transfer. This flat dispersion has a dramatic effect on the possibility of orbital ordering. Such ordering corresponds to the condensation of the orbital bosons in the ground state. The chemical potential λ is determined by $\sum_\gamma \langle b^{\dagger t}_{i\gamma} b^t_{i\gamma}\rangle = \frac{1}{N}\sum_{\gamma k} n_B(\epsilon_\gamma(\boldsymbol{k})) = 1$, where n_B is the Bose distribution function. The flat dispersion implies that the chemical potential $-\lambda$ is *always* negative, implying no condensation and no long range orbital ordering at finite temperature.

The orbital liquid in ferromagnetic manganites can also be formulated in the slave-boson method where the orbital pseudo-spinon and holon are identified as a fermion and a boson, respectively [61]. This will be discussed in Sect. 7.1.4.

4.2 Manganite with Layered Structure

4.2.1 Phase Diagram

The manganites $R_{2-2x}A_{1+2x}Mn_2O_7$ with R=La, Pr, Nd, Sm, and A=Sr, Ca have a layered crystal structure where a (R, A)O layer and a pair of MnO_2 layers are stacked along the c axis, as shown in Fig. 4.1. These materials comprise a Ruddlesden-Popper series $(R, A)_{n+1}Mn_nO_{3n+1}$ with $n = 2$, and are commonly called bi-layered or double-layered perovskite manganites. As was the case for the manganites with the pseudo-cubic perovskite structure, the formal valence of the Mn ions is 3+(4+) at $x = 0$ ($x = 1$). These compounds were first synthesized in 1980's [62], and, following the discovery of CMR in $La_{1.2}Sr_{1.8}Mn_2O_7$ [63], further work has brought to light other exotic phenomena [64]. The most remarkable difference as compared with $R_{1-x}A_x$MnO$_3$ is an anisotropy in the magnetic and electrical properties due to the layered structure, e.g, the electric resistivity in the *ab* plane is about 100 times smaller than that along the *c* axis in La$_{1.2}$Sr$_{1.8}$Mn$_2$O$_7$ at 300K. Recent experimental and theoretical work has lead to the recognition that these compounds are idea for the study of the orbital degree of freedom and its correlations with the spin ground state. In this subsection, we introduce $R_{2-2x}A_{1+2x}$Mn$_2$O$_7$ as a spin-charge-orbital coupled compound.

A magnetic and structural phase diagram of La$_{2-2x}$Sr$_{1+2x}$Mn$_2$O$_7$ has been established over a wide range of x ($0.30 \leq x \leq 0.50$), as shown in Fig. 4.11 [65], and has recently been expanded to larger x values ($0.5 \leq x$) (Fig. 4.12) [66]. The variety of magnetic phases is briefly summarized:

(I) $0.32 \leq x \leq 0.38$: At low temperatures, La$_{2-2x}$Sr$_{1+2x}$Mn$_2$O$_7$ exhibits FM order, although AFM order appears at intermediate temperatures, around $x \sim 0.3$. The easy axis is changed from the *c* axis to the *ab* plane with increasing x.

(II) $0.38 \leq x \leq 0.66$: The neighboring MnO$_2$ layers cant starting at $x \sim 0.39$. The spin alignment within a MnO$_2$ layer remains FM. The canting angle reaches 180° at $x \sim 0.48$. This is called the A-type AFM

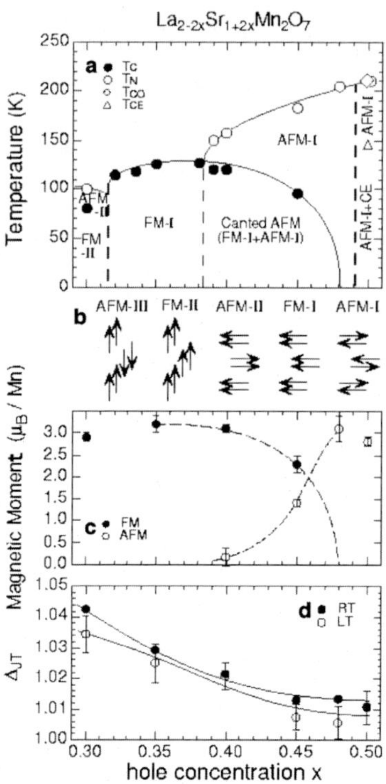

Fig. 4.11. (**a**) Magnetic phase diagram of La$_{2-2x}$Sr$_{1+2x}$Mn$_2$O$_7$ ($0.3 < x < 0.5$). (**b**) Several different magnetic structures appearing in the phase diagram. (**c**) Hole concentration dependence of FM and AFM magnetic moments at 10 K . (**d**) Hole concentration dependence of the lattice distortion Δ_{JT}. [65]

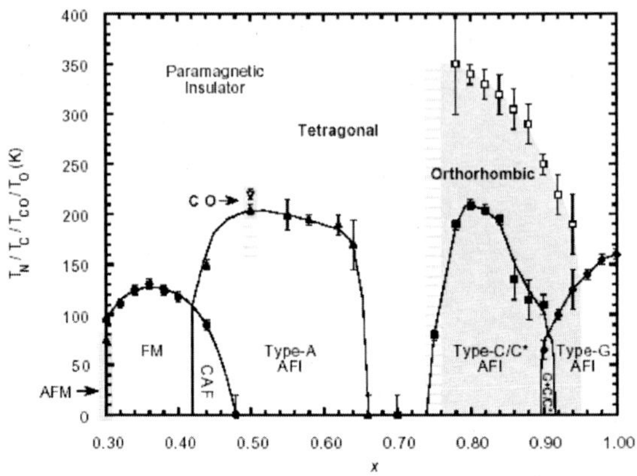

Fig. 4.12. Magnetic crystallographic phase diagram of $La_{2-2x}Sr_{1+2x}Mn_2O_7$. Solid data points represent magnetic transitions (AF Néel temperature T_N, ferromagnetic Curie temperature T_c, charge ordering temperature T_{CO}, and orbital ordering temperature T_O). Open data points represent crystallographic transitions. [66]

order, in analogy to that in manganites with the pseudo-cubic perovskite structure.

(III) $x = 0.5$: Around $x = 0.5$ in phase (II), charge and orbital order (75K $\leq x \leq$ 210K) and CE-type AFM order (75K $\leq x \leq$ 145K) appear. This is similar to $R_{0.5}A_{0.5}MnO_3$. A-type AFM order appears at the charge and orbital ordering temperature (210K), i.e., the A-type and CE-type AFM order parameters coexist between 75K and 145K.

(IV) $0.74 < x < 0.92$: There is C-type AFM order. A ferromagnetic rod lies parallel to the b axis. This magnetic structure may be correlated with the orthorhombic crystal structure which is observed around this hole concentration and which may stabilize the $3d_{3z^2-r^2}$-type orbital order.

(V) $0.88 < x < 1$: $La_{2-2x}Sr_{1+2x}Mn_2O_7$ has G-type AFM order where the spin alignment between NN Mn ions in MnO_2 bi-layers is antiparallel.

Up to now, there is no evidence of the long range magnetic order between the phases (III) and (IV). Below we focus on the magnetic and orbital structures in the phases (I) and (II), since the CMR effect is observed near the magnetic transition temperature and because there have been intensive studies of these phases. As shown in Fig. 4.11(d). The changes in magnetic structure are accompanied by systematic changes in the crystal lattice, i.e., in the anisotropy of a MnO_6 octahedra. We define

$$\Delta_{\mathrm{JT}} = \frac{d_{\mathrm{Mn-O(1)}} + d_{\mathrm{Mn-O(2)}}}{2d_{\mathrm{Mn-O(3)}}}, \tag{4.74}$$

where $d_{\mathrm{Mn-O}}$ is the distance between the nearest neighboring Mn and O ions and where the positions of the O ions are denoted in Fig. 4.13. This correlation between Δ_{JT} and magnetic structure is discussed later from the e_g orbital view point.

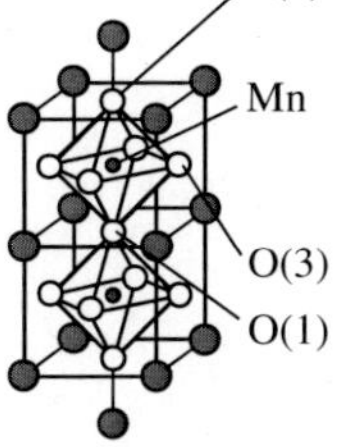

Fig. 4.13. Lattice structure and positions of O and Mn ions in $La_{2-2x}Sr_{1+2x}Mn_2O_7$.

4.2.2 Stability of Orbital and Magnetic Structure

In the manganites with cubic perovskite structure, competition between itinerant ferromagnetism and carrier localization accompanied by the antiferromagnetism dominates their unique magnetotransport. It was shown that the magnetic transition temperature is systematically varied as a function of the average ionic radius $\langle r_A \rangle$ of cations at the perovskite A-site, reflecting the tolerance factor t (see (4.59)) for the perovskite structure, mentioned previously. This factor controls the magnitude of the charge carrier hopping integral, although the conventional double exchange scenario cannot explain the drastic changes in the magnetic phase diagram by rather tiny modifications of $\langle r_A \rangle$ [51, 67]. On the other side, in the layered manganites, the tolerance factor t is not the driving force behind changes in the Curie, T_C, and the A-type AFM ordering Néel, T_N, temperatures. The values of t are located in a narrow region of $0.95 < t < 1.01$ while the temperatures T_C and T_N are distributed randomly.

The orbital degree of freedom is the key ingredient which determines the magnetic structure of the double layered manganites. As mentioned in Sect. 1.2 in connection with the high T_c cuprates, it is the electrostatic potential which determines the nature of the $3d$ orbitals in manganites. This observation might be justified by the fact that the manganites at $x = 0$ are classified as charge-transfer insulators for which the ionic model provides a good starting point. The energy-level structure based on the electron covalency shows a similar tendency to that given by the ionic model, because the e_g orbitals are the anti-bonding resulted from the Mn e_g and O $2p$ orbitals mixing. The electrostatic energy required to bring an electron from infinity to the site i in a solid, i.e. the Madelung site potential,

$$V(\boldsymbol{r}_j) = \sum_{i \neq j} \frac{Z_i e^2}{|\boldsymbol{r}_j - \boldsymbol{r}_i|}, \tag{4.75}$$

for a point charge $Z_i e$ at site i and with position $\boldsymbol{r}_i$. The electrostatic potential of an electron in the $3d_{3z^2-r^2}$ and $3d_{x^2-y^2}$ orbitals at site j can be estimated from

$$V_{3z^2-r^2} = \frac{1}{2}\left[V(\boldsymbol{r}_j + r_d \hat{z}) + V(\boldsymbol{r}_j - r_d \hat{z})\right], \tag{4.76}$$

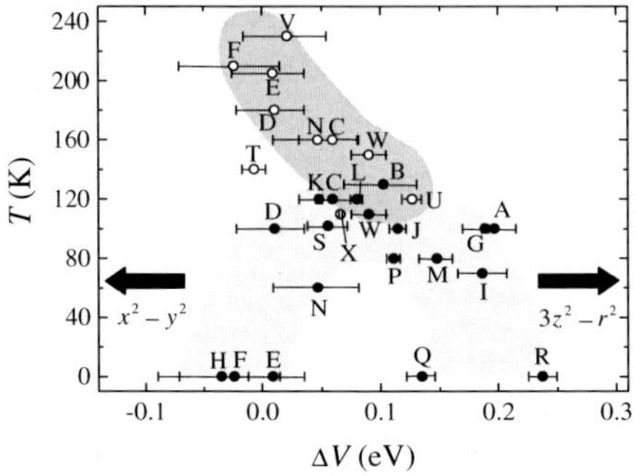

Fig. 4.14. Curie temperature T_C and Néel temperature T_N as functions of ΔV. Filled and open circles indicate T_C and T_N, respectively. Note that in the region with large positive (negative) ΔV, the $3d_{3z^2-r^2}$ ($3d_{x^2-y^2}$) orbital is occupied by an electron [68].

and

$$V_{x^2-y^2} = V(\boldsymbol{r}_j + r_d\hat{x}), \tag{4.77}$$

respectively. Here the radius r_d(=0.42Å) is that of the radial charge density maximum for a Mn $3d$ orbital, and $\hat{z}(\hat{x})$ is the unit vector in the $z(x)$ axis. The difference of these potentials

$$\Delta V = V_{3z^2-r^2} - V_{x^2-y^2}, \tag{4.78}$$

represents the relative stability of the two orbitals. With increasing ΔV, the electronic energy of a $3d_{3z^2-r^2}$ orbital decreases.

The correlation between the magnetic transition temperature, T_N and T_C, and ΔV has been examined for a number of double layered manganites (Fig. 4.14) [68]. It is clearly shown that both T_C and T_N scale very well with ΔV; T_N increases with decreasing ΔV and there is an optimal value of $\Delta V(\sim 0.08\,\mathrm{eV})$ for T_C. This observation of the clear correlations between spin and orbital structures in double layered manganites is well understood from the microscopic point of view [68,69]. For the A-type AFM structure, the FM spin alignment in a MnO_2 layer is caused by the double exchange interaction, while the AFM alignment between MnO_2 bi-layers reflects the superexchange interaction between the t_{2g} orbitals (J_{AF}) as discussed in (4.8). When the $3d_{x^2-y^2}$ orbital is lowest, the double exchange interaction in a layer (between the bi-layers) becomes stronger (weaker) and results in A-type AFM order. Mixing of the two orbitals is essential for FM interaction to overcome the AFM superexchange for all bond directions.

4.2.3 Experiments for Spin and Orbital Correlation

The clear correlation between the orbital and magnetic structures pointed above is confirmed in several physical observations in double layered manganites. We survey the experimental results supporting the strong spin-orbital correlation.

Magnetostriction. A strong spin-orbital correlation is suggested for $La_{2-2x}Sr_{1+2x}Mn_2O_7$ in magnetoelastic phenomena. The temperature and hole concentration dependence of the lattice striction ($\Delta L/L$) has been systematically investigated [70]. The gigantic change of this striction, i.e. magnetostriction, is observed at the magnetic ordering temperature. As shown in Fig. 4.15, this magnetostriction is anisotropic and its sign reverses around $x \sim 0.36$ in the FM phase; $\Delta L/L$ along the c axis increases (decreases) below T_C with $x = 0.3 - 0.36$ ($x = 0.4$). At $x = 0.45$, where the A-type AFM ordering appears at low temperatures, $\Delta L/L$ along the c axis monotonically deceases with decreasing temperature and shows no drastic change at the Néel temperature. Similar features are observed in the temperature and x dependence of the Mn-O bond lengths for the MnO_6 octahedra measured by the neutron diffraction experiments. The change of the distortion coordinate $\Delta Q(20K) - Q$(high temperature), where

$$Q = 2\left(d_{\mathrm{Mn-O(1)}} + d_{\mathrm{Mn-O(2)}} - 2d_{\mathrm{Mn-O(3)}}\right), \tag{4.79}$$

reverses sign at around $x = 0.36$ [71]. These phenomena are explained from the spin-orbital correlation. In the compounds which exhibit FM ordering below T_C ($0.3 < x < 0.4$), the magnetostriction occurs in order that orbital mixing is optimal for the FM structure. For $La_{2-2x}Sr_{1+2x}Mn_2O_7$ with $x = 0.3, 0.32$ ($x = 0.4$) the $3d_{3z^2-r^2}$ orbital component is larger (smaller) than the $3d_{x^2-y^2}$ character above T_C, while the $3d_{3z^2-r^2}$ component is reduced (enhanced) below T_C. In $La_{2-2x}Sr_{1+2x}Mn_2O_7$ with x=0.45 which exhibits the A-type AFM order at low temperatures, the $3d_{x^2-y^2}$ orbital is already stabilized above T_N. The orbital polarization is even larger below T_N, since the A-type AFM order and the $3d_{x^2-y^2}$ orbital order occur cooperatively.

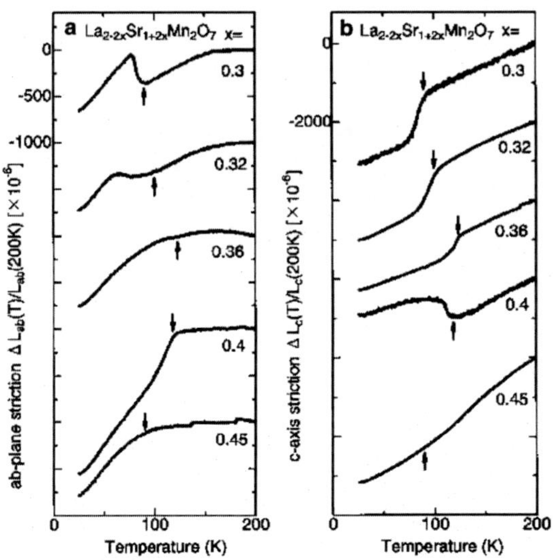

Fig. 4.15. Temperature dependence of the striction along (**a**) the ab plane and (**b**) the c axis for $La_{2-2x}Sr_{1+2x}Mn_2O_7$. Arrows indicate the magnetic transition temperature [70].

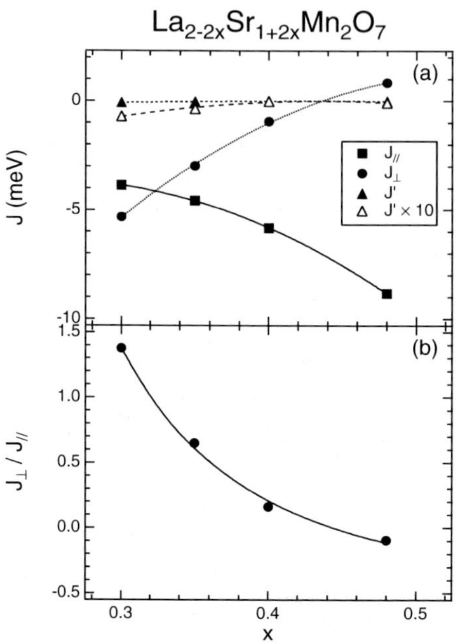

Fig. 4.16. The exchange interactions obtained by an analysis of the dispersion relations and the scattering cross sections obtained by inelastic neutron scattering [72].

Spin Waves. The dynamical magnetism of $La_{2-2x}Sr_{1+2x}Mn_2O_7$ is determined by the orbital degrees of freedom. Spin waves have been observed by inelastic neutron scattering experiments for the series $La_{2-2x}Sr_{1+2x}Mn_2O_7$ [72,73]. Two spin wave modes are observed in the $[hk0]$ direction, i.e., acoustic and optical branches corresponding to the in-phase and out-of phase motions of the Mn spins in the MnO_2 bi-layers. The spin wave dispersion relations change systematically with the hole concentration x. The spin stiffness constant monotonically increases with x for $x \geq 0.3$. On the other hand, the splitting between the acoustic and optical branches decreases. Finally, at $x = 0.48$ (A-type AFM phase), the two branches merge into a single branch.

Both the dispersion relation and the differential cross section are well fitted by the Heisenberg model where three kinds of the exchange interactions between Mn spins are introduced; $J_{\|}$ in a MnO_2 plane, $J_\perp$ between the NN planes, and J' between bi-layers . The x dependences of the exchange interactions are presented in Fig. 4.16. Upon increasing x beyond 0.3, $|J_{\|}|$ increases while $|J_\perp|$ rapidly approaches zero and then changes sign from negative to positive. On the other hand, $|J'|$ decreases with increasing x, and its value is 100∼1000 times smaller than $|J_{\|}|$ and $|J_\perp|$.

These systematic change of the interactions correlates with the distortion of the MnO_6 octahedra denoted by Δ_{JT} (Fig. 4.11(d)), which decreases from 1.035 at $x = 0.30$ to 1.005 at $x = 0.50$ [65]. With increasing x the $3d_{x^2-y^2}$ orbital becomes more stable. The net exchange interaction between Mn ions is a sum of the FM double-exchange and AFM superexchange interactions

(J_{AF}) acting between t_{2g} spins. With increasing stability of the $3d_{x^2-y^2}$ orbital, the FM interaction becomes stronger in the *ab* plane and weak along the *c* axis, and as a result J_{AF} overcomes the FM interaction along the *c* axis, as shown, in the region of $x = 0.4 - 0.48$.

Magnetic Compton Scattering. In comparison with the magnetic state, an observation of the orbital state is more difficult. One of the experimental probe is the resonant x-ray scattering which is introduced in the next section in detail. This is the x-ray diffraction experiment where one detects the orbital superlattice reflection due to the staggered type orbital order by utilizing the resonant scattering. In general, nature of the uniformly occupied orbital and a change of the orbital component are hard to detect by even this method because of a large scattering from electrons of inner orbitals. The magnetic Compton scattering is recently developed as a probe to detect the uniform component of the occupied orbital. This is a component of the Compton scattering depending on the magnetic moment. The spin dependent term is obtained by measurement of the Compton profile where the sample magnetization or the x-ray helicity are reversed. This profile reflects a difference of the projected momentum distribution function e.g.

$$J(p_z) = \int dp_x dp_y \left(\rho_\uparrow(\boldsymbol{p}) - \rho_\downarrow(\boldsymbol{p})\right), \tag{4.80}$$

where $\rho_{\uparrow(\downarrow)}(\boldsymbol{p})$ are the momentum distribution function with spin $\uparrow$ ($\downarrow$), and z axis is chosen to be parallel to the scattering vector. The key point is that only electrons concerning to the magnetic moments reflects to the magnetic Compton profile, that is, contributions from the non-magnetic orbitals are canceled out in the subtraction. Then, the information for the spatial distribution of the Mn $3d$ e_g and t_{2g} orbitals are detected by this method.

This technique is recently applied to $La_{2-x}Sr_{1+2x}Mn_2O_7$ with $x = 0.35$ and 0.42, and the Compton profiles are analyzed by the molecular orbital calculation [74]. The electron populations of the two e_g orbitals ($\rho_{3z^2-r^2}, \rho_{x^2-y^2}$) are $(0.19, 0.46)$ for $x = 0.35$ and (0.14,0.44) for $x = 0.42$. This implies that the $3d_{x^2-y^2}$ component is dominant in both the two compounds and the population of the $3d_{3z^2-r^2}$ orbital is reduced with increasing x. These observations are consistent with the systematic change of the orbital state in $La_{2-x}Sr_{1+2x}Mn_2O_7$ suggested by other experiments and theories.

4.3 Resonant X-ray Scattering (RXS)

Resonant x-ray scattering (RXS) occurs when the incident x ray energy is tuned around an absorption edge of a certain ion in a solid. This method was developed in 1970's by Templton and Templton [75], in order to detect the anisotropy of the unoccupied electronic states. In 1998, the method was first used to study orbital and charge ordering in $La_{0.5}Sr_{1.5}MnO_4$ by Murakami and coworkers [76]. Following this, the technique has developed rapidly and several orbitally active compounds have been investigated.

4.3.1 Experiments

It is well recognized that the experimental observation of the orbital structure and excitations is difficult. Orbital ordering is often associated with an anisotropic lattice distortion, i.e., cooperative Jahn-Teller distortion. Thus, one may think that the observation of a lattice distortion is equivalent to that of orbital ordering. However, in a solid, there are an infinite number of degrees of freedom and there are several interactions which compete and co-operate with each other. That is to say, in a solid, it not always the case that there is a one to one correspondence between orbital ordering and a certain lattice distortion.

Orbital ordering was first direct observed in K_2CuF_4 from an analysis of the magnetic form factor of the occupied orbital, using the polarized neutron diffraction [77]. The measurement of the charge density distribution using the x-ray and/or electron diffraction can also be used to determine which orbital is occupied [78]. Recently, a determination [79] of the spatial charge distribution of the orbital ordered manganite $NdSr_2Mn_2O_7$ was made using a Rietveld refinement of x-ray powder diffraction data and the maximum entropy method.

The charge and spin orderings of the single layered perovskite $La_{0.5}Sr_{1.5}MnO_4$ were determined magnetic neutron diffraction experiments well before Murakami's RXS experiments. The ordering patterns are similar to those of $La_{0.5}Ca_{0.5}MnO_3$ where an alternate ordering of the orbitals has been proposed to explain the long range CE-type AFM ordering. A similar multi-component ordering of $La_{0.5}Sr_{1.5}MnO_4$ is to be expected. In Murakami and coworkers RXS experiments [76] the incident x-ray energy was tuned near to the Mn K-edge of 6.552KeV and the reflection at $(\frac{2l+1}{4}\ \frac{2l+1}{4}\ n)$ was de-

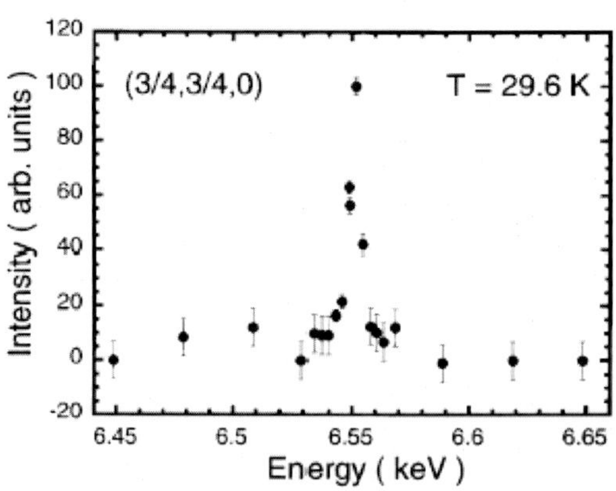

Fig. 4.17. Energy dependence of the scattering intensity for $La_{1.5}Sr_{0.5}MnO_4$ at the orbital order superlattice reflection (3/4 3/4 0) [76].

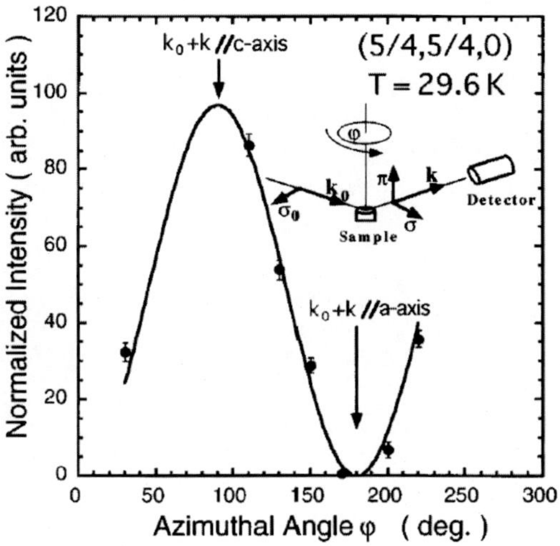

Fig. 4.18. Azimuthal angle dependence of the scattering intensity in $La_{1.5}Sr_{0.5}MnO_4$ at the orbital order superlattice reflection (5/4 5/4 0) [76].

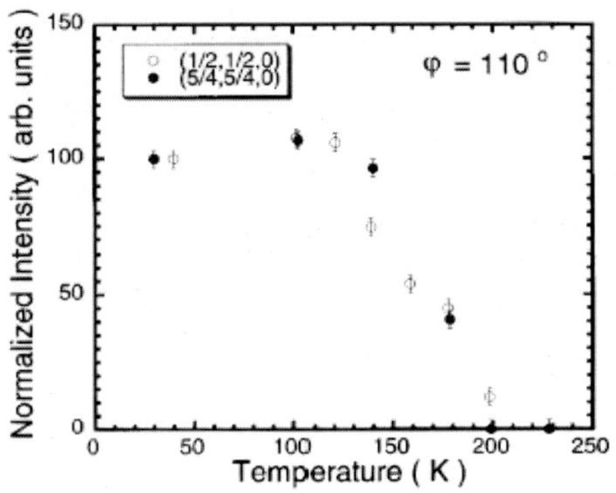

Fig. 4.19. Temperature dependence of the scattering intensity in $La_{1.5}Sr_{0.5}MnO_4$ at the orbital order superlattice reflection (5/4 5/4 0) and the charge order superlattice reflection (1/2 1/2 0) [76].

tected. The resonantly enhanced experimental results for the incident x-ray energy dependence of the scattering intensity are presented in Fig. 4.17.

The remarkable difference of RXS as compared with conventional x-ray scattering is seen in the azimuthal angle dependence around the scattering vector. The x ray coming from the synchrotron light source is highly polarized and the relative direction of electric vector to a sample is changed by rotating the sample around the scattering vector. The scattering intensity shown in Fig. 4.18 has a maximum (minimum) at $\varphi = 90°$ (180°) where the incident x-ray polarization is parallel to the *ab* plane (*c* axis) of the sample. The results are well fitted by $\sin^2 \varphi$, as expected from a phenomenological model assuming *orbital* ordering. The reflection at $(\frac{2l+1}{2}\ \frac{2l+1}{2}\ 0)$, corresponding to *charge* ordering, was also seen. Both the reflections at $(\frac{2l+1}{4}\ \frac{2l+1}{4}\ 0)$ and $(\frac{2l+1}{2}\ \frac{2l+1}{2}\ 0)$ appear around 200K and show an almost identical temperature dependence

Table 4.1. A list of experimental studies of RXS. A): A broad enhancement of the RXS intensity of the (100) reflection is observed at 141K. The second transition occurs at 77K. B): The orbital ordering is considered to occur at the Néel temperature T_N=181K. The RXS intensity is measured at 100K. C): A broad enhancement of the RXS intensity is observed at 7.6K. D): The orbital ordering is considered to occur at 25K. The RXS intensity is measured at 12K.

Materials	Orbital Ordering Temperature (K)
$LaMnO_3$	780
$La_{0.88}Sr_{0.12}MnO_3$	145
$Nd_{0.5}Sr_{0.5}MnO_3$	150
$Pr_{1-x}Ca_xMnO_3$ (x=0.4-0.5)	245
$La_{1.5}Sr_{0.5}MnO_4$	210
$LaSr_2Mn_2O_7$	210
$La_{1-x}Sr_xMnO_3$ (Artificial Lattice)	-
$YTiO_3$	-
$LaTiO_3$	-
YVO_3	77, 141 A)
V_2O_3	181 B)
$KCuF_3$	43
DyB_2C_2	25
HoB_2C_2	5.4
UPd_3	7.6 C)
CeB_6	3.2
NpO_2	25 D)

(Fig. 4.19). This implies that the charge and orbital order parameters couple to each other.

The method was then immediately applied to $LaMnO_3$ [80]. Independent of Murakami and coworkers' studies are RXS experiment and theory for V_2O_3 [81,82]. Nowadays, RXS is widely used to study the orbital degree of freedom in a number of correlated materials [83]; e.g. manganites, other transition-metal oxides, and f-electron systems. Simultaneously, theories of RXS from the several view points have been developed. The published experimental RXS studies of orbital states are listed in Table 4.1.

4.3.2 Scattering Cross Section

Perturbation theory, applied to electron-photon interaction, is used to calculate the differential cross section for x ray scattering from electrons. The Hamiltonian for the electron-photon coupled system is

$$\mathcal{H} = \mathcal{H}_{el} + \mathcal{H}_{ph} + \mathcal{H}_{el-ph}, \tag{4.81}$$

where $\mathcal{H}_{el}$, $\mathcal{H}_{ph}$ and $\mathcal{H}_{el-ph}$ correspond to the electrons, photons and their interactions. Specifically

$$\mathcal{H}_{el-ph} = \mathcal{H}_j + \mathcal{H}_\rho = -\frac{e}{c}\int \mathrm{d}\boldsymbol{r}\boldsymbol{j}(\boldsymbol{r})\cdot\boldsymbol{A}(\boldsymbol{r}) + \frac{e^2}{2mc^2}\int \mathrm{d}\boldsymbol{r}\rho(\boldsymbol{r})\boldsymbol{A}(\boldsymbol{r})^2. \tag{4.82}$$

where e is the unit of charge and m is the electron mass. The first term on the right hand side couple to the charge current $\boldsymbol{j}(\boldsymbol{r})$ while the second involves $\rho(\boldsymbol{r})$, the charge density. The photons are reflected by $\boldsymbol{A}(\boldsymbol{r})$ in both cases. The S matrix calculated to second order in $\boldsymbol{A}(\boldsymbol{r})$ provides the differential scattering-cross section for x-ray scattering [85, 86], i.e.,

$$\frac{\mathrm{d}^2\sigma}{\mathrm{d}\Omega\mathrm{d}\omega_f} = \sigma_T\frac{\omega_f}{\omega_i}\sum_f\left|S_1\boldsymbol{e}_{k_i\lambda_f}\cdot\boldsymbol{e}_{k_f\lambda_f} + \sum_{\alpha\beta}(\boldsymbol{e}_{k_i\lambda_i})_\alpha S_{2\alpha\beta}(\boldsymbol{e}_{k_f\lambda_f})_\beta\right|^2 \times\delta(\varepsilon_f + \hbar\omega_f - \varepsilon_i - \hbar\omega_i), \tag{4.83}$$

where

$$S_1 = \langle f|\rho_{\boldsymbol{k}_i-\boldsymbol{k}_f}|i\rangle, \tag{4.84}$$

and

$$S_{2\alpha\beta} = \frac{m}{e^2}\sum_m\left[\frac{\langle f|\boldsymbol{j}_{-\boldsymbol{k}_i\alpha}|m\rangle\langle m|\boldsymbol{j}_{\boldsymbol{k}_f\beta}|i\rangle}{\varepsilon_i - \varepsilon_m - \hbar\omega_f + \mathrm{i}\eta} + \frac{\langle f|\boldsymbol{j}_{\boldsymbol{k}_f\beta}|m\rangle\langle m|\boldsymbol{j}_{-\boldsymbol{k}_i\alpha}|i\rangle}{\varepsilon_i - \varepsilon_m + \hbar\omega_i + \mathrm{i}\eta}\right]. \tag{4.85}$$

This is the so-called Kramers-Heisenberg formula. We consider the scattering of an x ray with momentum $\boldsymbol{k}_i$, frequency $\omega_i = c|\boldsymbol{k}_i|$ and polarization λ_i, to $\boldsymbol{k}_f$, ω_f and λ_f. The polarization vector of x ray is $\boldsymbol{e}_{k_l\lambda_l}$ $(l = i, f)$, while $|i\rangle$, $|m\rangle$ and $|f\rangle$ are the electronic states in the initial, intermediate and final scattering states, with energies ε_i, ε_m and ε_f, respectively. The factor $\sigma_T = (e^2/mc^2)^2$ is the total scattering cross section for Thomson scattering and η in the denominators of the expression for S_2 is an infinitesimal positive constant. The Fourier transforms of the charge $\rho(\boldsymbol{r})$ and current $\boldsymbol{j}(\boldsymbol{r})$ densities are ρ_k and $\boldsymbol{j}_k$, respectively. If ρ_k and $\boldsymbol{j}_k$ are decomposed into charge and current operators defined at each site, then

$$\rho_{\boldsymbol{k}} = \sum_l \rho_{l\boldsymbol{k}}\mathrm{e}^{-\mathrm{i}\boldsymbol{k}\cdot\boldsymbol{r}_l}, \tag{4.86}$$

and

$$\boldsymbol{j}_{\boldsymbol{k}} = \sum_l \boldsymbol{j}_{l\boldsymbol{k}}\mathrm{e}^{-\mathrm{i}\boldsymbol{k}\cdot\boldsymbol{r}_l}. \tag{4.87}$$

The atomic scattering factor (ASF) at site l,

$$f_{l\alpha\beta}(\boldsymbol{k}_i, \boldsymbol{k}_f) = f_{0l}(\boldsymbol{k}_i, \boldsymbol{k}_f)\delta_{\alpha\beta} + \Delta f_{l\alpha\beta}(\boldsymbol{k}_i, \boldsymbol{k}_f), \tag{4.88}$$

where the first and second terms on the right hand side are said to be the "normal" and "anomalous" parts of ASF, respectively. These are defined in the same fashion as S_1 and $S_{2\alpha\beta}$ in (4.84) and (4.85) when $\rho_{\boldsymbol{k}}$ and $\boldsymbol{j}_{\boldsymbol{k}}$ are replaced by $\rho_{l\boldsymbol{k}}$ and $\boldsymbol{j}_{l\boldsymbol{k}}$, respectively.

When the incident x-ray energy $\hbar\omega_i$ is far from the energy difference of the initial and intermediate electronic states $\Delta\varepsilon_{mi} = \varepsilon_m - \varepsilon_i$, S_1 dominates the scattering. On the other hand, when $\hbar\omega_i \approx \Delta\varepsilon_{mi}$, the real part of the denominator in the second term of S_2 becomes small and the scattering diverges, i.e., there is resonant scattering. However the life time of the intermediate electronic state $|m\rangle$ limits the maximum scattering intensity [85]. This life time arises from the mixing of $|m\rangle$ with other states due to, e.g., the many body effects, and is reflected by the self-energy Σ. The denominator of the second term in S_2 is replaced by $\varepsilon_i - \varepsilon_m + \hbar\omega_i + i\Gamma$ where $\mathrm{Re}\Sigma$ is included in the definition of the ε_m and $\mathrm{Im}\Sigma + \eta$ is denoted by Γ. It is estimated that Δf, at the resonant condition, is of the order of 10 electrons per atom, which is comparable to f_0.

Consider a RXS experiment in which $\hbar\omega_i$ is tuned at the K edge of a transition-metal ion. In the intermediate scattering state, the $1s$ electron is excited to the $4p$ orbital at the same site. The current operator $\boldsymbol{j}_{lk}$ is given by

$$\boldsymbol{j}_{l\boldsymbol{k}} = -\frac{\mathrm{i}e}{2m} \sum_{\sigma} \boldsymbol{A}_{1s4p}(\boldsymbol{k}) c^{\dagger}_{l1s\sigma} c_{l4p\sigma} + \mathrm{h.c.}, \tag{4.89}$$

with

$$\boldsymbol{A}_{1s4p}(\boldsymbol{k}) = \int \mathrm{d}\boldsymbol{r} \mathrm{e}^{-\mathrm{i}\boldsymbol{k}\cdot\boldsymbol{r}} \left[\phi^*_{1s}(\boldsymbol{r}) \boldsymbol{\nabla}\phi_{4p}(\boldsymbol{r}) - \boldsymbol{\nabla}\phi^*_{1s}(\boldsymbol{r})\phi_{4p}(\boldsymbol{r})\right]. \tag{4.90}$$

The annihilation operator of an electron with orbital $1s(4p)$ and spin σ at site l is $c_{l1s(4p)\sigma}$, and $\phi_{1s(4p)}(\boldsymbol{r})$ is the Wannier function of the electron in the orbital $1s(4p)$. Note that the relevant region of the integral in (4.90) is determined by the average atomic radius $\langle r_{1s}\rangle$ of the $1s$ orbital which is much smaller than the wave length λ of an x ray. In the case of a Mn ion, the ratio $\lambda/\langle r_{1s}\rangle$ is about $1/100$. Thus the electric quadrupole transition strength is much smaller than for electric dipole transitions, and $\boldsymbol{A}_{1s4p}(\boldsymbol{k})$ is almost independent of the momentum of the x ray. For the detection of orbital ordering, RXS has the following advantages:

(1) The wave length of the x ray, which is tuned near the K absorption edge of a transition-metal ion, is shorter than a lattice constant of the perovskite unit cell. Thus, the diffraction experiments can be carried out for the orbital superlattice, unlike Raman scattering and optical reflection/absorption experiments.
(2) As shown by (4.85), S_2 is a tensor with respect to the incident and scattered x-ray polarizations, in contrast to S_1 which does not show any polarization dependence.
(3) By tuning the x-ray energy to the absorption edge, the scattering from a specific element, such as a Mn ion, can be identified.

4.3.3 Azimuthal Angle Dependence

Once the matrix elements of the ASF tensor are determined, the RXS scattering intensity can be calculated for a realistic experimental arrangement. One of the important experimental parameters is the azimuthal angle (φ) defined below. The conventional x-ray scattering intensity does not depend on this angle. That is to say, the φ dependent scattering intensity implies that some kinds of the anisotropic electronic structures are concerned in the resonant scattering.

Consider the RXS experimental arrangement appropriate to orbital ordered compounds (Fig. 4.20). It consists of a sample crystal (S) and the polarization analyzer comprising an analyzer crystal (A) and photon detector (D). The polarization scan is characterized by two rotation angles, i.e. the azimuthal angle (φ) and the analyzer angle (φ_A). The former is the rotation angle of the sample around the scattering vector $\boldsymbol{K} = \boldsymbol{k}_i - \boldsymbol{k}_f$, and the latter is that of the analyzer around an axis which is parallel to the scattered photon momentum. The incident x ray polarization is assumed to be perfectly polarized in the horizontal plane, i.e., has a σ-polarization. The direction of the electric vector of the incident x ray with respect to the crystalline axis is changed by an azimuthal rotation. Because of the tensor character of the ASF, the scattered x ray has both π- and σ-polarized components, which are separated by the analyzer scan. In this optical arrangement, the scattering intensity is given by [87]

$$I(\boldsymbol{k}_i, \boldsymbol{k}_f, \varphi, \varphi_A) = \sum_{\lambda} \left| \sum_{\lambda_f} M_{\lambda\lambda_f}(\varphi_A) A_{\lambda_f\lambda_i}(\boldsymbol{k}_i, \boldsymbol{k}_f, \varphi) \right|^2 , \tag{4.91}$$

where λ and $\lambda_{i(f)}$ $(= \pi, \sigma)$ indicate the polarization of the x ray, and where $M_{\lambda'\lambda}(\varphi_A)$ is the scattering matrix of the analyzer, i.e.,

$$M(\varphi_A) = F_A \begin{pmatrix} \cos\varphi_A & -\sin\varphi_A \\ \sin\varphi_A \cos 2\theta_A & \cos\varphi_A \cos 2\theta_A \end{pmatrix} , \tag{4.92}$$

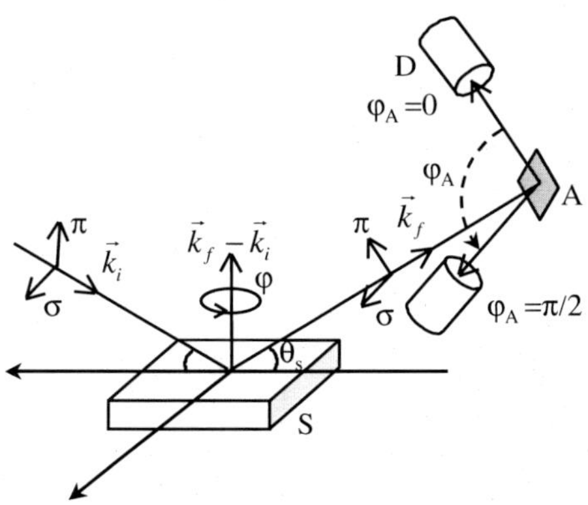

Fig. 4.20. A schematic experimental arrangement for RXS consisting of a sample crystal (S), an analyzer crystal (A) and a photon detector (D). The vectors $\boldsymbol{k}_i$ and $\boldsymbol{k}_f$ are the incident and scattered x-ray momentum, φ is the azimuthal angle, φ_A is the analyzer angle and θ_s is the scattering angle [87].

where $|F_A|$ and θ_A are the scattering factor and angle of the analyzer crystal. For simplicity, θ_A is fixed to be $\pi/4$. The scattering amplitude $A_{\lambda_f \lambda_i}(\boldsymbol{k}_f, \boldsymbol{k}_i, \varphi)$ in (4.91) is defined by

$$A_{\lambda_f \lambda_i}(\boldsymbol{k}_f, \boldsymbol{k}_i, \varphi) = \frac{e^2}{mc^2} \boldsymbol{e}_{\boldsymbol{k}_f \lambda_f} \left[U(\varphi) V F(\boldsymbol{k}_i, \boldsymbol{k}_f) V^\dagger U(\varphi)^\dagger \right] \boldsymbol{e}^t_{\boldsymbol{k} \lambda_i} , \quad (4.93)$$

where $F(\boldsymbol{k}_i, \boldsymbol{k}_f)$ is the structure factor given in the coordinate of the crystallographic axes $(\hat{a}, \hat{b}, \hat{c})$, and $\boldsymbol{e}_{\boldsymbol{k}_{i(f)} \lambda_{i(f)}}$ for the Cartesian coordinate in the laboratory system $(\hat{e}_1, \hat{e}_2, \hat{e}_3)$. The polarization vectors of the incident and the scattered x ray are given by

$$\boldsymbol{e}_{\boldsymbol{k}_i \sigma} = (1 \quad 0 \quad 0) , \quad (4.94)$$
$$\boldsymbol{e}_{\boldsymbol{k}_f \sigma} = (1 \quad 0 \quad 0) , \quad (4.95)$$
$$\boldsymbol{e}_{\boldsymbol{k}_i \pi} = (0 \quad \sin\theta \quad \cos\theta) , \quad (4.96)$$
$$\boldsymbol{e}_{\boldsymbol{k}_f \pi} = (0 \quad -\sin\theta \quad \cos\theta) . \quad (4.97)$$

The unitary matrix $U(\varphi)$, in (4.93), generates the azimuthal rotation of the sample around the $\hat{e}_3$ axis:

$$U(\varphi) = \begin{pmatrix} \cos\varphi & -\sin\varphi & 0 \\ \sin\varphi & \cos\varphi & 0 \\ 0 & 0 & 1 \end{pmatrix} , \quad (4.98)$$

and the matrix V is that for the transformation from a coordinate of the crystallographic axis $(\hat{a}, \hat{b}, \hat{c})$ to the laboratory system $(\hat{e_1}, \hat{e_2}, \hat{e_3})$. For example, in the case where the scattering vector is parallel to the (100) axis, the matrix V is given by

$$V = \begin{pmatrix} 0 & 0 & 1 \\ 1 & 0 & 0 \\ 0 & 1 & 0 \end{pmatrix} . \quad (4.99)$$

The azimuthal angle $\varphi = 0$ corresponds to the condition that the crystallographic c axis is included in the scattering plane. The structure factor

$$F_{\alpha\beta}(\boldsymbol{k}_i, \boldsymbol{k}_f) = N \sum_{l \in \mathrm{cell}} \mathrm{e}^{\mathrm{i}(\boldsymbol{k}_i - \boldsymbol{k}_f) \cdot \boldsymbol{r}_l} f_{l\alpha\beta}(\boldsymbol{k}_i, \boldsymbol{k}_f) , \quad (4.100)$$

where N is the number of unit cells. When the matrix V and the tensor elements of ASF at site l in a unit cell $f_{l\alpha\beta}(\boldsymbol{k}_i, \boldsymbol{k}_f)$ are identified, the scattering intensity is calculated as a function of the azimuthal angle φ.

Let us apply the above formulae to the orbital ordered $LaMnO_3$. Below 780K, two different orbitals are alternately aligned in the ab plane and uniformly along the c axis. These two orbital sublattices are denoted by A and B. A kind of the occupied orbital at each sublattice is specified by the orbital mixing angle, i.e. $\theta_{A(B)}$ defined in (4.1). Here we assume a condition $\theta_A = -\theta_B = \theta_T$. Consider the RXS at $(h\ k\ l) = (\frac{2l+1}{2}\ \frac{2l+1}{2} n)$ which is called "the orbital superlattice reflection". Because of the symmetry, the following

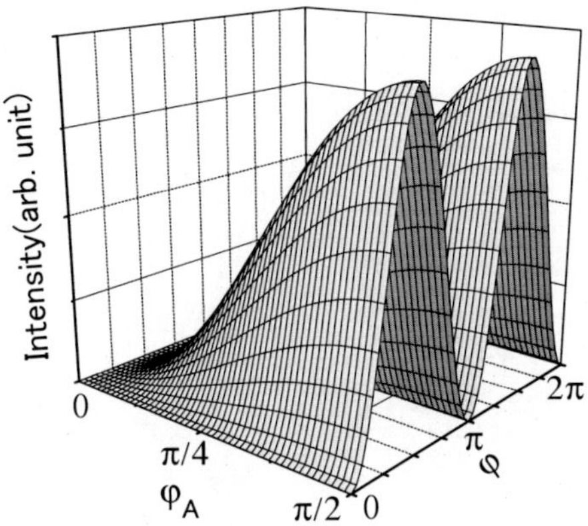

Fig. 4.21. The azimuthal angle dependence at the orbital superlattice reflection [87].

conditions for the ASF tensors Δf_A and Δf_B for Mn ions in the A and B orbital sublattices, respectively, are required;

$$\Delta f_{Axx} = \Delta f_{Byy}, \tag{4.101}$$

$$\Delta f_{Ayy} = \Delta f_{Bxx}, \tag{4.102}$$

$$\Delta f_{Azz} = \Delta f_{Bzz}. \tag{4.103}$$

The scattering intensity normalized by a factor $N^2\sigma_T|F_A|^2$ is obtained as

$$\tilde{I}(\varphi, \varphi_A = 0) = 0, \tag{4.104}$$

$$\tilde{I}(\varphi, \varphi_A = \pi/2) = \left|\Delta f_{-xx}\right|^2 \sin^2\varphi \cos^2\theta, \tag{4.105}$$

with

$$\Delta f_{-\alpha\alpha} = \frac{1}{2}(\Delta f_{A\alpha\alpha} - \Delta f_{B\alpha\alpha}). \tag{4.106}$$

The resulting $\tilde{I}(\varphi, \varphi_A)$ is shown in Fig. 4.21. The characteristics of the azimuthal angular dependence are:

(i) The scattering condition for $\varphi_A = 0$ corresponds to $\lambda_i = \lambda_f = \sigma$ (the $\sigma-\sigma$ scattering). This scattering intensity corresponds to $\Delta f_{-xx} + \Delta f_{-yy}$, and, reflecting the symmetry expressed as (4.103), the scattering intensity disappears.

(ii) The scattering condition at $\varphi_A = \pi/2$ implies $\lambda_i = \sigma$ and $\lambda_f = \pi$ (the $\sigma - \pi$ scattering). This scattering amplitude provides direct evidence that the scattering arises from the anisotropy of the anomalous part of the atomic scattering factor.

(iii) The function $\tilde{I}(\varphi, \varphi_A = \pi/2)$ is the square of the sinus of φ. The intensity is maximum (minimum) at $\varphi = (2n+1)\pi/2$ ($\varphi = n\pi$) where the electric field of incident x ray is parallel to the ab plane (the c axis). This is explained by the fact that $\Delta f_{-\alpha\alpha}$ is zero for $\alpha = z$ and is finite for $\alpha = x$ and y.

The different types of φ dependent scattering intensity are expected for other types of orbital ordered states. The azimuthal angle dependence thereby permits different types of the orbital ordering to be identified. It is experimentally confirmed, for several perovskite manganites, that the observed φ dependent intensity at the orbital superlattice reflection is well fitted by (4.105). The azimuthal angle dependence of the RXS intensity for other types of orbital ordered states has been done in $YTiO_3$, YVO_3 and so on.

4.3.4 Mechanism of RXS

One of the main issues for RXS as a probe to detect the orbital ordering, is the mechanism of scattering from orbital ordered state, i.e., mechanism for the anisotropic tensor elements of the atomic scattering factor. Consider RXS in the orbital ordered manganites. The initial x-ray is tuned around the Mn K edge and causes the dipole transitions from the Mn $1s$ orbital to the Mn $4p$ orbital in the intermediate scattering states. Thus, this transition itself does not have direct access to the $3d$ orbital which is responsible for the orbital ordering. Needed is some mechanisms by which bring about an anisotropy in the ASF for the orbital ordered state. One may think that the direct access is possible by the resonant scattering at the Mn L edge where the Mn $2p$ electron is excited to the Mn $3d$ orbital in the intermediate state. However, in this case, satisfying the diffraction condition is difficult, except for the orbital order with the large unit cell.

It is easily shown that covalency involving the Mn $3d$-O $2p$ orbitals, and the O $2p$-Mn $4p$ orbitals is not relevant to the anisotropy of the ASF. Consider the six O $2p$ orbitals in a cubic MnO_6 octahedra which are elongated along the Mn-O bond directions. The six linear combinations of these $2p$ orbitals (molecule orbitals) are defined as $\{\psi_{p_{3z^2-r^2}}, \psi_{p_{x^2-y^2}}, \psi_{p_{r^2}}, \psi_{p_x}, \psi_{p_y}, \psi_{p_z}\}$ which have the E_{gu}, E_{gv}, A_{1g}, T_{2gx}, T_{2gy}, and T_{2gz}, symmetries, respectively, in the O_h group. The $3d_{3z^2-r^2}$ and $3d_{x^2-y^2}$ and the $\psi_{p_{3z^2-r^2}}$ and $\psi_{p_{x^2-y^2}}$ wave functions couple with each other, respectively, as do the $4p_l$ $(l = x, y, z)$ and ψ_{p_l} wave functions. Therefore, the $3d$ e_g orbitals decouple from the three $4p$ orbitals. In the following, we introduce two possible mechanisms for RXS in the orbital ordered manganites and other $3d$ transition metal oxides.

Coulomb Mechanism. One of the mechanisms for RXS from orbital order is the Coulomb interaction between $3d$ and $4p$ electrons [88, 89]:

$$\mathcal{H}_{3d-4p} = \sum_{\gamma\gamma'\alpha\sigma\sigma'} V_{\gamma\gamma'\alpha} d^\dagger_{\gamma\sigma} d_{\gamma'\sigma} P^\dagger_{\alpha\sigma'} P_{\alpha\sigma'}, \tag{4.107}$$

where $P_{\alpha\sigma}$ is the annihilation operator of the Mn $4p$ electron with $\alpha(= x, y, z)$, and $V_{\gamma\gamma'\alpha}$ is the Coulomb interaction defined by

$$V_{\gamma\gamma'\alpha} = \begin{cases} F_0(3d,4p) + \dfrac{4}{35} F_2(3d,4p) \cos\left(\theta_\gamma - \dfrac{2\pi}{3} m_\alpha\right) & \text{for } \gamma' = \gamma \\ \dfrac{4}{35} F_2(3d,4p) \sin\left(\theta_\gamma - \dfrac{2\pi}{3} m_\alpha\right) & \text{for } \gamma' = \bar{\gamma} \end{cases}, \tag{4.108}$$

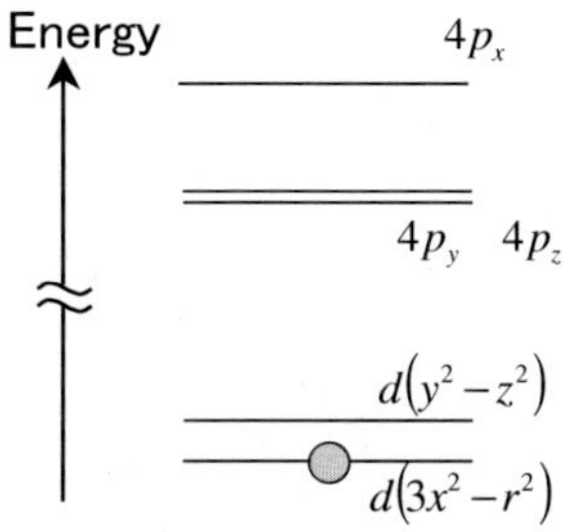

Fig. 4.22. A schematic energy diagram of the Mn $4p$ orbitals at the $3d_{3x^2-r^2}$ orbital occupied site for the Coulomb mechanism.

where $(m_x, m_y, m_z) = (1, 2, 3)$ and $\bar{\gamma}$ indicates the counterpart of γ. The Slater-integral for the $3d$ and $4p$ electrons are

$$F_0(3d, 4p) = F^0(3d, 4p), \tag{4.109}$$

$$F_2(3d, 4p) = \frac{1}{49} F^2(3d, 4p), \tag{4.110}$$

with

$$F^n(3d, 4p) = \int_0^\infty \mathrm{d}r_1 \mathrm{d}r_2 r_1^2 r_2^2 R_{32}(r_1)^2 R_{41}(r_2)^2 \frac{r_<^n}{r_>^{n+1}}, \tag{4.111}$$

where $R_{32}(r)$ and $R_{41}(r)$ are the radial parts of the wave functions for Mn $3d$ and Mn $4p$ electrons, respectively, and $r_<(r_>)$ is defined in (4.19).

Consider the orbital ordered state where the $3d_{3x^2-r^2}$ and $3d_{3y^2-r^2}$ orbitals are alternately aligned in the ab plane. At the $3d_{3x^2-r^2}$ orbital occupied site, the energy levels of the three $4p$ orbitals split as shown in Fig. 4.22. The anisotropy of the ASF arises when

$$(|\Delta f_{xx}|, |\Delta f_{yy}|, |\Delta f_{zz}|) = (f_s, f_l, f_l), \tag{4.112}$$

for a $3d_{3x^2-r^2}$ orbital occupied site, and when

$$(|\Delta f_{xx}|, |\Delta f_{yy}|, |\Delta f_{zz}|) = (f_l, f_s, f_l), \tag{4.113}$$

for a $3d_{3y^2-r^2}$ orbital occupied site. In the vicinity of the K-edge, the condition $f_l > f_s$ is satisfied, since the $4p_y$ and $4p_z$ ($4p_y$ and $4p_z$) orbitals dominate the K-edge for the $3d_{3x^2-r^2}$ ($3d_{3y^2-r^2}$) orbital occupied sites. Thus the RXS intensity at the orbital superlattice reflection point, being proportional to $|\Delta f_{3x^2-r^2\alpha\alpha} - \Delta f_{3y^2-r^2\alpha\alpha}|^2$, is finite when $\alpha = x$ and y.

In addition to the $3d - 4p$ Coulomb interaction, the inter-site Coulomb interaction between Mn $3d$ and O $2p$ electrons leads to an anisotropy of the ASF. Because of the strong hybridization between the Mn $3d$ and O $2p$ orbitals, the $|d^1_{3x^2-r^2}\rangle$ state strongly mixes with the $|d^1_{3x^2-r^2} d^1_{y^2-z^2} \underline{L}_{y^2-z^2}\rangle$ state. Here $\underline{L}_{y^2-z^2}$ indicates a state where one hole occupies the O $2p$ $\psi_{p_{y^2-z^2}}$ orbital. Thus the inter-site Coulomb interaction between a $2p$ hole and a $4p$ electron splits the three $4p$ orbitals. This interaction is given by

$$\mathcal{H}_{2p-4p} = \sum_{\gamma\alpha\sigma\sigma'} V^{inter}_{\gamma\alpha} p^{(h)\dagger}_{\gamma\sigma} p^{(h)}_{\gamma\sigma} P^\dagger_{\alpha\sigma'} P_{\alpha\sigma'}, \tag{4.114}$$

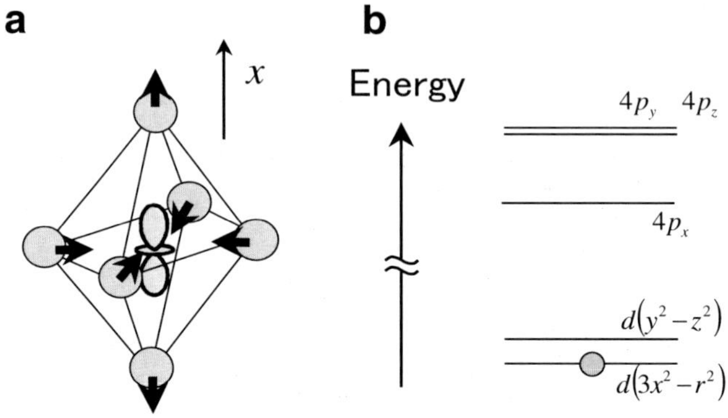

Fig. 4.23. (**a**) A distortion in a MnO_6 octahedron where $d_{3x^2-r^2}$ orbital is occupied by an electron. (**b**) A schematic energy diagram of the Mn 4p orbitals for a $3d_{3x^2-r^2}$ orbital occupied site and the Jahn-Teller mechanism.

with

$$V_{\gamma\alpha}^{inter} = -\varepsilon - \frac{\varepsilon\rho^2}{5}\cos\left(\theta_\gamma + m_\alpha\frac{2\pi}{3}\right) . \tag{4.115}$$

The annihilation operator, $p_{\gamma\sigma}^{(h)}$, for a hole in the ψ_{p_γ} orbitals ($\gamma = 3z^2 - r^2, x^2 - y^2$). The quantity $\varepsilon = Ze^2/a$ and $\rho = \langle r_{4p}\rangle/a$ with $Z = 2$ and the average radius of the Mn 4p orbital is $\langle r_{4p}\rangle$. The relative level structure caused by this interaction is the same as that shown in Fig. 4.22. Thus, the two mechanisms cause an anisotropy of the ASF, cooperatively.

Jahn-Teller Mechanism. Another Jahn-Teller mechanism leading to an anisotropy of the ASF has been proposed [90–92]. Consider a $d_{3x^2-r^2}$ orbital occupied site. The Jahn-Teller effect leads to a lattice distortion of the O_6 octahedron along the x direction see (Fig. 4.23(a)). Since the Mn-O bond lengths along the x directions are elongated, the hybridization between O $2p_x$ and Mn $4p_x$ orbitals for these bonds is weaker than for the other bonds. As a result, the three 4p orbitals split so that the energy level of the $4p_x$ orbital is lower than those of the $4p_y$ and $4p_z$ orbitals (see Fig. 4.23(b)), this because the 4p orbitals are anti-bonding orbitals. This tendency is opposite to that due to the Coulomb mechanism and is called the Jahn-Teller mechanism [90–92]. The anisotropy in the ASF at the K edge arises when

$$(|\Delta f_{xx}|, |\Delta f_{yy}|, |\Delta f_{zz}|) = (f_l, f_s, f_s), \tag{4.116}$$

for a $d_{3x^2-r^2}$ orbital occupied site, and when

$$(|\Delta f_{xx}|, |\Delta f_{yy}|, |\Delta f_{zz}|) = (f_s, f_l, f_s), \tag{4.117}$$

for a $d_{3y^2-r^2}$ orbital occupied site. Since the $4p_x$ and $4p_y$ orbitals dominate the K-edge at the $3d_{3x^2-r^2}$ and $3d_{3y^2-r^2}$ orbital occupied sites, we have the relation $f_l > f_s$.

As mentioned above, the RXS intensity is given by $|f_l - f_s|^2$. This is proportional to the square of the level splitting of the Mn $4p$ orbitals $\Delta(\equiv \varepsilon_x - \varepsilon_z)$, where $\varepsilon_{x(z)}$ is the energy of the Mn $4p_{x(z)}$ orbital, and does not depend on the sign of Δ [80]. Then, quantitative estimations for the anisotropy of the ASF are required in order to identify the dominant contributions to RXS from orbital order. It is, in general, difficult theoretically to treat quantitatively both the local correlation effects, such as the core hole potential and the $3d$-$4p$ Coulomb interactions, and the itinerant character of the $4p$ electrons on an equal footing. The mechanism responsible for the anisotropy of the ASF is still under debate.

4.3.5 Microscopic Calculations of the RXS Intensity

As mentioned above, the appropriate microscopic treatments are essentially required to describe the anomalous term of the scattering factor and the RXS intensity in orbital ordered state. This is because the anisotropy of the scattering factor is sensitive to both the local and itinerant nature of electrons. One of the method which is adopted in the calculation of the RXS intensity is based on the local density of state (DOS) of the $4p$ electron. Instead of the standard formula of RXS intensity presented in (4.85), the difference of the local DOS on the two orbital sublattices is calculated by, for example, the LSDA+U or KKR type band structure calculation method. This is the method being sensitive to the electron itinerant nature, and effects of the core hole is neglected in the intermediate scattering process. Another finite difference method is one of the ab initio cluster method, such as in the MnO_6 cluster. The wave functions of the intermediate states in RXS are obtained by solving the Schrödinger equation for the photo excited $4p$ electron under a certain Coulomb potential constructed from a given electron charge density. Thus, the core hole effects are taken into account as an effective potential, although this is also essentially based on the one-electron picture.

Alternate method introduced in this subsection is based on the projection method (the memory function method, the composite-operator method) in the Green's function formalism [94–96]. This is a non-perturbational method widely applied to the electronic structure of molecules and solids. This method is a reliable way of accounting for the itinerant and localized nature of the correlated electrons on an equal footing, and, of course, independent of the Coulomb/Jahn-Teller mechanisms. The detailed theoretical framework of this method is presented in Appendix C.

Consider a generic tight-binding model, for example, for $LaMnO_3$ where the Hamiltonian includes the Mn $3d$, $4p$, $1s$ and O $2p$ electron degrees of freedom and the several kinds of interactions between them, such as the on-site Coulomb interaction, the core hole potential, the electron transfers between different sites and so on. The relevant part of the ASF, $\Delta f_{l\alpha\alpha}$ at site l (4.88) is given by the retarded Green's function via

$$\mathrm{Im}\Delta f_{l\alpha\alpha} = -\frac{|A_{1s4p}|^2}{m} \sum_{\sigma} \mathrm{Im}G_{l\alpha\sigma}(\omega + \mathrm{i}\Gamma), \tag{4.118}$$

where $G_{l\alpha\sigma}(\omega)$ is the Fourier transform of

$$G_{l\alpha\sigma}(t) = \theta(t)\langle i|[J^{\dagger}_{l\alpha\sigma}(t), J_{l\alpha\sigma}(0)]|i\rangle. \tag{4.119}$$

The operator $J_{l\alpha\sigma}$ indicates the dipole transition from the $1s$ to $4p$ orbitals at site l defined by $J_{l\alpha\sigma} = P^{\dagger}_{l\alpha\sigma} s_{l\sigma}$ where $P_{l\alpha\sigma}$ and $s_{l\sigma}$ are the annihilation operators of Mn $4p$ and $1s$ electrons, respectively, with spin σ and the Cartesian coordinate α. In the memory function formalism, the imaginary part of the Green's function is given by a continued fraction [96]:

$$\mathrm{Im}G_{l\alpha\sigma}(\omega) = \beta\omega \mathrm{Im}\delta M^{(0)}(\omega)\Big|_{T\to 0}, \tag{4.120}$$

with

$$\delta M^{(n-1)}(\omega) = \frac{I^{(n)}}{\omega - \left(M_0^{(n)} + \delta M^{(n)}(\omega)\right) I^{(n)-1}}, \tag{4.121}$$

for $n \geq 1$, where $I^{(n)}$ is a normalization factor, i.e.,

$$I^{(n)} = \langle \psi_n^{\dagger} \psi_n \rangle_{\lambda}, \tag{4.122}$$

and $M_0^{(n)}$ is the static part of the self-energy

$$M_0^{(n)} = \langle (\mathrm{i}\partial_t \psi_n^{\dagger})\psi_n \rangle_{\lambda}. \tag{4.123}$$

Here, we define

$$\langle A^{\dagger}(t)B(0)\rangle_{\lambda} = T\int_0^{\beta} \mathrm{d}\lambda \langle A^{\dagger}(t)B(\mathrm{i}\lambda)\rangle, \tag{4.124}$$

for temperature T. The ψ_n are operator products (the composite operators) appearing in the equation of motion of the operator $J_{l\alpha\sigma}$. An appropriate choice of a set of ψ_n is essential in this formalism and depends on the issue. For example, in the present RXS calculation for $LaMnO_3$, one of the suitable choice is

$$(\psi_1, \psi_2, \psi_3) = \left(J_{l\alpha\sigma}, J_{l\alpha\sigma}\delta n^{d}_{l\gamma}, J_{l\alpha\sigma}\delta n^{dp}_{l\gamma}\right), \tag{4.125}$$

where $\delta A = A - \langle A \rangle$, $n^d_{l\gamma} = \sum_\sigma d^{\dagger}_{l\gamma\sigma} d_{l\gamma\sigma}$ and $n^{dp}_{l\gamma} = \sum_\sigma d^{\dagger}_{l\gamma\sigma} p_{l\gamma\sigma}$. $d_{l\gamma\sigma}$ and $p_{l\gamma\sigma}$ are the annihilation operators for Mn $3d$ and O $2p$ orbitals with spin σ and orbital γ. ψ_2 and ψ_3 indicate the $1s \to 4p$ dipole excitations associated with the $3d$ charge fluctuation and the charge transfer excitation between Mn $3d$ and O $2p$ orbitals, respectively. These operators are treated as single quantum variables in this formalism.

The advantages of the memory functional method for this problem are: (1) The local many body excitations arising around a Mn ion are treated properly by the composite operators ψ_n of (4.125). For example, due to the operator

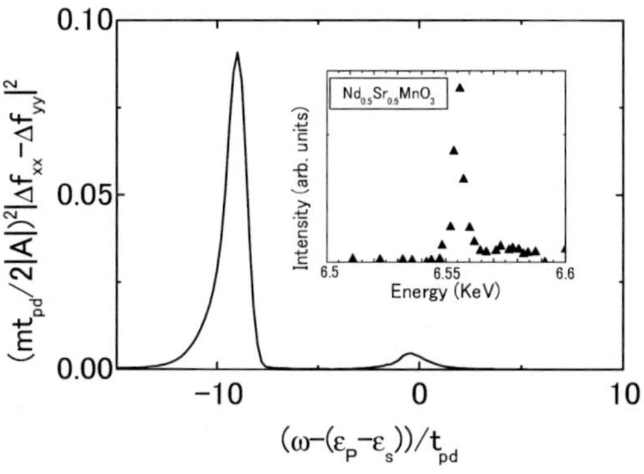

Fig. 4.24. The energy dependence of the scattering intensity of RXS in the $[3d_{3x^2-r^2}/3d_{3y^2-r^2}]$-type orbital ordered state [93]. The inset is the experimental results of the scattering intensity for $Na_{0.5}Sr_{0.5}MnO_3$ [84].

product $\psi_2 = J_{l\alpha\sigma}\delta n_{i\gamma}^{dp}$, the so-called well-screened state $|\underline{1s}3d^5 4p^1 \underline{L}\rangle$ is taken into account properly. Here, L indicates the ligand O $2p$ orbitals and the underline implies the one hole occupied state. It is well known that this state dominates the Mn K edge rather than the simple poorly screened state $|\underline{1s}3d^4 4p^1\rangle$. (2) The itinerant effects of the $4p$ electrons are taken into account in $\delta M^{(n)}(\omega)$ where the loop approximation in the diagrammatic technique is applied. These effects are crucial to description of the energy dependence of the ASF in manganites where the band width of the itinerant $4p$ electrons is a large.

The calculated RXS intensity is shown in Fig. 4.24 associated with the experimental data for $Nd_{0.55}Sr_{0.45}MnO_3$ [84]. This projection method is applied to the orbital ordered manganite. The calculation is based on the Coulomb mechanism and the lattice distortions are not taken into account. A sharp peak structure near the lower edge of the spectrum (the K edge) and a small intensity above the edge are attributed to the itinerant $4p$ band nature and the local electron correlation, respectively. Both structures are in seen in the experimental spectra for several manganites with orbital order.

4.4 Orbital Excitation

Orbital Wave. The orbital degree of freedom is a quantum variable of an electron, as are spin and charge degrees of freedom, and hence this degree of freedom has its own dynamics called "orbital excitations". When compared with the static properties of orbitals, very little is known about orbital excitations and experimental probes which might detect them. The collective excitation of the orbital degree of freedom, or "orbital wave" is the analog of a spin wave in a magnetically ordered system. A quantized orbital wave is called "orbiton".

Collective excitations of an orbital degree of freedom were first proposed theoretically by Cyrot and Lyon-Caen [38], and Komarov et al. [100]. Several kinds of collective orbital excitations have been investigated theoretically. These are classified as follows;

(1) the orbital wave in a model in which the orbital space has a continuous symmetry, such as SU(2) [38]. The excitation in the orbital ordered state is essentially the same as the spin wave in the Heisenberg model, i.e. the gapless Goldstone mode. A particularly interesting example is so-called one-dimensional SU(4) spin-orbital model [97–99];

$$\mathcal{H} = J \sum_{\langle ij \rangle} \left(\boldsymbol{S}_i \cdot \boldsymbol{S}_j + \frac{1}{4} \right) \left(\boldsymbol{T}_i \cdot \boldsymbol{T}_j + \frac{1}{4} \right) . \tag{4.126}$$

This model corresponds to (4.33) with $J_1 = J_2$, i.e. the exchange interaction $J = 0$, and has a SU(4) symmetry where the generators are $\boldsymbol{S}$, $\boldsymbol{T}$ and $\boldsymbol{ST}$. The ground state is a singlet with a periodicity $4a$. The elementary excitations, i.e. the spin-orbital coupled excitations, are gapless and soften at the momentums $\boldsymbol{k} = \pi/2$ and π. These low lying excitations contribute to the thermodynamics, such as the specific heat.

(2) Orbital excitations in a model where the anisotropy in the pseudo-spin space is taken into account [100]. The orbital waves calculated for $LaMnO_3$ based on the Hamiltonian (4.27) are introduced in this section in detail [40]. The spin-orbital excitations are also investigated in the anisotropic orbital model [101,102]:

$$\mathcal{H} = J \sum_{\langle ij \rangle} \left[4 \; \boldsymbol{S}_i \cdot \boldsymbol{S}_j \left(\tau_i^l - \frac{1}{2} \right) \left(\tau_j^l - \frac{1}{2} \right) + \left(\tau_i^l + \frac{1}{2} \right) \left(\tau_j^l + \frac{1}{2} \right) \right], \tag{4.127}$$

which is obtained from (4.27) by assuming $J_1 = J_2$. The theoretical results of this excitation are introduced in Chap. 7.

(3) the orbital-lattice coupled excitation [103,104]. The dispersion relation of the collective excitations are dominated by the orbital-phonon interaction and the electronic exchange interaction. The vibronic collective excitation associated with the Jahn-Teller effect has been called a "vibron" [48].

Here we focus on the orbital excitation in $LaMnO_3$ with an A-type AFM order and a $[3d_{3x^2-r^2}/3d_{3y^2-r^2}]$-type orbital order. This type of the orbital order parameter is represented by the expectation value of the pseudo-spin operators as

$$\langle \boldsymbol{T}_k \rangle = \left(\frac{\sqrt{3}}{4} \delta_{\boldsymbol{k}=(\pi\pi 0)}, 0, -\frac{1}{4} \delta_{\boldsymbol{k}=(000)} \right) , \tag{4.128}$$

where $\langle \boldsymbol{T}_k \rangle$ is the Fourier transform of $\langle \boldsymbol{T}_i \rangle$. The orbital excitations are defined as deviations of $\boldsymbol{T}_i$ from the above value. When an excitation occurs at a Mn site, this excitation propagates through the interaction, J, which couples orbitals at different sites and which results in the dispersion of the orbital wave. Assume the lattice degree of freedom being frozen for the orbiton. This adiabatic approximation may be justified because the characteristic energy of the orbital wave, J, is larger than the phonon frequency of the Jahn-Teller mode ($\sim$ 700cm^{-1}). Then, start with the model Hamiltonian

$$\mathcal{H} = \mathcal{H}_J + \mathcal{H}_{JT}, \tag{4.129}$$

where $\mathcal{H}_J$ and $\mathcal{H}_{JT}$ are defined by (4.31) and (4.51), respectively. The lattice distortions described by Q_{i2} and Q_{i3} are treated as static variables in this calculation. The spin operator $\boldsymbol{S}_i$ in $\mathcal{H}_J$ is also replaced by a classical number for the A-type AFM phase, since the band width of the spin waves in $LaMnO_3$ is of the order of 30meV. In the xz plane of the pseudo-spin space, a rotated frame is introduced and the new z axis is taken to be parallel to the $3d_{3x2-r^2}$ and $3d_{3y^2-r^2}$ orbitals. The dispersion relation of the orbital wave is obtained by applying the SU(2) Holstein-Primakoff transformation within the linear spin wave approximation,

$$\begin{aligned} \widetilde{T}_{ix} &= \frac{1}{2}T_{ix} \pm \frac{\sqrt{3}}{2}T_{iz} \sim \frac{1}{2} - a_i^\dagger a_i, \\ \widetilde{T}_{iz} &= \mp\frac{\sqrt{3}}{2}T_{ix} + \frac{1}{2}T_{iz} \sim \frac{1}{2}\left(a_i^\dagger + a_i\right), \end{aligned} \tag{4.130}$$

where the upper and lower signs are for the $3d_{3x^2-r^2}$- and $3d_{3y^2-r^2}$-orbital occupied sites, respectively, and a_i is the annihilation operator for an orbiton. The results shown in Fig. 4.25 is the dispersion relations of the orbital waves for $LaMnO_3$ [40, 105]. The four branches of the orbital wave are attributed to the four inequivalent Mn ions in the unit cell. The excited states at the point Γ in the Brillouin zone are classified by the irreducible representations of these excitations, A_{1g}, B_{3g}, B_{1g} and B_{2g}. For example, the lowest branch is explicitly given by $\psi_{B_{3g}} = \frac{1}{2}(\psi_1 - \psi_2 - \psi_3 + \psi_4)$ where ψ_l ($l = 1-4$) indicates the site at which the e_g orbital occupation is changed, i.e., this at $\boldsymbol{r}_i$ for $l=1$, $\boldsymbol{r}_i + a\hat{\boldsymbol{x}}$ for $l=2$, $\boldsymbol{r}_i + c\hat{\boldsymbol{z}}$ for $l=3$, and $\boldsymbol{r}_i + a\hat{\boldsymbol{x}} + c\hat{\boldsymbol{z}}$ for $l=4$. As shown in Fig. 4.25, there is a gap in the orbital wave attributed to the fact

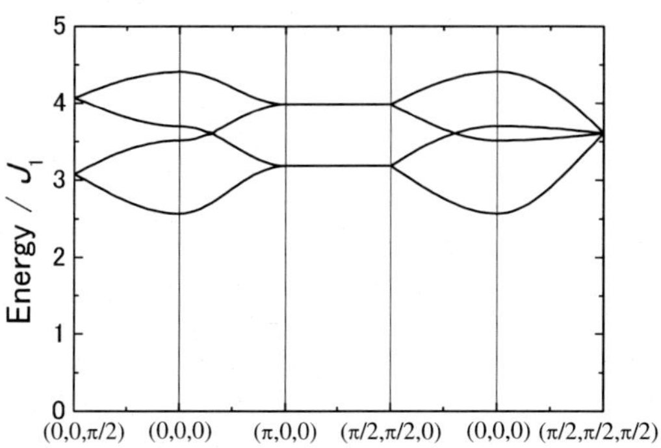

Fig. 4.25. Dispersion relation of the orbital wave in the C-type orbital ordered state with an A-type AFM structure. [105].

that (1) The model Hamiltonian does not have a continuous symmetry in the pseudo-spin space. The symmetry is lowered furthermore in the A-type AFM phase where the cubic symmetry is broken. (2) The coupling with the Jahn-Teller type lattice distortion acts as an external field in the pseudo-spin space.

Raman Scattering from Orbital Wave. The orbital wave is a new elementary excitation in the orbital ordered state and may contribute to several optical, thermodynamic, and transport properties. Then, the issue is how to measure experimentally this excitation. The optical absorption/reflection is not available as a probe to detect the orbiton because the two e_g orbitals have a same parity, except for utilizing the optical transition associated with phonons with the odd-parity symmetry. Raman scattering is a candidate as a probe which can detect orbital waves [105,106]. It is known that the magnons in antiferromagnets are observed by the Raman scattering as the two-magnon scattering through the exchange of up and down spins between the NN magnetic ions. The orbiton is also expected to be excited through the inter-site exchange processes in the orbital ordered state. The most distinct point in the orbiton scattering from the magnon Raman scattering is that the total orbital quantum number, e.g. $\sum_i T_{iz}$ is not conserved due to the electron transfer between the different orbitals in the NN sites. That is to say, the one-orbiton scattering is possible in the e_g orbital ordered state. The Raman scattering cross section is expressed by the dynamical correlation function for the orbiton operator $a_{\boldsymbol{k}}$ introduced in (4.130) at momentum $\boldsymbol{k} = 0$. The polarizations of the incident and scattered photons determines the direction of the bond where the electronic exchanges occur. Thus, the polarization analyses are crucial to identify the scatterings from the orbiton.

Recently, peak structures in the Raman scattering spectra were found around 120–170cm^{-1} in $LaMnO_3$ [107]. The experimental results are shown in the insets of Fig. 4.26. The relative intensity of the peak structures depends strongly on the polarization configuration. Two magnon Raman scattering is not expected in the ab plane where spins align ferromagnetically. The multiphonon excitations may be at a disadvantage since (1) The Raman energy shift cannot be reproduced by a combination of the observed Raman and infrared phonon modes. (2) A substantial temperature dependence of the energy shift is observed, in contrast to a negligible temperature dependence of the observed phonon mode energies. As shown in Fig. 4.26, the polarization and energy dependences and relative intensity of the Raman scattering peaks are well explained by the theoretical calculations based on orbital wave scattering. A snapshot of the orbital wave (orbiton) is presented in Fig. 4.27.

Although Raman scattering is a suitable tool by which to detect the orbital waves, information for the orbiton dispersion relation obtained by this method is restricted. The resonant inelastic x-ray scattering (RIXS) and the (nonresonant) inelastic x-ray scattering (IXS) are the possible methods to detect the dispersion relation. These experimental methods have al-

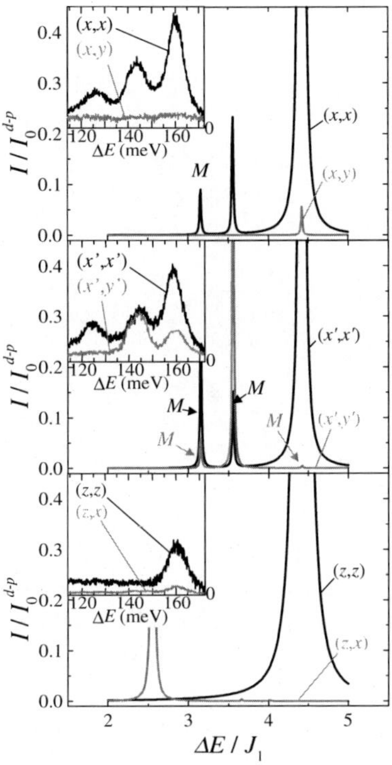

Fig. 4.26. The Raman scattering spectra from the orbital wave in the A-type AFM phase [105]. Insets show the experimental Raman spectra for $LaMnO_3$ at 9 K [107].

ready been used as a momentum resolved probe to detect bulk excitations in solids [108, 109], and have been rapidly developed due to the recent availability of third-generation synchrotron light sources. RIXS is an inelastic version of the resonant x-ray scattering (RXS) introduced in Sect. 4.3. The scattering intensity is largely enhanced at the resonant condition. Unfortunately, up to now, the energy resolution is limited to be of the order of 0.1eV which

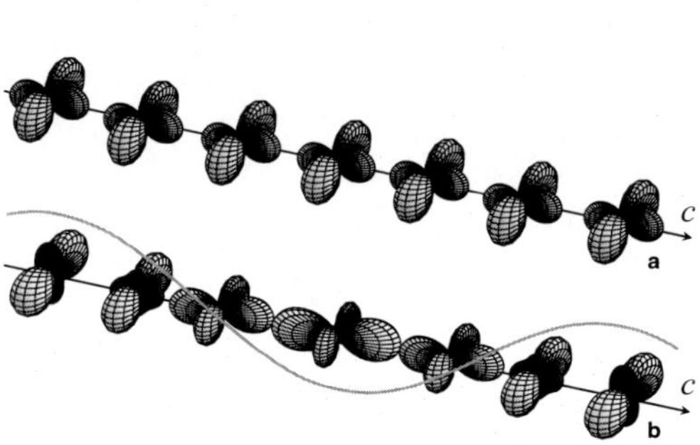

Fig. 4.27. (**a**) Electron cloud in the ground state, (**b**) a snapshot of orbital wave (orbiton).

is comparable or larger than the orbiton energy. Actually, by this method, the electronic excitations across the Mott gap have been detected in the insulating cuprates $Ca_2CuO_2Cl_2$ [110] and the orbital ordered $LaMnO_3$ [111]. On the other hand, in the usual IXS experiments, the energy resolution is much higher than RIXS being enough to detect the orbiton. However, the scattering intensity is proportional to the square of the charge valence Z^2. That is, the expected scattering intensity for orbiton is much weaker than that for phonon. The further developments of the synchrotron light source and the related experimental devices are expected.

4.5 Other Orbital-Related Topics

Phase Separation. Electronic phase separation in manganites and other transition metal oxides has been discussed in early 90's by Nagaev [112]. Some anomalous properties in doped manganites are interpreted in the FM-AFM two phase scenario. After the discovery of CMR, a number of experimental and theoretical studies of the phase separation has been done. The existence of the phase separated regime between the hole undoped AFM and hole rich FM phases in the so-called "double exchange model" was shown by Dagotto and coworkers [23]. This is associated with a charge segregation, i.e. a system is separated into the hole rich FM region and the poor AFM one. In the thermodynamic sense, appearance of the FM phase due to the homogeneous hole doping in the AFM phase implies the negative compressibility, i.e. $\kappa = n^{-2}(\partial n/\partial \mu) < 0$ due to the larger band width of the FM phase than that of the AFM one.

The phase separation is also found in the theoretical model with the orbital degree of freedom [113,114]. As mentioned in Sect. 4.1.4, there are two kinds of ferromagnetic interactions, i.e. the superexchange interaction and the double exchange one described by $\mathcal{H}_J$ (4.27) and $\mathcal{H}_t$ (4.25), respectively. These two interactions favor the different types of the orbital states; the staggered type orbital order for the hole poor region and the uniform orbital order and/or disorder for the hole rich one. Between the two phases, the two interactions compete with each other and a phase separated regime appears. The main mechanism of this phase separation is similar to the spin case, i.e. the thermodynamic instability occurs by doping of holes into the staggered type orbital order with smaller band width comparing with that in the uniform orbital state. In addition, due to the anisotropy in the orbital space induces the discontinuous change of the orbital state and further promote the phase separation. This kind of the orbital induced phase separation is suggested by a number of experiments, mainly through observations of the two phases associated with different type/amplitude of lattice distortions.

One of the issue in the phase separation idea is that this is the charge-segregated type or not. This is because the energy cost of large charged cluster is expected to be large due to the long-range Coulomb interactions. A micrometer scale mixture of the charge ordered insulating and FM metallic regions observed in $La_{5/8-y}Pr_yCa_{3/8}MnO_3$ is interpreted to have a same carrier concentration $x = 1/2$ [115]. Mechanism of this kind of phase separation with equal electronic density is discussed theoretically in connection with the quenched disorder, although more detail theoretical studies and comparison with experiments are still up-to-date.

Other Exotic Orbital States. Here we briefly mention other unusual orbital states observed experimentally and proposed theoretically in manganites. Mechanism of the long range orbital order observed in a number of half-doped manganites (see Fig. 4.8) is still open question. This is not just an interest for the multi-component complicated order but may be related to mechanism of CMR which appears near the transition between the FM metal and the spin-charge-orbital ordered insulator. The $3d_{3x^2-r^2}$ and $3d_{3y^2-r^2}$ type orbitals are alternately aligned along the [1 -1 0] direction and uniformly along the [110] one in the Mn^{3+} sublattice. Two key experimental results for this issue are the following; (1) the orbital and charge orders occur at the same temperature ($T_{OO} = T_{CO}$) which is far above the CE-type AFM Néel temperature in some manganites, and (2) RXS experiments indicate that, above T_{OO}, the orbital correlation length is shorter than that for charge [116]. The result (1) indicates that the magnetic interactions, such as J_{AF} in (4.8), are not responsible for this orbital order, and the CE-type AFM order is a consequence of this order. Even for the charge order, on which the orbital order is realized, a simple Coulombic interaction scenario is not applicable; Mn ions with the same valence are aligned along the c axis. The electronic interactions between Mn^{3+} and Mn^{4+} ions as well as the long ranged phononic one are supposed to play key roles in the present charge-orbital ordering issue. This order may be related to the paired striped state observed by the electron diffraction experiments in $La_{1/3}Ca_{2/3}MnO_3$ [117] where the $3d_{3x^2-y^2}$ orbital rod running along the [110] direction forms a pair with the $3d_{3y^2-r^2}$ orbital rod.

The orbital order with the complex wave function, i.e. the y component of the pseudo-spin $\langle T_y \rangle$ are proposed theoretically [118, 119]. The eigen wave function of this operator is explicitly

$$|\psi_\pm^C\rangle = \frac{1}{\sqrt{2}} \left(|d_{3z^2-r^2}\rangle \pm \mathrm{i} |d_{x^2-y^2}\rangle \right) . \tag{4.131}$$

The remarkable points of this complex orbital are (1) this represents the magnetic octupole which breaks the time-reversal symmetry, and (2) the electron charge distribution has the cubic symmetry in contrast to the real mixed orbitals

$$|\psi_\pm^R\rangle = \frac{1}{\sqrt{2}} \left(|d_{3z^2-r^2}\rangle \pm |d_{x^2-y^2}\rangle \right) , \tag{4.132}$$

which are the eigen states of T_x. The orbital ordered state with the $\langle T_y \rangle$ order parameter has been proposed in the f-electron systems with quadrupole moments, such as CeB_6. This order is not expected to realize in the Mott insulating limit, $LaMnO_3$, since the operator T_y does not appear in the superexchange term of the Hamiltonian $\mathcal{H}_J$ (4.27) and the Jahn-Teller coupling stabilizes the real orbital. Possibilities of the uniform and staggered type complex orbital orders are suggested theoretically in moderately hole doped ferromagnetic manganites. The cubic charge distribution is compatible with the experimentally observed almost isotropic structural and magnetic properties, although the direct evidence of this orbital order is not reported experimentally up to now.

CMR. The main issue of study in manganites is, of course, origin of CMR, a gigantic reduction of the electric resistivity applying magnetic field. A giant magnetoresistance has been first observed in 1970's near the transition between the paramagnetic metallic phase to the ferromagnetic one. This phenomena is understood by the theoretical calculation based on the double exchange interaction. On the other hand, the recently observed CMR occurs at vicinity of the paramagnetic insulating phase to the FM metallic one; this is recognized to be the metal-insulator transition induced by applying magnetic field. Therefore, the central problem of CMR is to reveal nature of the insulating phase located at vicinity of the FM metallic one. It is first suggested by Millis et al. that the paramagnetic insulating behavior and a gigantic change of the resistivity at the Curie temperature are not understood within the simple double exchange scenario [13]. It is considered that the interplay of the double exchange interaction and the Jahn-Teller effects under the orbital degree of freedom reproduces the behavior of the resistivity [120, 121]. That is, the self-trapped state due to the Jahn-Teller coupling, termed the Jahn-Teller polaron is realized above T_C. By applying magnetic field, an itinerant nature of electron overcomes the carrier localization. This scenario may be applied to the manganites with lightly and moderately hole doped regions.

In the higher doped region close to $x = 0.5$, one of the interpretations is that CMR is a consequence of the first order phase transition between competing phases, such as the FM metallic phase and the insulating one associated with the charge, orbital and spin order. Short range correlations of the lattice distortion, indicating the charge and orbital short range order, are observed in the insulating phase above the Curie temperature. This is measured as the diffuse scattering in the x-ray and neutron scatterings [122, 123]. The scattering intensity grows with decreasing temperature, and abruptly disappears at T_C; there is clear correlation between the temperature dependence of the diffuse scattering and that of the electric resistivity. The correlation length is estimated to be about 10Å$\sim$ 30Å. These results suggest one possibility that the FM metallic state strongly competes with the insulating charge-orbital ordered one associated with lattice distortion observed near half doped manganites, and a collapse of the insulating phase implies the insulator-metal transition. Raman scattering experiments also indicate the dynamical charge/orbital ordering above T_C and suggest an important role of directional ordering of orbital through the measurement of the anisotropic diffuse scattering [124]. The phase separation scenario with spin and orbital inhomogeneity may be closely related to this picture.

4.6 Summary

Study of the manganites with perovskite structure has been started since more than 40 years ago. The recent diverse experimental and theoretical researches from the modern physics view points shed light on this antiquated materials and provides a new meaning of a spin-charge-orbital-lattice coupled system with electron correlation. More than 10 years have passed since discovery of the CMR; nevertheless, even nowadays, new phenomena are continuously reported experimentally and exotic concepts are proposed theoretically. Some of the fundamental issues of manganites including origin of CMR are still under debate and arose much interest of researchers. We should finally remark that the studies of manganites do not only make a large progress of the fundamental condensed matter physics, but also promote the technological applications of the correlated electron systems. The multi-functional devices with fast speed, colossal response and high capacity, by utilizing the multi degrees of freedom, e.g. spin, charge, orbital and lattice, are certainly expected in the next generation, in spite of the present semiconductor based devices.

References

1. J. G. Bednorz, K. A. Müller: Z. Phys. **64**, 189 (1986)
2. G. H. Jonker, H. van Santen: Physica **16**, 337 (1950)
3. G. H. Jonker: Physica **22**, 707 (1956)
4. K. Chahara, T. Ohono, M. Kasai, Y. Kanke, Y. Kozono: Appl. Phys. Lett. **62**, 780 (1990)
5. R. von Helmolt, J. Wecker, R. Holzapfel, L. Schultz, K. Samwer: Phys. Rev. Lett. **71**, 2331 (1993)
6. Y. Tokura, A. Urushibara, Y. Moritomo, T. Arima, A. Asamitsu, G. Kido, N. Furukawa: Jour. Phys. Soc. Jpn. **63**, 3931 (1994)
7. S. Jin, T. H. Tiefel, M. McCormack, R. A. Fastnacht, R. Ramesh, L. H. Chen: Science **264**, 413 (1994)
8. C. Zener: Phys. Rev. **82**, 403 (1951)
9. P. W. Anderson, H. Hasegawa: Phys. Rev. **100**, 675 (1955)
10. P. G. de Gennes: Phys. Rev. **118**, 141 (1960)
11. C. W. Searle, S. T. Wang: Can. Jour. Phys. **48**, 2023 (1970)
12. Y. Tomioka, A. Asamitsu, Y. Moritomo, Y. Tokura: Jour. Phys. Soc. Jpn. **64**, 3626 (1995)
13. A. J. Millis, P. B. Littlewood, B. I. Shrainman: Phys. Rev. Lett. **74**, 5144 (1995)
14. G. Matsumoto: Jour. Phys. Soc. Jpn. **29**, 606 (1970)

15. J. B. A. A. Elemans, B. van Laar, K. R. van der Veen, B. O. Loopstra: Jour. Soli. Stat. Chem. **3**, 238 (1971)
16. Z. Jirák, S. Krupička, V. Nekvasil, E. Pollert, G. Villeneuve, F. Zounová: Jour. Mag. Mag. Mat. **15-18**, 519 (1980)
17. E. O. Wollan, W. C. Koehler: Phys. Rev. **100**, 545 (1955)
18. Y. Tokura: *Colossal Magnetoresistance Oxides*, (Amsterdam, Gordon and Breach 2000)
19. T. Kaplan, S. D. Mahanti: *Physics of Manganites* (Gordon and Breach 1999)
20. M. Imada, A. Fujimori, Y. Tokura: Rev. Mod. Phys. **70** 1039 (1998)
21. Y. Tokura, N. Nagaosa: Science **288** 462 (2000)
22. T. Chatterji: *Colossal Magnetoresistive Manganites*, (Kluwer Academic Pub, London 2004)
23. E. Dagotto: *Nanoscale Phase Separation and Colossal Magnetoresistance* (Springer-Verlag, New York 2003)
24. A. Fujimori, A. E. Bocquet, T. Saitoh, T. Mizokawa, Jour. Elec. Spec. Relat. Phenom. **62** 141 (1993)
25. T. Saitoh, A. E. Bocquet, T. Mizokawa, H. Namatame, A. Fujimori, M. Abbate, Y. Takeda, Y. Takano: Phys. Rev. B **51**, 13942 (1995)
26. A. Chainani, M. Mathew, D. D. Sarma: Phys. Rev. B **47**, 15397 (1993)
27. J. -H. Park, C. T. Chen, S. -W. Cheong, W. Bao, G. Meigs, V. Chakarian, Y. U. Idzerda: Phys. Rev. Lett. **76**, 4215 (1996)
28. Y. Okimoto, T. Katsufuji, T. Ishikawa, A. Urushibara, T. Arima, Y. Tokura: Phys. Rev. Lett. **75**, 109 (1995)
29. J. H. Jung, K. H. Kim, D. J. Eom, T. W. Noh, E. J. Choi, J. Yu, Y. S. Kwon, Y. Chung: Phys. Rev. B **55**, 15489 (1997)
30. L. F. Feiner, A. M. Oleś: Phys. Rev. B **59**, 3295 (1999)
31. R. Shiina, T. Nishitani, H. Shiba: Jour. Phys. Soc. Jpn. **66**, 3159 (1997)
32. J. H. Van Vleck: Rev. Mod. Phys. **25**, 220 (1953)
33. J. B. Goodenough: Phys. Rev. **100**, 564 (1955)
34. J. Kanamori: J. Phys. Chem. Solids **10**, 87 (1959)
35. L. M. Roth: Phys. Rev. **149**, 306 (1966)
36. K. I. Kugel, D. I. Khomskii: Sov. Phys.- JETP Letter **15**, 446 (1972)
37. L. I. Korovin, E. K. Kudinov: *Sov. Phys. -Solid State* **16**, 1666 (1975)
38. M. Cyrot, C. Lyon-Caen: Jour. Physique **36**, 253 (1975)
39. S. Inagaki: Jour. Phys. Soc. Jpn. **39**, 596 (1975)
40. S. Ishihara, J. Inoue, S. Maekawa: Phys. Rev. B **55**, 8280 (1997)
41. K. Kubo, N. Ohata: Jour. Phys. Soc. Jpn. **33**, 21 (1972)
42. N. Furukawa: Jour. Phys. Soc. Jpn. **64**, 2734 (1996)
43. H. A. Jahn, E. Teller: Proc. Roy. Soc. A **161**, 220 (1937)
44. H. C. Longuet-Higgins: Adv. Spectrosc. **2**, 429 (1962)
45. M. V. Berry: Proc. R. Soc. London **A 392**, 45 (1984)
46. G. Delacrétaz, E. R. Grant, R. L. Whetten, L. Wöste, J. W. Zwanziger: Phys. Rev. Lett. **56**, 2598 (1986)
47. G. A. Gehring, K. A. Gehring: Rep. Prog. Phys. **38**, 1 (1975)
48. M. D. Kaplan, B. G. Vekhter: *Cooperative Phenomena in Jahn-Teller Crystals* (Plenum Press, New York) (1995)
49. J. Kanamori: Jour. Appl. Phys. Suppl. **31**, 14S (1960)
50. S. Okamoto, S. Ishihara, and S. Maekawa: Phys. Rev. B **65**, 144403 (2002)
51. H. Y. Hwang, S. -W. Cheong, P. G. Radaelli, M. Marezio, and B. Batlogg: Phys. Rev. Lett. **75**, 914 (1995)
52. Y. Tokura: J. Appl. Phys. **79**, 5288 (1996)
53. H. Kuwahara, T. Okuda, Y. Tomioka, A. Asamitsu, Y. Tokura: Phys. Rev. Lett. **82**, 4316 (1999) and Mat. Res. Soc. Sym. Proc. **494**, 83 (1998)

54. R. Maezono, S. Ishihara, N. Nagaosa: Phys. Rev. B **57**, R13993 (1998)
55. S. Yunoki, A. Moreo, E. Dagotto: Phys. Rev. Lett. **81**, 5612 (1998)
56. T. Hotta, Y. Takada, H. Koizumi, E. Dagotto: Phys. Rev. Lett. **84**, 2477 (2000)
57. S. Okamoto, S. Ishihara, S. Maekawa: Phys. Rev. B **61**, 14647 (2000)
58. D. D. Sarma, N. Shanthi, S. R. Krishnakumar, T. Saitoh, T. Mizokawa, A. Sekiyama, K. Kobayashi, A. Fujimori, E. Weschke, R. Meier, G. Kaindl, Y. Takeda, M. Takano: Phys. Rev. **B53**, 6873 (1996)
59. M. C. Martin, G. Shirane, Y. Endoh, K. Hirota, Y. Moritomo, Y. Tokura: Phys. Rev. **B 53**, (1996)
60. S. Ishihara, M. Yamanaka, N. Nagaosa: Phys. Rev. B **56**, 686 (1997)
61. R. Kilian, G. Khaliullin: Phys. Rev. B **58**, R11841 (1998)
62. R. A. M. Ram, P. Ganguly, C. N. R. Rao: J. Sol. Stat. Chem. **70**, 82 (1987)
63. Y. Moritomo, A. Asamitsu, H. Kuwahara, Y. Tokura: *Nature (London)* **380**, 141 (1996)
64. T. Kimura, Y. Tomioka, H. Kuwahara, A. Asamitsu, M. Tamura, Y. Tokura: Science **274**, 1698 (1996)
65. M. Kubota, H. Fujioka, K. Hirota, K. Ohoyama, Y. Moritomo, H. Yoshizawa, Y. Endoh: Jour. Phys. Soc. Jpn. **69**, 1606 (2000)
66. C. D. Ling, J. E. Millburn, J. F. Mitchell, D. N. Argyriou, J. Linton, H. N. Bordallo: Phys. Rev. B **62**, 15096 (2000)
67. S. -W. Cheong, H. Y. Hwang: in *Colossal Magnetoresisitance Oxides* edited by Y. Tokura (Gordon and Breach, 2000)
68. S. Okamoto, S. Ishihara, S. Maekawa: Phys. Rev. B **63**, 104401 (2001); T. Akimoto, Y. Moritomo, K. Ohoyama, O. O. Kamoto, S. Ishihara, S. Maekawa and A. Nakamura: Phys. Rev. B **59**, R14153 (1999)
69. R. Maezono, N. Nagaosa: Phys. Rev. B **61**, 1825 (2000)
70. T. Kimura, Y. Tomioka, A. Asamitsu, Y. Tokura: Phys. Rev. Lett. **81**, 5920 (1998)
71. M. Medarde, J. F. Mitchell, J. E. Millburn, S. Short, J. D. Jorgensen: Phys. Rev. Lett. **83**, 1223 (1999)
72. K. Hirota, S. Ishihara, H. Fujioka, M. Kubota, H. Yoshizawa, Y. Moritomo, Y. Endoh, S. Maekawa: Phys. Rev. B **65**, 064414 (2001)
73. T. G. Perring, D. T. Adroja, G. G. Chaboussant, G. Aeppli, T. Kimura, Y. Tokura: Phys. Rev. Lett. **87**, 217201 (2001)
74. A. Koizumi, S. Miyaki, Y. Kakutani, H. Koizumi, N. Hiraoka, K. Makoshi, N. Sakai, K. Hirota, Y. Murakami: Phys. Rev. Lett. **86**, 5589 (2001)
75. D. H. Templeton, L. K. Templeton: Acta. Crystallogr. A **38**, 62 (1982)
76. Y. Murakami, H. Kawada, H. Kawata, M. Tanaka, T. Arima, H. Moritomo, Y. Tokura: Phys. Rev. Lett. **80**, 1932 (1998)
77. Y. Ito, J. Akimitsu: Jour. Phys. Soc. Jpn. **40**, 1333 (1976)
78. J. M. Zuo, M. Kim, M. O'Keefe, J. C. Spence: Nature **401**, 49 (1999)
79. M. Takata, E. Nishibori, K. Kato, M. Sakata, Y. Moritomo: Jour. Phys. Soc. Jpn. **68**, 2190 (1999)
80. Y. Murakami, J. P. Hill, D. Gibbs, M. Blume, I. Koyama, M. Tanaka, H. Kawata, T. Arima, Y. Tokura, K. Hirota, Y. Endoh: Phys. Rev. Lett. **81**, 582 (1998)
81. M. Fabrizio, M. Altarelli, M. Benfatto: Phys. Rev. Lett. **80**, 3400 (1998) : Phys. Rev. Lett. **81**, E4030 (1998)
82. L. Paolasini, C. Vettier, F. de Bergevin, F. Yakhou, D. Mannix, A. Stunault, W. Neubeck: Phys. Rev. Lett. **82**, 4719 (1999)
83. S. Ishihara, S. Maekawa: Rep. Prog. Phys. **65** 561 (2002)

84. K. Nakamura, T. Arima, A. Nakazawa, Y. Wakabayashi, Y. Murakami: Phys. Rev. B **60**, 2425 (1999)
85. J. J. Sakurai: *Advanced Quantum Mechanics* (Addison-Wesley, Reading) (1987)
86. M. Blume: Jour. Appl. Phys. **57**, 3615 (1985)
87. S. Ishihara, S. Maekawa: Phys. Rev. B **58**, 13442 (1998)
88. S. Ishihara, S. Maekawa: Phys. Rev. Lett. **80**, 3799 (1998)
89. S. Ishihara, S. Maekawa: Phys. Rev. B **62**, 5690 (2000)
90. I. S. Elfimov, V. I. Anisimov, G. A. Sawatzky: Phys. Rev. Lett. **82**, 4264 (1999)
91. M. Benfatto, Y. Joly, C. R. Natoli: Phys. Rev. Lett. **83**, 636 (1999)
92. M. Takahashi, J. Igarashi, P. Fulde: Jour. Phys. Soc. Jpn. **68**, 2530 (1999)
93. S. Ishihara, S. Maekawa: Jour. Mag. Mag. Mat. **21**, 233 (2001)
94. D. Forster: *Hydrodynamic Fluctuations, Broken Symmetry and Correlation Functions* (Addison-Wesley, Massachusetts) (1990)
95. P. Fulde: *Electron Correlations in Molecules and Solids* (Spinger-Verlag, Berlin) (1993)
96. H. Matsumoto, M. Sasaki, S. Ishihara, M. Tachiki: Phys. Rev. B **46**, 3009 (1992)
97. Y. Yamashita, N. Nishitani, K. Ueda: Phys. Rev. B **58**, 9114 (1998)
98. Y. Q. Li, M. Ma, D. N. Shi, F. C. Zhang: Phys. Rev. B **60**, 12781 (1999)
99. I. Affleck: Nucl. Phys. B **265**, 409 (1986)
100. A. G. Komarov, L. I. Korovin, E. K. Kudinov: Sov. Phys. Solid State **17** 1531 (1976)
101. L. F. Feiner, A. M. Oleś, J. Zaanen: Phys. Rev. Lett. **78**, 2799 (1997)
102. G. Khaliullin, V. Oudovenko: Phys. Rev. B **56**, R14243 (1997)
103. V. Perebeinos, P. B. Allen: Phys. Stat. Sol. **215**, 607 (1999)
104. J. van den Brink: Phys. Rev. Lett. **87**, 217202 (2001)
105. S. Okamoto, S. Ishihara, S. Maekawa: Phys. Rev. B **66**, 014435 (2002)
106. J. Inoue, S. Okamoto, S. Ishihara, W. Koshibae, Y. Kawamura, S. Maekawa: Physica B **237-238**, 51 (1997)
107. E. Saitoh, S. Okamoto, K. T. Takahashi, K. Tobe, K. Yamamoto, T. Kimura, S. Ishihara, S. Maekawa, Y. Tokura: Nature **410**, 180 (2001)
108. D. L. Ederer, J. H. McGuire: *Raman emission by x-ray scattering* (World Scientific, Singapore) (1996)
109. A. Kotani, S. Shin: Rev. Mod. Phys. **73**, 203 (2001)
110. M. Z. Hasan, E. D. Isaacs, Z. -X. Shen, L. L. Miller, K. Tsutsui, T. Tohyama, S. Maekawa: Science **288**, 1811 (1999)
111. T. Inami, S. Ishihara, H. Kondo, J. Mizuki, T. Fukuda, S. Maekawa, H. Nakao, T. Matsumura, K. Hirota, Y. Murakami, Y. Endoh: **67**, 045108 (2003)
112. E. L. Nagaev: Phys. Stat. Sol. (b) **186**, 9 (1994)
113. S. Yunoki, A. Moreo, E. Dagotto: Phys. Rev. Lett. **81**, 5612 (1998)
114. S. Okamoto, S. Ishihara, S. Maekawa: Phys. Rev. B **61**, 451 (2000)
115. M. Uehara, S. Mori, C. H. Chen, S. -W. Cheong: Nature **399**, 560 (1999)
116. M. von Zimmermann, J. P. Hill, D. Gibbs, M. Blume, D. Casa, B. Keimer, Y. Murakami, Y. Tomioka, Y. Tokura: Phys. Rev. Lett. **83**, 4872 (1999)
117. S. Mori, C. H. Chen, S. -W. Cheong: Nature **392**, 473 (1998)
118. J. van den Brink, D. Khomskii: Phys. Rev. B **63**, 140416 (2001)
119. A. Takahashi, H. Shiba: Jour. Phys. Soc. Jpn. **69**, 3328 (2000)
120. A. J. Millis, B. I. Shrainman, R. Mueller: Phys. Rev. Lett. **77**, 175 (1996)
121. H. Röder, J. Zang, A. R. Bishop: Phys. Rev. Lett. **76**, 1356 (1996)
122. S. Shimokura, N. Wakabayashi, H. Kuwahara, Y. Tokura: Phys. Rev. Lett. **83**, 4389 (1999)

123. L. Vasiliu-Doloc, S. Rosenkranz, R. Osborn, S. K. Shinha, J. Lynn, J. Mesot, O. H. Seek, G. Preosti, A. J. Fedro, J. M. Mitchell: Phys. Rev. Lett. **83**, 4393 (1999)
124. E. Saitoh, Y. Tomioka, T. Kimura, Y. Tokura: Jour. Phys. Soc. Jpn. **69**, 2403 (2000)

5. Titanates and Vanadates

5.1 Introduction

In addition to the colossal magnetoresistive manganites introduced in Sect. 4.1, the transition metal oxides with light transition metal ions, such as Ti and V, are recognized to be an alternative class of materials with orbital degrees of freedom. Except for compounds where the formal valence of Ti (V) is 4+ (5+), such as $BaTiO_3$, TiO_2 and V_2O_5, the electrons occupy the partially degenerate transition metal ion $3d$ orbitals. For example, with a cubic crystalline field, the five fold degeneracy of the $3d$ orbitals is split into the e_g and t_{2g} orbitals. The latter, with relatively low in energy, i.e. the $3d_{xy}$, $3d_{yz}$ and $3d_{zx}$ orbitals, are partially filled by electrons in Ti and V ions.

One of the best known compounds of such light transition metal oxides is V_2O_3 which has been studied since the 1960s. It is considered to be the prototypical material exhibiting a metal-insulator transition. This transition can be induced by changing the temperature, the pressure and by the substitution of ions. Distorted VO_6 octahedra are located in the corundum-type crystal structure. The crystalline field with the trigonal symmetry further splits the triply degenerate t_{2g} orbitals into the a_{1g} and e_g orbitals. One of the two electrons at each V site occupies the doubly degenerate e_g orbitals. Castellani et al. suggested in 1970s that one electron in the a_{1g} orbital forms a kind of dimerized states involving a vertical pair of V ions, and that the e_g orbital degree of freedom is active [1]. The effective spin-orbital Hamiltonian, which is similar to (4.27) was derived and the stable spin and orbital orders were investigated. Since late 1990s, V_2O_3 has been re-visited; the spin and orbital orders and their excitations have been investigated by the neutron scattering [2], resonant x-ray scattering [3], inelastic x-ray scattering [4] and so on. Based on these new experimental results, a variety of spin and orbital states have been proposed theoretically.

Another class of light transition metal oxides with orbital degree of freedom are the titanates and vanadates with perovskite crystal structure, $R_{1-x}A_xTiO_3$ and $R_{1-x}A_xVO_3$ (R: trivalent cations, A: divalent cations). The parent compounds with $x = 0$, e.g., $YTiO_3$ and $LaVO_3$, are the Mott-Hubbard insulators. The formal valence of the Ti and V ions is 3+, and the three t_{2g} orbitals are filled by one and two electrons, respectively. Through the substitution of the R by A ions, holes are introduced into the system and a metal-insulator transition occurs for a certain value of x. In the metallic phase close to the phase boundary there are anomalous electric, magnetic and optical properties. Some of these phenomena are interpreted as being due to critical fluctuations in the spin, orbital and charge degrees of freedom. In this chapter, the electronic states in the hole doped and undoped titanates and vanadates are reviewed from the view point of the orbital degree of freedom. Also see the review [5].

5.2 Orbital States

One of the topics in perovskite titanates and vanadates is the t_{2g} orbital states in the end materials, $R\mathrm{TiO_3}$ and $R\mathrm{VO_3}$. The fundamental properties of titanates and vanadates and a variety of the recent studies for the orbital degree of freedom in these compounds are reviewed.

5.2.1 Perovskite Titanates

The crystal structure of $R\mathrm{TiO_3}$ is a distorted perovskite with orthorhombic cell of *Pbnm* space group. The structural, magnetic and orbital properties of $R\mathrm{TiO_3}$ are systematically investigated by changing the ionic radius r_R of the R ion in passing from La to Y [6]. The smaller is the radius r_R, the larger is the rotation of the $\mathrm{TiO_6}$ octahedron referred to as a "$\mathrm{GdFeO_3}$-type distortion". This distortion bends a Ti-O-Ti bond; the bond angle is 157° in $\mathrm{LaTiO_3}$, and 140° (in the ab plane) and 144° (along the c axis) in $\mathrm{YTiO_3}$. As mentioned in Sect. 4.1.4, the large bond bending implies the small overlapping of the Mn $3d$ and O $2p$ wave functions, that is, the small electron transfer integral between nearest neighbor (NN) Ti sites. $\mathrm{YTiO_3}$ is a Mott-Hubbard type insulator with the optical gap of about 1eV [7]. Ferromagnetic (FM) order occurs at 30K and the observed saturated FM moment is about 80% of that expected for $S = 1/2$. With decreasing r_R, the Curie temperature T_C monotonically decreases, and the system eventually exhibits a G-type antiferromagnetic (AFM) order (AFM spin alignments between all NN bonds). For $\mathrm{LaTiO_3}$, which is a Mott insulator with small optical gap ($\sim$ 0.2eV), T_N reaches 145K. A similar phase change is found in the mixed crystal of $\mathrm{Y_{1-x}La_xTiO_3}$. The experimental phase diagram is shown in Fig. 5.1. Recently the series $R\mathrm{TiO_3}$ has stimulated much experimental and theoretical interest. However, some issues are still open questions. One of the serious questions are why the smaller distorted $\mathrm{LaTiO_3}$ shows the G-type AFM order instead of the FM one expected from the orbitally degenerate systems. A detailed crystal structure has been obtained by the x-ray diffraction. The magnitude of the Jahn-Teller lattice distortion is quantified by the distribution of Ti-O bond lengths denoted by $\Delta d_{\mathrm{Ti-O}}/\langle d_{\mathrm{Ti-O}}\rangle$ where $\Delta d_{\mathrm{Ti-O}}$ is the difference between the longest and shortest Ti-O bond lengths and $\langle d_{\mathrm{Ti-O}}\rangle$ is the average. It is found that $\Delta d_{\mathrm{Ti-O}}/\langle d_{\mathrm{Ti-O}}\rangle$, is about 3% in $\mathrm{YTiO_3}$ and 0.5% in

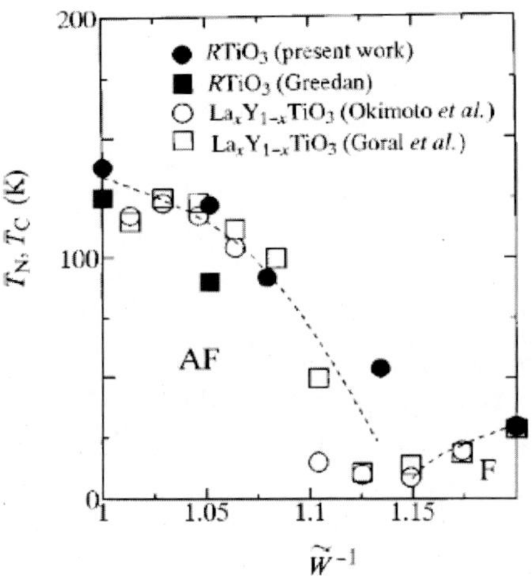

Fig. 5.1. Phase diagram of $RTiO_3$. $\widetilde{W}$ is the ratio of the band width W of $RTiO_3$ to that of $LaTiO_3$. [7].

$LaTiO_3$. Thus, a magnitude of the Jahn-Teller distortion with E_g symmetry, i.e. the Q_2 and Q_3 distortions defined in (4.38) and (4.39), is much smaller than that in $LaMnO_3$ ($\sim 15\%$).

A detailed information of the orbital structure in $YTiO_3$ has been obtained by utilizing the recent several experimental probes. The polarized neutron scattering experiment is the magnetic neutron diffraction where the difference of the scattering cross sections with the up and down neutron spins, i.e. the polarization ratio, is measured. Because the magnetically active electrons occupy the most outer orbitals, i.e. the t_{2g} orbitals, the observed magnetic structure factor reflects the shape of the occupied t_{2g} orbitals. This method was first applied to the observation of the orbital order in K_2CuF_4 by Akimitsu et al. and is recently utilized to determine the orbital order in $YTiO_3$ [8]. Through the comparison with the model calculation, the occupied orbital wave functions are obtained as in the inequivalent four Ti ions in a unit cell,

$$|\psi_1\rangle = c_1|d_{zx}\rangle + c_2|d_{xy}\rangle, \tag{5.1}$$

$$|\psi_2\rangle = c_1|d_{yz}\rangle - c_2|d_{xy}\rangle, \tag{5.2}$$

$$|\psi_3\rangle = c_1|d_{zx}\rangle - c_2|d_{xy}\rangle, \tag{5.3}$$

$$|\psi_4\rangle = c_1|d_{yz}\rangle + c_2|d_{xy}\rangle, \tag{5.4}$$

with $c_1/c_2 = 1.2$, where the four Ti ions are located at $(0\ \frac{1}{2}\ 0)$, $(\frac{1}{2}\ 0\ 0)$, $(0\ \frac{1}{2}\ \frac{1}{2})$ and $(\frac{1}{2}\ 0\ \frac{1}{2})$ (see Fig. 5.2). The resonant x-ray scattering (RXS), introduced in Sect. 4.3, suggests also the similar type of the orbital order [9]. The RXS intensities are observed in the three orbital superlattice reflection points, i.e. (100), (001), and (011). Through comparing the relative magnitudes of the intensities, the type of the orbital order is identified. For example, in the orbital order with $c_1 = 0$, the intensity at (001) almost disappears and that at (100)

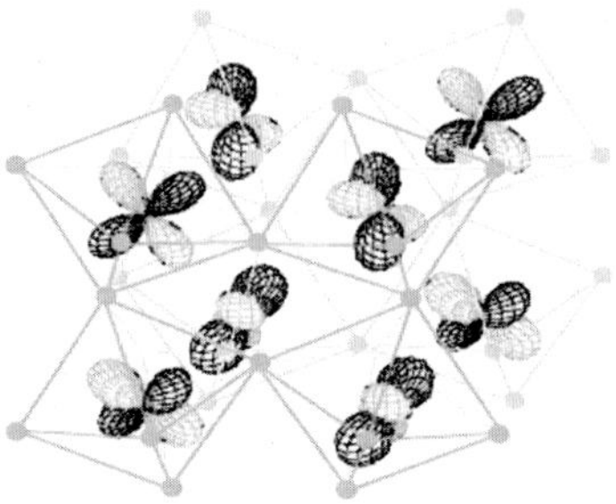

Fig. 5.2. Schematic structure of the experimentally proposed orbital order in $YTiO_3$.

becomes maximum. The obtained four sets of RXS data are consistently explained by the orbital wave function given by (5.4) with $c_1/c_2 \sim 1$. The ^{47}Ti nuclear magnetic resonance (NMR) spectrum, where the occupied orbital reflects on the electric hyperfine interaction, is explained by the above type of the orbital order [10]. These wave functions are also predicted by the theoretical calculations based on the Hartree-Fock approximation and the ab-initio band structure calculation with the local density approximation (LDA)+U methods [11, 12]. On the contrary, the ferromagnetic spin wave obtained by the inelastic neutron scattering shows almost isotropic dispersion relation. In a simple sense, in the long range orbital order, the magnetic exchange interactions in the *ab* plane and the *c* axis are expected to be anisotropic as seen in $LaMnO_3$. A connection between these two results, i.e. existence of the orbital order and the isotropic spin wave which seem to be contradict with each other, is still open for question.

The orbital states in other side of RTiO$_3$ series, $LaTiO_3$, is further controversy; the orbital is ordered or disordered? This state of things may be attributed to some unexpected experimental results; a weak or almost no RXS intensity, the almost isotropic spin wave dispersion, the G-type AFM order being unexpected in a simple orbital order, a small magnetic moment $0.45\mu_B$ and so on [13]. In order to resolve these anomalous experimental results, the orbital liquid state where the the quantum orbital fluctuation disturbs the orbital order is theoretically proposed. Through the recent neutron and x-ray diffraction experiments, the structural changes were found out at T_N=146K and are interpreted to lift the orbital degeneracy. The NMR spectra show a nuclear quadrupole splitting which is explained by the existence of a quadrupole moment of $3d$ electrons. An experimentally proposed orbital order is approximately [14, 15]

$$|\psi_1\rangle = a|d_{zx}\rangle + b|d_{zx}\rangle + c|d_{xy}\rangle, \tag{5.5}$$
$$|\psi_2\rangle = b|d_{zx}\rangle + a|d_{yz}\rangle + c|d_{xy}\rangle, \tag{5.6}$$
$$|\psi_3\rangle = a|d_{zx}\rangle + b|d_{zx}\rangle - c|d_{xy}\rangle, \tag{5.7}$$
$$|\psi_4\rangle = b|d_{zx}\rangle + a|d_{yz}\rangle - c|d_{xy}\rangle, \tag{5.8}$$

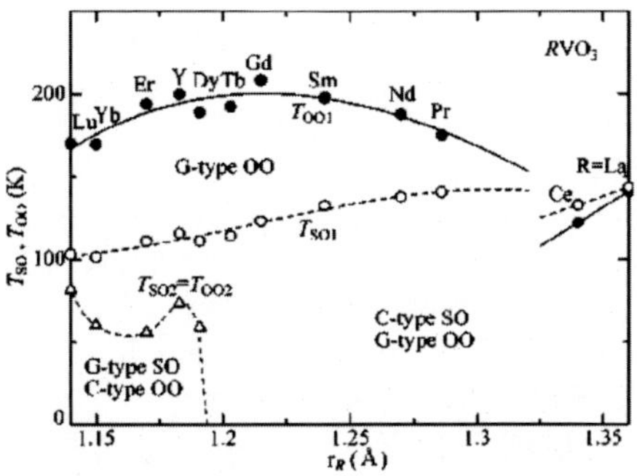

Fig. 5.3. Spin-orbital phase diagram of $R\mathrm{VO}_3$. [16].

with $a \sim b \sim c \sim 1/\sqrt{3}$. However, a connection between these wave functions and the listed unusual experimental results are still under debate.

5.2.2 Perovskite Vanadates

It has been recognized since 1970s that $R\mathrm{VO}_3$ is a prototypical Mott-Hubbard insulator. As well as a $R\mathrm{VO}_3$ series, the structural, magnetic and orbital properties of $R\mathrm{TiO}_3$ are systematically investigated by changing the R ion (see Fig. 5.3 [16]). The optical gap of LaVO_3 is about 1eV, and C-type AFM order occurs at T_N=143K. Slightly below T_N, the structural phase transition occurs at T_{OO}=141K which is caused by the G-type staggered orbital order. There are two kinds of orbitals in this order

$$|\psi_1\rangle = |d_{zx}\rangle|d_{xy}\rangle, \tag{5.9}$$

$$|\psi_2\rangle = |d_{yz}\rangle|d_{xy}\rangle. \tag{5.10}$$

The $3d_{xy}$ orbital is occupied at all V sites while the $3d_{yz}$ and $3d_{zx}$ orbitals are alternately occupied in all direction NN bonds by an electron as shown in Fig. 5.4.

On the other hand, in YVO_3, the spin and orbital ordering temperatures are inverted; The G-type orbital ordering occurs at T_{OO1}=200K followed by C-type AFM order with T_{N1} =115K. Upon further lowering the temperature, the spin and orbital ordered patterns are changed, i.e., the G-type AFM with

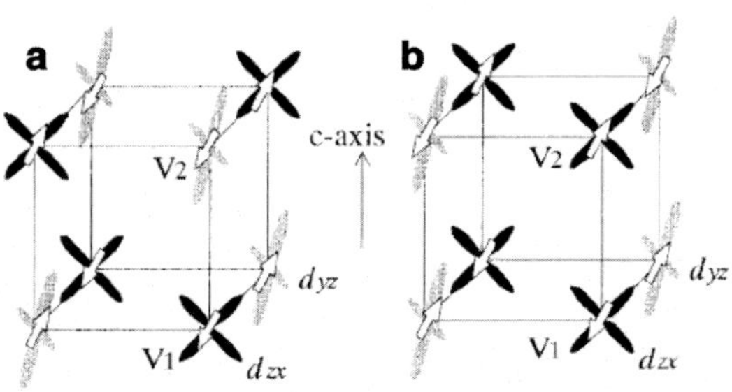

Fig. 5.4. Structures of (**a**) the C-type AFM and G-type orbital orders, and (**b**) the G-type AFM and C-type orbital orders in $R\mathrm{VO}_3$ [20].

the C-type orbital order is observed below $T_{N2} = T_{OO2}$ =77K. Here, the two orbital wave functions represented by (5.10) are alternately aligned in the ab plane and uniformly along the c axis. These sequential spin and orbital phase transitions were first suggested through the observation of a temperature induced magnetization reversal phenomena [17]. A net magnetization caused by the tilting of the antiferromgnetic moments is abruptly reversed at $T_{N2} = T_{OO2}$. The origin of this phenomena is discussed in connection with the spin-orbital transition. More recently the material has been examined by the non-resonant [18] and resonant x-ray scatterings [19]. The azimuthal angle dependence (see Sect. 4.3.3) of RXS intensity, by which the orbital pattern is identified, is changed above and below T_{OO2} indicating a change of the orbital order. It is confirmed that the orbital orders in $LaVO_3$ and the low-temperature G-type AFM phase of YVO_3 are consistent with the observed anisotropic optical properties. On the other hand, the optical spectrum in the intermediate-temperature C-AFM phase of YVO_3, are not fully explained based on the simple G-type orbital order [20, 21].

5.3 Metal–Insulator Transition

One of the most attractive phenomena of perovskite titanates and vanadates is the metal-insulator transition caused by doping of holes in the end compounds. With increasing a hole concentration x in $Y_{1-x}Ca_xTiO_3$ [22], the ferromagnetic long rang order disappears around $x = 0.15$. Above this concentration, the insulating behavior still retains, although a magnitude of the electrical resistivity gradually decreases; the non-magnetic insulating phase is realized. Then, the metallic feature appears at $x (= x_c) = 0.4$. In a $La_{1-x}Sr_xTiO_3$ series, the metal-insulator transition occurs at much smaller hole concentration (x_c=0.05) [23]. This smaller critical concentration in comparison with $Y_{1-x}Ca_xTiO_3$ is ascribed to the barely insulating nature in $LaTiO_3$ due to the smaller $GdFeO_3$-type distortion. It is unclear that the orbital order is retained or not in the non-magnetic insulating and metallic phases. In the metallic side at vicinity of the transition, the system shows several prototypical behavior of correlated metals [23]. The electric resistivity shows a T^2 behavior at low temperatures and this coefficient A increases as the system approaches the phase boundary from the metallic side. The effective mass estimated from the electronic specific heat coefficient (γ) and the Pauli paramagnetic susceptibility (χ) also increase divergently. At the vicinity of the transition, γ reaches approximately 20mJ/molK^2 which is much larger than that for conventional metal. The relation of A to γ termed the Kadowaki-Woods relation is about $C = A/\gamma^2 = 1.0\times\ 10^{-5}\mu\Omega$ cm (mol K /mJ)^2 in $La_{1-x}Sr_xTiO_3$. This follows the universal value for the several heavy electron systems. Another ratio $R_W = \chi/\gamma$ termed the Wilson ratio provides a magnitude of the electron correlation; $R_W = 1$ in the free electron case, and $R_W = 2$ in the strong coupling limit. In both $Y_{1-x}Ca_xTiO_3$

and $La_{1-x}Sr_xTiO_3$, R_W is approximately 2 which is insensitive to the hole concentration. These properties realized in titanates, i.e. the critical behavior of A, m^* and χ near the metal-insulator transition, the universal value of C and so on, are explained by the fermi-liquid theory, and indicate that this metal-insulator transition is not attributed to the carrier reduction but to the mass enhancement due to the electron correlation.

Nature of the metal-insulator transition is systematically examined in $La_{1-x}Sr_xVO_3$ since early 1970s. As mentioned above, at $x = 0$, the C-type AFM order appears at T_N =143K which is slightly higher than the structural phase transition temperature ($T_s(= T_{OO})$=141K) where the G-type orbital order occur. Between T_N and T_s, the system is the AFM phase without orbital order. With increasing x for $La_{1-x}Sr_xVO_3$ starting from $x = 0$, T_s decreases more rapidly than T_N, and this intermediate phase becomes distinct. Finally, at $x(= x_c) = 0.176$, the metal-insulator transition occurs with disappearing of the structural phase transition. Above x_c, the system is an AFM metallic state with orbital disorder. The metallic behaviors at a vicinity of x_c are rather different from those observed in titanates [24]. The temperature dependence of the electric resistivity has a component being proportional to $T^{1.5}$, unlike titanates, and the coefficient of this component monotonically increases by approaching x_c. The linear temperature coefficient of the specific heat γ gradually reduces from $x \sim 0.25$ with decreasing x. These discrepancies between titanates and $La_{1-x}Sr_xVO_3$ may be attributed to existence of the AFM metallic phase neighboring of the Mott insulating phase.

5.4 Electronic State and Model Hamiltonian

The insulating titanates/vanadates with perovskite structure are classified to the Mott-Hubbard insulators, unlike $LaMnO_3$ and La_2CuO_4 being the charge-transfer type insulators. Compare the light and heavy $3d$ transition metal oxides. There are some tendencies for the electronic structures in a series of the oxides. (1) the on-site Coulomb interaction U increases with the atomic number Z from Ti to Cu, and (2) the charge transfer gap Δ decreases with increasing Z. These are obtained by the cluster analysis of the valence band photoemission spectra. In $LaTiO_3$, and $LaVO_3$, U and Δ values are approximately (U,Δ)=(4eV,7eV) and (4.5eV,7eV), respectively, indicating the Mott-Hubbard type insulators. A boundary between the Mott-Hubbard type to the charge-transfer type seems to be located around $LaCrO_3$ in the so-called the Zaanen-Sawatzky-Allen diagram. For the heavy transition metal oxides, (U, Δ)=(5.5eV,4eV) for $LaMnO_3$ and (7.5eV,0) for $LaCuO_3$ implying the charge-transfer types. Thus, when we consider the electronic model for titanates and vanadates, up to several eV, the multi-orbital Hubbard model consisting of the Ti(V) $3d$ electrons without O $2p$ degree of freedom provides a good starting points.

Another difference of titanates and vanadates from manganites and other heavy transition metal oxides is seen in the electron hopping integral. As explained in the chapter on manganites (Sect. 4.1), the electron hopping between the NN transition-metal ions in the perovskite crystal occurs through the O $2p$ orbitals between them. In an ideal crystal lattice, the $3d_{xy}$ orbital couples with $2p_x$ ($2p_y$) along the $x(y)$ direction and decouples with $2p_z$. Therefore, for example, for a bond along the x direction, an electron hops between the d_{xz} orbitals and the d_{xy} orbitals, which are called "active orbitals", but not between the d_{yz} "inactive orbitals". Explicitly, the transfer integral between sites i orbital γ and j with γ' is given as

$$t_{ij}^{\gamma\gamma'} = t\delta_{\gamma\gamma'}(\delta_{\gamma a_l} + \delta_{\gamma b_l}), \tag{5.11}$$

where l indicates a direction of the bond connecting i and j. a_l, b_l and c_l correspond to the two active and the single inactive orbitals for direction $l(= x, y, z)$, respectively. The hopping integral t is a second order process involving the hopping integral between t_{2g} and O$2p$ orbitals, i.e., to a "π bond", denoted by $t_{pd\pi}$ which is about 2 times smaller than that for a σ bond $t_{pd\sigma}$. In principle, there is a finite hopping integral $t_{dd\delta}$ between the NN inactive orbitals through the direct overlap with the t_{2g} orbitals, however, the magnitude is much smaller than t and this is usually neglected.

With considering the above two characters in the t_{2g} orbital systems, we derive the effective Hamiltonian for titanates and vanadates as done for manganites in Sect. 4.1. Start from the generalized Hubbard Hamiltonian for a three dimensional cubic lattice where triply degenerate t_{2g} orbitals are introduced at each Ti or V site:

$$\mathcal{H} = \varepsilon_d \sum_{i\sigma\gamma} d_{i\gamma\sigma}^\dagger d_{i\gamma\sigma} + \sum_{\langle ij\rangle\gamma\gamma'\sigma} \left(t_{ij}^{\gamma\gamma'} d_{i\gamma\sigma}^\dagger d_{j\gamma'\sigma} + \text{h.c.} \right) + \mathcal{H}_{el-el}, \tag{5.12}$$

with

$$\begin{aligned}\mathcal{H}_{el-el} &= U \sum_{i\gamma} n_{i\gamma\uparrow} n_{i\gamma\downarrow} + U' \frac{1}{2} \sum_{i\gamma\neq\gamma'} n_{i\gamma} n_{i\gamma'} \\ &+ J \sum_{i\gamma>\gamma'\sigma\sigma'} d_{i\gamma\sigma}^\dagger d_{i\gamma'\sigma'}^\dagger d_{i\gamma\sigma'} d_{i\gamma'\sigma} + I \sum_{i\gamma\neq\gamma'} d_{i\gamma\uparrow}^\dagger d_{i\gamma\downarrow}^\dagger d_{i\gamma'\downarrow} d_{i\gamma'\uparrow},\end{aligned} \tag{5.13}$$

where $d_{i\gamma\sigma}^\dagger$ is an annihilation operator for the t_{2g} electron at site i with spin $\sigma(=\uparrow,\downarrow)$ and orbital $\gamma(= xy, yz, zx)$. Definitions of the electron-electron interaction parameters U, U', I and J are given in Sect. 4.1. The relativistic spin-orbit interaction and the electron-lattice interactions are considered later. Consider the insulating titanates RTiO$_3$ where the formal valence of Ti ion is 3+. The t_{2g}^1 states of a Ti ion are six fold degenerate. The t_{2g}^2 and t_{2g}^0 states are induced by the hopping of electrons. In the independent electron picture there is a 15 ($={}_6C_2$) fold degeneracy of the t_{2g}^2 electron configuration which splits into the multiplets due to the electron-electron interaction $\mathcal{H}_{el-el}$, i.e., there are the ${}^1A_1$ state with the energy $E_{A_1}(t_{2g}^2) = U + 2I$,

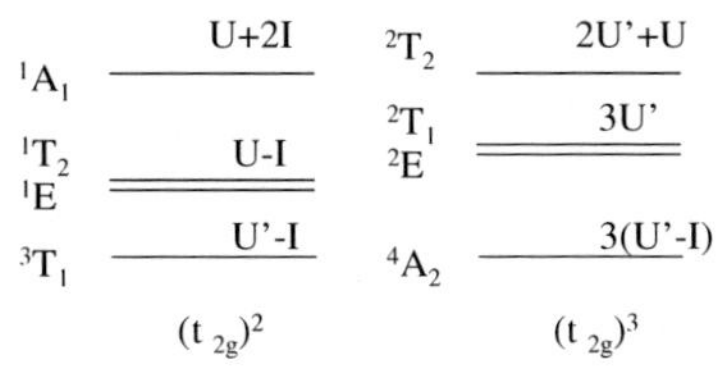

Fig. 5.5. Energy level diagrams of the t_{2g}^2 and t_{2g}^3 electron configurations.

${}^1T_2$ with $E_{T_2}(t_{2g}^2) = U' + I$, ${}^1E$ with $E_E(t_{2g}^2) = U - I$ and ${}^3T_1$ with $E_{T_1}(t_{2g}^2) = U' - I$. The energy diagram is given in Fig. 5.5. The spin triplet state ${}^3T_1$ is the ground state due to the Hund rule coupling.

As mentioned previously, the values of U for titanates and vanadates are smaller than that estimated for the manganites with perovskite structure. However, the transfer integral t is also smaller in titanates since the π bond is weaker. It is reasonable to assume that the electron configuration of a Ti ion is basically t_{2g}^1 but that virtual electron hopping mixes in the t_{2g}^2 configuration. The effective Hamiltonian in a suitably restricted Hilbert space provides a good starting point by which to describe the low energy electronic states of titanates. The so-called J term which is second order in the transfer integral are labeled by the intermediate states in exchange processes [25], i.e.,

$$\mathcal{H}_J = \mathcal{H}_{T_1} + \mathcal{H}_{T_2} + \mathcal{H}_E + \mathcal{H}_{A_1}, \tag{5.14}$$

with

$$\mathcal{H}_{T_1} = -J_{T_1} \sum_{\langle ij \rangle} \left(\frac{3}{4} n_i n_j + \boldsymbol{S}_i \cdot \boldsymbol{S}_j \right) \left(B^l - C^l + D^l \right), \tag{5.15}$$

$$\mathcal{H}_{T_2} = -J_{T_2} \sum_{\langle ij \rangle} \left(\frac{1}{4} n_i n_j - \boldsymbol{S}_i \cdot \boldsymbol{S}_j \right) \left(B^l + C^l + D^l \right), \tag{5.16}$$

$$\mathcal{H}_E = -J_E \sum_{\langle ij \rangle} \left(\frac{1}{4} n_i n_j - \boldsymbol{S}_i \cdot \boldsymbol{S}_j \right) \left(\frac{2}{3} A^l - \frac{2}{3} C'^l \right), \tag{5.17}$$

$$\mathcal{H}_{A_1} = -J_{A_1} \sum_{\langle ij \rangle} \left(\frac{1}{4} n_i n_j - \boldsymbol{S}_i \cdot \boldsymbol{S}_j \right) \left(\frac{1}{3} A^l + \frac{2}{3} C'^l \right). \tag{5.18}$$

The prefactor J_Γ with $\Gamma = (T_1, T_2, E, A_1)$ is $J_\Gamma = t^2/E_\Gamma(t_{2g}^2)$, and $\boldsymbol{S}_i$ is a $S = 1/2$ spin operator at site i. The five orbital operators A^l, B^l, C^l, C'^l and D^l in the Hamiltonian involve eight operators $O_{i\Gamma\gamma}$ which in turn are defined in terms of the Gell-Mann matrices which are generators of the SU(3) Lee algebra [26]:

$$\lambda_1 = \begin{pmatrix} 0 & 1 & 0 \\ 1 & 0 & 0 \\ 0 & 0 & 0 \end{pmatrix}, \quad \lambda_2 = \begin{pmatrix} 0 & -\mathrm{i} & 0 \\ \mathrm{i} & 0 & 0 \\ 0 & 0 & 0 \end{pmatrix}, \quad \lambda_3 = \begin{pmatrix} -1 & 0 & 0 \\ 0 & 1 & 0 \\ 0 & 0 & 0 \end{pmatrix}, \tag{5.19}$$

$$\lambda_4 = \begin{pmatrix} 0 & 0 & 1 \\ 0 & 0 & 0 \\ 1 & 0 & 0 \end{pmatrix}, \qquad \lambda_5 = \begin{pmatrix} 0 & 0 & \mathrm{i} \\ 0 & 0 & 0 \\ -\mathrm{i} & 0 & 0 \end{pmatrix}, \qquad \lambda_6 = \begin{pmatrix} 0 & 0 & 0 \\ 0 & 0 & 1 \\ 0 & 1 & 0 \end{pmatrix}, \tag{5.20}$$

$$\lambda_7 = \begin{pmatrix} 0 & 0 & 0 \\ 0 & 0 & -\mathrm{i} \\ 0 & \mathrm{i} & 0 \end{pmatrix}, \qquad \lambda_8 = \frac{1}{\sqrt{3}} \begin{pmatrix} 1 & 0 & 0 \\ 0 & 1 & 0 \\ 0 & 0 & -2 \end{pmatrix}. \tag{5.21}$$

By using these matrices, we define the basic orbital operators by

$$O_{i\Gamma\gamma} = \frac{-1}{\sqrt{2}} \sum_{\sigma\alpha\beta} d^{\dagger}_{i\alpha\sigma} (\lambda_l)_{\alpha\beta} d_{i\beta\sigma}, \tag{5.22}$$

where the identification between the $\Gamma\gamma$, which correspond to an irreducible representation, on the left hand side and the label l on the right hand side is: $(\Gamma\gamma; l) = (Eu; 8)$, $(Ev, 3)$, $(T_2x; 6)$, $(T_2y; 4)$, $(T_2z; 1)$, $(T_1x; 7)$, $(T_1y; 5)$ and $(T_1z; 2)$. The operators $O_{iE\gamma}$ and $O_{iT_2\gamma}$ describe the electric quadrupole moments and $O_{iT_1\gamma}$ describes the magnetic dipole moment, i.e. the orbital angular momentum. In terms of the above operators,

$$A^l = 2\left(\frac{2}{3} - \sqrt{\frac{2}{3}} O^l_{iEu}\right)\left(\frac{2}{3} - \sqrt{\frac{2}{3}} O^l_{jEu}\right) + 4 O^l_{iEv} O^l_{jEv}, \tag{5.23}$$

$$B^l = \left(\frac{2}{3} - \sqrt{\frac{2}{3}} O^l_{iEu}\right)\left(\frac{2}{3} - \sqrt{\frac{2}{3}} O^l_{jEu}\right) - 2 O^l_{iEv} O^l_{jEv}, \tag{5.24}$$

$$C^l = 2\left(O_{iT_2l} O_{jT_2l} + O_{iT_1l} O_{jT_1l}\right), \tag{5.25}$$

$$C'^l = 2\left(O_{iT_2l} O_{jT_2l} - O_{iT_1l} O_{jT_1l}\right), \tag{5.26}$$

$$\begin{aligned} D^l = & \left(\frac{1}{3} + \sqrt{\frac{2}{3}} O^l_{iEu}\right)\left(\frac{2}{3} - \sqrt{\frac{2}{3}} O^l_{jEu}\right) \\ & + \left(\frac{2}{3} - \sqrt{\frac{2}{3}} O^l_{iEu}\right)\left(\frac{1}{3} + \sqrt{\frac{2}{3}} O^l_{jEu}\right), \end{aligned} \tag{5.27}$$

where the $O^l_{iE\gamma}$ are defined by

$$\begin{pmatrix} O^l_{iEu} \\ O^l_{iEv} \end{pmatrix} = \begin{pmatrix} \cos\frac{2\pi}{3} m_l & \sin\frac{2\pi}{3} m_l \\ -\sin\frac{2\pi}{3} m_l & \cos\frac{2\pi}{3} m_l \end{pmatrix} \begin{pmatrix} O_{iEu} \\ O_{iEv} \end{pmatrix}, \tag{5.28}$$

with $m_l = (1, 2, 3)$ for $l = (x, y, z)$. As was $\mathcal{H}_J$ for manganites, the orbital part of this Hamiltonian has no continuous SU(3) symmetry and the orbital quantum numbers are not conserved quantities. When the pair hopping term described by C'^l is neglected, it *is* the case that number of electrons of each orbital type is conserved. There are other representations for the spin-orbital Hamiltonian for the t^1_{2g} systems derived in [27–29] (also see Chap. 7).

The effective electronic Hamiltonian for vanadates with perovskite structure RVO_3 is derived in the similar way. Consider the ground state of the t_{2g}^2 configuration, i.e. the ${}^3A_1$ states. In the independent electron scheme, the excited states of the t_{2g}^3 configuration are $20(={}_6C_3)$ fold degenerated. These are split into the multiplets: ${}^4A_2$ with the energy $E_{A_2}(t_{2g}^3) = U' - I$, ${}^2E$ with $E_E(t_{2g}^3) = U' + I$ ${}^2T_1$ with $E_{T_1}(t_{2g}^3) = U' + I$, and ${}^2T_2$ with $E_{T_2}(t_{2g}^3) = U' + 4I$, see Fig. 5.5, where the ${}^4A_2$ state with $S = 3/2$ is the ground state. Accounting for the virtual excitation of the t_{2g}^1 and t_{2g}^3 states, the effective Hamiltonian for the t_{2g}^2 states

$$\mathcal{H}_{d^2} = \mathcal{H}_{A_1} + \mathcal{H}_E + \mathcal{H}_{T_1} + \mathcal{H}_{T_2}, \tag{5.29}$$

where

$$\mathcal{H}_{T_1} = -J_{T_1} \sum_{\langle ij \rangle} (2 + \boldsymbol{S}_i \cdot \boldsymbol{S}_j) \left(-\frac{1}{4} A^l + \frac{1}{2} B^l - C^l \frac{1}{2} D^l \right), \tag{5.30}$$

$$\mathcal{H}_{T_2} = -J_{T_2} \sum_{\langle ij \rangle} (1 - \boldsymbol{S}_i \cdot \boldsymbol{S}_j) \left(-\frac{1}{4} A^l + \frac{1}{2} B^l + C^l \frac{1}{2} D^l \right), \tag{5.31}$$

$$\mathcal{H}_E = -J_E \sum_{\langle ij \rangle} (1 - \boldsymbol{S}_i \cdot \boldsymbol{S}_j) \left(\frac{2}{3} A^l + \frac{2}{3} C^{\prime l} \right), \tag{5.32}$$

$$\mathcal{H}_{A_1} = -J_{A_1} \sum_{\langle ij \rangle} (1 - \boldsymbol{S}_i \cdot \boldsymbol{S}_j) \left(\frac{1}{3} A^l + \frac{2}{3} C^{\prime l} \right), \tag{5.33}$$

where the prefactor $J_\Gamma = t^2/(E_\Gamma(t_{2g}^3) - 2E_{{}^1T_1}(t_{2g}^2))$ and the spin operator $\boldsymbol{S}_i$ corresponds to $S = 1$. The orbital parts A^l, B^l, C^l, $C^{l'}$ and D^l are again given by (5.23)-(5.27). The similarity between the Hamiltonians for t_{2g}^1 and t_{2g}^2 systems follows from electron-hole symmetry, i.e., the high spin d^2 and d^1 states are equivalent with respect to the orbital degree of freedom. A similar Hamiltonian, in the restricted Hilbert space, has been derived in [30]. Magnitudes of the exchange interactions are estimated from the analysis of the spin wave dispersion relation and the optical spectra. J_{A_1} is of the order of 10meV.

Spin Orbit Interaction. The effects of the relativistic spin-orbit coupling (LS coupling) is different for the e_g and t_{2g} systems. This basic interaction Hamiltonian is

$$\mathcal{H}_{ls} = \sum_i \xi \boldsymbol{l}_i \cdot \boldsymbol{s}_i, \tag{5.34}$$

where

$$\xi = \frac{\hbar^2}{2m^2c^2r} \frac{\mathrm{d}V}{\mathrm{d}r}, \tag{5.35}$$

The electron potential is V, and $\boldsymbol{s}_i$ and $\boldsymbol{l}_i$ are the spin and angular momentums of the i-th electron. When we focus on the lowest LS multiplet, the effective Hamiltonian reduces to

$$\mathcal{H}_{ls} = \lambda \boldsymbol{L} \cdot \boldsymbol{S}, \tag{5.36}$$

where λ is the LS coupling constant. Consider this interaction for the e_g orbitally degenerated system. The operator $\boldsymbol{L}(= \mathrm{i}\hbar \boldsymbol{r} \times \boldsymbol{\nabla})$ is an axial vector and with T_{1g} symmetry. The matrix elements of $\boldsymbol{L}$ between the e_g^1 states have the symmetries $E_g \times T_{1g} \times E_g$ which reduces to $3T_{1g} + T_{2g}$. Since this does not include A_{1g}, the matrix elements are zero, even for an orbitally degenerated system. On the other hand, for the t_{2g}^1 systems, the matrix elements have the symmetries $T_{2g} \times T_{1g} \times T_{2g}$ which reduces to $A_{1g} + A_{2g} + 2E_g + 4T_{1g} + 3T_{2g}$ and includes A_{1g}. Therefore, the three fold degeneracy of the t_{2g} system is lifted by the spin-orbit interaction. In order to evaluate the matrix elements of the LS coupling Hamiltonian, the three t_{2g} wave functions are identified as p_x, p_y and p_z orbitals. The p orbital, with $L = 1$, leads to a 6 $(= 3 \times 2)$ fold degeneracy of the t_{2g}^1 states. This is split into a quartet with $|\boldsymbol{J}(\equiv \boldsymbol{L} + \boldsymbol{S})| = 3/2$ with energy $E_{3/2} = -\lambda/2$, and a, $|\boldsymbol{J}| = 1/2$, doublet for which $E_{1/2} = \lambda$. The eigen-states are explicitly

$$\psi_{3/2} = \left\{ \frac{1}{\sqrt{2}} \left(|d_{yz\uparrow}\rangle \pm \mathrm{i} |d_{zx\uparrow}\rangle \right), \ \frac{1}{\sqrt{6}} \left(2|d_{xy\uparrow}\rangle \pm |d_{yz\downarrow}\rangle - \mathrm{i}|d_{zx\downarrow}\rangle \right) \right\}, \tag{5.37}$$

and

$$\psi_{1/2} = \left\{ \frac{1}{\sqrt{3}} \left(|d_{xy\uparrow}\rangle \pm |d_{yz\downarrow}\rangle + \mathrm{i}|d_{zx\downarrow}\rangle \right) \right\}, \tag{5.38}$$

where the quantization axis is taken to be the z axis.

The value of λ is about $20 \sim 25$meV estimated from free Ti and V ions. In the case where the orbital degeneracy is lifted by other reasons, such as the lattice distortion, the orbital angular momentum $\langle \boldsymbol{L} \rangle$ in the ground state is quenched in the zero-th order of λ. $\langle \boldsymbol{L} \rangle$ is recovered by the second order perturbation of λ between the ground and excited states. When the level splitting ΔE is chosen to be about 1eV, for example, the effective interaction amplitude is approximately given by $\lambda^2/\Delta E$ which is of the order of 0.5meV.

Jahn-Teller Coupling. It is expected that, in the t_{2g} orbital systems, the electron-lattice interaction is weaker than that in the e_g orbital system, since the electronic clouds of the t_{2g} orbitals are *not* elongated along the Ti(V)-O bonds. The appropriate Jahn-Teller coupling Hamiltonian

$$H_{JT} = g_{E_g} \sum_{i\gamma = u,v} O_{iE_\gamma} Q_{iE_{g\gamma}} + g_{T_{2g}} \sum_{i\gamma = x,y,z} O_{iT_{2\gamma}} Q_{iT_{2g\gamma}}, \tag{5.39}$$

where $O_{iE\gamma}$ and $O_{iT_2\gamma}$ are the orbital operators given in (5.22), and g_{E_g} and $g_{T_{2g}}$ are the coupling constants. The O vibration modes $Q_{iE_g u}(= Q_3)$ and $Q_{iE_g v}(= Q_2)$, see (4.38) and (4.39), respectively, correspond to O motions along the Ti-O bond directions. In the second term, on the other hand, for

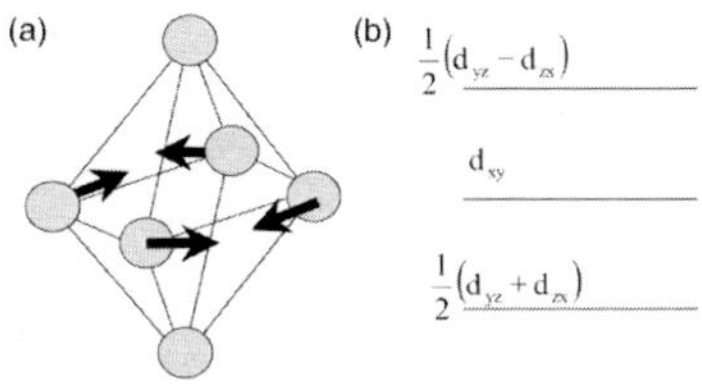

Fig. 5.6. (**a**) The oxygen vibrations for the Q_{xy} mode. (**b**) Energy level diagram of the t_{2g} orbitals split by the Jahn-Teller coupling to the Q_{xy} mode where the coupling constant is assumed to be positive.

the $Q_{iT_{2g}x}(=Q_{yz})$, $Q_{iT_{2g}y}(=Q_{zx})$ and $Q_{iT_{2g}z}(=Q_{xy})$ modes, the O ions vibrate in the plane of the Ti(V)O_4 squares. Explicitly Q_{xy} is

$$Q_{xy}=\frac{1}{2}\left[q_y(\mathrm{d}\hat{x})-q_x(\mathrm{d}\hat{y})-q_y(-\mathrm{d}\hat{x})+q_x(-\mathrm{d}\hat{y})\right], \tag{5.40}$$

and Q_{yz} and Q_{zx} are given in the similar way. $q_l(\boldsymbol{r})$ for $l=(x,y,z)$ indicates the vibration of the O ion at the position $\boldsymbol{r}$ along the direction l. The origin for $\boldsymbol{r}$ is taken to be the position of the Ti(V) ion. A schematic picture of the modes Q_{xy} is shown in Fig. 5.6. As seen from (5.39), theses modes mix the three t_{2g} orbitals. It is worth noting that, in contrast to the e_g orbital systems, the triple degeneracy is not completely lifted through the interaction with $Q_{iE_g\gamma}$, e.g., under the Q_2 or Q_3 distortion, the three t_{2g} orbitals are split into a $3d_{xy}$ singlet along with a, $3d_{yz}$ and $3d_{zx}$, doublet. In contrast, a $Q_{iT_{2g}}$ distortion lifts the degeneracy completely, i.e., the distortion Q_{xy} splits the triplet t_{2g} leading to the, non-degenerate, orbitals $\frac{1}{\sqrt{2}}(3d_{yz}+3d_{zx})$, $\frac{1}{\sqrt{2}}(3d_{yz}-3d_{zx})$ and $3d_{xy}$ as shown in Fig. 5.6(b). In the LDA+U method, the Jahn-Teller coupling constant of the E_g symmetry is estimated to be g_{E_g} ~15meV from the Jahn-Teller stabilization energy (~ 27meV) for $LaVO_3$ which is comparable to the exchange interactions in vanadates.

Some Consequences of Spin and Orbital Models. A number of theoretical study for the spin and orbital states in titanates and vanadates has been done based on the modified Hubbard Hamiltonian (5.12) with (5.13) and the effective spin-orbital one (5.14) and (5.29) associated with the spin-orbit interaction, the Jahn-Teller coupling, as well as the ab-initio band structure calculation.

The ferromagnetic "mixed"-type orbital order in $YTiO_3$ (see (5.4)) are obtained by the Hartree-Fock calculation in the $p-d$ type Hamiltonian, the effective spin-orbital Hamiltonian and the band structure calculation based on the local spin density approximation (LSDA) and the generalized gradient approximation (GGA) [11, 12, 29]. The key ingredients for this type of the orbital order may be the $GdFeO_3$-type distortion and the Jahn-Teller type one; the FM phase with the mixed orbital order realizes by introducing these lattice distortion. This is interpreted to be ascribed to the σ bond character of the hopping integral between NN Ti sites. The orbital order associated with the G-type AFM in $LaTiO_3$ may also require the additional ingredients to the pure exchange model; the relativistic spin-orbit interaction promotes the G-type AFM order and the Jahn-Teller distortions with D_{3d} symmetry, represented as $Q_{xy} + Q_{yz} + Q_{zx}$ [31], favors the orbital order discussed in Sect. 5.2. The quantum orbital liquid state in AFM phase and the orbital fluctuation in FM phase proposed based on the effective spin-orbital model [28] are introduced in Chap. 7, in more detail.

The C-AFM and G-type orbital orders observed in $LaVO_3$ are well reproduced by the electronic exchange model Hamiltonian such as (5.29). The AFM magnetic order in the *ab* plane and the FM one along the *c* axis are responsible for the uniform order of the d_{xy} orbital and the staggered alignment of the $3d_{zx}$ and $3d_{yz}$ orbitals, respectively. On the other hand, the G-AFM order associated with the C-type orbital one is not stable within the mean field calculation in the exchange Hamiltonian (5.29). The cooperative Jahn-Teller distortion with the E_g symmetry being compatible with the C-type orbital order may stabilized this spin and orbital ordered phase.

5.5 Summary

In conclusion, the perovskite titanates and vanadates are recognized to be prototypical compounds where the t_{2g} orbital physics is examined theoretically and experimentally. There is a number of distinct properties from the e_g orbital degenerate systems, i.e. the three-fold orbital degeneracy, the orthogonality of the electron hopping, the relativistic spin-orbit coupling, the Jahn-Teller distortion with the T_{2g} symmetry, and so on. As a result, it is expected that a number of cooperative and/or competitive factors bring about a variety of exotic phenomena. A number of important issues in titanates and vanadates are still up-to-date and further progress of researches are required. Another interests are in the metal-insulator transition appearing by doping of holes. Several observed anomalies near the transition are well arranged within the fermi-liquid theory. One of the remaining issues is roles of the orbital order/fluctuation on the metal-insulator transition. This problem may open a new research field in the correlated metallic systems.

References

1. C. Castellani, C. R. Natoli, J. Ranninger: Phys. Rev. B **9**, 189 (1976)
2. W. Bao, C. Broholm, G. Aeppli, P. Dai, J. M. Honig, P. Metcalf: Phys. Rev. Lett. **78**, 507 (1997)
3. L. Paolasini, C. Vettier, F. de Bergevin, F. Yakhou, D. Mannix, A. Stunault, W. Neubeck: Phys. Rev. Lett. **82**, 4719 (1999)
4. E. D. Isaacs, P. M. Platzman, P. Metcalf, J. M. Honig: Phys. Rev. Lett. **76**, 4211 (1996)
5. M. Imada, A. Fujimori, Y. Tokura: Rev. Mod. Phys. **70** 1039 (1998)
6. D. A. MacLean, H.-N. Ng, J. E. Greedan: Jour. Sol. Stat. Chem. **30**, 35 (1979)
7. T. Katsufuji, Y. Taguchi, Y. Tokura: Phys. Rev. B **56**, 10145 (1997)

8. H. Ichikawa, J. Akimitsu, M. Nishi, K. Kakurai: Physica B **281&282**, 482 (2000)
9. H. Nakao, Y. Wakabayashi, T. Kiyama, Y. Murakami, M. v. Zimmermann, J. P. Hill, D. Gibbs, S. Ishihara, Y. Taguchi, Y. Tokura: Phys. Rev. B **66**, 184419 (2002)
10. M. Itoh, M. Tsuchiya, H. Tanaka, K. Motoya: Jour. Phys. Soc. Jpn. **68**, 2783 (1999)
11. T. Mizokawa, A. Furjimori: Phys. Rev. B **54**, 5368 (1996); T. Mizokawa, D. I. Khomskii, G. A. Sawatzky: Phys. Rev. B **60**, 7309 (1999)
12. H. Sawada, N. Hamada, K. Terakura: Physics B **237-238**, 46 (1997); H. Sawada, K. Terakura: Phys. Rev. B **58**, 6831 (1998)
13. B. Keimer, D. Casa, A. Ivanov, J. W. Lynn, M. v. Zimmermann, J. P. Hill, D. Gibbs, Y. Taguchi, Y. Tokura: Phys. Rev. Lett. **85**, 3946 (2000)
14. T. Kiyama, M. Itoh, Phys. Rev. Lett. **91** 167202 (2003)
15. M. Cwik, T. Lorenz, J. Baier, R. Müller, G. André, F. Bourée, F. Lichtenberg, A. Freimuth, R. Schmitz, E. Müller-Hartmann, M. Braden: Phys. Rev. B **68** 060401 (2003)
16. S. Miyasaka, Y. Okimoto, M. Iwama, Y. Tokura: Phys. Rev. B **68**, 100406 (2003)
17. Y. Ren, T. T. M. Palstra, D. I. Khomskii, E. Pellegrin, A. A. Nugroho, A. A. Menovski, G. A. Sawatzky: Nature, **396**, 441 (1998)
18. G. R. Blake, T. T. M. Palstra, Y. Ren, A. A. Nugroho, A. A. Menovsky: Phys. Rev. Lett. **87**, 245501 (2001)
19. M. Noguchi, A. Nakazawa, S. Oka, T. Arima, Y. Wakabayashi, H. Nakao, Y. Murakami: Phys. Rev. B **62**, R9271 (2000)
20. S. Miyasaka, Y. Okimoto, Y. Tokura: Jour. Phys. Soc. Jpn. **71**, 2086 (2002)
21. Y.Motome, H. Seo, Z. Fang, N. Nagaosa: Phys. Rev. Lett. **90** 146602 (2003)
22. T. Katsufuji, Y. Okimoto, Y. Tokura: Phys. Rev. Lett. **75**, 3497 (1995)
23. Y. Tokura, Y. Taguchi, Y. Okada, Y. Fujishima, T. Arima, K. Kumagai, Y. Iye: Phys. Rev. Lett. **70**, 2126 (1993)
24. S. Miyasaka, T. Okuda, Y. Tokura: Phys. Rev. Lett. **85**, 5388 (2000)
25. S Ishihara, T. Hatakeyama, S. Maekawa: Phys. Rev. B **65**, 064442 (2002)
26. M. Gell-Mann, Y. Ne'eman: *The eightfold way*, (Benjamin, New York, 1964)
27. K. I. Kugel, D. I. Khomskii: Sov. Phys. Solid State **17**, 285 (1975)
28. G. Khaliullin, S. Maekawa: Phys. Rev. Lett. **85**, 3950 (2000)
29. M. Mochizuki, M. Imada: Jour. Phys. Soc. Jpn. **69**, 1982 (2000)
30. G. Khaliullin, P. Horsch, A. M. Oleś: Phys. Rev. Lett. **86**, 3879 (2001)
31. M. Mochizuki, M. Imada: Phys. Rev. Lett. **91**, 167203 (2003)

6. Cobaltates

Cobalt oxides have been studied extensively because of their interesting transport and magnetic properties, as well as in connection with a variety of potential applications. For example, $La_{1-x}Sr_xCoO_3$ [1, 2] is a candidate material for high-performance electrodes in solid-oxide fuel-cells and $NaCo_2O_4$ [3–12] has potential as a thermoelectric material. Recently, superconductivity [13] has been discovered in water-intercalated $Na_{0.35}CoO_2{\cdot}1.3H_2O$.

In these cobalt oxides, a cobalt ion is caged in an octahedron of oxygen ions. Usually, the electronic state of a transition metal ion is fixed, that is to say the total-spin quantum-number and the number of electrons in the $3d$ orbitals have more or less precise values. In the cobalt oxides, however, the electronic state of a Co ion can have both different valence states and even when the valence is fixed there exist different spin states. In some cases, the ionic state of a Co ion changes with temperature. This transition is accompanied with change of the transport and magnetic properties. The interplay of the properties and the transition of ionic state is a main issue of the cobalt oxides, i.e., the problem of low-, high- and intermediate-spin states.

6.1 Low-, High- and Intermediate-Spin States

Let us first focus on an isolated CoO_6 octahedron and examine the electronic properties. Due to the effect of the crystal field, the $3d$ orbitals split into two fold, e_g, and three fold t_{2g} degenerate levels with the splitting defined to be $10Dq$. There exist six (five) electrons in the $3d$ shell of the Co^{3+} (Co^{4+}) ion. If the crystal field is strong enough, all the electrons go into t_{2g} orbitals (see Fig. 6.1(a)). Suppose that two of the electrons go into the e_g orbitals. In this case, the electronic state costs an energy $10Dq \times 2$ and is unstable. The stability of the electronic state also depends on the spin degree of freedom. The Hund's rule coupling J_H favors the electronic configurations $t_{2g}^4 e_g^2$ with $S = 2$ for Co^{3+} and $t_{2g}^3 e_g^2$ with $S = 5/2$ for Co^{4+}. The Hund's rule coupling and the crystal-field splitting $10Dq$ compete with each other. When neither coupling dominates, the electronic configurations $t_{2g}^5 e_g^1$ with $S = 1$ and $t_{2g}^4 e_g^1$ with $S = 3/2$ may also be possible for Co^{3+} and Co^{4+}, respectively. It is interesting to note that the states shown in Fig. 6.1(c) have the same orbital degree of freedom as those described in Chap. 4. The electronic states of

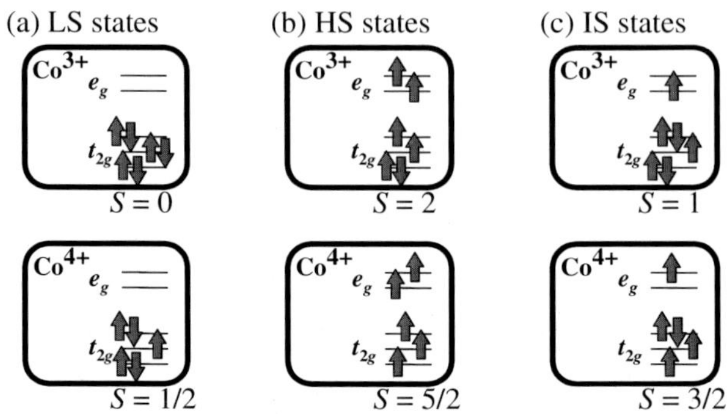

Fig. 6.1. Schematic representation of the electronic states of cobalt ions, Co^{3+} and Co^{4+}. The horizontal lines indicate the energy levels of e_g and t_{2g} orbitals. The arrow represents a spin of an electron. (**a**) Low-spin states, (**b**) high-spin states, and (**c**) intermediate-spin states. The magnitude of spin S is indicated.

a Co^{3+} (Co^{4+}) ion with the electronic configurations, $t_{2g}^6 e_g^0$ ($t_{2g}^5 e_g^0$), $t_{2g}^4 e_g^2$ ($t_{2g}^3 e_g^2$), and $t_{2g}^5 e_g^1$ ($t_{2g}^4 e_g^1$) are called the low-spin (LS), high-spin (HS) and intermediate-spin (IS) states, respectively.

The competition between the LS, HS, and IS states leads to the unique transport and magnetic behavior of $LaCoO_3$ which have been studied extensively. The history dates back more than half a century. It has been shown that $LaCoO_3$ has no long-range magnetic order down to 4 K in an early neutron diffraction study [14, 15] and is a nonmagnetic insulator with LS state of Co^{3+} ions. With increasing temperature, however, the spin state of the Co ions changes, with the magnetic susceptibility showing a maximum at around 100 K (see Fig. 6.2), and signals a transition in the spin state of the Co^{3+} ions. Polarized neutron scattering [16] and Knight shift [17] measurements have confirmed that the abrupt increase of magnetic susceptibility is due to a change in the spin state of the Co^{3+} ions. An anomalous temperature dependence has also been observed [18–21] in the electrical resistivity and the thermopower (see Figs. 6.3 and 6.4): In the temperature region below 100 K, the electrical resistivity is large. With increasing temperature, however, the resistivity is drastically suppressed and the magnitude becomes compa-

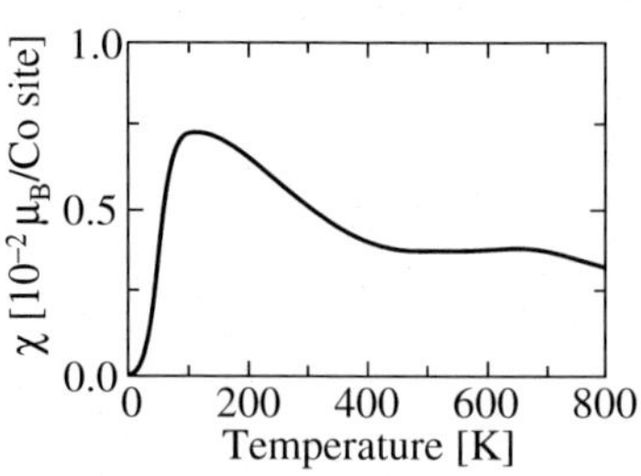

Fig. 6.2. Temperature dependence of the magnetic susceptibility (χ) in $LaCoO_3$. This sketch is taken from [19].

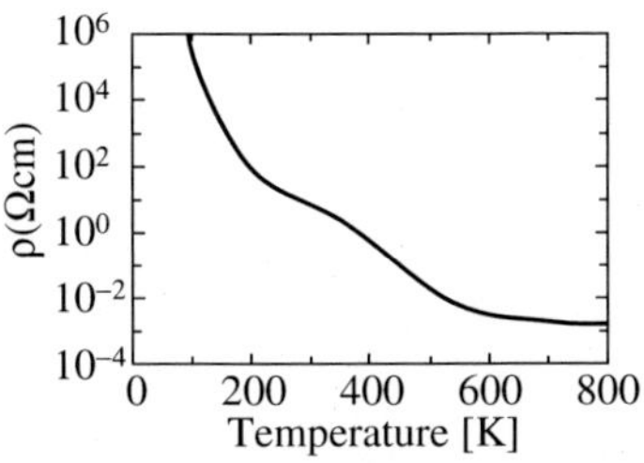

Fig. 6.3. Temperature dependence of the electrical resistivity (ρ) in $LaCoO_3$. This sketch is taken from [19].

rable with that of a good metal above 600 K. In the high temperature region (>600K), the thermopower is no larger than more usual metals, but with decreasing temperature, this starts to increase. There is a plateau in the range $\sim$ 200 K – 400 K and a clear maximum at around 70K. The relation between these anomalies and the LS – HS spin transition has been much debated. The role of the other, "$S = 1$ state", was also examined. Heikes et al. [22] took into account the $S = 1$ state when examining the electronic properties of $LaCoO_3$. The possibility of stabilizing the IS state and its role in the various anomalies has attracted much attention since 1990s. X-ray and electron spectroscopy measurements have been performed to study the local spin-state of $LaCoO_3$ [23–25]. Abbate et al. [23, 24] have focused attention on the local LS-to-HS transition. The importance of IS state in the local spin-state transition have been emphasized by Saitoh et al. [25]. Based on band-structure calculations using the so-called LDA+U method, i.e., a generalization of the local density approximation, the relative stability of the IS state and the HS state was discussed by Korotin et al. [30]. They find that a strong $p - d$ hybridization stabilizes the IS state. They suggest that the orbital ordering of the IS state plays an important role in this stabilization. The orbital ordering is schematically shown in Fig. 6.5. This mechanism, i.e., the spin-state transition stabilized in part by orbital ordering, has been examined experimentally by Yamaguchi et al. [18]. They report the existence of local lattice distortions which accompany the spin transition, this supporting the IS state model. There are several [26–28] theoretical studies, based on a Hartree-Fock mean-field approximation, which examine the electronic and magnetic states. Numerical diagonalization has also been used to the problem [29]. The im-

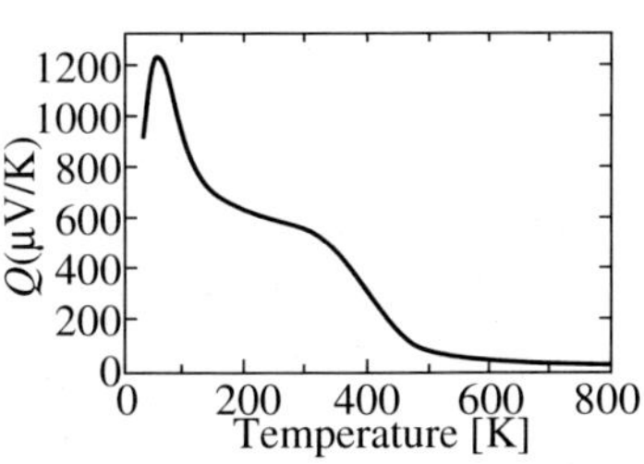

Fig. 6.4. Temperature dependence of the thermopower (Q) in $LaCoO_3$. This sketch is taken from [21].

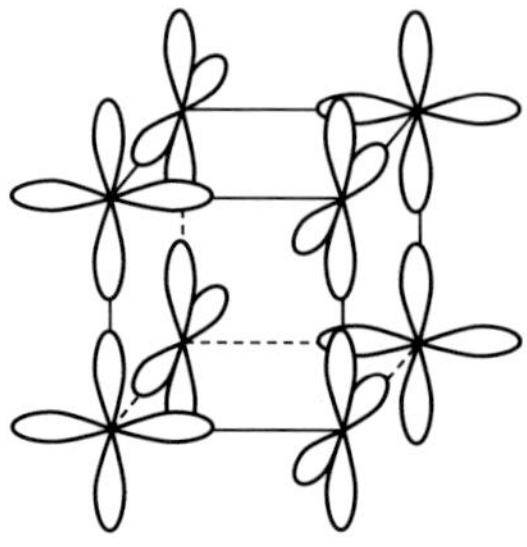

Fig. 6.5. Schematic representation of the orbital ordering in $LaCoO_3$ proposed by Korotin et al. [30]

portance of the spin-orbit interaction for the "$S = 1$ state" has been pointed out. In fact, based upon electron-spin-resonance measurements, Noguchi et al. [32] have suggested that the "$S = 1$ state" be considered as a triplet state formed from HS state through the effects of the spin-orbit interaction.

Let us consider again the local electronic structure of LS, HS and IS states. In the LS state all the $3d$ electrons occupy the t_{2g} orbitals. On the other hand, two electrons go into the e_g orbitals in the HS state while a single electron does so in the IS state. The e_g orbitals extend towards, while the t_{2g} orbitals point away from, the oxygen ions. Between the Co $3d$ states are the O $2p$ orbitals and the hybridization between these depends on their spatial configuration (see Fig. 6.6). Specifically, due to symmetry this hybridization can vanish. (Examples are shown in Fig. 6.6 (d) $\sim$ (f).) Even when the hybridization matrix elements are finite they depend upon the spatial configuration of orbitals. Specifically, the e_g and $2p$ orbitals point towards each other and have a larger overlap than do the t_{2g} and $2p$ equivalents which point away form each other. (See Fig. 6.6(a)–(c).) The coupling between $3d$ orbitals is then a second order processes in these matrix elements with the result that the transfer-matrix-element between adjacent t_{2g} orbitals is smaller than that for e_g orbitals. Consequently, the electrons in t_{2g} orbitals have a tendency to localize. That the electrons in e_g orbitals can gain kinetic energy through hybridization helps stabilize the HS and IS states. At the same time, various interactions such as the Hund's rule coupling and crystal fields compete

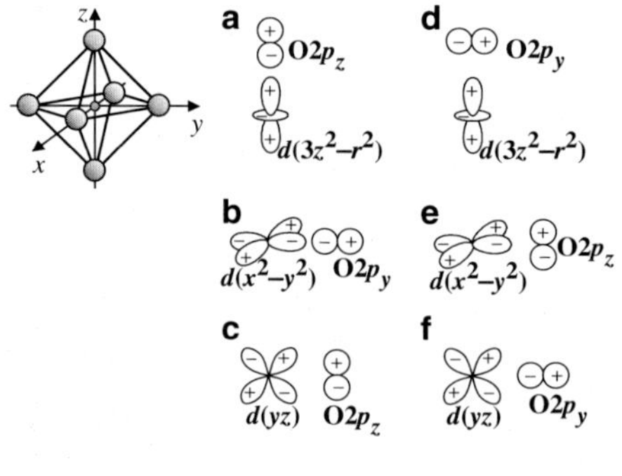

Fig. 6.6. Configurations of the $3d$ orbitals of a Co ion and $2p$ orbitals of the surrounding oxygens.

each other. Through this subtle competition, unique magnetic and transport properties are realized in the cobalt oxides. Therefore, it is often said that the cobalt oxides are a "melting-pot" of interactions.

6.2 Thermoelectric Materials and Cobalt Oxides

In this and the following sub-sections are discussed the thermo-electric response of $NaCo_2O_4$ and related cobalt oxides. At least above room temperature, these systems exhibit neither a phase transition nor an anomaly of the type described in the previous sub-sections. These cobalt oxides, however, have attracted attention as the potential new thermoelectric materials [3].

Thermoelectric materials generate electric energy from heat without pollution essentially. Potential applications include the utilization of waste heat and as power sources for deep-space probes etc. [33].

The in-plane resistivity of the layered-hexagonal oxide $NaCo_2O_4$ shows metallic behavior; the electrical resistivity (ρ) at 300 K is 200 $\mu\Omega$cm, i.e., that of a good metal. Although the thermopower of good metals is usually small, this oxide shows a large thermopower, increasing with temperature to reach 100 μV/K at 300 K.

A real advantage of the cobalt oxide is the large thermopower at higher temperatures. Recently large high temperature thermopowers have also been observed in many related compounds [4–7]. The thermoelectric performance of a material is quantified by the figure of merit ZT ($= Q^2T/\rho\kappa$ where T and κ are temperature and thermal conductivity, respectively). The maximum ZT in sintered samples of $NaCo_2O_4$ is around $\sim$1000K [8], and the large value $ZT \geq 1$ had been measured [9] at $T \geq$ 873K for layered-hexagonal oxide $Ca_2Co_2O_5$.

Traditionally semiconductors have been the most used thermoelectric materials. The electronic structure of semiconductors is well described by the band theory. The design of conventional thermoelectric material is a compromise between low electrical resistivity and large thermopower. Both depend on carrier density with the thermopower increasing logarithmically with decreasing carrier density. However a low carrier density also implies a larger resistivity. The optimum carrier density is estimated to be $10^{18} \sim 10^{19}$ cm^{-3}. Present thermoelectric materials are synthesized with this restriction on the carrier density.

For conventional thermoelectric materials, attention is focused on the charge degree of freedom of electrons. However, as will be seen in the next sub-section, thermopower does not always reflect these degrees of freedom. It has to do with the entropy transported by an electric current. From this point of view, let us ask what are the key ingredients to be used in the design of new thermoelectric materials.

Let us focus on the electronic structure of $NaCo_2O_4$. The lamellar structured crystal $NaCo_2O_4$ has hexagonal layer of CoO_2 sheets as shown in

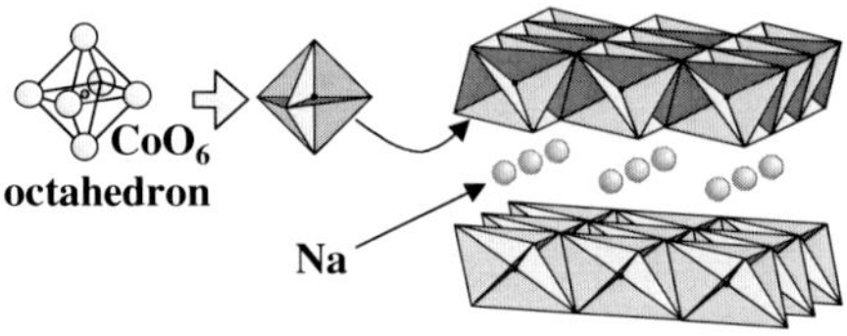

Fig. 6.7. Crystal structure of $NaCo_2O_4$.

Fig. 6.7. Similarly $Ca_2Co_2O_5$, (Bi,Pb)-Sr-Co-O systems and other related cobalt oxides also contain CoO_2 layers.

That the Coulomb interaction is important to the understanding of $NaCo_2O_4$ has been shown theoretically and experimentally [3, 7, 34–37]. It has been observed [34, 35] that there exist Co^{3+} and Co^{4+} sites with the configurations of t_{2g}^6 and t_{2g}^5, respectively. Band calculations [37] show that the conduction band is in a t_{2g} manifold and that the e_g manifold is reasonably well separated from these states. The effective on-site Hubbard U and the band width are estimated to be 5∼8 eV and 1.6 eV, respectively. The bands derived from these t_{2g} orbitals are narrower than in other metallic transition metal oxides with perovskite structure. This is a reflection of the crystal structure: The CoO_6 octahedra in the hexagonal layer share edges with each other, so that the angle of the Co-O-Co bond is about 90 degrees. In the 90-degree configuration, the hopping-matrix-elements of adjacent $3d$ orbitals due to σ bond are zero, and consequently, the $3d$ electrons have small hopping-matrix-elements due to a π bond with neighboring sites. The Hamiltonian for these $3d$ electrons may be written as follows;

$$H = H_{\text{inter}} + H_{\text{intra}} \tag{6.1}$$

$$\begin{aligned} H_{\text{inter}} &= -\sum_{\langle ij\rangle}\sum_{\gamma\gamma'}\sum_{\sigma} t_{ij}^{\gamma\gamma'}\left(d_{i\gamma\sigma}^{\dagger}d_{j\gamma'\sigma} + \text{h.c.}\right) \\ H_{\text{intra}} &= U\sum_{i}\sum_{\gamma} n_{i\gamma\uparrow}n_{i\gamma\downarrow} + U'\sum_{i}\sum_{\gamma\neq\gamma'} n_{i\gamma}n_{i\gamma'} \\ &\quad + J_H\sum_{i}\sum_{\gamma\neq\gamma'}\sum_{\sigma\sigma'} d_{i\gamma\sigma}^{\dagger}d_{i\gamma'\sigma'}^{\dagger}d_{i\gamma\sigma'}d_{i\gamma'\sigma}, \end{aligned} \tag{6.2}$$

where the indices γ and γ' reflect the three t_{2g} orbitals and where other notations as specified in Chapt. 1. In contrast to semi-conductors, the electronic properties of the cobalt oxides are characterized by strong correlated ($t_{ij}^{\gamma\gamma'} << U, U'$) t_{2g} electrons.

Before discussing the thermo-electronic behavior of $NaCo_2O_4$ in more detail, let us review the thermo-electric response of a solid [38].

6.3 Thermoelectric Effect

Thermodynamic systems are either in equilibrium or are not. In turn, non-equilibrium systems might be either linear or non-linear. The thermoelectric effects which are of interests here correspond to an approximation where the deviations from the thermodynamic equilibrium values are small and are linear in the driving fields.

It is known that not only the electric field $\nabla\varphi$ but also the gradient of the temperature ∇T drive the electric current, where φ is the electrostatic potential. When an electric current flows, for $T \neq 0$ there is a concurrent heat current. This situation is reflected by the following linear response equations,

$$\begin{cases} j_e = l^{11}\nabla\varphi + l^{12}\nabla T, \\ j_{\text{heat}} = l^{21}\nabla\varphi + l^{22}\nabla T, \end{cases} \tag{6.3}$$

where j_e and j_{heat} are the electric and heat currents, respectively. The linear dynamical laws of thermoelectric response normally exhibit Onsager reciprocity. However, (6.3) will violate the Onsager reciprocity theorem relation $l^{12} = l^{21}$, since the generalized forces, which correspond to $\nabla\varphi$ and ∇T, and the currents, j_e and j_{heat}, are not chosen appropriately. Let us derive expressions for the dynamic thermoelectric linear response which do not have this defect.

Shown in Fig. 6.8 is a finite but very small area in a thermo-electric system. For simplicity, it is assumed that all flows and forces are parallel to the x-direction. The low- and high-temperature sides are labeled a and b, respectively, and between these there exist differences in the electro-static potential $\Delta\varphi$ and temperature ΔT, where n and ΔU denote the amount of transferred particles and energy.

The system is irreversible, i.e., there is a net generation of entropy. The change in the local entropy density $\Delta \mathrm{s}_a$ on side a is expressed as

$$\Delta \mathrm{s}_a = -\frac{\Delta U}{T} + \frac{\mu(T)}{T}n, \tag{6.4}$$

where $\mu(T)$ is the chemical potential at temperature T. On side b,

$$\Delta \mathrm{s}_b = \frac{\Delta U}{T+\Delta T} + \frac{\mu(T+\Delta T) - e\Delta\varphi}{T+\Delta T}n, \tag{6.5}$$

where e (> 0) is the absolute value of electric charge. The rate of change of the entropy is therefore,

$$\begin{aligned} \frac{\mathrm{d}\mathrm{s}}{\mathrm{d}t} &= \left[\frac{\mathrm{d}(\Delta \mathrm{s}_a + \Delta \mathrm{s}_b)}{\mathrm{d}t}\right] \\ &= \frac{\mathrm{d}(\Delta U)}{\mathrm{d}t}\left(-\frac{\Delta T}{T^2}\right) + \frac{\mathrm{d}n}{\mathrm{d}t}\left[\frac{e}{T}\Delta\varphi - \Delta T\frac{\partial}{\partial T}\left(\frac{\mu}{T}\right)\right]. \end{aligned} \tag{6.6}$$

Note that $\mathrm{d}n/\mathrm{d}t$ is the rate at which particles move through the boundary of a and b, i.e., is the particle current. In the same way, $d(\Delta U)/dt$ is the

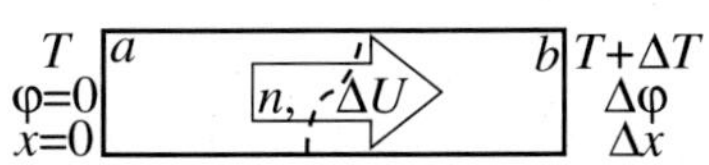

Fig. 6.8. Thermo-electric system.

energy current. The generalized forces are identified from the coefficients which multiply the energy and number currents in (6.6). These are written as,

$$\frac{e}{T}\Delta\varphi - \Delta T\frac{\partial}{\partial T}\left(\frac{\mu}{T}\right) = \Delta x\left[\frac{e}{T}\frac{\partial\varphi}{\partial x} - \frac{\partial}{\partial x}\left(\frac{\mu}{T}\right)\right] \tag{6.7}$$

for the particle current and

$$-\frac{\Delta T}{T^2} = \Delta x\frac{\partial}{\partial x}\left(\frac{1}{T}\right) \tag{6.8}$$

for the energy current, respectively. Finally, the appropriate form for the linear dynamical laws are

$$\begin{cases} j_1 = M^{11}\left[\frac{e}{T}\frac{\partial\varphi}{\partial x} - \frac{\partial}{\partial x}\left(\frac{\mu}{T}\right)\right] + M^{12}\frac{\partial}{\partial x}\left(\frac{1}{T}\right), \\ j_2 = M^{21}\left[\frac{e}{T}\frac{\partial\varphi}{\partial x} - \frac{\partial}{\partial x}\left(\frac{\mu}{T}\right)\right] + M^{22}\frac{\partial}{\partial x}\left(\frac{1}{T}\right), \end{cases} \tag{6.9}$$

where the j_l are the particle, (l=1), and energy, (l=2), flux densities. Here, the quantities M^{lm} are called the kinetic coefficients and satisfy the Onsager reciprocity theorem $M^{12} = M^{21}$.

The kinetic coefficients M^{lm} are determined from the experimental data through the relevant phenomenological laws, i.e., Ohm's law of electrical conduction,

$$j_e\,(= -ej_1) = \frac{1}{\rho}\left(-\frac{\partial\varphi}{\partial x}\right) \tag{6.10}$$

(where j_e is the electric current and ρ is the electrical resistivity), Fick's law of diffusion,

$$j_1 = -D\frac{\partial\tilde{n}}{\partial x} \tag{6.11}$$

(where $\tilde{n}$ is particle density and D is the diffusion coefficient), Fourier's law of heat conduction,

$$j_{\text{heat}}\,(= j_2 - \mu j_1) = -\kappa\frac{\partial T}{\partial x} \tag{6.12}$$

(where j_{heat} is the heat-current flux-density and κ is the thermal conductivity). The final law corresponds to the Seebeck effect which is to be discussed below.

When the temperature and concentration gradients are not allowed ($\partial T/\partial x$=0 and $\partial\mu/\partial x$=0), the first equation in (6.9) gives

$$j_1 = M^{11}\frac{e}{T}\frac{\partial\varphi}{\partial x} \tag{6.13}$$

and the relation

$$\frac{1}{\rho} = e^2 \frac{M^{11}}{T} \tag{6.14}$$

is obtained. In the case that $\partial T/\partial x$=0 and $\partial \varphi/\partial x$=0, the first equation in (6.9) gives

$$j_1 = -M^{11} \frac{1}{T} \frac{\partial \mu}{\partial x} = -M^{11} \frac{1}{T} \frac{\partial \mu}{\partial \tilde{n}} \frac{\partial \tilde{n}}{\partial x} \tag{6.15}$$

and we find

$$D = M^{11} \frac{1}{T} \frac{\partial \mu}{\partial \tilde{n}}. \tag{6.16}$$

The thermal conductivity κ is expressed as

$$\kappa = \frac{1}{T^2} \left[M^{22} - \frac{(M^{12})^2}{M^{11}} \right] \tag{6.17}$$

with the condition that j_1=0 and $\partial \varphi/\partial x$=0.

An expression for the thermopower (Q) is obtained from the first (6.9) for an open circuit (j_1=0), i.e.,

$$-\left(\frac{\partial \varphi}{\partial x} - \frac{1}{e} \frac{\partial \mu}{\partial x} \right) = \left(-\frac{1}{eT} \frac{M^{12}}{M^{11}} + \frac{\mu}{eT} \right) \frac{\partial T}{\partial x} \equiv Q \frac{\partial T}{\partial x}. \tag{6.18}$$

This defines the *absolute* Seebeck coefficient, i.e., the thermopower is defined to be the ratio of the gradients of electro-chemical potential ($\varphi - \mu/e$) and the temperature, i.e.,

$$Q = -\frac{1}{eT} \frac{M^{12}}{M^{11}} + \frac{\mu}{eT}. \tag{6.19}$$

When we apply a temperature gradient to a solid, the hot electrons will spread into the cold end of the sample due to the diffusive force originating from the temperature gradient. Consequently, electrons tend to pile up at the cold end. The resulting charge imbalance results an electric field. The force due to the electric field will compete with the diffusive force. The difference between the forces appears as a net electro-motive force between the ends of the sample. This is the physics implied by (6.18). The steady state ($j_1 = 0$) thermo-electric response follows from the linear dynamical law, (6.9), i.e.,

$$0 = M^{11} \left[\frac{e}{T} \frac{\partial \varphi}{\partial x} - \frac{\partial}{\partial x} \left(\frac{\mu}{T} \right) \right] + M^{12} \frac{\partial}{\partial x} \left(\frac{1}{T} \right). \tag{6.20}$$

The steady state reflects the balance among the various terms. There are not only electric and diffusive forces but an interference between the electric and heat currents (the last term of the left hand side of (6.20)). This is the nature of the thermo-electric effect.

We are now in the position to expose an interesting insight into the physical meaning of the thermopower. Given that the heat $\Delta \mathsf{q}$ is related to the entropy change by $T\Delta \mathsf{s}$, then the relation between heat, J_{heat}, and entropy, J_s, currents is

$$J_{\text{heat}} = TJ_s. \tag{6.21}$$

On the other hand, the heat current-density is defined by

$$J_{\text{heat}} = j_2 - \mu j_1. \tag{6.22}$$

Inserting (6.9) into (6.22), the entropy current-density can be expressed as

$$\begin{aligned} J_s &= \frac{1}{T}\left(\frac{M^{12}}{M^{11}} - \mu\right) j_1 + \frac{1}{T}\left[M^{22} - \frac{(M^{12})^2}{M^{11}}\right]\frac{\partial}{\partial x}\left(\frac{1}{T}\right) \\ &= -eQj_1 + \frac{1}{T}\left(-\kappa\frac{\partial T}{\partial x}\right). \end{aligned} \tag{6.23}$$

This expression shows that there are two contributions to the flow of entropy: one is the entropy carried by the Fourier heat and the other is carried by electric current $(-ej_1)$. An electric current is a flow of charge carriers. Equation (6.23) demonstrates why the thermopower is often said to be the entropy of electric current and to measure the entropy per carrier.

It is often said that the thermopower reflects the sign and the density of charge carriers in metals and semiconductors. However a little thought about (6.23) should convince the reader that the thermopower is not *solely* determined by carrier charge degrees of freedom. The thermopower is, in fact, a direct probe of the charge carrier entropy. An electron has not only charge but also spin and orbital are those required degrees freedom. Although these degrees of freedom exist for all charge carriers, in metals and degenerate semiconductors, they will not always contribute to their entropy, since the electronic state is well described by a degenerate electron gas.

To understand this physics in more detail, we need a microscopic formulation for the thermopower. In the next sub-section, we derive the thermoelectric response (6.9) in terms of linear response theory, using the methods of statistical mechanics.

6.4 Linear Response Theory for Thermoelectric Systems

A real macroscopic body can be divided into a set of sub-systems which are relatively small but yet macroscopic. These sub-systems interact with each other, i.e., the extensive parameters (such as energy) flow to and from the neighboring sub-systems and the interesting extensive parameters undergo spontaneous fluctuating *even* in the thermal equilibrium state. Evidently, the smaller the sub-system, the larger are the spontaneous fluctuations. The fluctuation will then dissipate through the spontaneous flow of the relevant extensive parameters to or from the surrounding sub-systems. The Onsager reciprocity theorem follows if these spontaneous fluctuations obey the same dynamical laws as do the irreversible process described in the previous sections. With this assumption, the phenomenological law (6.9) can be obtained using linear response theory based on statistical mechanics.

The formula for the linear response to an external perturbation is known as the Kubo formula. Take H to be the Hamiltonian and $\mathrm{e}^{-\eta t}\mathcal{F}$ to be a perturbation, where t is the time and η is an infinitesimal quantity so that the perturbation turns on adiabatically. The expectation value $\langle B\rangle$ of a physical quantity B is given by,

$$\langle B\rangle = -\int_0^{\infty} \mathrm{d}t\mathrm{e}^{-\eta t} \int_0^{1/k_B T} \mathrm{d}\tau \mathrm{Tr}\left[\rho_0 \frac{\partial \mathcal{F}(-t-\mathrm{i}\tau)}{\partial t} B\right], \tag{6.24}$$

where the symbol "Tr" denotes trace, i.e., is a summation over any complete set of states and

$$\mathcal{F}(-t-\mathrm{i}\tau) = \mathrm{e}^{-\mathrm{i}\mathcal{H}(t+\mathrm{i}\tau)}\mathcal{F}\mathrm{e}^{\mathrm{i}\mathcal{H}(t+\mathrm{i}\tau)}. \tag{6.25}$$

Here $\mathcal{H} = H - \mu\hat{N}$, $\hat{N}$ the total particle number operator, for the grand canonical ensemble or simply $\mathcal{H} = H$ for the canonical formalism. The density matrix ρ_0 is defined as

$$\rho_0 = \mathrm{e}^{-\mathcal{H}/k_B T}/\mathrm{Tr}\left[\mathrm{e}^{-\mathcal{H}/k_B T}\right], \tag{6.26}$$

where k_B is the Boltzmann constant.

When the perturbation is given by a mechanical disturbance such as an external electric field, the linear response theory corresponds directly to (6.24). However, we must deal with the thermoelectric response where the temperature gradient is the source of the electric current. Of course, the Hamiltonian for the perturbation corresponding to the thermal disturbance does not exist.

It is interesting to note that in (6.24) the perturbation term $\mathcal{F}$ appears as the time derivative $\partial\mathcal{F}(-t)/\partial t$. For the present problem this implies the change of the energy per unit time, i.e., corresponds to the production of heat. The change of heat $\Delta \mathsf{q}$ causes the change of the entropy $T\Delta \mathsf{s}$. Therefore,

$$\partial\mathcal{F}/\partial t = T\mathrm{d}\mathsf{s}/\mathrm{d}t. \tag{6.27}$$

Inserting (6.6) and (6.27) into (6.24), we obtain as a microscopic expressions for the kinetic coefficients

$$M^{lm} = T\int_0^{\infty} \mathrm{d}t \int_0^{1/k_B T} \mathrm{d}\tau \mathrm{Tr}\left[\rho_0 \hat{\jmath}_m(-t-\mathrm{i}\tau)\hat{\jmath}_l\right], \tag{6.28}$$

where the $\hat{\jmath}_l$ are particle currents (l=1) and energy flux (l=2) operators in the Heisenberg representation. The expectation value of the operator $\hat{\jmath}_l$ is just the current density, i.e., $j_l = \langle \hat{\jmath}_{l\gamma}\rangle$. In the expression (6.28), the relation $M^{12} = M^{21}$ is evident.

From the microscopic viewpoint the thermopower is given by (6.28) which makes manifest the fact that the entropy of carriers gives rise to the thermopower.

6.5 Thermopower in Strongly Correlated Systems: Approach from High Temperature Side

Let us apply linear response theory to the thermoelectric response of the cobalt oxides. Although linear response theory gives an explicit expression for the thermopower, the underlying mechanism is not always clear. For high temperatures, it is possible to obtain a simpler and more easily interpreted result. In this sub-section we derive this expression.

When the trace is expressed in terms of energy eigenstates, the kinetic coefficients

$$M^{lm} = \pi k_B \sum_{nn'} \langle n|\rho_0 \hat{\jmath}_m|n'\rangle\langle n'|\hat{\jmath}_l|n\rangle \delta(E_n - E_{n'}), \tag{6.29}$$

where E_n and $|n\rangle$ are the eigenvalues and eigenstates of the Hamiltonian H, respectively, i.e, $H|n\rangle = E_n|n\rangle$.

For high temperatures, i.e., $t_{ij}^{\gamma\gamma'} \ll k_BT \ll U$, the density matrix

$$\rho_0 \equiv \frac{\mathrm{e}^{-(H_{\mathrm{inter}}+H_{\mathrm{intra}})/k_BT}}{\mathrm{Tr}\left[\mathrm{e}^{-(H_{\mathrm{inter}}+H_{\mathrm{intra}})/k_BT}\right]} \simeq \frac{\mathrm{e}^{-H_{\mathrm{intra}}/k_BT}}{\mathrm{Tr}\left[\mathrm{e}^{-H_{\mathrm{intra}}/k_BT}\right]}. \tag{6.30}$$

In the approximation to the right, the kinetic energy H_{inter} is neglected since $t_{ij}^{\gamma\gamma'}/k_BT \sim 0$. With this the approximate energy eigenstates correspond to simple real space configurations. Shown in Fig. 6.9 are the electronic states of a cobalt ion which involve the triplet t_{2g}. In Fig. 6.9(a), the states 1, 4 and 5 correspond to a Co^{3+} ion, while the states 2 and 3 reflect Co^{4+} ion. Another combination with Co^{3+} and Co^{4+} ions is shown in Fig. 6.9(b). In Fig. 6.9(c), the state 3 involves four electrons while the other combinations correspond to Co^{3+} ions. By comparing the energies of these states, it is seen that the state (c) costs an energy U more than (a) or (b). For $U \gg k_BT$, the states such as Fig. 6.9(c) do not contribute to (6.29) and (6.30) because of the Boltzmann factor $\exp(-U/k_BT)$. As a result, (6.29) is reduced to

$$M^{lm} = \pi k_B \sum_{nn'}{}' \langle n|\rho_0 \hat{\jmath}_m|n'\rangle\langle n'|\hat{\jmath}_l|n\rangle \delta(E_n - E_{n'}), \tag{6.31}$$

where $\sum'$ denotes the summation over the states which consist of only Co^{4+} and Co^{3+} sites. In (6.31), H_{intra} in the numerator of ρ_0 simplifies to $N_{(\mathrm{I})}\varepsilon_{\mathrm{I}} + (L - N_{(\mathrm{I})})\varepsilon_{\mathrm{II}}$, where ε_{I} and $\varepsilon_{\mathrm{II}}$ are the energies of Co^{4+} and Co^{3+} sites, respectively, and $N_{(\mathrm{I})}$ is the number of Co^{4+} sites. Finally, (6.29) becomes

$$\begin{aligned} M^{lm} = \pi k_B & \frac{\mathrm{e}^{-[N_{(\mathrm{I})}\varepsilon_{\mathrm{I}}+(L-N_{(\mathrm{I})})\varepsilon_{\mathrm{II}}]/k_BT}}{\mathrm{Tr}\left[\mathrm{e}^{-H_{\mathrm{intra}}/k_BT}\right]} \\ & \times \sum_{nn'}{}' \langle n|\hat{\jmath}_m|n'\rangle\langle n'|\hat{\jmath}_l|n\rangle \delta(E_n - E_{n'}). \end{aligned} \tag{6.32}$$

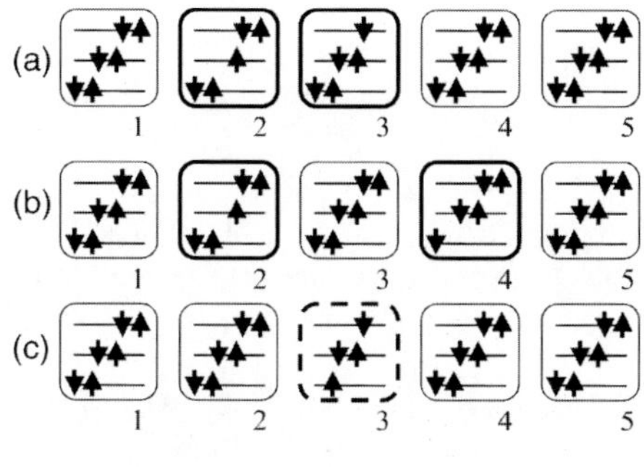

Fig. 6.9. Configurations of electrons. (**a**)–(**c**) have the same numbers of electrons. However, state (**c**) costs more energy than (**a**) or (**b**) by U (see the text).

It is seen that M^{12}/M^{11} in the expression for the thermopower (6.19) becomes temperature independent, and the thermopower at high temperatures finally is given by the simple expression

$$Q = \frac{\mu}{eT}. \tag{6.33}$$

Thermodynamics tells us that μ/T is given by the partial derivative with respect to the number of particles N of the entropy s, so from (6.33)

$$Q = -\frac{1}{e}\left(\frac{\partial s}{\partial N}\right)_{E,V}, \tag{6.34}$$

where E and V are the internal energy and the volume of the electron system, respectively. In other words, (6.34) states that the thermopower is the entropy per carrier.

In the high temperature limit $k_B T/t \to \infty$, (6.34) is valid. In this limit, the kinetic energy and inter-site interaction are neglected. In this situation, the entropy or μ/T are obtained by enumerating all possible states.

There exist two kinds of sites (I) and (II), i.e., Co^{4+} and Co^{3+}. If L and $N_{(\mathrm{I})}$ are the number of unit cells and of sites of type (I), the total number of ways of arranging the two kinds of the sites and their internal degrees of freedom results in a total number of the possible states

$$g = [g_{(\mathrm{I})}]^{N_{(\mathrm{I})}}[g_{(\mathrm{II})}]^{L-N_{(\mathrm{I})}}\frac{L!}{N_{(\mathrm{I})}!(L-N_{(\mathrm{I})})!}, \tag{6.35}$$

where $g_{(\mathrm{I})}$ and $g_{(\mathrm{II})}$ are the degeneracies of (I) and (II), respectively. Knowing that the entropy is given by $s = k_B \ln g$, (6.34) is written as

$$Q = -\frac{k_B}{e}\frac{\partial \ln g}{\partial N}. \tag{6.36}$$

Using Stirlings approximation, (6.36) becomes

$$Q = \frac{k_B}{e}\left(\ln g_{(\mathrm{I})} - \ln g_{(\mathrm{II})}\right) - \frac{k_B}{e}\ln\left(\frac{x}{1-x}\right), \tag{6.37}$$

with $x = N_{(\mathrm{I})}/L$, i.e., is the concentration of the Co^{4+} sites.

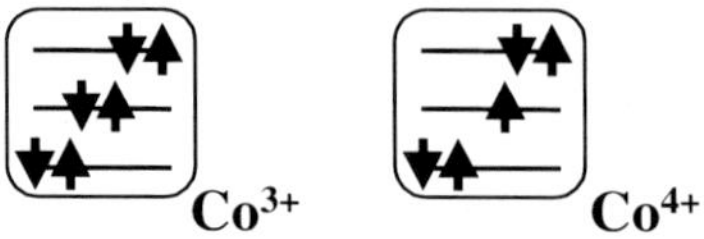

Fig. 6.10. The schematic representation of the configuration of the t_{2g} electrons in Co^{3+} and Co^{4+}.

There are two terms on right hand side of (6.37). The first reflects the degeneracies $g_{(\mathrm{I})}$ and $g_{(\mathrm{II})}$ while the second involves only the hole concentration x. Clearly the spin and orbital degrees of freedom are accounted for in the first term.

Let us apply the theory to $NaCo_2O_4$. Figure 6.10 shows the schematic representation of the t_{2g} electrons for Co^{3+} and Co^{4+} sites. We can see that the configuration of the electrons is uniquely determined for Co^{3+}. Removing an electron from Co^{3+} results in Co^{4+} and there are six ways to do this. It follows, Co^{4+} has a six-fold degeneracy. Since the average valence of a cobalt ion in the stoichiometric compound is +3.5, the ratio of Co^{3+} to Co^{4+} ions is unity, i.e., $x = 0.5$. Thus $g_{(\mathrm{I})} = 1$ and $g_{(\mathrm{II})} = 6$ and with x=0.5, (6.37) gives 154 μV/K for the high-temperature value of the thermopower. This result clearly shows why $NaCo_2O_4$ has the large thermopower with positive sign regardless of the carrier-density. The last term in (6.37) vanishes for x=0.5, and it is solely the spin and orbital degrees of freedom reflecting the strong Coulomb interaction ($U >> t$) which gives rise to the large thermopower.

The approximation that the kinetic energy of electrons is neglected, was used to obtain (6.37). The kinetic term must be accounted for finite temperatures. In addition, the contribution M^{12}/M^{11} in (6.19) cannot be neglected. The theory is extended to the finite temperature region in the next subsection.

6.6 Spin and Orbital States and the Thermopower

We will examine the thermopower in the strongly correlated electron systems using a simplified model, and focus our attention on the role of spin and orbital degeneracies [39] which gave a large thermopower in the previous sub-section. To this end, we use numerical diagonalization for a finite size system. With this the atomic parameters such as Hund's-rule coupling and the energy-level splitting between orbitals are fully taken into account.

We examine the thermopower and the electrical resistivity in the following model (see Fig. 6.11);

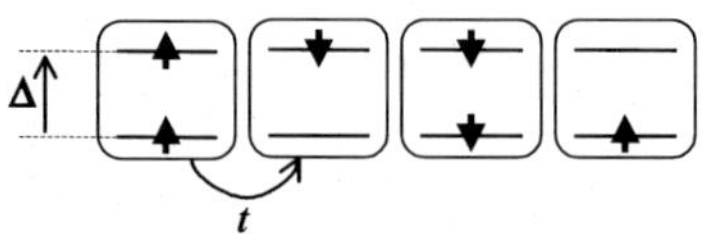

Fig. 6.11. Schematic representation of the model given by Hamiltonian, (6.38).

$$H = -t\sum_{i\nu\sigma}(c^\dagger_{i,\nu,\sigma}c_{i+1,\nu,\sigma} + c^\dagger_{i+1,\nu,\sigma}c_{i,\nu,\sigma}) + J\sum_{i,\nu}\boldsymbol{S}_{i\nu}\cdot\boldsymbol{S}_{i+1,\nu} + J_H\sum_i \boldsymbol{S}_{i,a}\cdot\boldsymbol{S}_{i,b} + \sum_{i,\nu}\Delta_\nu n_{i,\nu}, \tag{6.38}$$

where ν $(=a,b)$ and σ $(=\uparrow,\downarrow)$ are the indices for orbital and spin states, respectively, and where J_H denotes the Hund's-rule coupling. The number of electrons N is taken to be larger than the number of unit cells L. The energy-level splitting is parameterized by $\Delta = \Delta_a - \Delta_b$. The effect of the strong Coulomb interaction is taken into account by excluding double occupation of a given orbital. The chemical potential μ is calculated as a function of temperature, using

$$N = \frac{\sum_{N_e}\mathrm{Tr}[\mathrm{e}^{-(H-\mu N_e)/k_BT}N_e]}{\sum_{N_e}\mathrm{Tr}[\mathrm{e}^{-(H-\mu N_e)/k_BT}]}, \tag{6.39}$$

for the average number of electrons. The summation is performed over all possible numbers of electrons.

The conservation law for the particle and energy densities gives an expression for the current and energy flux operators in real space. The particle density and energy density at site r are defined as

$$n_r = \sum_{\nu\sigma} c^\dagger_{r,\nu\sigma}c_{r,\nu\sigma}, \tag{6.40}$$

$$h_r = -\frac{t}{2}\sum_{\delta=\pm1,\nu\sigma}(c^\dagger_{i,\nu,\sigma}c_{i+\delta,\nu,\sigma} + c^\dagger_{i+\delta,\nu,\sigma}c_{i,\nu,\sigma}) + \frac{J}{2}\sum_{\delta=\pm1,\nu}\boldsymbol{S}_{r,\nu}\cdot\boldsymbol{S}_{r+\delta,\nu} + J_H\sum_r \boldsymbol{S}_{r,a}\cdot\boldsymbol{S}_{r,b} + \sum_{r,\nu}\Delta_\nu n_{r,\nu}, \tag{6.41}$$

respectively. The appropriate conservation laws are

$$-\mathrm{i}[n_r, H] = \sum_\delta \delta^{-1}\hat{j}_{1,(r,r+\delta)}, \tag{6.42}$$

$$-\mathrm{i}[h_r, H] = \sum_\delta \delta^{-1}\hat{j}_{2,(r,r+\delta)}, \tag{6.43}$$

where $\sum_{\delta>0}\hat{j}_{1,(r,r+\delta)}$ and $\sum_{\delta>0}\hat{j}_{2,(r,r+\delta)}$ define the current and energy flux operators in the forward direction, respectively. The Fourier transform of $\hat{j}_{l,q}$ is defined via

$$\sum_{\delta>0}\hat{j}_{l,(r,r+\delta)} = \frac{1}{L}\sum_q \hat{j}_{l,q}\exp[\mathrm{i}qr], \tag{6.44}$$

where L denotes the number of unit cells.

We calculate the M^{lm} in (6.28) for a canonical ensemble. We diagonalize the Hamiltonian (6.38) on a periodic-chain with 6 unit-cells and 10 electrons using Householder method. We then calculate the thermopower Q and

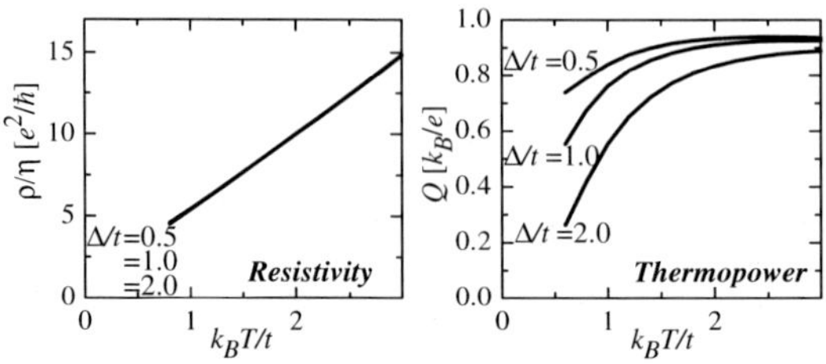

Fig. 6.12. Thermopower and resistivity in the model given by (6.38) with L=6 and N=10 in periodic boundary condition, respectively. The resistivity is in units of η. J/t=0.25 and J_H/t=−10 are used.

electrical resistivity ρ. The results are shown in Fig. 6.12. The electrical resistivity and thermopower are increasing functions of temperature k_BT/t, but the thermopower saturates at high temperatures. The Δ dependence of thermopower and resistivity is interesting: The thermopower increases monotonically with decreasing Δ/t, whereas the resistivity is almost independent of Δ/t.

The energy-level splitting Δ controls the orbital degree of freedom, i.e., for small Δ the orbitals are degenerate and can carry entropy. The Δ dependence of the thermopower of the results, Fig. 6.12, are easily understood from (6.37). In the model (6.38), there are two kinds of sites (I) and (II) with degeneracies $g_{(\mathrm{I})}$ and $g_{(\mathrm{II})}$ which originate from the spin and orbital degrees of freedom. When t, Δ, $J_H \ll k_BT$, electrons are equally distributed on both orbitals for (I), and $g_{(\mathrm{I})}$ is estimated to be 4 due to the spin and orbital degrees of freedom. For (II), both orbitals are occupied by electrons, so that $g_{(\mathrm{II})}$ is 4. For a somewhat lower temperature when, t, $\Delta \ll k_BT \ll J_H$, $g_{(\mathrm{II})}$ is 3 because the electrons in (II) forms a spin-1 state. Lowering the temperature so that $t \ll k_BT \ll \Delta$, J_H corresponding to a larger energy-level splitting Δ, and the electrons in (I) tend to occupy the lowest energy orbitals. In other words, the contribution of the orbital degree of freedom to $g_{(\mathrm{I})}$ decreases with increasing Δ with $g_{(\mathrm{I})}$ estimated to be 2 for $\Delta/t \to \infty$.

The underlying physics is easily understood. Suppose that the state (II) is on the site i and the state (I) is on the site $i+1$. When an electron moves from i to $i+1$, the sites i and $i+1$ become (I) and (II), respectively. In the process, the charge $-e$ moves from i to $i+1$ and the degeneracy is exchanged. This displacement of the degeneracy corresponds to a motion of entropy. This is reflected as a large thermopower since the thermopower is a direct probe of the entropy per carrier, this in contrast to the resistivity.

We have seen that the strong Coulomb interaction induces the spin and orbital degrees of freedom which enhance the thermopower. This realization indicates a new path by which to develop novel high performance thermoelectric materials.

6.7 Thermopower of the Degenerate Electron Gas

The above discussion has emphasized how the spin and orbital degrees of freedom can be used to enhance the thermopower. It has been shown that these spin and orbital degrees of freedom are a consequence of a strong Coulomb interaction ($U >> t$). For comparison, the thermoelectric response of a weakly interacting, ($U << t$), electron gas is summarized in this sub-section.

Using linear response theory, the thermopower Q and the electrical conductivity σ of an electron gas with impurities and adiabatic phonons are related via [47];

$$Q = \frac{\pi^2}{3}\left(\frac{k_B^2 T}{e}\right)\left[\frac{\mathrm{d}\ln\sigma(\epsilon)}{\mathrm{d}\epsilon}\right]_{\epsilon=E_F} . \tag{6.45}$$

In the relaxation time approximation, the electrical conductivity

$$\sigma\left(E_F\right) = \frac{e^2}{3}\left(v_F\right)^2 \mathcal{N}\left(E_F\right)\tau, \tag{6.46}$$

or

$$\sigma\left(E_F\right) = \frac{ne^2\tau}{m^*}, \tag{6.47}$$

where v_F, $\mathcal{N}(E_F)$ and τ denote the Fermi velocity, the density of state at the Fermi energy E_F and the life time of an electron, respectively. Here, n is the density of electrons and m^* is the effective mass of an electron.

With a series of assumption, (6.45) is obtained from the Boltzmann equation. This result has a wide application. In the Fermi liquid theory, the interactions between simply electrons renormalize the effective mass and the life time. Consequently, the thermopower is determined by the density of states at the Fermi energy or the electron density. For example, the large thermopower in the Kondo systems [48] is seen to follow from the enhancement of the density of state at the Fermi energy. From (6.45)-(6.47), it is clear that the electrical resistivity and thermopower both depend on the carrier density. The thermopower is enhanced with decreasing carrier density, but this unfortunately results in an increased electrical resistivity.

As has already been stated, it is evident from (6.23) that the thermopower is nothing but the entropy transported by charge carriers. For a Fermi liquid the spin and orbital degrees of freedom are suppressed and only the charge degree of freedom plays an important role.

6.8 Summary and Discussion – The Path to New Thermoelectric Materials

In this section, we have shown that the thermopower can be enhanced with little effect on resistivity when, due to strong Coulomb interactions, there is a large degeneracy in spin and orbital degrees of freedom. This is completely different from the physics of conventional thermoelectric materials based on semiconductors.

It should however be noted that the electron-phonon and superexchange interactions can lead to the long-range order which suppress the spin and orbital degeneracies for temperatures lower than that at which short-range correlations set in and that this might be considerably higher than the critical temperature T_c for long-range order. Such correlations are detrimental to a substantial spin and orbital thermopower especially in the bipartite lattices. However, for a triangular lattice there is a frustration of the spin and orbital degrees of freedom and this helps prevent the electron-phonon and superexchange interactions from lifting the degeneracies. Thus frustrated lattices favor a large thermopower.

The carrier doping is usually needed in order to have a good conductivity. Although the effects of the spin and orbital degrees of freedom in the thermopower have been also observed for some other doped transition metal oxides, the effects are lost with increasing carrier concentration [49,50] since Coulomb correlations are less important as the doping is increased. Upon doping, sufficient strong Coulomb interaction (U) is required to retain its internal degrees of freedom which are reflected in the thermopower. The layered-hexagonal structure of $NaCo_2O_4$ with the Co-O-Co bond angle of about 90 degrees provides the narrow electron band width (W) [37], i.e., U/W is large and the effects of Coulomb interaction are important.

In this sense, small angle TM–O–TM bonds (TM denotes transition metal ion) leading to a narrow band, strong correlation of the electrons and frustration are considered to be the key ingredients for potential transition metal oxide thermoelectric materials.

Finally, transition metal oxides, other than cobalt oxides, may also provide large thermopowers. The thermopower derived from the first term in (6.37) for several transition metal oxides is listed in Table 6.1.

Table 6.1. Possible thermopower induced by spin and orbital in several transition metal oxides. The first column shows the combination of ions (I) and (II). The second column shows the ratio of the local degeneracies $g_{(I)}$ and $g_{(II)}$.

(I) and (II)	$g_{(I)}/g_{(II)}$	$(k_B/e)\ln(g_{(I)}/g_{(II)})$
$Ti^{4+}(d^0)$, $Ti^{3+}(d^1)$	1/6	-154 μV/K
$V^{4+}(d^1)$, $V^{3+}(d^2)$	6/9	-35 μV/K
$Cr^{4+}(d^2)$, $Cr^{3+}(d^3)$	9/4	70 μV/K
$Mn^{4+}(d^3)$, $Mn^{3+}(d^4)$	4/10	-79 μV/K

References

1. H. Uchida, S. Arisaka, M. Watanabe: Solid State Ionics **135** 347 (2000)
2. X. Chen, S. Wang, Y. L. Yang, L. Smith, N.J. Wu, B.-I. Kim, S. S. Perry, A.J. Jacobson, A. Ignatiev: Solid State Ionics **146** 405 (2002)
3. I. Terasaki, Y. Sasago, K. Uchinokura: Phys. Rev. B **56**, R12685 (1997)
4. T. Itoh, I. Terasaki: Jpn. J. Appl. Phys. **39**, 6658 (2000)
5. I. Terasaki: Proceedings of the 18th International Conference on Thermoelectrics, Baltimore, USA, (1999)
6. A. C. Masset, C. Michel, A. Maignan, M. Hervieu, O. Toulemonde, F. Studer, B. Raveau, J. Hejtmanek: Phys. Rev. B **62**, 166 (2000)
7. T. Yamamoto, K. Uchinokura, I. Tsukada: Phys. Rev. B **65**, 184434 (2002); T. Yamamoto, I. Tsukada, K. Uchinokura, M. Takagi, T. Tsunobe, M. Ichihara, K. Kobayashi: Jpn. J. Appl. Phys. **39**, L747 (2000); T. Yamamoto: Ph.D. thesis, University of Tokyo, 2001
8. M. Ohtaki, E. Maeda: J. Jpn. Soc. Powder and Powder Metallurgy **47** 1159 (2000)
9. R. Funahashi, I. Matsubara, H. Ikuta, T. Takeuchi, U. Mizutani, S. Sodeoka: Jpn. J. Appl. Phys. **39**, L1127 (2000)
10. Y. Miyazaki, M. Onoda, T. Oku, M. Kikuchi, Y. Ishii, Y. Ono, Y. Morii, T. Kajitani: J. Phys. Soc. Jpn. **71**, 491 (2002)
11. R. Ishikawa, Y. Ono, Y. Miyazaki, T. Kajitani: Jpn. J. Appl. Phys. **41**, L337, (2002)
12. Y. Ono, R. Ishikawa, Y. Miyazaki, T. Kajitani: J. Phys. Soc. Jpn. **70**, Suppl. A, 235 (2001); Y. Ono, R. Ishikawa, Y. Miyazaki, Y. Ishii, Y. Morii, T. Kajitani: J. Solid State Chem. **166**, 177 (2002)
13. K. Takada, K. Takada, H. Sakurai, E. Takayama-Muromachi, F. Izumi, R.A. Dilanian, T. Sasaki: Nature **422**, 53 (2003)
14. W. C. Koehler, E. O. Wollan, J. Phys. Chem. Solids **2**, 100 (1957)
15. N. Menyuk, K. Dwight, P. M. Raccah: J. Phys. Chem. Solids **28**, 549 (1967)
16. K. Asai, P. Gehring, H. Chou, G. Shirane: Phys. Rev. B **40**, 10982 (1989)
17. M. Itoh, M. Sugahara, I. Natori, K. Motoya: J. Phys. Soc. Jpn. **64**, 3967 (1995)
18. S. Yamaguchi, Y. Okimoto, Y. Tokura: Phys. Rev. B **55**, R8666 (1997)
19. Y. Tokura, Y. Okimoto, S. Yamaguchi, H. Taniguchi, T. Kimura, H. Takagi: Phys. Rev. B **55**, R1699 (1998)
20. M. Imada, A. Fujimori, Y. Tokura: Rev. Mod. Phys. **70**, 1039 (1998)
21. M. A. Señarís, J. B. Goodenough: J. Solid State Chem. **116**, 224 (1995)
22. R. R. Heikes, R. C. Miller, R. Mazelsky: Physica (Amsterdam) **30**, 1600 (1964)
23. M. Abbate, J. C. Fuggle, A. Fujimori, L. H. Tjeng, C. T. Chen, R. Potze, G. A. Sawatzky, H. Eisaki, S. Uchida: Phys. Rev. B **47**, 16124 (1993)
24. M. Abbate, R. Potze, G. A. Sawatzky, A. Fujimori: Phys. Rev. B **49**, 7210 (1994)
25. T. Saitoh, T. Mizokawa, A. Fujimori, M. Abbate, Y. Takeda, M. Takano: Phys. Rev. B **55**, 4257 (1997)
26. H. Takahashi, F. Munakata, M. Yamanaka: Phys. Rev. B 53, 3731 (1996); *ibid.* **57**, 15211 (1998)
27. T. Mizokawa, A. Fujimori: Phys. Rev. B **54**, 5368 (1996)
28. M. Zhuang, W. Zhang, N. Ming: Phys. Rev. B **57**, 10705 (1998); *ibid.* **57**, 10710 (1998)
29. K. Tsutsui, J. Inoue, S. Maekawa: Phys. Rev. B **59**, 4549 (1999)
30. M. A. Korotin, S. Yu. Ezhov, I. V. Solovyev, V. I. Anisimov, D. I. Khomskii, G. A. Sawatzky: Phys. Rev. B **54**, 5309 (1996)
31. J. Androulakis, N. Katsarakis, J. Giapintzakis: Phys. Rev. B **64**, 1774401 (2001)

32. S. Noguchi, S. Kawamata, K. Okuda, H. Nojiri, M. Motokawa: Phys. Rev. B **66**, 094404 (2002)
33. G. Mahan, B. Sales, J. Sharp: Phys. Today, **50**, No.3, 42 (1997)
34. T. Tanaka, S. Nakamura, S. Iida: Jpn. J. Appl. Phys. **33**, L581 (1994)
35. R.Ray, A. Ghoshray, K. Ghoshray, S. Nakamura: Phys. Rev. B **59**, 9454 (1999)
36. Y. Ando, N. Miyamoto, K. Segawa, T. Kawata I. Terasaki: Phys. Rev. B **60** 10580 (1999)
37. D. J. Singh: Phys. Rev. B **61**, 13397 (2000)
38. As the typical text books, one can find: H. B. Callen: *Thermodynamics and an Introduction to Thermostatistics, 2nd Edition* (J. Wiley, New York, 1985) D. K. Kondepudi, I. Prigogine: *Modern Thermodynamics: From Heat Engines to Dissipative Structures* (J. Wiley, New York, 1998) G. D. Mahan: *Many-Particle Physics, 3ed Edition* (Plenum Press, New York, 2000) C. Kittel: *Elementary Statistical Physics* (J. Wiley & Sons, Inc. New York, 1958) A. Katchalsky, Peter F. Curran: *Nonequilibrium thermodynamics in biophysics* (Harvard University Press, Cambridge, Mass. 1965) T.C. Harman, J.M. Honig: *Thermoelectric and thermomagnetic effects and applications* (McGraw-Hill, New York 1967)
39. The importance of the spin and orbital degrees of freedom on the thermopower of strongly correlated systems has been studied by many authors. See the references [40–46].
40. P. M. Chaikin, G. Beni: Phys. Rev. B **13**, 647 (1976)
41. J.-P. Doumerc: J. Sold State Chem. **109**, 419 (1994)
42. W. Koshibae, K. Tsutsui, S. Maekawa: Phys. Rev. B **62**, 6869 (2000)
43. W. Koshibae, S. Maekawa: Phys. Rev. Lett. **87**, 236603 (2001)
44. A. Oguri, S. Maekawa: Phys. Rev. B **41**, 6977 (1990)
45. D.B. Marsh, P.E. Parris: Phys. Rev. B **54**, 7720 (1996); *ibid.* **54**, 16602 (1996)
46. G. Pálsson, G. Kotliar: Phys. Rev. Lett. **80**, 4775 (1998)
47. M. Jonson, G. D. Mahan: Phys. Rev. B **21**, 4223 (1980); *ibid.* **42**, 9350 (1990)
48. The role of the Kondo effect in the thermoelectric response has been studied for many years. One may find the following articles and the references there in: N. E. Bickers, D. L. Cox, J. W. Wilkins: Phys. Rev. Lett. **54**, 230 (1985), S. Maekawa, S. Kashiba, M. Tachiki, S. Takahashi: J. Phys. Soc. Jpn. **55**, 3194 (1986) and G. D. Mahan: Solid State Phys. **51**, 81 (1998)
49. A. B. Kaiser, C. Uher: Studies of High Temperature Superconductors, Vol. 7, edited by A. V. Narlikar (New York: Nova Science Publishers), p. 353 (1991)
50. T. Okuda, K. Nakanishi, S. Miyasaka, Y. Tokura: Phys. Rev. B **63**, 113104 (2001)

7. Quantum Effects in Orbitally Degenerate Systems

In a classical approach to orbital physics, one usually assumes that a long-range coherence of the orbital polarization sets in below the cooperative orbital/Jahn-Teller transition temperature, and that the spin-exchange interactions on every bond are fixed by the Goodenough-Kanamori rules [1–4]. Implicit to this picture is that the orbital splittings are large enough so that we can consider orbital populations as classical numbers. Such a classical treatment of orbitals is certainly justified when orbital order is driven by strong cooperative lattice distortions that lead to a large splitting of the – initially degenerate– orbital levels. In this limit the orbital excitations are more or less localized high-energy quadrupole moment (or crystal-field) transitions, and therefore they effectively just renormalize the spin degrees of freedom but otherwise do not have much effect on the physical properties of the system at low energy scales.

Quantum effects, however, might start to dominate the ground state properties and elementary excitations when classical order is frustrated by some interaction that opposes the tendency of the orbitals to order. The reason that this might very well be the case in some regions of the phase diagram of the manganites is that orbitals strongly interact with spins via the superexchange and that orbital order is frustrated by the kinetic energy of carriers in a metallic system. But the actual situation in manganites is still subject to debate; while in the undoped case the orbital excitations are rather high in energy and therefore quite decoupled from the spins, there are many indications that orbital dynamics play an essential role in the physics of doped manganites as discussed below.

Orbital frustration is a very general property of cubic perovskite compounds. Titanium and vanadium oxides, in which oxygen octahedra are only weakly distorted, are of particular interest. In this case the orbital frustration and related quantum effects are more pronounced because the degeneracy of t_{2g} level is larger, which enhances quantum effects. In addition to the frustration that is present in the e_g-system, there also exists the possibility to form quantum singlets among t_{2g} orbitals. Therefore quantum tunneling between different local orbital configurations may occur. The cubic titanate $LaTiO_3$ is the most representative example: orbitals in this Mott insulator remain disordered even at low temperature, which can be explained by the formation of a coherent quantum liquid state. The cubic vanadates are also interesting in this respect, although they are different because of the larger spin value. In this case the spin-orbital frustration is resolved with the help of low-dimensional orbital fluctuations.

The origin of the orbital frustration is quite different from that in a conventional spin systems without orbital degeneracy. In the latter case, spin interactions are of the same, e.g. Heisenberg form for all the bonds, and it is a special lattice geometry that usually brings about spin frustrations; a typical example is the Heisenberg model on triangular or pyrochlore lattices. In such frustrated lattices, there are large number of classical spin configurations with the same energy. The distinct feature of the orbitally degenerate models is that frustration occurs for conventional, e.g. simple cubic, lattices. The reason is that the very form of the orbital interaction depends on the bond direction in the crystal. Because of the spatial anisotropy of the orbital wavefunctions, interactions on different bonds require the population of different orbital states and hence compete. A bond directionality of the interactions is a special feature of orbital models which leads to a large degeneracy of the low energy states and pronounced quantum effects. In this way, the orbital degeneracy provides a new root for frustrated quantum models in three dimensions, in addition to the conventional one driven by geometrical frustration.

7.1 Systems with e_g Orbital Degeneracy

In this section we discuss some theoretical models and several physical examples which show that there is a strong dynamical interplay between orbital fluctuations and spin and charge degrees of freedom. Both spin exchange and charge motion are highly sensitive to the state of the orbital bonds, and the basic idea is that the interaction energy cannot be optimized simultaneously for all the bonds: this leads to peculiar frustrations and quantum resonances among orbital bonds. Of course a classical treatment in such cases gives very poor estimates of energies, and the quantum dynamics of the coupled orbital-spin-charge system becomes of crucial importance.

7.1.1 Spin-Orbital Model

Quantum fluctuations of the orbitals originate mainly from two kind of interactions. One possibility involves the superexchange interactions and will be discussed first. The second mechanism –the very effective frustration of orbitals by doped holes– we describe later on.

Let us take as an example a version of the superexchange model for e_g-electrons (proposed by Kugel and Khomskii [4]), neglecting Hund's rule corrections for simplicity. On a three-dimensional cubic lattice it takes the form:

$$H = \sum_{\langle ij \rangle} \left(\boldsymbol{S}_i \cdot \boldsymbol{S}_j + \frac{1}{4} \right) \hat{J}_{ij}^{(\gamma)}, \tag{7.1}$$

$$\hat{J}_{ij}^{(\gamma)} = J \left(T_i^{(\gamma)} T_j^{(\gamma)} - \frac{1}{2} T_i^{(\gamma)} - \frac{1}{2} T_j^{(\gamma)} + \frac{1}{4} \right),$$

with $J = 4t^2/U$, where t and U are, respectively, the hopping integral and on-site Coulomb repulsion in the Hubbard model for two-fold degenerate e_g-electrons at half filling. The structure of $T_i^{(\gamma)}$ depends on the index γ which specifies the orientation of the bond $\langle ij \rangle$ relative to the cubic axes a, b and c:

$$T_i^{(a/b)} = \frac{1}{4}(-\sigma_i^z \pm \sqrt{3}\sigma_i^x), \qquad T_i^{(c)} = \frac{1}{2}\sigma_i^z, \tag{7.2}$$

where σ^z and σ^x are the Pauli matrices. (Here, a convention is used such that $\sigma^z = 1$ (-1) corresponds to a $|x^2 - y^2\rangle$ $(|3z^2 - r^2\rangle)$ state). Physically, the $T^{(\gamma)}$ operators describe the dependence of the spin-exchange interaction on the orbital occupation, and the main feature of this model –as is suggested by the very form of Hamiltonian (7.1)– is the strong interplay between spin and orbital degrees of freedom. It was recognized first in [5] that this simple model contains rather nontrivial physics: the classical Néel state in (7.1) (where $\langle \boldsymbol{S}_i \cdot \boldsymbol{S}_j \rangle = -1/4$) is infinitely degenerate in the orbital sector thus strongly frustrating the orbital order; this extra degeneracy must be lifted by some mechanism.

Before discussing this mechanism, let us first elaborate on this degeneracy and the importance of quantum effects in this case. We first notice that the effective spin exchange constant in this model is definite positive $(\langle J_{ij}^{(\gamma)} \rangle \geq 0)$ for any configuration of orbitals, where its value can vary from zero to J, depending on the orientation of orbital pseudospins. We therefore expect a simple two-sublattice antiferromagnetic, G-type, spin order. There is however a problem: a classical G-type ordering has cubic symmetry and can therefore not lift the orbital degeneracy, not even locally. In more formal terms, the spin part $(\boldsymbol{S}_i \cdot \boldsymbol{S}_j + 1/4)$ of the Hamiltonian (7.1) simply becomes zero in this state for all bonds, so that these orbitals effectively do not interact –they are completely uncorrelated– and hence retain the full rotational freedom on every lattice site. In other words, we gain no energy from the orbital interactions that are present in the model. This shows that, from the point of view of the orbitals, the classical Néel state is energetically a very poor state for the system.

The mechanism for developing intersite orbital correlations (and hence to gain energy from orbital ordering) must involve a strong deviation in the spin configuration from the Néel state –a deviation from $\langle \boldsymbol{S}_i \cdot \boldsymbol{S}_j \rangle = -\frac{1}{4}$. This implies an intrinsic tendency of the system to develop low-dimensional spin fluctuations which can most effectively be realized by an ordering of orbitals as shown in Fig. 7.1. In this situation the effective spin interaction is quasi one-dimensional along the chains in the c-direction so that quantum spin fluctuations are very much enhanced and quantum energy is gained from

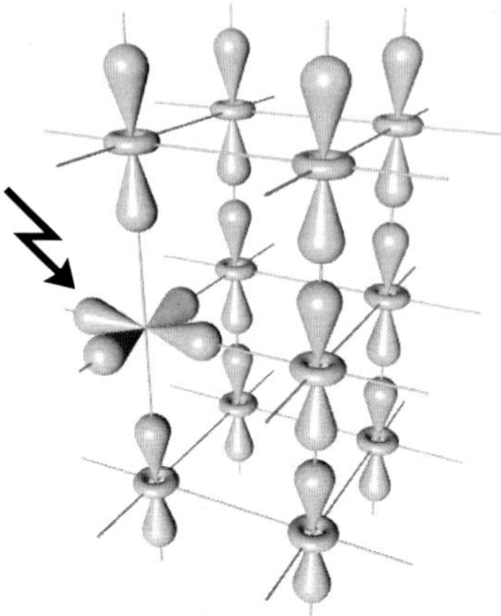

Fig. 7.1. Illustrated is $|3z^2 - r^2\rangle$-orbital order which leads to weakly coupled AF spin chains ($J_c = J$, $J_\perp = J/16$). This type of orbital ordering provides the largest energy gain due to quantum spin fluctuations. An orbital flip (indicated by an arrow) modulates the strength of the neighboring exchange bonds, breaking the c-chain. In the classical Néel state, such orbital excitations cost no energy. Due to the presence of strong quasi one-dimensional spin fluctuations, however, a finite orbital gap opens through the order from disorder mechanism, thus stabilizing this structure. (After [9]).

the bonds along the chain. Here $\langle \boldsymbol{S}_i \cdot \boldsymbol{S}_j + \frac{1}{4}\rangle < 0$, so that the effective orbital (pseudospin) exchange is indeed ferromagnetic, which leads to the orbital structure shown in Fig. 7.1. At the same time the cubic symmetry is explicitly broken, as fluctuations of spin bonds are different in different directions. This leads to a finite splitting of the e_g-levels, and therefore an orbital gap is generated. One can say that in order to stabilize the ground state, orbital order and spin fluctuations support and enhance each other –a situation that is very similar to Villain's order from disorder phenomena known previously in connection with frustrated spin systems [6].

More quantitatively, the expectation value of exchange coupling (4) between $|3z^2 - r^2\rangle$ orbitals (consider $\sigma_i^z = -1$ in (7.1,7.2)) is $J_c = J$ along the c-axis, and it is only small in the ab-plane: $J_\perp = J/16$. Exchange energy is mainly accumulated in c-chains and can be approximated as $J_c\langle \boldsymbol{S}_i \cdot \boldsymbol{S}_j + \frac{1}{4}\rangle_c + 2J_\perp\langle \boldsymbol{S}_i \cdot \boldsymbol{S}_j + \frac{1}{4}\rangle_\perp \simeq -0.16J$ per site (using $\langle \boldsymbol{S}_i \cdot \boldsymbol{S}_j\rangle_c = 1/4 - \ln 2$ for 1D and assuming $\langle \boldsymbol{S}_i \cdot \boldsymbol{S}_j\rangle_\perp \sim 0$). On the other hand $(x^2 - y^2)$-type ordering results in two dimensional magnetic structure ($J_{a,b} = 9J/16, J_c = 0$), and has a much smaller energy gain $\simeq -0.09J$.

From the technical point of view, it is obvious that a conventional expansion about the classical Néel state would fail to remove the orbital degeneracy: only quantum fluctuations can lead to orbital correlations. This is precisely the reason why in a linear spin-wave approximation one does not generate an orbital gap, and low-energy singularities appear [5, 7]. The problem was resolved in Refs. [8,9]: the singularities vanish once quantum spin fluctuations are explicitly taken into account in the calculations of the orbiton spectrum. These fluctuations generate a finite gap for orbital, as well as for composite spin-orbital, excitations and in this way the spin fluctuations remove the orbital frustration. The long-range spin-orbital order indicated in Fig. 7.1 is stable against residual interactions because of this orbital gap, and because of the small, but finite, coupling between spin chains.

The Hamiltonian (1) can be studied within the following scheme. (*i*) Rewrite (1) in the form $H = H_{sp} + H_{\rm orb} + H_{\rm int}$. Here the first two terms describe spin and orbital sectors separately. Treating spin-orbital coupling on the mean-field level:

$$H_{sp} = \sum_{\langle ij \rangle} \langle \hat{J}_{ij}^{(\gamma)} \rangle \left(\boldsymbol{S}_i \cdot \boldsymbol{S}_j + \frac{1}{4} \right), \tag{7.3}$$

$$H_{\rm orb} = \sum_{\langle ij \rangle} \left\langle \boldsymbol{S}_i \cdot \boldsymbol{S}_j + \frac{1}{4} \right\rangle \delta(\hat{J}_{ij}^{(\gamma)}), \tag{7.4}$$

where $\delta(A) = A - \langle A \rangle$. It is important to examine the stability of the mean-field, spin-orbital separated state against fluctuations generated by dynamical coupling between spin and orbital excitations. This coupling is represented by

$$H_{\rm int} = \sum_{\langle ij \rangle} \delta(\hat{J}_{ij}^{(\gamma)}) \delta(\boldsymbol{S}_i \cdot \boldsymbol{S}_j). \tag{7.5}$$

(*ii*) Assume the G-type antiferromagnetic spin order and uniform $(3z^2 - r^2)$- or $(x^2 - y^2)$-type ordering of the orbitals, and employ a spin (orbital) wave representation for $\boldsymbol{S}_i$ ($\boldsymbol{\sigma}_i$) operators. (*iii*) Calculate spin-orbit interaction corrections to the excitation spectrum and to the order parameters. Since latter quantities enter in coupling constants in (7.3),(7.4), all steps have to be performed self-consistently.

Consider $(3z^2 - r^2)$-type orbital order which results in a highly anisotropic quasi-1D magnetic structure. Let us discuss first the mean-field results, which follow from (7.3),(7.4). In linear spin-wave approximation, one expands the spin operators in (7.3) in terms of Holstein-Primakoff bosons as $S_i^z = 1/2 - a_i^\dagger a_i$, $S_i^+ \simeq a_i$ on spin-up lattice sites, while $S_i^z = a_i^\dagger a_i - 1/2$ and $S_i^+ \simeq a_i^\dagger$ on spin-down sublattice. The linearized bosonic Hamiltonian can easily be diagonalized in a momentum space by the Bogoliubov transformation $a_{\boldsymbol{k}} = u_{\boldsymbol{k}} \alpha_{\boldsymbol{k}} + v_{\boldsymbol{k}} \alpha_{-\boldsymbol{k}}^\dagger$. This gives

$$H_{SW} = E_{SW} + \sum_{\boldsymbol{k}} \omega_{1,\boldsymbol{k}} \alpha_{\boldsymbol{k}}^\dagger \alpha_{\boldsymbol{k}}, \tag{7.6}$$

where $E_{SW} = \sum_{\boldsymbol{k}} (\omega_{1,\boldsymbol{k}} - J_1)/2$ stands for zero-point spin fluctuation energy, and $\omega_{1,\boldsymbol{k}} = J_1 \sqrt{1 - \lambda_{1,\boldsymbol{k}}^2}$ is the spinwave dispersion. The momentum dependence of $\omega_{1,\boldsymbol{k}}$, as well as of the Bogoliubov transformation coefficients $u_{\boldsymbol{k}} = \{(s+1)/2\}^{1/2}$ and $v_{\boldsymbol{k}} = -\{(s-1)/2\}^{1/2} \mathrm{sign}(\lambda_{1,\boldsymbol{k}})$ with $s = (1 - \lambda_{1,\boldsymbol{k}}^2)^{-1/2}$, are determined by a function $\lambda_{1,\boldsymbol{k}} = (J_c \cos k_z + J_\perp \cos k_x + J_\perp \cos k_y)/J_1$. Here $J_1 = (J_c + 2J_\perp)$, and the spin exchange constants

$$J_c = \frac{J}{4} \langle 1 - \sigma_i^z - \sigma_j^z + \sigma_i^z \sigma_j^z \rangle^{(c)}, \tag{7.7}$$

$$J_\perp = \frac{J}{4} \langle 1 + \frac{1}{2}\sigma_i^z + \frac{1}{2}\sigma_j^z + \frac{1}{4}\sigma_i^z \sigma_j^z + \frac{3}{4}\sigma_i^x \sigma_j^x \rangle^{(a,b)}$$

are determined by orbital correlations. The orbital state, on the other hand, is governed by $H_{\rm orb}$, (7.4). Expanding the pseudospins in the orbital exchange operator $\hat{J}_{ij}^{(\gamma)}$ in terms of Holstein-Primakoff bosons, $\sigma_i^z = 2b_i^\dagger b_i - 1$, $\sigma_i^x \simeq b_i + b_i^\dagger$, and again keeping only terms quadratic in the boson operators b,$b^\dagger$, one obtains the orbital wave Hamiltonian

$$H_{OW} = E_{OW} + \sum_{\boldsymbol{k}} \omega_{2,\boldsymbol{k}} \beta_{\boldsymbol{k}}^\dagger \beta_{\boldsymbol{k}}. \tag{7.8}$$

The orbital zero-point energy $E_{OW} = \sum_{\boldsymbol{k}}(\omega_{2,\boldsymbol{k}} - J_2)/2$ and the orbital dispersion $\omega_{2,\boldsymbol{k}} = J_2\sqrt{1 - 2\lambda_{2,\boldsymbol{k}}}$ are controlled by spin fluctuations via the parameters $\kappa_\gamma = -\langle \boldsymbol{S}_i \cdot \boldsymbol{S}_j + \frac{1}{4}\rangle^{(\gamma)}$ as follows:

$$J_2 = 2J\left(\kappa_c - \frac{1}{4}\kappa_\perp\right), \quad \lambda_{2,\boldsymbol{k}} = \frac{3}{4}\frac{\kappa_\perp}{4\kappa_c - \kappa_\perp}(\cos k_x + \cos k_y). \tag{7.9}$$

Note that all κ_γ, and hence $\omega_{2,\boldsymbol{k}}$, vanish in a classical Néel state for spins, making it explicit that orbital order is stabilized by spin fluctuations. Conversely, the spin coupling constants J_c and $J_\perp$, which control the spin fluctuations, are determined by orbital correlations.

Self-consistent numerical calculations [8] show that the orbital pseudospin is almost saturated (the mixture of the $|x^2 - y^2\rangle$ state is about one percent only). Coupling between chains $J_\perp$ is weak, $J_\perp/J_c \simeq 0.052$ only, as expected, but sufficient to support 3D long-range spin order with $\langle S^z\rangle = 0.226$ and to produce a quite large magnon dispersion in the *ab*-plane (see thin dashed lines in Fig. 7.2). Orbital excitations are gapful, since the orbital ordering is not associated with the breaking of any continuous symmetry. Quantitatively, we see that the orbital gap (of the order of $t^2/U = J/4$) is smaller than the spin-wave bandwidth. The softness of the orbital excitations is related to the fact that the orbital degeneracy in the model (1) can be lifted only by quantum effects in the spin sector.

Now, what happens when we switch on the dynamical coupling between spin and orbital excitations? The latter is represented by $H_{\rm int}$ (7.5), which in terms of spin ($\alpha_{\boldsymbol{k}}$) and orbital ($\beta_{\boldsymbol{k}}$) wave excitations is

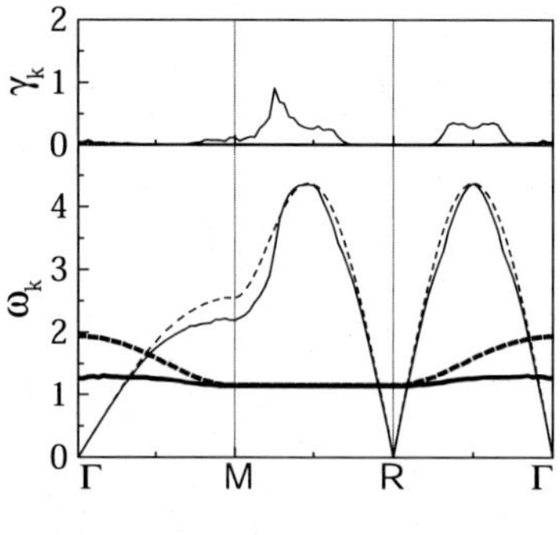

Fig. 7.2. Dispersion $\omega_{\boldsymbol{k}}$ and damping $\gamma_{\boldsymbol{k}}$ (in units of t^2/U) of the spin and orbital waves along the direction $\Gamma \to M(\pi,\pi,0) \to R(\pi,\pi,\pi) \to \Gamma$ in the Brillouin zone, calculated in the self-consistent mean-field approximation (dashed lines), and further renormalized by including interaction effects (solid lines). Thin (thick) lines correspond to the spin (orbital) excitation. For orbital waves $\gamma_{\boldsymbol{k}}$ is almost indistinguishable from the zero line. (After [8]).

$$H_{\rm int} = -V \sum_{\boldsymbol{kp}} \left\{ f_{\boldsymbol{k},\boldsymbol{p}} \alpha^\dagger_{\boldsymbol{k}} \alpha_{\boldsymbol{p}} + \frac{1}{2} g_{\boldsymbol{k},\boldsymbol{p}} (\alpha^\dagger_{\boldsymbol{k}} \alpha^\dagger_{-\boldsymbol{p}} + \alpha_{-\boldsymbol{k}} \alpha_{\boldsymbol{p}}) \right\} (\beta_{\boldsymbol{q}} + \beta^\dagger_{-\boldsymbol{q}}), \qquad (7.10)$$

where only the lowest order terms are kept. Here $V = \sqrt{3}J/8$, and $\boldsymbol{q} = \boldsymbol{k} - \boldsymbol{p}$. The matrix elements are

$$f_{\boldsymbol{k},\boldsymbol{p}} = [\eta_{\boldsymbol{q}}(u_{\boldsymbol{k}} u_{\boldsymbol{p}} + v_{\boldsymbol{k}} v_{\boldsymbol{p}}) + (\eta_{\boldsymbol{k}} + \eta_{\boldsymbol{p}})(u_{\boldsymbol{k}} v_{\boldsymbol{p}} + v_{\boldsymbol{k}} u_{\boldsymbol{p}})] \times (1 - 2\lambda_{2,\boldsymbol{k}})^{-1/4}, \qquad (7.11)$$

$$g_{\boldsymbol{k},\boldsymbol{p}} = [\eta_{\boldsymbol{q}}(u_{\boldsymbol{k}} v_{\boldsymbol{p}} + v_{\boldsymbol{k}} u_{\boldsymbol{p}}) + (\eta_{\boldsymbol{k}} + \eta_{\boldsymbol{p}})(u_{\boldsymbol{k}} u_{\boldsymbol{p}} + v_{\boldsymbol{k}} v_{\boldsymbol{p}})](1 - 2\lambda_{2,\boldsymbol{k}})^{-1/4},$$

with $\eta_{\boldsymbol{k}} = (\cos k_x - \cos k_y)/2$. Note that while the interaction involves two magnons, a single orbital wave can be created or absorbed with an amplitude which remains finite even at zero momentum. This reflects a special feature of the orbital pseudospin which is not a conserved quantity.

Physically, the interaction (7.10) accounts for processes where spin exchange is accompanied by the simultaneous orbital transition $|3z^2 - r^2\rangle \leftrightarrow |x^2 - y^2\rangle$, thus enhancing the $|x^2 - y^2\rangle$ orbital component in the ground state. Alternatively, $H_{\rm int}$ reflects the fluctuating nature of the operator $\hat{J}^{(\gamma)}_{ij}$ in (7.1): The spin exchange "constant" in an orbitally degenerate system is not a constant anymore, but a dynamical quantity. The first term in $H_{\rm int}$ accounts for the scattering of magnons by orbital waves. The creation of a pair of magnons, accompanied by an orbital wave excitation is also allowed (see the second term in (7.10)). Both processes renormalize magnon energies. In a second order perturbation theory, an excited magnon may couple to an intermediate state containing a magnon and orbital waves as shown in Fig. 7.3(a). In other words, a single spin flip couples to a composite excitation consisting of both spin and orbital fluctuations. The orbital excitation may decay into two magnons (Fig. 7.3(b)), resulting in a renormalization of orbital wave energy. Such second order corrections to the spin and orbital excitations and order parameters has been calculated in Ref. [8]. It was shown that the dynamical spin-orbit coupling results in the following: (i) It enhances quantum fluctuations in both subspaces thus reducing the staggered moment to $\langle S^z \rangle = 0.191$, and increasing the weight of the $|x^2 - y^2\rangle$ orbital component (which is about 6 percent). The latter effect is also reflected in a larger value of the ratio $J_\perp/J_c = 0.072$. The spin dynamics are quasi one dimensional: spin correlations along the chains $\langle \boldsymbol{S}_i \cdot \boldsymbol{S}_j \rangle_c = -0.42$, while $\langle \boldsymbol{S}_i \cdot \boldsymbol{S}_j \rangle_\perp = -0.10$ only. Such a strong anisotropy of the spin bonds is

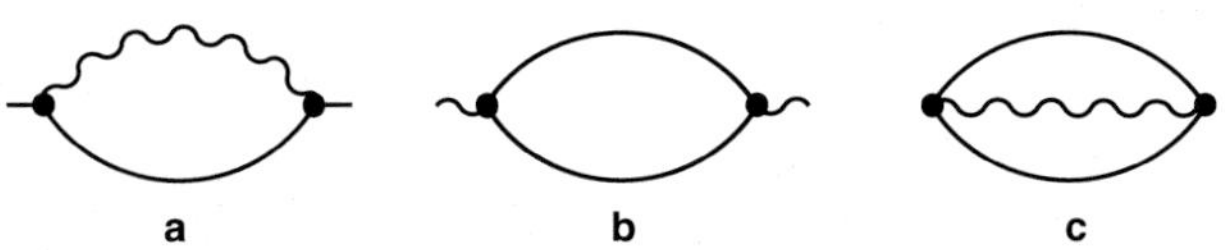

Fig. 7.3. Interaction corrections to the spin (**a**) and orbital (**b**) excitations, and to the ground state energy (**c**). Straight lines (wavy lines) represent spin (orbital) waves.

precisely what supports quasi 1D orbital order which, in its turn, supports low dimensionality of the spin sector. (*ii*) Spin and orbital excitations are both softened by interaction effects. This is more pronounced in particular momentum directions, see Fig. 7.2. The orbital gap still remains well defined. (*iii*) Spin waves get a finite damping. Orbital waves are almost undamped since the density of spin states inside the orbital gap is small. (*iv*) Joint spin-orbital fluctuations significantly lower the ground state energy. The latter is given by $E_0 = E_{SW} + E_{OW} + \langle H_{\text{int}} \rangle$, where the first two terms stand for the independent spin and orbital fluctuations, and the interaction correction calculated from the last diagram in Fig. 7.3, is

$$\langle H_{\text{int}} \rangle = -V^2 \sum_{\boldsymbol{kp}} \frac{g^2_{\boldsymbol{k},\boldsymbol{p}}}{\omega_{1,\boldsymbol{k}} + \omega_{1,\boldsymbol{p}} + \omega_{2,\boldsymbol{k}-\boldsymbol{p}}}. \tag{7.12}$$

This is nothing but a second order perturbation theory result for the last term in H_{int}. It accounts for the zero-point energy stemming from composite fluctuations involving simultaneously two magnons and an orbital wave. The ground state energy $E_0 = -0.17J$ per site is found, which is close to our estimation above, based upon physical considerations.

Consider $(x^2 - y^2)$-type ordering, one can show that it is unstable against fluctuations. It turns out that orbital excitations around this mean-field state are gapless at the Γ point, $\omega_{2,\boldsymbol{q}} \sim q$. In addition, the spin-orbit interaction vertex remains finite at $q = 0$, and orbital waves can therefore not be considered as Goldstone modes. As a consequence there are divergencies in perturbation theory indicating that the $(x^2 - y^2)$-type ordered state is not appropriate, as we already showed above from energy considerations.

We may conclude that the infinite degeneracy of the classical Néel state in the model (1) is removed by reducing the effective dimensionality of the spin system. Quantum spin fluctuations then generate an orbital excitation gap through a spin-orbit coupling mechanism. Orbital degeneracy in the model (1) should manifest itself as a strong reduction of the Néel temperature and by a soft quasi-1D spin structure.

In general the Kugel-Khomskii model is a very instructive example of how an apparently three dimensional system may by itself develop low dimensional quantum fluctuations. It is quite interesting and amusing that these quantum fluctuations do not only coexist with a weak three dimensional staggered spin moment, but that, by generating an orbital gap and intersite orbital correlations, they are actually of vital importance to the stabilization of this long-range order. The physical origin of this peculiar situation is, of course, the strong spatial anisotropy of the e_g-orbital wave functions. Because of this anisotropy it is impossible to optimize all the bonds simultaneously; this results in orbital frustration. The best the system can finally do, is to create some specific arrangement of strong and weak bonds which effectively reduces the dimensionality of the spin interactions. In this way the system can gain quantum energy. At the same time tunneling between different orbital configurations is suppressed: the spin fluctuations produce an energy gap for the rotation of orbitals.

7.1.2 Orbital-Only Model

Let us consider for the moment that the spins are frozen in a ferromagnetic configuration, and ask how the orbitals would behave in this case. The model (7.1) then reads as

$$H_{\mathrm{orb}} = A \sum_{\langle ij \rangle} T_i^{(\gamma)} T_j^{(\gamma)}, \tag{7.13}$$

where $T_i^{(\gamma)}$ is given by (7.2) and $A = J/2$. This limit has actually been considered in Ref. [10] as a model system to describe orbital dynamics in an undoped orbitally degenerate ferromagnet and for the ferromagnetic insulating phase of the underdoped manganites. In the latter case this mapping is not quite satisfactory since the ferromagnetic insulator is in fact stabilized by frozen-in doped holes that strongly affect the orbitals in their neighborhood, as will be discussed later. The model (7.13) is clearly anisotropic and in that respect underlines the important difference between orbital excitations and conventional spin dynamics. Another very closely related model is a so-called "cubic" model defined as follows:

$$H_{\mathrm{cub}} = A \left(\sum_{\langle ij \rangle_a} T_i^x T_j^x + \sum_{\langle ij \rangle_b} T_i^y T_j^y + \sum_{\langle ij \rangle_c} T_i^z T_j^z \right), \tag{7.14}$$

where along each a, b and c crystallographic axis only one of the respective pseudospin operators T^x, T^y and T^z is active. In this "cubic" model the bond anisotropy in Hamiltonian (7.13) is taken to the extreme. Although the model is not directly applicable to the manganites, it does contain the essential bond anisotropy and degeneracy that makes orbital models so different from conventional spin Hamiltonians in which interactions on different bonds are the same and act cooperatively, e.g. Heisenberg Hamiltonian. Already in the 1970's [11], this model has been proposed and discussed in the context of orbital frustrations. It is interesting to note that it is precisely this "cubic" model which represents magnetic anisotropy energy for the cubic titanates [12].

We discuss the two models in (7.13,7.14) in parallel as they display very similar peculiarities in the low-energy limit. Since the pseudospins in both models interact antiferromagnetically for all bonds, a staggered orbital ordered state is expected to be the ground state of the system. In the three-dimensional system at the orbital degeneracy point, however, linear spin-wave theory leads to a gapless two-dimensional excitation spectrum. This results in an apparent instability of the ordered state at any finite temperature [10], an outcome that sounds at least counter intuitive. Actually, the problem is even more severe: a close inspection shows that the interaction corrections to the orbiton excitations diverge even at zero temperature, making manifest the fact that the linear spin-wave expansion about a classical staggered orbital, Néel-like, state is not adequate in this case.

The origin of these problems was clarified in Refs. [12, 14, 15]. By symmetry, there are only a finite number of directions (three equivalent cubic axes), one of which will be selected by a staggered pseudospin order parameter. Since this breaks only a discrete symmetry, the excitations about the ordered state must have a gap. A linear spin wave theory fails however to give the gap, because (7.13,7.14) acquire a rotational symmetry in the limit of classical spins. This results in an infinite degeneracy of classical states, and an accidental pseudo Goldstone mode appears, which is however not a symmetry property of the original quantum models (7.13,7.14). This artificial gapless mode leads to low-energy divergences that arise because the coupling constant for the interaction between orbitons does not vanish in the zero momentum limit, as it would happen for a true Goldstone mode. Hence the interaction effects are non-perturbative.

At this point the order from disorder mechanism comes again into play: a particular classical state is selected so that the fluctuations about this state maximize the quantum energy gain, and a finite gap in the excitation spectra opens, this since in the ground state of the system the rotational invariance is broken. To explore this point explicitly, we calculate quantum corrections to the ground state energy as a function of the angle θ between c-axis and the staggered moment. Assuming the latter is perpendicular to b-axis, we rotate globally the pseudospin quantization axes as $T^z \to T^z \cos\theta - T^x \sin\theta$, $T^x \to T^x \cos\theta + T^z \sin\theta$, and perform then a conventional spin wave expansion $T_i^z = 1/2 - a_i^\dagger a_i$, $T_i^x \simeq a_i + a_i^\dagger$ about classical Néel state, where the staggered moment is oriented now along the new z direction. As the rotation of the quantization axes changes the form of the Hamiltonian, one observes that the magnon excitation spectrum has an explicit θ-dependence. Specifically, in case of the model (7.13) we obtain:

$$\omega_{\mathbf{p}}(\theta) = 2A\Big[1 - \gamma_{\mathbf{p}} - \frac{1}{\sqrt{3}}\eta_{\mathbf{p}} \sin 2\theta - \lambda_{\mathbf{p}}(1 - \cos 2\theta)\Big]^{1/2} , \tag{7.15}$$

Here, $\gamma_{\mathbf{p}} = (c_x + c_y)/2$, $\eta_{\mathbf{p}} = (c_x - c_y)/2$, $\lambda_{\mathbf{p}} = (2c_z - c_x - c_y)/6$, and $c_\alpha = \cos p_\alpha$. Calculating the zero point magnon energy $E(\theta) = -\sum_{\mathbf{p}}(A - \omega_{\mathbf{p}}(\theta)/2)$, one obtains an effective potential for the staggered moment, which at small θ is harmonic: $E(\theta) = const + \frac{1}{2}K_{\rm eff}\theta^2$, with an effective “spring” constant $K_{\rm eff} = A\kappa$. The parameter κ is given by

$$\kappa = \frac{1}{3}\sum_{\mathbf{p}}\Big[\frac{2\gamma_{\mathbf{p}}}{(1-\gamma_{\mathbf{p}})^{1/2}} - \frac{\eta_{\mathbf{p}}^2}{(1-\gamma_{\mathbf{p}})^{3/2}}\Big] \approx 0.117 . \tag{7.16}$$

The physical meaning of the above calculation is that zero point quantum fluctuations, generated by interactions between the spin waves, are enhanced when the staggered moment stays near a symmetric position (one of three cubic axes), and this leads to the formation of an energy profile of cubic symmetry. A breaking of this discrete symmetry results then in the magnon gap, which should be about $\sqrt{K_{\text{eff}}/M}$ in the harmonic approximation, where the "effective inverse mass" $1/M$ is order of value of the magnon bandwidth $W = 2\sqrt{2}A$. More quantitatively, the potential $E(\theta)$ can be associated with an effective uniaxial anisotropy term, $2K_{\text{eff}} \sum_{\langle ij \rangle_c} T_i^z T_j^z$, generated by spin-wave interactions in the symmetry broken phase. This low energy effective anisotropy leads to the magnon gap $\Delta = 2\sqrt{AK_{\text{eff}}} = 2A\sqrt{\kappa} \sim 0.7A$. The excitation gap is comparable to the full magnon bandwidth as $\Delta/W \simeq 0.24$. Almost the same gap/bandwidth ratio for the model (7.13) was obtained, using the equation of motion method, in [15].

The excitation spectrum of "cubic" model (7.14) can be analyzed in a similar way, either by calculating a quantum energy that depends on the order parameter direction, or using the equation of motion method [12]. Both methods lead to a large excitation gap of the order of A for this model. The gap is generated by interactions between the pseudospin excitations.

Although fluctuations may considerably reduce the order parameter in these models, long-range order still survives at finite temperatures; it is stabilized by the quantum physics which leads to the opening of a finite excitation gap. It should be noticed however, that while nature of the ground state and excitations gap are more of less determined for the models (7.13,7.14), further studies are required in order to fully characterize the excitation spectra. Of particular interest are damping effects, as we expect substantial incoherent features in the orbiton modes because of the strong interaction between low-energy ($\sim \Delta$) orbitons.

What can we learn from the above examples? The calculation of the excitation spectrum in systems with orbital degeneracy is somewhat involved even in the half filled, insulating limit. This is due to the peculiar frustration of superexchange interactions, which leads to infrared divergencies when linear spin wave theory is used [5, 7, 10]. That such divergencies occur in lowest order approximations is a universal feature of e_g-orbital models on cubic lattice –it reflects the special symmetry properties of e_g- orbital pseudospins. To calculate the excitation spectrum a more careful treatment of quantum effects is therefore required.

The main message is that orbital ordering, unlike spin ordering, is not accompanied by a simple Goldstone mode. Rather, collective orbiton excitations always have a finite gap and we also expect a substantial incoherent damping over all momentum space. This distinct feature of the orbitons has to be kept in mind when interpreting experimental data. Of course the precise way in which an orbiton gap is generated, depends on the model, but an order from disorder scenario seems to be a rather common mechanism by which to remove the frustration of classically degenerate orbital configurations.

7.1.3 Orbital-Charge Coupling, Orbital Polarons

We turn now to doping effects in manganites. A doped hole in a Mott insulator strongly interacts with a variety of low-energy degrees of freedom, leading to polaronic effects. In a pure spin-charge $t-J$ model, as are used for the cuprates, a hole breaks the spin bonds, and more importantly, frustrates spin order when it hops around. In manganites, the presence of orbital degeneracy introduces new degrees of freedom which are involved in the dynamics of holes. There are in general two channels of orbital-charge coupling, which we consider now following Ref. [16].

The first mechanism, which acts via the electron-transfer term, is similar to the spin-charge coupling mentioned above. This interaction of holes with the orbital degrees of freedom changes the character of the hole motion: the scattering of orbital excitations leads to a suppression of the coherent quasiparticle weight and to incoherent charge dynamics. We study this effect in a framework of the orbital t-J Hamiltonian [16, 17]. At low temperatures and intermediate doping levels, the double-exchange mechanism induces a parallel alignment of spins, leaving us with the orbital t-J Hamiltonian:

$$H_{tJ} = -\sum_{\langle ij\rangle}\left(t^{(\gamma)}_{\alpha\beta}\hat{c}^{\dagger}_{i\alpha}\hat{c}_{j\beta} + \text{h.c.}\right) + \frac{J}{2}\sum_{\langle ij\rangle} T^{(\gamma)}_i T^{(\gamma)}_j . \tag{7.17}$$

The nearest-neighbor spatial bond directions are denoted by $\gamma \in \{a, b, c\}$ as usual. We use constrained operators $\hat{c}^{\dagger}_{i\alpha} = c^{\dagger}_{i\alpha}(1-n_i)$ which create an e_g electron at site i in orbital α only under the condition that the site is empty. The first term in (7.17) describes the inter-site transfer of constrained e_g electrons. The transfer amplitude depends upon the orientation of orbitals at a given bond as is reflected by the transfer matrices

$$t^{(a/b)}_{\alpha\beta} = t\begin{pmatrix} 1/4 & \mp\sqrt{3}/4 \\ \mp\sqrt{3}/4 & 3/4 \end{pmatrix}, \quad t^{(c)}_{\alpha\beta} = t\begin{pmatrix} 1 & 0 \\ 0 & 0 \end{pmatrix}.$$

Chosen here is a representation with respect to the orbital basis $\alpha \in \{|3z^2-r^2\rangle, |x^2-y^2\rangle\}$. Due to the non-diagonal structure of $t^{(a/b)}_{\alpha\beta}$, the orbital quantum numbers are not conserved by Hamiltonian (7.17) – inter-site transfer processes induce fluctuations in the orbital sector. The second term in (7.17) accounts for processes involving the virtual occupation of sites by two e_g electrons. This superexchange mechanism leads to an inter-site coupling between orbital pseudospins of strength $J = 4t^2/U$, where U is the on-site repulsion between spin-parallel e_g electrons. The pseudospin operators $T^{(\gamma)}$ are defined in (7.2). Jahn-Teller phonons mediate an additional interaction between orbital pseudospins which is of the same form as the superexchange term. The numerical value of J has to be chosen such as to reflect both effects.

In the following we study the influence of orbitals on the motion of the doped holes employing an "orbital wave" approximation. Starting from the assumption that long-range orbital order has developed we use a slave-fermion representation of the electron operators in the transfer term of the Hamiltonian:

$$c^\dagger_{i\alpha} = b^\dagger_{i\alpha} f_i.$$

Within this picture, the orbital pseudospin is assigned to bosonic orbitons and charge to fermionic holons. The lattice is divided into two sublattices which are ascribed different preferred pseudospin directions. As an example, one can choose a C-type orbital order, that is

$$|\uparrow\rangle \equiv \left(|3z^2 - r^2\rangle + |x^2 - y^2\rangle\right)/\sqrt{2} \tag{7.18}$$

on sublattice A, and

$$|\downarrow\rangle \equiv \left(|3z^2 - r^2\rangle - |x^2 - y^2\rangle\right)/\sqrt{2} \tag{7.19}$$

on sublattice B within ab planes repeating itself along the c-direction; this closely resembles the type of order observed experimentally in $LaMnO_3$. In analogy to conventional spin-wave theory, excitations around this ground state can be treated by employing the following mapping of the orbiton operators $b_{i\alpha}$ onto "orbital-wave" operators β_i:

$$b_{i\uparrow} = \begin{cases} 1, & \text{sublattice A,} \\ \beta_i, & \text{sublattice B,} \end{cases} \qquad b_{i\downarrow} = \begin{cases} \beta_i, & \text{sublattice A,} \\ 1, & \text{sublattice B.} \end{cases}$$

In the momentum representation the transfer term in Hamiltonian (7.17) then becomes

$$H_t = \sum_{\boldsymbol{k}} \omega_{\boldsymbol{k}} f^\dagger_{\boldsymbol{k}} f_{\boldsymbol{k}} + \sum_{\boldsymbol{kp}} \left[\gamma_{\boldsymbol{k}} \beta^\dagger_{\boldsymbol{p}} + \gamma_{\boldsymbol{k}+\boldsymbol{p}} \beta_{-\boldsymbol{p}}\right] f^\dagger_{\boldsymbol{k}} f_{\boldsymbol{k}+\boldsymbol{p}}. \tag{7.20}$$

Here $\omega_{\boldsymbol{k}} = -t(c_x + c_y - 2c_z)/2$ and $\gamma_{\boldsymbol{k}} = t[(2-\sqrt{3})c_x + (2+\sqrt{3})c_y + 2c_z]/2$ with $c_\gamma = \cos k_\gamma$. The first term in (7.20) describes the coherent motion of holes within a band of width $W^{\rm coh}_{\rm hl} = 4t$. The second term describes the interaction of holes with excitations of the orbital background. The dynamics of the latter is controlled in general by superexchange and Jahn-Teller interactions. To simplify the calculations, we approximate orbital excitations by dispersionless, non-propagating orbitons of energy J:

$$H_J = J \sum_{\boldsymbol{k}} \beta^\dagger_{\boldsymbol{k}} \beta_{\boldsymbol{k}}. \tag{7.21}$$

This is actually quite realistic for manganites in which orbitons are expected to be gapped because of: (i) the quantum orbiton gap as discussed previously, and (ii) the Jahn-Teller coupling induced level splitting.

Scattering of orbital excitations leads to a suppression of the coherent quasiparticle weight and a simultaneous widening of the incoherent holon bandwidth. In analogy to studies of spin systems [46, 47], we analyze these effects by employing a self-consistent Born approximation for the self energy of holes – within this method all non-crossing diagrams of the self energy are summed up to infinite order while crossing diagrams are neglected. Restricting ourselves to the case of a single hole moving at the bottom of the band, we obtain the following expression for the holon self energy:

$$\Sigma(\mathrm{i}\omega) = t^2 \sum_{\boldsymbol{p}} \gamma_{\boldsymbol{p}}^2 \, G(\mathrm{i}\omega - J, \boldsymbol{p}). \tag{7.22}$$

The Matsubara frequencies are defined as $\mathrm{i}\omega = \mathrm{i}(2n+1)\pi T$, where T denotes temperature and n an integer number.

Our first aim is to study the loss of coherency in the hole motion. Employed is a dominant-pole approximation [46]: We split the holon propagator in (7.22) into its coherent and incoherent parts,

$$G(\mathrm{i}\omega, \boldsymbol{k}) = \frac{a_{\boldsymbol{k}}}{\mathrm{i}\omega - \tilde{\omega}_{\boldsymbol{k}}} + G^{\mathrm{inc}}(\mathrm{i}\omega, \boldsymbol{k}), \tag{7.23}$$

where $a_{\boldsymbol{k}}$ denotes the quasiparticle weight and $\tilde{\omega}_{\boldsymbol{k}}$ the not-yet-known renormalized holon dispersion. Keeping only the coherent part and using $a_{\boldsymbol{k}} = [1 - (\partial/\partial\omega)\Sigma'(\omega)]^{-1}$ with $\Sigma'(\omega) = \mathrm{Re}[\Sigma(\mathrm{i}\omega \to \omega + \mathrm{i}0^+)]$, the following recursion relation for the quasiparticle weight is obtained:

$$a_{\bar{\boldsymbol{q}}} = \left[1 + t^2 \sum_{\boldsymbol{p}} \gamma_{\boldsymbol{p}}^2 \frac{a_{\boldsymbol{p}}}{(\tilde{\omega}_{\bar{\boldsymbol{q}}} - \tilde{\omega}_{\boldsymbol{p}} - J)^2}\right]^{-1}, \tag{7.24}$$

with $\bar{\boldsymbol{q}} = (0, 0, \pi)$ at the bottom of the band. Equation (7.24) can be approximately solved by expanding the integrand around $\bar{\boldsymbol{q}}$, which yields

$$a_{\bar{q}} = \begin{cases} 1 - \dfrac{1}{\sqrt{2\pi^2}} \left(\dfrac{t}{J}\right)^{1/2} & \text{for} \quad J \gg t, \\ \sqrt[4]{2\pi^2} \left(\dfrac{J}{t}\right)^{1/4} & \text{for} \quad J \ll t. \end{cases} \tag{7.25}$$

In the limit $J/t \to \infty$ a coherent hole motion with $a_{\bar{q}} = 1$ is recovered. In contrast to the spin $t - J$ problem in cuprates, the hole mobility in this limit is still possible due to the presence a small but finite non-diagonal hopping matrix elements which do not conserve orbital pseudospin. Therefore the "string" effect [18] that occurs in the spin model, is less severe in the orbital t-J model. This peculiar property of a hole motion in orbital systems has been emphasized by several authors, see Refs. [16, 19, 20]. In the opposite limit $J/t \to 0$, however, the holon quasiparticle weight is completely lost, which indicates a strong scattering of holes by orbital fluctuations, and the excitation spectrum of a doped hole shows only a broad continuum in a momentum space. On general grounds, it is also expected that mobile holes will fill in the orbiton gap, and will eventually destroy orbital order in a similar way as they melt the spin order in cuprates.

Next we turn to a study of the renormalization of the total holon bandwidth, accounting for incoherent processes as well. Inserting $G(\mathrm{i}\omega, \boldsymbol{k}) =$

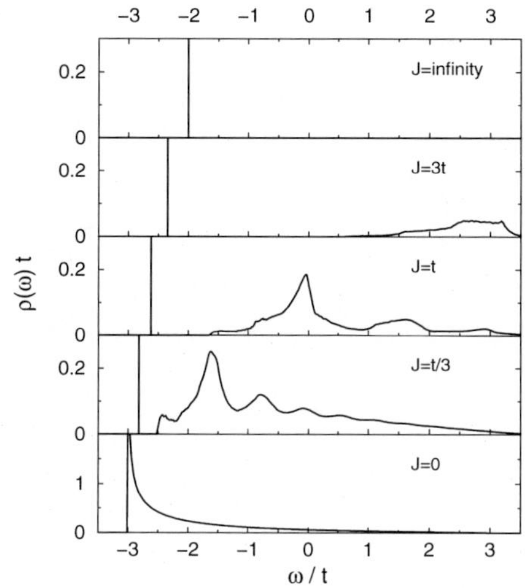

Fig. 7.4. Spectral function of a hole moving at the bottom of the band. Different values of J/t are used: The spectrum is completely incoherent for $J = 0$, having a lower bound at $\omega_{\min} = -3t$. With increasing values of J/t, the spectral weight is shifted from the incoherent part of the spectrum to a coherent quasiparticle peak (denoted by a vertical line). In the limit $J/t \to \infty$, the quasiparticle peak is at $\omega_{\mathrm{QP}} = -2t$ and has accumulated all spectral weight. (After [16]).

$[\mathrm{i}\omega - \omega_{\boldsymbol{k}} - \Sigma(\mathrm{i}\omega)]^{-1}$ into (7.22) leads to a recursion relation for the holon self energy:

$$\Sigma(\mathrm{i}\omega) = t^2 \sum_{\boldsymbol{p}} \frac{\gamma_{\boldsymbol{p}}^2}{\mathrm{i}\omega - \omega_{\boldsymbol{p}} - J - \alpha\Sigma(\mathrm{i}\omega - J)} . \tag{7.26}$$

The factor $\alpha = (z-1)/z$ partially accounts for the constraint that forbids more than one orbital excitation per site – the hole may therefore not return to a previously visited site unless it is reabsorbed by an excitation. Equation (7.26) has been numerically solved in [16] in order to determine the spectral function $\rho_{\bar{\boldsymbol{q}}}(\omega) = -\frac{1}{\pi}\mathrm{Im}[\mathrm{G}(\mathrm{i}\omega \to \omega + \mathrm{i}0^+, \bar{\boldsymbol{q}})]$ at the bottom of the band. The result is shown in Fig. 7.4.

Different values of J/t are used. In the limit $J/t \to 0$, the spectrum is completely incoherent and extends down to $\omega_{\min} = -3t$ corresponding to a holon bandwidth of $W_{\mathrm{hl}} = 2|\omega_{\min}| = 6t$. In this limit the hole creates its own disorder and effectively moves within an orbital-liquid state characterized by strong orbital fluctuations. In the opposite case $J/t \to \infty$, all spectral weight accumulates in a quasiparticle peak (denoted by a vertical line) at $\omega_{\mathrm{QP}} = -2t$ which corresponds to a bandwidth of $W_{\mathrm{hl}} = 2|\omega_{\mathrm{QP}}| = 4t$. The orbital state is here static and excitations are completely suppressed. At finite values of J/t, the total spectral weight is divided between the coherent and incoherent parts. The latter is separated from the quasiparticle peak by the orbital excitation energy J. Processes in which the hole creates more than one orbital excitation are reflected by a succession of peaks in the incoherent spectrum. For $J = t/3$ which is realistic for the manganites, the quasiparticle peak accounts for $\approx 65\%$ of the spectral weight and the width of the holon band is

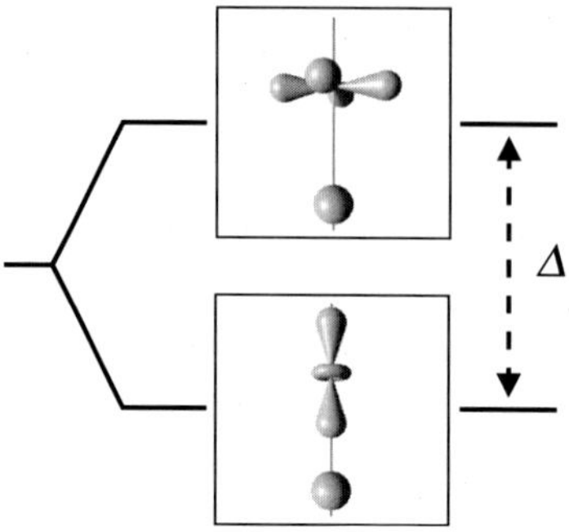

Fig. 7.5. Polarization of e_g-levels on sites next to a hole: Bond stretching phonons and Coulomb interactions induce an energy splitting $\Delta = \Delta^{\rm ph} + \Delta^{\rm ch}$. Here the sphere indicates the location of a hole (Mn^{4+}-ion) adjacent to a Mn^{3+}-ion. (After [16]).

$$W_{\rm hl} \approx 5.7t. \tag{7.27}$$

Comparing this with $W_{\rm hl} = 6t$ for the orbital-liquid state, we find a reduction of about 5%. It is therefore concluded that a disorder-order crossover in the orbital sector has only a secondary effect on the kinetic energy of charge carriers.

It turns out, however, that the spatial anisotropy of e_g-orbitals provides yet another, very important channel of orbital-charge coupling that is specific to the case of manganites. Notice that, unlike the cuprates, holes remain localized up to rather high doping levels, and at the same time induce an isotropic ferromagnetic state. The complete breakdown of metalicity at hole concentrations below $x_{\rm crit} \approx 0.15$ - 0.2 occurs despite the fact that ferromagnetism is fully sustained, and is sometimes even stronger, in this regime [21,22]. This seems very surprising when compared with the standard double-exchange picture. To explain this puzzle, the concept of orbital polarons was introduced in Ref. [16].

The important point is that in an orbitally degenerate Mott-Hubbard system there also exists a specific coupling between holes and orbitals that stems from the polarization of e_g-orbitals in the neighborhood of a hole. In this way the cubic symmetry at manganese ions close to the hole is lifted, see Fig. 7.5. The displacement of oxygen ions, and also the Coulomb force exerted by the positively charged hole, and hybridization of electrons, causes a splitting of orbital levels on sites adjacent to a hole. This splitting is comparable in magnitude to the kinetic energy of holes so that the orbital-hole binding energy can be large enough for holes and surrounding orbitals to form a bound state. A strong reduction of the bandwidth comes from this mechanism as orbital-hole bound states begin to form. We discuss now the two most important mechanisms leading to such a level splitting: (1) the displacement of oxygen ions that move towards the empty site; and (2) the splitting of e_g states induced by the Coulomb force between a positive hole and negative electrons. The magnitude of the sum $\Delta = \Delta^{\rm ph} + \Delta^{\rm ch}$ is estimated as follows: The former phonon contribution $\Delta^{\rm ph}$ originates from the coupling of holes to the lattice breathing mode Q_1 and of e_g electrons to two Jahn-Teller modes

Q_2 and Q_3:

$$H_{\mathrm{el-ph}} = -\sum_i \left(g_1 Q_{1i} n_i^h + g_2 Q_{2i} \sigma_i^x + g_3 Q_{3i} \sigma_i^z + \frac{K}{2} \boldsymbol{Q}_i^2 \right), \qquad (7.28)$$

where n_i^h denotes the number operator for holes and the Pauli matrices $\sigma_i^{x/z}$ act on the orbital subspace. The coupling constants are g_1 and $g_2 \approx g_3$ and K is the lattice spring constant. Hamiltonian (7.28) mediates an interaction between empty and occupied sites. The effective Hamiltonian describing this coupling is obtained by integrating (7.28) over oxygen displacements $\boldsymbol{Q}_i = (Q_{1i}, Q_{2i}, Q_{3i})$. For a given bond along the z direction this yields $H^z = -\frac{1}{2}\Delta^{\mathrm{ph}}\, n_i^h \sigma_j^z$ with

$$\Delta^{\mathrm{ph}} = g_1 g_2 \sqrt{2}/(3K) \approx (g_1/g_2) E_{\mathrm{JT}}. \qquad (7.29)$$

A lower bound for this quantity is given by the Jahn-Teller energy, i.e., $\Delta^{\mathrm{ph}} \geq E_{\mathrm{JT}} \approx 0.2$ eV, assuming that the coupling to the breathing mode is at least as strong as coupling to the Jahn-Teller modes. Next we estimate the contribution to the e_g-level splitting that follows from the Coulomb interaction between a positively charged hole and an e_g electron on a neighboring site. The magnitude of Δ^{ch} is assessed by taking into account the covalence of Mn $3d$ and O $2p$ orbitals, which gives

$$\Delta^{\mathrm{ch}} \approx \frac{3}{4} \gamma^2 e^2 / R_{\mathrm{Mn-Mn}}. \qquad (7.30)$$

The covalency factor $\gamma = t_{pd}/\Delta_{pd}$ can be obtained from the transfer amplitude and the charge gap between Mn and O sites, $t_{pd} \approx 1.8$ eV and $\Delta_{pd} \approx 4.5$ eV. [23] Together with a lattice spacing of $R_{\mathrm{Mn-Mn}} = 3.9$Å this leads to $\Delta^{\mathrm{ch}} \approx 0.4$ eV. In total the polarization of e_g levels on sites next to a hole yields an energy splitting $\Delta \approx 0.6$ eV. This is comparable in magnitude to the transfer amplitude t and indicates that the direct coupling of charge and orbital degrees of freedom are of importance in manganites.

For a given bond along the z direction the interaction reads as $H^z = -\frac{1}{2}\Delta\, n_i^h \sigma_j^z$. This is precisely the new orbital-charge coupling channel which is of course absent in pure spin-charge models. The splitting of e_g-levels effects all six sites surrounding a hole, and the analogous expressions for the x and y directions can easily be derived by a rotation of the interaction Hamiltonian in the orbital space. The complete orbital-charge coupling Hamiltonian for the cubic system is then

$$H_{\mathrm{ch-orb}} = -\Delta \sum_{\langle ij \rangle} n_i^h T_j^{(\gamma)}, \qquad (7.31)$$

with orbital pseudospin operators given by (7.2). The interactions in Hamiltonian (7.31) describe the tendency of the system to form orbital polarons. For low enough hole concentrations the polaron consists of a bound state between a central hole with surrounding e_g-orbitals pointing towards the hole as is shown in Fig. 7.6.

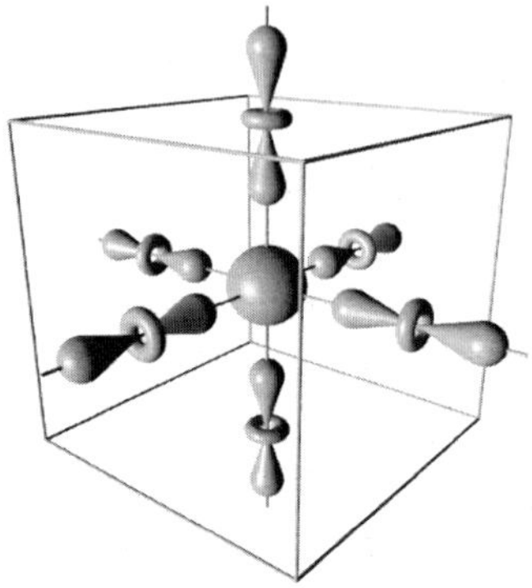

Fig. 7.6. Orbital polaron in the strong-coupling limit: Six e_g- states point towards a central hole. (After [16]).

In conventional lattice-polaron theory, the binding energy is a function of the coupling constant g and the stiffness of the lattice which is controlled by the spring constant K: The energy gain stemming from the interaction between charge carriers and the lattice competes against the deformation energy of the crystal. In the case of orbital polarons, the underlying picture is very similar. Here the coupling constant is given by the orbital-charge interaction energy Δ, while the energy scale J is a measure of the "stiffness" of the orbital sector. These two quantities are expected to determine the binding energy of the orbital polaron.

In order to estimate the polaron binding energy, consider a static hole placed in an orbital ordered state, and calculate the reduction of the total energy due to the interaction Hamiltonian (7.31). Orbitals on sites adjacent to the hole will cant towards directions shown in Fig. 7.6, by angles that are determined by competing effects of J and orbital-change Δ coupling. Accounting for contributions from the polarization of nearest neighbor orbitals surrounding a hole, the following expression for the polaron binding energy can be obtained (see for details [16]):

$$E_b^{\mathrm{orb}} = 3\left[\sqrt{\Delta^2 + J^2} - J\right] . \tag{7.32}$$

In the limit $J/\Delta \to \infty$, the orbital state is very stiff and cannot be polarized by the hole; the binding energy then vanishes as $3\Delta^2/(2J)$. On the other hand, the polarization is complete in the limit $J/\Delta \to 0$, yielding a maximum value $E_b^{\mathrm{orb}} = 3\Delta$ for the binding energy.

We note that the functional form of (7.32) differs from the conventional lattice-polaron case where $E_b^{\mathrm{ph}} = g^2/(2K)$. This is due to the fact that there exists an upper limit for the orbital polarization in which the orbitals around a hole have been fully reoriented to point towards the empty site (see Fig. 7.6); technically the existence of this upper bound is reflected by the hard-core nature of the orbital operators in (7.2). The familiar form for the binding energy $\propto \Delta^2/J$ is recovered only for the weak-coupling limit in which orbital distortions around the hole are small. Here it can be considered that J is the counterpart of the lattice stiffness K in conventional polaron theory.

The structure of the orbital polaron adds a large amplitude of virtual excursions of e_g-electrons into the empty site. Thus, besides minimizing the interaction energy of Hamiltonian (7.31), the polaron also allows for a lower kinetic energy. We note that these virtual hopping processes locally enhance the magnetic moments of core and e_g-spins via the double-exchange mechanism in all three directions and gives a large effective spin to the orbital polaron. This naturally explains the development of ferromagnetic clusters experimentally observed at temperatures above T_C at low doping levels [24]. At finite hole densities these clusters start to interact, thereby inducing a global ferromagnetic state. Clearly, the orbital state of such a ferromagnetic insulator must be very complex due to the presence of frozen-in orbital polarons. The relevant model must contain the interactions given by both (7.31) and (7.13) as a first step, and further be complemented by, at least, the Coulomb forces between holes. It was suggested [16] that such a state may have orbital/Jahn-Teller glass features, which reduces the long-range component of static Jahn-Teller distortions. The orbital excitations are expected therefore to have a large broadening in momentum space. These issues clearly deserve more theoretical work, particularly with the perspective of future experiments aim at the observation of orbital excitations in doped manganites.

The formation of orbital polarons competes with kinetic energy of holes that tends to delocalize the charge carriers, but also competes with the fluctuation rate $\propto xt$ of orbitals: the faster the orbitals fluctuate, the less favorable it is to form a bound state in which orbitals have to give up part of their fluctuation energy. In other words, the binding energy of the polaron decreases at higher doping levels. When this orbital polaron picture is combined with that of conventional lattice polarons [25], the transition from ferromagnetic metal to ferromagnetic insulator can be well explained [16].

Once the orbital polaron is formed it will be trapped by even a weak disorder. Near commensurate fillings, say about 1/8, they may form a polaron superlattice, optimizing simultaneously the charge and orbital configuration energies, as discussed in [26]. Commensuration through phase separation – which is certainly relevant in manganites – is also possible, but we will not discuss this complex issue here.

7.1.4 Orbital Liquids, Anomalous Transport

An almost universal feature of manganites is that at higher doping concentrations (around $x \approx 1/3$) a ferromagnetic metallic state emerges. In this doping regime there is a strong competition between the tendency of orbitals to order locally and the kinetic energy of the charge carriers that tends to destroy long range orbital order. This is comparable to the situation in high-T_c superconductors, where the long-range antiferromagnetic order of, in this case, spins is frustrated by mobile charge carriers. In an analogous way the mobile holes in optimally doped manganites can melt the orbital order: in the ferromagnetic metallic phase of the manganites the orbitals are "rotating" and may form resonating orbital bonds. In other words, orbitals are disordered due to quantum fluctuations, forming an orbital liquid, similar in spirit to the RVB state of the spin systems discussed in Chap. 3.

One possible manifestation of the orbital liquid state in manganites is observed anomalies in optical spectra as we discuss now. The appearance of a ferromagnetic metallic state is explained in a framework of double exchange physics. But if one takes the double exchange model as the starting point to explain the properties of the metallic state, it is quite surprising that in experiment one finds that the e_g-electrons do not behave as conventional spin-polarized carriers in a uniform ferromagnetic phase at all. For such carriers one expects that optical spectral weight is mostly accumulated into a low-frequency Drude peak, but the weight of optical spectra in manganites robustly extends up to $\sim$ 1eV, even at very low temperatures [27].

This experimental finding is remarkable for at least two reasons. First, the energy scale extends to the electron-volt range, which rules out a purely phononic origin of the incoherence; and second, the incoherent spectral weight is very large in magnitude, even at low temperatures.

Other experimental studies show that collective as well as local lattice distortions are absent in metallic manganites at low temperatures, which may be an indication that orbital fluctuations are strong. Based upon this observation, several authors [28–30] attributed the incoherent structure of the optical conductivity to the orbital degrees of freedom. While the study [29] is based on a simple band picture, a more elaborate treatment of both orbital degeneracy and on-site correlations was suggested in Refs. [28] and [30], employing slave-fermion and slave-boson methods, respectively. In this works the orbital liquid, a quantum disordered state of e_g-quadrupole moments in metallic manganites, was discussed. Such quantum disorder of orbitals is caused by the motion of holes which mixes up dynamically all possible orbital configurations.

At first glance this idea seems to be very similar to the concept of the moving holes that cause spin disorder in cuprates. There are two differences, however. The first one is that in a ferromagnetic system the orbital superexchange interactions are rather frustrated from the very beginning and this actually favors the orbital liquid, as we discussed previously. The effective two-dimensionality of the pseudospin fluctuations is, in fact, emphasized in [28] as an important disorder mechanism. While indeed enhancing quantum effects, this effect alone would still not be sufficient to destroy the pseudospin order, as one can see from the behavior of orbital models discussed above. Rather, it acts cooperatively with the frustrations induced by charge motion. The second, and most important, difference from cuprates is the strong tendency for holes in manganites to localize and form orbital/lattice polarons. It is the restricted charge mobility that confines the orbital liquid metallic state of manganites to a rather small region of the phase diagram.

The optical conductivity in the metallic phase of manganites has been calculated in [30], taking the idea of an orbital liquid as a starting point. The strongly correlated nature of the e_g-electrons in the metallic system can conveniently be accounted for by employing a slave-boson representation of electron operators:

$$c_{i\alpha}^\dagger = f_{i\alpha}^\dagger b_i.$$

Here the orbital pseudospin is carried by fermionic orbitons $f_{i\alpha}$, and charge by bosonic holons b_i. In comparison with slave-fermion treatment of the orbital t-J model, this representation is most adapted to describe the correlated Fermi-liquid state of manganites since it naturally captures both the coherent and incoherent features of excitations. On the other hand, short-range orbital correlations are not fully captured in this representation. This way, slave-boson and slave-fermion treatments are complementary to each other.

Introducing further mean-field parameters $\chi = t^{-1}\sum_{\alpha\beta} t_{\alpha\beta}^{(\gamma)}\langle f_{i\alpha}^\dagger f_{i\beta}\rangle$ and $x = \langle b_i^\dagger b_j\rangle$, where x is the concentration of holes and $\chi \approx \frac{1}{2}$, the two types of quasiparticles in the Hamiltonian (7.17) can be decoupled:

$$H_{\text{orb}} = -\left(x + \frac{\chi J}{2t}\right)\sum_{\langle ij\rangle} t_{\alpha\beta}^{(\gamma)}\left(f_{i\alpha}^\dagger f_{j\beta} + \text{h.c.}\right), \tag{7.33}$$

$$H_{\text{hl}} = -\chi t\sum_{\langle ij\rangle}\left(b_i^\dagger b_j + \text{h.c.}\right). \tag{7.34}$$

Diagonalizing the above expressions in the momentum representation one obtains

$$H_{\text{orb}} = \sum_{\boldsymbol{k}\nu} \xi_{\boldsymbol{k}}^\nu f_{\boldsymbol{k}\nu}^\dagger f_{\boldsymbol{k}\nu}, \quad H_{\text{hl}} = \sum_{\boldsymbol{k}} \omega_{\boldsymbol{k}} b_{\boldsymbol{k}}^\dagger b_{\boldsymbol{k}},$$

with index $\nu = \pm$ and dispersion functions

$$\xi_{\boldsymbol{k}}^\pm = (xt + \frac{1}{2}J\chi)\Big[-\epsilon_0(\boldsymbol{k}) \pm \sqrt{\epsilon_1^2(\boldsymbol{k}) + \epsilon_2^2(\boldsymbol{k})}\Big],$$

$$\omega_{\boldsymbol{k}} = 6\chi t\Big[1 - \frac{1}{3}\epsilon_0(\boldsymbol{k})\Big],$$

where $\epsilon_0(\boldsymbol{k}) = c_x + c_y + c_z$, $\epsilon_1(\boldsymbol{k}) = (c_x + c_y)/2 - c_z$, $\epsilon_2(\boldsymbol{k}) = \sqrt{3}(c_x - c_y)/2$ with $c_\gamma = \cos k_\gamma$. The essence of this slave-particle mean-field treatment is that orbital and charge fluctuations are assigned different energy scales. This is reflected by the bandwidths of orbiton and holon quasiparticles, respectively:

$$W_{\text{orb}} = 6(xt + \frac{1}{2}J\chi), \qquad W_{\text{hl}} = 12t\chi. \tag{7.35}$$

The former quantity W_{orb} sets the energy scale of orbital fluctuations – the terms proportional to xt and J describe fluctuations induced by the motion of holes and by the coupling between pseudospins, respectively. The latter quantity $W_{\rm hl}$ finally defines the itineracy of holes in the orbital-liquid state.

The calculations of optical conductivity are somewhat involved and include several contributions stemming from: (i) Optical transitions within the coherent orbiton bands $\xi_{\boldsymbol{k}}^{\pm}$ resulting in a small coherent Drude peak, (ii) Weak interband transitions between the orbiton bands contributing at finite energy, and (iii) Highly incoherent optical transitions in which several elementary (orbiton and holon) particles do participate (see for details [30]). Latter contribution accounts for the composite nature of charge carriers in correlated systems reflected by the orbiton-holon convolution, and leads to finite optical conductivity at large frequencies. The small Drude peak as well as a broad optical absorption spectrum, that extends up to the bare bandwidth, are well reproduced by these calculations. The physical picture is that the charge carriers scatter strongly on the dynamical disorder caused by the fluctuations of orbital bonds that are due to the correlated rotation of orbitals. The fact that the anomalous transport properties in the ferromagnetic metallic phase can be described consistently, supports the validity of the orbital-liquid framework. It is worth also to mention that the same idea of fluctuating orbital bonds has successfully been used to explain anomalous magnon dispersions in ferromagnetic manganites [31]. An alternative description of the orbital liquid using slave-fermion formalism is given in Sect. 4.1.5.

7.2 Systems with t_{2g} Orbital Degeneracy

Systems with t_{2g} orbitals may exhibit different and even more interesting phenomena because of their larger degeneracy. Indeed, experimental developments indicate the limitations of the standard Goodenough-Kanamori picture of static orbitals in t_{2g} electron systems. It has been argued that orbital quantum fluctuations might sometimes be of crucial importance, and a quantum version of the orbital physics is needed. New concepts such as the three dimensional orbital liquid in $LaTiO_3$ [32,33], one dimensional orbital chains showing Heisenberg spin chain like dynamics in cubic vanadates [34] have been discussed. It is not accidental that these ideas emerged from a study of titanates and vanadates having t_{2g}^1 and t_{2g}^2 electronic configurations, respectively. This is because of the large, threefold degeneracy of t_{2g} orbitals, which enhances quantum effects. Another crucial point is that the JT coupling is relatively weak for t_{2g} systems. Indeed, distortion of octahedra for the titanates and vanadates is much smaller when compared with typical JT distortions in manganites with e_g orbitals. Since the JT related orbital splittings scale as square of the distortions, one may think that t_{2g} orbital states would be much less affected by the electron-lattice coupling, and that the intrinsic dynamics of coupled spin-orbital system are, as a first approximation, governed by electronic superexchange interactions. Effects of lattice distortions (which are always present) can then be accounted for in a next step.

In this section we discuss the orbital ground state and the orbital quantum dynamics of the cubic titanates $ReTiO_3$ (Re=La,...,Y), in which magnetoactive 3d electron of Ti^{3+} ion occupies the threefold degenerate crystal field level of t_{2g} symmetry. This class of Mott insulators are very interesting: titanates show a nearly continuous transition from an antiferromagnetic state in $LaTiO_3$ with an unusually small staggered moment to a saturated ferromagnet in $YTiO_3$ [35]. Even more surprisingly, spin wave data shows that both antiferromagnetic (AF) and ferromagnetic (F) titanates are highly isotropic having the same exchange couplings in all cubic directions [32, 36]. This is in sharp contrast with what might be expected on the basis of conventional picture of orbital ordering which commonly results in a strong spatial anisotropy of the spin exchange bonds, and leads one to anticipate unusual orbital states.

7.2.1 Spin-Orbital Model

We begin with a general discussion of the t_{2g} -superexchange interaction. In terms of fermions $a_{i,\sigma}$, $b_{i,\sigma}$, $c_{i,\sigma}$ corresponding to t_{2g} levels of yz, zx, xy symmetry, respectively, the hopping term in the Hubbard is

$$-t\left(a^{\dagger}_{i,\sigma}a_{j,\sigma} + b^{\dagger}_{i,\sigma}b_{j,\sigma} + \text{h.c.}\right) \tag{7.36}$$

for the bonds along c-direction in a cubic crystal. The motivation for a,b,c nomenclature for orbital flavor index is obvious: each t_{2g} orbital is orthogonal to one of the cubic axes, and cannot move in that particular direction. Therefore, only two orbitals relevant for each direction as is easily seen from the hopping Hamiltonian (see also Fig. 7.7). On a- and b-axis bonds, (b, c) and (c, a) orbital pairs have to be used, respectively, in the derivation of superexchange Hamiltonian. Correlation energies of the doubly occupied d^2 virtual states of Ti-ion depend on the Coulomb repulsion parameters U (for electrons on the same orbital), U' (for electrons on different orbitals), and on the Hund's exchange constant J_H. One usually assumes the relation $U^{'} = U - 2J_H$ which holds in the atomic limit. With this assumption, the following energetics of d^2 states can be obtained [37]: $U - 3J_H$, if d^2 intermediate configuration has the high spin (triplet) state; $U + 2J_H$ for low spin (singlet) state if two electrons occupy the symmetric, $(a_i a_i + b_i b_i + c_i c_i)/\sqrt{3}$ orbital wave-function, whereas all other low spin states have correlation energy $U - J_H$.

In order to obtain the superexchange interaction, one eliminates the doubly occupied virtual states perturbatively. For t_{2g} systems, this has been done by Kugel and Khomskii [38] neglecting the Hund's coupling (J_H) splitting of the excited multiplet. These authors analyzed the mean-field solutions and pointed out the high degeneracy of the classical ground states (see also Chap. 5). This emphasizes importance of quantum effects which are addressed below.

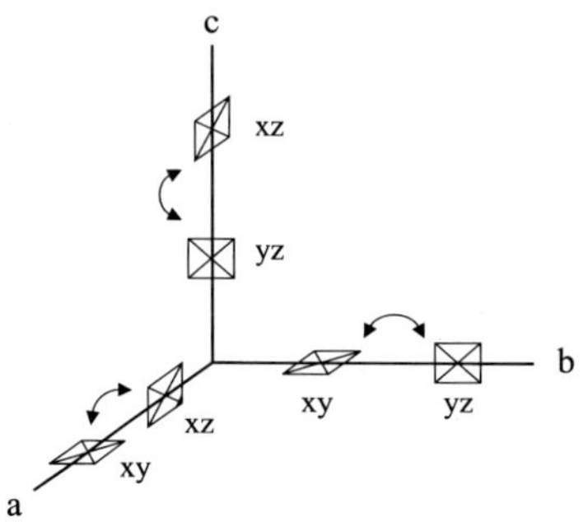

Fig. 7.7. On every bond of a cubic crystal, two out of three t_{2g} orbitals are equally involved in the superexchange and may resonate. The same two orbitals select also a particular component of angular momentum. (After [12]).

The exchange interaction taking into account Hund's coupling effects in the intermediate states is derived in Refs. [12, 13], in a form which is similar to the case of e_g orbital superexchange, see (7.1):

$$H = \sum_{\langle ij \rangle} \left[\left(\boldsymbol{S}_i \cdot \boldsymbol{S}_j + \frac{1}{4} \right) \hat{J}_{ij}^{(\gamma)} + \frac{1}{2} \hat{K}_{ij}^{(\gamma)} \right] , \tag{7.37}$$

where the orbital operators $\hat{J}_{ij}^{(\gamma)}$ and $\hat{K}_{ij}^{(\gamma)}$ depend on bond directions $\gamma (= a, b, c)$. They are given by the following expressions:

$$\hat{J}_{ij}^{(\gamma)} = J \left[\frac{r_1 + r_2}{2} A_{ij}^{(\gamma)} - \frac{r_2 - r_3}{3} B_{ij}^{(\gamma)} - \frac{r_1 - r_2}{4} (n_i + n_j)^{(\gamma)} \right], \tag{7.38}$$

$$\hat{K}_{ij}^{(\gamma)} = J \left[\frac{r_1 - r_2}{2} A_{ij}^{(\gamma)} + \frac{r_2 - r_3}{3} B_{ij}^{(\gamma)} - \frac{r_1 + r_2}{4} (n_i + n_j)^{(\gamma)} \right], \tag{7.39}$$

where $J = 4t^2/U$, as usual. The coefficients $r_1 = 1/(1-3\eta)$, $r_2 = 1/(1-\eta)$ and $r_3 = 1/(1+2\eta)$ follow from the Hund's splitting of the excited t_{2g}^2 multiplet via $\eta = J_H/U$, as discussed above. The operators $A_{ij}^{(\gamma)}$, $B_{ij}^{(\gamma)}$ and $n_i^{(\gamma)}$ can conveniently be written in terms of constrained particles a_i, b_i, c_i (orbitons with $n_{ia} + n_{ib} + n_{ic} = 1$). Namely,

$$A_{ij}^{(c)} = n_{ia} n_{ja} + n_{ib} n_{jb} + a_i^\dagger b_i b_j^\dagger a_j + b_i^\dagger a_i a_j^\dagger b_j, \tag{7.40}$$

$$B_{ij}^{(c)} = n_{ia} n_{ja} + n_{ib} n_{jb} + a_i^\dagger b_i a_j^\dagger b_j + b_i^\dagger a_i b_j^\dagger a_j, \tag{7.41}$$

$$n_i^{(c)} = n_{ia} + n_{ib} \tag{7.42}$$

for the pair along the c axis. Similar expressions are obtained for the exchange bonds along the a- and b-axis by replacing orbitons (a, b) in (7.40)-(7.42) by (b, c) and (c, a) pairs, respectively. It should be noticed that a_i, b_i, c_i particles have hereafter only an orbital quantum number, since a spin component of the original d-electron is represented, in (7.37), by spin one-half operators, as usual. Such a separation of the orbital and spin sectors is a general property of Mott insulators, and it is taken to the extreme in a classical Goodenough-Kanamori picture which treats the mutual coupling of two sectors as a static interplay between spin and orbital order parameters. However, spin-orbital separation does not occur in the one-dimensional case as we will discuss soon; the idea of spin-orbital confinement at intermediate and high energy scales is also crucial to understand anomalous properties of titanates.

Another representation of the orbital exchange operators is via the angular momentum operators of the t_{2g} level [38, 39], using the following relations:

$$l_x = \mathrm{i}(c^\dagger b - b^\dagger c), l_y = \mathrm{i}(a^\dagger c - c^\dagger a), l_z = \mathrm{i}(b^\dagger a - a^\dagger b). \tag{7.43}$$

The orbital operators, e.g.,

$$A_{ij}^{(c)} = [(1 - l_x^2)_i(1 - l_x^2)_j + (l_x l_y)_i(l_y l_x)_j] + [x \leftrightarrow y]. \tag{7.44}$$

Similar expressions for the a and b bonds can be obtained by permutations of angular momentum components, that is by replacing (l_x, l_y) by (l_y, l_z) and (l_z, l_x), respectively.

Finally, it is also instructive to rewrite the orbital interactions in terms of auxiliarly pseudospins. For instance,

$$A_{ij}^{(\gamma)} = 2\Big(\boldsymbol{\tau}_i \cdot \boldsymbol{\tau}_j + \frac{1}{4}n_i n_j\Big)^{(\gamma)}, \tag{7.45}$$

Here $\boldsymbol{\tau}^{(\gamma)}$ is a pseudospin one-half operating on subspace of orbital doublet $(\alpha, \beta)^{(\gamma)}$ relevant for γ-bond, i.e., the pseudospin $\boldsymbol{\tau}^{(c)}$ operates on the subspace spanned by (a, b) pair of orbitons, while $\boldsymbol{\tau}^{(a)}$ and $\boldsymbol{\tau}^{(b)}$ act on (b, c) and (c, a) doublets, respectively.

The interactions as given by (7.38–7.39) look rather complicated. However, the dominant part of the Hamiltonian, obtained by neglecting Hund's coupling terms (consider the $\eta \to 0$ limit, for which $r_{1,2,3} = 1$), has a simple and transparent structure:

$$H = J\sum_{\langle ij\rangle} 2\Big(\boldsymbol{S}_i \cdot \boldsymbol{S}_j + \frac{1}{4}\Big)\Big(\boldsymbol{\tau}_i \cdot \boldsymbol{\tau}_j + \frac{1}{4}n_i n_j\Big)^{(\gamma)}. \tag{7.46}$$

An unessential energy shift ($= -J$) is not shown here. The Hamiltonian (7.46) contains the most relevant features of t_{2g} spin-orbital systems, and represents a highly frustrated many-body problem. The major role of the Hund's rule corrections and the weak orbital-lattice coupling, which are always present in a realistic case, is to select a particular spin-orbital state among the low energy, almost degenerate states of the Hamiltonian (7.46).

Comparing (7.46) with its e_g counterpart, (7.1), one may notice substantial differences in the orbital interactions. The orbital part in (7.46) is much more symmetric due to the fact that two orbitals equally participate in superexchange process for every direction and may form pseudospins that are coupled in a Heisenberg-like fashion. This introduces more orbital dynamics when compared with e_g orbital case.

To develop some intuition about the model (7.46), let us discuss some simpler examples. Consider, say, an exchange pair along the c-direction. If both ions are occupied by active orbitals ($n_i^{(c)} = n_j^{(c)} = 1$), one obtains an interaction of the form $2J\left(\boldsymbol{S}_i \cdot \boldsymbol{S}_j + \frac{1}{4}\right)\left(\boldsymbol{\tau}_i \cdot \boldsymbol{\tau}_j + \frac{1}{4}\right)$ which shows perfect symmetry between spin and orbital pseudospin. The pair has sixfold degeneracy in the lowest energy state: both the spin triplet/orbital singlet and the spin singlet/orbital triplet states gain the same quantum exchange energy equal to $-J/2$. In other words, the spin exchange constant may have equally strong ferromagnetic and antiferromagnetic nature depending on the symmetry of the orbital wavefunction. This is at odds with the classical Goodenough-Kanamori picture in which ferromagnetic spin exchange occurs only at finite Hund's coupling and hence is smaller by factor of J_H/U. In this respect, t_{2g} superexchange clearly differs from that in the e_g spin-orbital model (1), in which spin exchange interaction is positive, since the formation of orbital singlets is impossible.

When such t_{2g} pairs forms a 1D chain, one obtains a model which has been investigated in context of spin-orbital physics as the SU(4) model [40–42]. A large amount of of quantum energy (of order of $-0.4J$) is gained in this model due to the resonance between the local configurations spin singlet/orbital triplet and spin triplet/orbital singlet. As a result, the low-energy excitations are of a composite spin-orbital nature. Figure 7.8(a) illustrates the idea of correlated spin-orbital fluctuations. In fact, the one-dimensional version of this model has been solved by the Bethe ansatz [43] and the excitation spectrum consists of three gapless mixed spin-pseudospin modes. Critical exponents of various correlation functions have also been determined by conformal field theory analysis of the Bethe ansatz solution [44].

For the real three dimensional perovskite titanates, the situation is more complicated since the SU(4) spin-orbital resonance along one direction necessarily frustrates interactions for the remaining two directions. This is because the pseudospins $\tau^{(\gamma)}$ and densities $n^{(\gamma)}$ in equation (7.46) are bond dependent and require different orbital pairs, see Fig. 7.7. The quantum numbers, say, $\tau^{(c)} = 1/2$ and $n^{(c)} = 1$ imply that $\tau^{(a,b)} = 0$ because of the local constraint. Thus orbital interactions on different bonds compete, a situation which might be called "cubic" frustration. "Cubic" frustration is very general and and is a distinct property of spin-orbital and orbital-only models. It leads to infinite degenerate classical ground states. As we have seen in connection with the e_g orbital problem, this degeneracy can be removed by quantum spin fluctuations through the formation of quasi one dimensional orbital ordering.

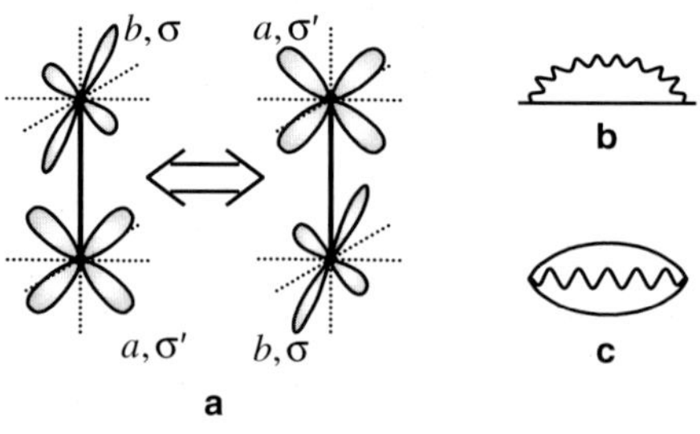

Fig. 7.8. (**a**) Spin-orbital resonance between two (left and right) states of the c-bond pair. Electrons exchange both their spin and orbital quantum numbers during this process. (**b**) Self-consistent Born approximation for the fermionic orbital self-energy. (**c**) Diagram contributing to the bond order parameter χ. Solid (wavy) lines represent orbitons (magnons). (After [33]).

Crucial to that idea was that the bond directional geometry of $(3z^2 - r^2)$ orbital. The t_{2g} orbitals are however not bond oriented (they are all planar), and not much spin fluctuation energy can be gained by any pattern of static orbital orderings. There is a special symmetry (see Fig. 7.7) for the t_{2g} orbitals such that every exchange bond has two equivalent orbitals which may form orbital singlet or triplet bonds. Orbital bonds may form however only if the spin deviation from the classical Néel state occurs simultaneously: The operator $(\boldsymbol{S}_i \cdot \boldsymbol{S}_j + 1/4)$ in (7.46) has a finite value, providing an effective dynamical coupling for orbitals. Thus one again arrives at the idea of a composite spin-orbital excitation, which consists of spin fluctuations accompanied by the formation of dynamical orbital bonds. Such local excitations, composed of a magnon and an orbital fluctuation, removes the orbital frustration dynamically. A disordered orbital state is suggested by this picture: Spin-orbital resonances for the three different directions of a cubic lattice involve different pairs of orbitals, thus all three t_{2g} states are equally populated on every site on average. This best optimizes the overall quantum energy of the t_{2g} spin-orbital superexchange system on a cubic lattice.

One should notice that spins and orbitals in the three dimensional model (7.46) are not equivalent. The spin sector is of simple Heisenberg form, thus interactions on different bonds cooperate and support spin long-range order in the ground state. One can say therefore that it is the "cubic frustration" of the orbital sector which brings about the unusual quantum physics in the three dimensional system. Although the spin sector itself is free of "cubic frustration", the latter nevertheless strongly affects the spin degrees of freedom through the dynamical coupling between the two sectors: As short-wavelength magnons are actively involved in the local resonance of the exchange bonds, a staggered long-range spin order parameter is expected but with a considerably reduced magnitude. Another, more formal observation is that spin sector is effectively half-filled $(n_{i,\uparrow} + n_{i,\downarrow} = 1)$; on the other hand, the orbital occupation number $n_i^{(\gamma)}$ is not conserved but rather fluctuates around the average value 2/3. Because of the threefold degeneracy, one has a less severe constraint $n_{i,a} + n_{i,b} + n_{i,c} = 1$ in orbital sector, hence quantum zero-point fluctuations are stronger in this sector. While true $SU(4)$ mixed excitations cannot develop in a cubic system because of the spin-orbital asymmetry, one still may think in terms of a virtual $SU(4)$ resonance which can be viewed in the present context as a local excitation composed of a magnon and an orbital fluctuation.

To implement the idea of spin-orbital resonance in the model Hamiltonian (7.46), it is convenient to introduce the three-particle bond variable, $\hat{\chi}_\alpha^{(ij)} = \alpha_i^\dagger \alpha_j (s_i^\dagger + s_j)$. Here, α denotes an appropriate fermionic orbiton, say an a or b orbitons for c -bond pair (ij), while $s^\dagger$ is a magnon excitation above the Néel state. This is an analog to $SU(4)$ flavor bond variable [40] in a spin ordered state. The physical picture is, that while the large energy $(\sim 4t^2/U)$ behaviour of the model (7.46) is governed by a coupled spin-orbital dynamics, the underlying weak magnetic order results in a separation of low energy spin excitations forming coherent magnons. The bare magnon dispersion is simply given by the Heisenberg interaction with an average exchange constant $\bar{J} = \langle \hat{J}_{ij}^{(\gamma)} \rangle$, which will be calculated below. In contrast, the orbital dynamics are exclusively due to the three-particle resonance, since the spin-only prefactor $\langle \mathbf{S}_i \cdot \mathbf{S}_j + 1/4 \rangle$ is almost zero (~ -0.05) in 3D. We may write (7.46) as $H = H_{\rm sp} + H_{\rm int} + \text{const.}$, where H_{sp} describes bare magnons, and

$$H_{\rm int} = -\frac{1}{2} \sum_{\langle ij \rangle} \sum_{\alpha\beta} \hat{\chi}_\alpha^{(ij)} \hat{\chi}_\beta^{(ji)} . \tag{7.47}$$

The strategy is then to decouple $H_{\rm int}$ in terms of bond variables. The solution with uniform bond amplitude χ gives $H_{\rm int} \Rightarrow H_a + H_b + H_c + \text{const.}$, where

$$H_a = -\chi \cdot \sum_{\langle ij \rangle_{bc}} \left\{ a_i^\dagger a_j \left(s_i^\dagger + s_j \right) + \text{h.c.} \right\} . \tag{7.48}$$

Here the summation is restricted to the (b, c) planes, and we have chosen $J = 4t^2/U$ as an energy scale. Similar expressions hold for b and c orbitons. Since the filling factor is only 1/3 per orbital flavor, the orbital sector is far from any nesting condition, thus a mean-field uniform solution is in the same spirit as large N -theories [45] and is believed to be a reasonable starting point to describe a disordered orbital state. Equation (7.48) describes an effective hopping of orbitons accompanied by simultaneous magnon excitations. Thus, we have mapped the superexchange for an undoped system to an effective χ-$\bar{J}$ model, which in momentum space is

$$H_{\chi-\bar{J}} = \sum_{\mathbf{kp}} \sum_{\alpha} M_{\mathbf{kp}}^{(\alpha)} (\alpha_{\mathbf{k+p}}^\dagger \alpha_{\mathbf{k}} s_{\mathbf{p}} + \text{h.c.}) + \sum_{\mathbf{p}} \omega_{\mathbf{p}} s_{\mathbf{p}}^\dagger s_{\mathbf{p}} . \tag{7.49}$$

Here $M_{\mathbf{kp}}^\alpha = -4\chi(\gamma_{\mathbf{k+p}}^{(\alpha)} u_{\mathbf{p}} + \gamma_{\mathbf{k}}^{(\alpha)} v_{\mathbf{p}})$, while the form-factors are $\gamma_{\mathbf{k}}^{(a)} = (c_y + c_z)/2$, $\gamma_{\mathbf{k}}^{(b)} = (c_z + c_x)/2$, $\gamma_{\mathbf{k}}^{(c)} = (c_x + c_y)/2$, where $c_\lambda = \cos k_\lambda$. The bare magnon dispersion $\omega_{\mathbf{p}} = 3\bar{J}(1 - \gamma_{\mathbf{p}}^2)^{1/2}$ with $\gamma_{\mathbf{p}} = (c_x + c_y + c_z)/3$, and

$$\begin{aligned} u_{\mathbf{p}} &= \frac{1}{\sqrt{2}} \left[(1 - \gamma_{\mathbf{p}}^2)^{-1/2} + 1 \right]^{1/2} , \\ v_{\mathbf{p}} &= -\frac{1}{\sqrt{2}} \left[(1 - \gamma_{\mathbf{p}}^2)^{-1/2} - 1 \right]^{1/2} \text{sign}(\gamma_{\mathbf{p}}) \end{aligned} \tag{7.50}$$

are conventional Bogoliubov transformation coefficients for magnons and a cubic lattice. It is noticed that the coupling constant in (7.49) vanishes in the $p = 0$ limit (spin conservation is respected), and it is the short-wavelength magnons which are important. The model is similar to the t-J model for the doped Néel state of cuprates (see [46,47] for comparison), yet the parameters χ and $\bar{J}$ have to be self-consistently determined. It is noticed, that there are three branches of 2D fermions, while the dimensionality of magnon sector is three in the present case.

We calculate first the fermionic spectrum within a self-consistent Born approximation. Unlike holes in the t-J model, an orbiton creates a magnon at the site on which it arrives, and eliminates a magnon when leaving a site (see (7.48)). Therefore orbiton motion contains a coherent component even in the Ising limit for the magnons. The limit is used to simplify the momentum integration in the self-energy, Fig. 7.8(b). One then obtains $Re\Sigma_\omega^{(\alpha)}(\mathbf{k}) = f_\omega - \kappa_\omega(\gamma_\mathbf{k}^{(\alpha)})^2$. Here, the frequency dependent coefficient

$$\kappa_\omega = (4\chi)^2 \int_0^\infty d\xi \frac{\rho(\xi)}{\xi + 3\bar{J} - \omega} , \tag{7.51}$$

where $\rho(\xi) = \langle \rho_\mathbf{k}(\xi) \rangle_\mathbf{k}$ is fermionic density of states ($\langle ... \rangle_\mathbf{k}$ implies an average over Brillouin zone). The function f_ω has similar structure:

$$f_\omega = (4\chi)^2 \int_{-\infty}^0 d\xi \frac{\tilde{\rho}(\xi)}{\omega + 3\bar{J} - \xi} , \tag{7.52}$$

where $\tilde{\rho}(\xi) = \langle \rho_\mathbf{k}(\xi)(\gamma_\mathbf{k}^{(\alpha)})^2 \rangle_\mathbf{k}$. One observes there is a two-sublattice structure to the orbiton dispersion (imposed by spin order) as expected from the analogy with the t-J model [47]. This is reflected in the momentum dependence of the self-energy, hence the orbiton dispersion is given by $(\gamma_\mathbf{k}^{(\alpha)})^2$, implying a coherent motion of orbitons within spin-up/down sublattices, similar to the case of hole motion in a quantum antiferromagnet. We estimate the fermionic (unrenormalized) mass as $m \simeq 1/\kappa_0$, and the density of states to be $\rho(0) \simeq 1/\pi\kappa_0$. Each orbiton forms its own 2D fermi-surface (FS). For, say, the a orbitons the two Fermi surfaces are centered at $(0,0)$ and (π,π), are almost circular, and lie in the (k_y, k_z) plane. Two dimensional nature of orbiton bands is a natural consequence of the fact that the t_{2g} orbital can only move along two directions. Assuming a constant density of states within an interval of width $W_{\rm orb} = 1/\rho(0)$, we find $\kappa_0 \simeq 4\chi/\sqrt{\pi}$, and $W_{\rm orb} \simeq 4\sqrt{\pi}\chi$. The energy scale for orbital fluctuations is $W_{\rm orb}$. Further, we estimate the bond amplitude $\chi = \langle a_i^\dagger a_j (s_i^\dagger + s_j) \rangle$. In the Ising limit, for intermediate magnons, a straightforward calculation of diagram Fig. 7.8(c) gives $\chi \simeq (4/3\sqrt{\pi})\langle(\gamma_\mathbf{k}^{(a)})^2\rangle_{FS}$. For $\langle n^{(a)} \rangle = 1/3$ we estimate $\langle(\gamma_\mathbf{k}^{(a)})^2\rangle_{FS} \simeq 0.3$. The spin stiffness in the present model is controlled by $\bar{J}$. The mean-field value 2/9 for $\bar{J}$ follows from (2). This number is reduced to $\bar{J} \simeq 0.16$ due to the corrections shown in Fig. 7.9(a). Physically this reduction is a consequence of the fact that the $SU(4)$ resonance induces a ferromagnetic component into spin interactions.

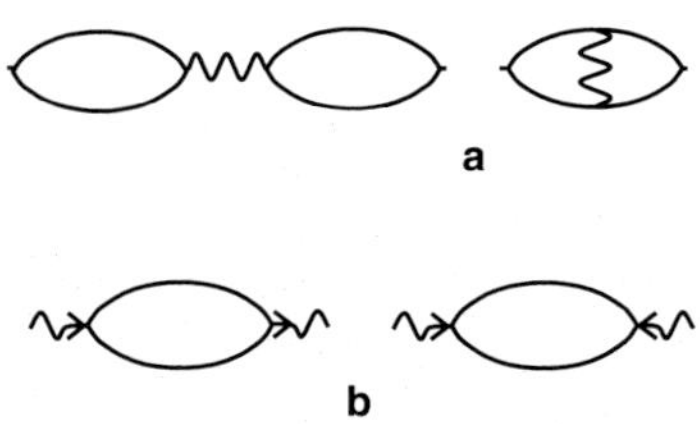

Fig. 7.9. (**a**) Interaction corrections to the spin exchange constant $\bar{J}$. (**b**) Magnon self-energies.

Thus, we have fixed the basic energy scales (in units of $J = 4t^2/U$): $W_{\rm orb} \simeq 1.6$ for orbital sector, and $3\bar{J} \simeq 0.48$ for magnon bandwidth. The spin stiffness is relatively small because the Pauli principle is relevant only if electrons occupy the same orbital, and the probability for this is reduced by the large orbital degeneracy. The principal energy gain stems from the magnon-orbital resonance: $E_{\rm int} = -6\chi^2 \simeq -0.31$. (The spin-only contribution is negligible: $E_{SW} \simeq -0.02$). This number should be compared with the energy gain associated with statically ordered planar orbitals: $\simeq -0.17$. This comparison shows the importance of quantum fluctuations in formation of the spin and orbital ground states of the model.

We discuss now the physical consequences of the theory in context of experimental data on $LaTiO_3$. Keimer et al. [32] reported a dynamical quenching of t_{2g} orbital angular moments in $LaTiO_3$ which is not associated by any detectable orbital/Jahn-Teller ordering. The magnon spectra are found to fit a spin one-half, nearly isotropic Heisenberg model on a cubic lattice. Keimer et al. suggested a picture of strongly fluctuating orbitals to explain their results. These observations seem to be consistent with present superexchange theory, in which "cubic frustration" phenomena results in a quantum disordered state for orbitals. The observed magnon dispersions of cubic symmetry are very natural in above picture with orbitals fluctuating faster than magnons. The energy scale $4t^2/U \simeq 100$ meV, required to fit $\bar{J}_{exp} = 15.5 \pm 1$meV, is also reasonable in view of the values $U \sim 5$eV, and $t \sim 0.3$eV [35]. This scale implies the orbital fluctuation energy in $LaTiO_3$ is $W_{\rm orb} \sim 160$meV.

Keimer et al. observed a very small anisotropy gap, indicating that effects of conventional relativistic spin-orbit coupling are strongly suppressed. This is in fact an unavoidable consequence of the theory, since the ground state has no orbital degeneracy. We may calculate the spectral density of orbital angular momentum fluctuations. These correspond to the fermionic orbiton excitations ("orbiton Stoner continuum"). The spectral density of the local susceptibility of angular moments $l_i^z = {\rm i}(a^\dagger b - b^\dagger a)_i$ vanishes in the static, $\omega = 0$ limit: $S(\omega) = 2\rho^2(0)\omega$. This implies a complete quenching of orbital angular moments, so they disappear from low-energy physics. Also, the ratio $\Lambda_{so}/W_{\rm orb} \sim 10^{-1}$ ($\Lambda_{so} \sim 19$ meV [48]) means that the relativistic spin-orbit coupling $\Lambda_{so}(\boldsymbol{Sl})$ corrections are small.

Despite the three dimensionality of magnon spectra, the staggered moment in $LaTiO_3$ is small, i.e. $0.45\mu_B$ [32] - an obvious problem for a conventional spin-wave picture. The present theory resolves this difficulty: The intensity of spin Bragg peak is partially taken away by a quantum magnon-orbiton resonance and redistributed over a finite frequency region. In other words, fluctuating orbitals generate additional quantum spin fluctuations in the ground state. A spin moment reduction can be calculated following the method as used for the t-J model [49].

Accounting for lowest order interaction corrections (Fig. 7.9(b)) to the magnons, one finds (at T=0):

$$\delta S^z_{\rm int} = \frac{3}{2}(4\chi)^2 \sum_{\mathbf{kk}'} n_{\mathbf{k}}(1-n_{\mathbf{k}'})\Big\{\frac{A_{\mathbf{kk}'}}{(\omega_{\mathbf{p}}+\xi_{\mathbf{k}'}-\xi_{\mathbf{k}})^2} - \frac{B_{\mathbf{kk}'}}{\omega_{\mathbf{p}}(\omega_{\mathbf{p}}+\xi_{\mathbf{k}'}-\xi_{\mathbf{k}})}\Big\}, \qquad (7.53)$$

where $\xi_{\mathbf{k}}$ is the a orbiton dispersion, $\mathbf{k}' = \mathbf{k}+\mathbf{p}$, $A_{\mathbf{kk}'} = \lambda_+^2/(1+\gamma_{\mathbf{p}}) + \lambda_+\lambda_-/(1-\gamma_{\mathbf{p}}^2)^{1/2}$, $B_{\mathbf{kk}'} = \lambda_+^2\gamma_{\mathbf{p}}/(1+\gamma_{\mathbf{p}})$, and $\lambda_\pm = \gamma_{\mathbf{k}}^{(a)} \pm \gamma_{\mathbf{k}'}^{(a)}$. By averaging first the matrix elements A, B in (5) over the Brillouin zone (they are rather regular functions), we evaluate $\delta S^z_{\rm int} \simeq 0.15$. (Another estimation, using the Ising limit for magnons, gives 0.13). Adding also a conventional 3D spin-wave correction $\delta S^z_{SW} \simeq 0.075$, one then obtains a staggered moment $0.55\mu_B$ in fair agreement with experiment.

It should be noticed here that calculation of temperature dependence of the spin order parameter is more subtle issue. Because of 2D-nature of the t_{2g} electron motion, long-range spin order in the model (7.46) is possible only at zero temperature [50], thus the complete model (7.37) including the Hund's coupling terms has to be considered at finite temperature calculations. Such calculations have not yet been performed thus far.

In conclusion, due to the large degeneracy and special geometry of the orbitals, the t_{2g}-superexchange model (7.46) most likely has an orbitally disordered ground state. The observed anomalies of $LaTiO_3$ find a natural explanation within this picture. In the metallic manganites, e_g orbital disorder is enforced by the presence of mobile carriers. Apparently, insulating $LaTiO_3$ illustrates the idea that frustrated t_{2g} orbitals can also create orbital disorder, without doping and well before the Mott transition to a metallic state is reached.

7.2.2 Orbital-Only Model

Let us return now to the original superexchange model (7.37), and ask the question: What kind of orbital state would optimize the superexchange energy if we fix the spin state to be ferromagnetic as in $YTiO_3$? Even though spin fluctuations are frozen in this case and composite spin-orbital excitation is "switched off" in the ground state, the previous discussion of orbital-only models (7.13,7.14) suggests that we may still expect highly nontrivial physics. The reason is again the "cubic frustration" phenomena which is still fully operative in the orbital sector and leads to an infinite degeneracy of classical ground states, as is the case for the e_g models.

If spins are all aligned ferromagnetically (i.e., $\boldsymbol{S}_i \cdot \boldsymbol{S}_j = 1/4$), the t_{2g} superexchange model (7.37) reduces to an orbital-only Hamiltonian of very simple form:

$$H_{\rm orb} = \frac{1}{2}\sum_{\langle ij\rangle} A_{ij}^{(\gamma)} \, . \tag{7.54}$$

Hereafter, we use $r_1 J$ as a unit of energy, and drop a constant energy shift $(= -r_1 J)$. For convenience, we show again the orbital operators:

$$A_{ij}^{(c)} = n_{ia}n_{ja} + n_{ib}n_{jb} + a_i^\dagger b_i b_j^\dagger a_j + b_i^\dagger a_i a_j^\dagger b_j \tag{7.55}$$

$$= [(1 - l_x^2)_i(1 - l_x^2)_j + (l_x l_y)_i(l_y l_x)_j] + [x \leftrightarrow y] \tag{7.56}$$

for the pair along the c-axis. Interactions on other bonds are obtained by permutations of the a, b, c orbiton labels or angular momentum components l_z, l_y, l_x, as has been explained earlier.

It is useful to look at the structure of $H_{\rm orb}$ from different points of view: (i) On a given bond, the operator $A_{ij}^{(\gamma)}$ acts within a particular doublet of equivalent orbitals. Spin-like physics, that is the formation of orbital singlets is therefore possible. Thus quantum fluctuations are stronger here in comparison with previous orbital-only models (7.13),(7.14). (ii) On the other hand, interactions on different bonds are competing: they involve different doublets, thus frustrating each other. This brings about Potts model like frustrations, from which the high degeneracy of the classical orbital configurations follows. This property is exactly the same as in e_g orbital case. (iii) Finally, we observe in $A_{ij}^{(\gamma)}$ a pseudospin $l = 1$ interaction of pure biquadratic form (see (7.56)). If $\boldsymbol{l}$ was a classical vector, its sign could be changed at any site independently without effecting the energy. Such a local (so-called Z_2) symmetry and the associated degeneracy of the classical states tell us that angular momentum ordering, if any, must be of pure quantum origin.

The above points (i-iii) govern the underlying physics of the orbital Hamiltonian. We need to find classical states which maximize the zero point energy gain when we switch on the quantum fluctuations. In other words, certain classical orbital patterns will be selected and stabilized by quantum effects. Normally, these orderings are expected to be along symmetric orientations of the crystal this reflecting the symmetry of the underlying interactions. By inspection of the global structure of interactions $A_{ij}^{(\gamma)}$ in above equations, one observes that the first two terms are definitely positive. However, the last two terms (which change the "color" of orbitals) can be made negative on *all the bonds simultaneously*, if: (i) on every bond, two particular components of $\boldsymbol{l}_i$ and $\boldsymbol{l}_j$ are antiparallel, and (ii) the remaining third components are parallel. For c-bonds the rule reads as: $l_{iz}l_{jz}$ and $l_{ix}l_{jx}$ are both negative, while the l_y components are parallel. (In terms of orbitons: the pairs c_i and c_j and a_i and a_j are in antiphase, but b_i and b_j have the same phase). There are only two topologically different arrangements (called (a) and (b)) on a cubic lattice, which can accommodate this curious mixture of "2/3 antiferro" plus "1/3 ferro" correlations (see Fig. 7.10).

For technical reasons, and also to simplify the physical picture it is useful to introduce new quantization axes. This is done in two steps. First, we

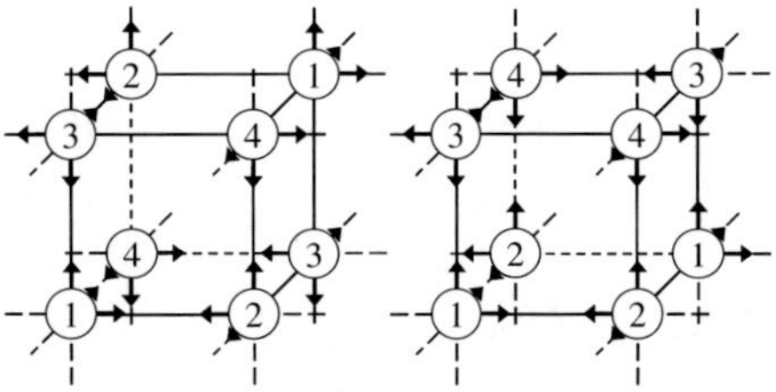

Fig. 7.10. Arrangement of the local quantization axes in states (**a**) and (**b**). Each state has four sublattices denoted by numbers. Arrows show a direction of the quantization axes at each site. They represent also a snapshot of local correlations of orbital angular momentum: on every bond, two out of three components l_α are correlated antiferromagnetically. (After [51]).

introduce local, sublattice specific, quantization axes (see Fig. 7.10):

$$\begin{aligned} &1:(x,y,z)\to(x,y,z), &&2:(x,y,z)\to(-x,-y,z),\\ &3:(x,y,z)\to(-x,y,-z), &&4:(x,y,z)\to(x,-y,-z). \end{aligned} \tag{7.57}$$

After the resulting sign transformations of the $l_{i\alpha}$ and orbitons, one obtains a negative sign in front of the last two terms of $A_{ij}^{(\gamma)}$, thereby "converting" the interactions in the new axes to a fully ferromagnetic one. From now on, the sublattice structure will no longer enter in the excitation spectrum. With the above observations, it is also clear that all the components of $\boldsymbol{l}$ are equally needed in order to optimize all the three directions. We anticipate therefore that the cubic diagonals are "easy" (or "hard") axes for $\boldsymbol{l}$ fluctuations/orderings (recall that the Hamiltonian has no rotational symmetry for $\boldsymbol{l}$ vector). This suggests the second step in our transformations: Rotate the quantization axes such that the new z-axis ($\tilde{z}$) is along one of the cubic diagonals (say, [111]). This is done as follows: $\boldsymbol{l}_i = \hat{R}\,\tilde{\boldsymbol{l}}_i$ (and $\mathbf{t} = \hat{R}\,\tilde{\mathbf{t}}$ for orbitons, where $\mathbf{t}$ denotes the (a,b,c) orbital triplet). The matrix $\hat{R} = \hat{R}_z(\pi/4)\hat{R}_y(\theta_0)\hat{R}_z(-\pi/4)$ with $\tan\theta_0 = \sqrt{2}$. The rotation $\hat{R}_\alpha(\theta)$ is around α-axis by angle θ. More explicitly, $\hat{R}$ is given by

$$\hat{R} = \frac{1}{\sqrt{3}}\begin{pmatrix} c+s & c-s & 1\\ c-s & c+s & 1\\ -1 & -1 & 1 \end{pmatrix}, \tag{7.58}$$

with $c = 1/2$ and $s = \sqrt{3}/2$. In terms of wave functions, this transformation reads as follows:

$$\begin{aligned} \psi_{\tilde{c}} &= \frac{1}{\sqrt{3}}(d_{yz}+d_{xz}+d_{xy}),\\ \psi_{\tilde{a}} &= \frac{1}{\sqrt{3}}[(c+s)d_{yz}+(c-s)d_{xz}-d_{xy}],\\ \psi_{\tilde{b}} &= \frac{1}{\sqrt{3}}[(c-s)d_{yz}+(c+s)d_{xz}-d_{xy}]. \end{aligned} \tag{7.59}$$

After the above transformations, the interactions $A_{ij}^{(\gamma)}$ in the orbital Hamiltonian (7.54) are of the following form (a symbol "tilde", denoting local coordinate axes, is implied below):

$$\begin{aligned}3A_{ij}^{(\gamma)} &= \frac{2}{3}(1 - Q_{iz}Q_{jz}) - \frac{1}{2}l_{iz}l_{jz} \\ &+\frac{1}{2}(Q_xT_{-1} + T_{-1}Q_x - T_0T_1 - T_1T_0)_{ij} \\ &+\frac{1}{2}(Q_xQ_x - T_0T_0 + T_{-1}T_{-1})_{ij}^{(\gamma)} \\ &+\frac{2}{3}Q_{iz}(T_0 + cT_1)_j^{(\gamma)} + \frac{2}{3}(T_0 + cT_1)_i^{(\gamma)}Q_{jz} \\ &+\frac{1}{2}l_{iz}(l_x + l_y)_j^{(\gamma)} + \frac{1}{2}(l_x + l_y)_i^{(\gamma)}l_{jz} \\ &-\frac{1}{2}(l_x + l_y)_i^{(\gamma)}(l_x + l_y)_j^{(\gamma)}.\end{aligned} \tag{7.60}$$

Here, Q_z and Q_x represent the quadrupole momentum operators of e_g symmetry (of $(3z^2 - r^2)$- and $(x^2 - y^2$-type, respectively). Further, $T_0 = T_z$ and $T_{\pm 1} = T_y \pm T_x$, where T_z, T_y and T_x represent the quadrupole operators of t_{2g} (xy, xz and yz, respectively) symmetry. These quadrupole momentum operators are expressed in terms of both the angular momentum and the orbiton operators as follows:

$$Q_z = \frac{1}{2}(\boldsymbol{l}^2 - 3l_z^2) = n_c - \frac{1}{2}(n_a + n_b), \tag{7.61}$$

$$Q_x = l_x^2 - l_y^2 = n_b - n_a, \tag{7.62}$$

$$T_z = l_xl_y + l_yl_x = -(a^\dagger b + b^\dagger a), \tag{7.63}$$

$$T_x = l_yl_z + l_zl_y = -(b^\dagger c + c^\dagger b), \tag{7.64}$$

$$T_y = l_xl_z + l_zl_x = -(c^\dagger a + a^\dagger c). \tag{7.65}$$

The bond γ-dependence of the quadrupole operators is obtained by changing the angular momentum operators in their definition as follows:

$$l_x^{(\gamma)} = \begin{cases} -cl_x \pm sl_y, & \text{for } \gamma = a(b) \\ l_x, & \text{for } \gamma = c \end{cases} \tag{7.66}$$

$$l_y^{(\gamma)} = \begin{cases} -cl_y \mp sl_x, & \text{for } \gamma = a(b) \\ l_y. & \text{for } \gamma = c \end{cases} \tag{7.67}$$

Explicitly,

$$Q_x^{(\gamma)} = \begin{cases} -cQ_x \mp sT_z, & \text{for } \gamma = a(b) \\ Q_x, & \text{for } \gamma = c \end{cases} \tag{7.68}$$

$$T_z^{(\gamma)} = \begin{cases} -cT_z \pm sQ_x, & \text{for } \gamma = a(b) \\ T_z, & \text{for } \gamma = c \end{cases} \tag{7.69}$$

$$T_x^{(\gamma)} = \begin{cases} -cT_x \mp sT_y, & \text{for } \gamma = a(b) \\ T_x, & \text{for } \gamma = c \end{cases} \tag{7.70}$$

$$T_y^{(\gamma)} = \begin{cases} -cT_y \pm sT_x, & \text{for } \gamma = a(b) \\ T_y. & \text{for } \gamma = c \end{cases} \quad (7.71)$$

Finally, $T_{\pm 1}^{(\gamma)} = T_y^{(\gamma)} \pm T_x^{(\gamma)}$.

It should be noted that among the eight operators, namely, the five quadrupole and the three angular momentum operators, only four operators are independent. This is because of the local constraint imposed on the a_i, b_i and c_i particles.

Though it looks somewhat complicated, an obvious advantage of the transformed form of the interaction $A_{ij}^{(\gamma)}$ is that it is well structured, such that the full Hamiltonian can be divided into two parts, i.e., $H_{\text{orb}} = H_{Ising} + H_{trans}$. Here, H_{Ising} represents the unfrustrated, "longitudinal" part of the interaction, reflected by the first line of $A_{ij}^{(\gamma)}$ in (7.60) that depends solely on z-components of operators. It can be rewritten also as follows:

$$H_{Ising} = \frac{1}{6} \sum_{\langle ij \rangle} \Big[l_{iz}^2 + l_{jz}^2 - \frac{3}{2} l_{iz}^2 l_{jz}^2 - \frac{1}{2} l_{iz} l_{jz} \Big]. \quad (7.72)$$

All bonds act cooperatively in this interaction supporting the ordering of either the Q_z or l_z variables. A "transverse" part of the Hamiltonian, H_{trans} is responsible for the fluctuations of transverse components l_x, l_y and quadrupole moments of various symmetry. In terms of a, b, c orbitons, H_{Ising} promotes a condensation of an appropriate orbiton, while H_{trans} gives the dispersion of orbitons and their interactions.

Now we are ready to discuss possible orbital orderings. Equation (7.72) makes it explicit that there are the two classical minima corresponding to two different "ferro-type" states denoted by I and II: (I) Quadrupole ordering with finite $\langle l_z^2 \rangle$ (but $\langle l_z \rangle = 0$), and (II) Magnetic ordering with a finite expectation value of the angular momentum ($l_{iz} = 1$ classically). We notice that the last, magnetic term in (7.72) is generated by the quantum commutation rules when we rotated H_{orb}; this makes explicit that the "Z_2 symmetry" is classical and emphasizes the quantum origin of the orbital magnetism. We will see below that states I and II are degenerate even at the quantum level. Noticing that an arbitrary cubic diagonal could be taken as $\tilde{z}$, and having in mind also the two structures in Fig. 7.10, one obtains a multitude of degenerate states. This makes any orderings very fragile.

To be more specific about the last statement, we calculate the excitation spectrum. Introduce first a quadrupole order parameter (we restore now "tilde" denoting operators in a rotated basis): $Q = \langle (\tilde{l}_x^2 + \tilde{l}_y^2)/2 - \tilde{l}_z^2 \rangle = \langle n_{\tilde{c}} - (n_{\tilde{a}} + n_{\tilde{b}})/2 \rangle$, and the magnetic moment $m_l = \langle \tilde{l}_z \rangle = \mathrm{i} \langle \tilde{b}^\dagger \tilde{a} - \tilde{a}^\dagger \tilde{b} \rangle$ (we set $\mu_B = 1$). State I corresponds to the condensation of the orbital $\tilde{c}$ (an equal mixture of the original a, b, c states, see (7.59)), while the ordering orbital in the state II is an imaginary one $(\tilde{a} - \mathrm{i}\tilde{b})/\sqrt{2}$. Let us call the remaining two orbitals α and β, which are the excitations of the model. Specifically, $\alpha = \tilde{a}$, $\beta = \tilde{b}$ for the state I, and $\alpha = [\tilde{a} + \mathrm{i}\tilde{b} + (1+\mathrm{i})\tilde{c}]/2$, $\beta = [\tilde{b} - \mathrm{i}\tilde{a} - (1-\mathrm{i})\tilde{c}]/2$

for the state II. Note that latter doublet can be regarded as a magnons of orbital origin, describing fluctuations of the orbital magnetism. Remarkably, in terms of these doublet excitations, the linearized Hamiltonian in momentum space has the same form for both phases. In the orbital wave approximation for quadrupole ordered state, the condensed orbiton operator is approximated as $\tilde{c} = \sqrt{(1 - n_\alpha - n_\beta)} \simeq 1 - n_\alpha/2 - n_\beta/2$. Then, the angular and quadrupole momentum operators in (7.60) are expanded in powers of operators α, β, keeping only terms up to second order, e.g. $Q_z = 1 - 3(n_\alpha + n_\beta)/2$, $l_x = \mathrm{i}(\beta - \beta^\dagger) + \ldots$, $T_x = -(\beta + \beta^\dagger) + \ldots$, etc. After somewhat lengthy but straightforward algebra one obtains a linearized orbital wave Hamiltonian:

$$H_{OW} = \sum_{\boldsymbol{k}} \Big[n_{\alpha\boldsymbol{k}} + n_{\beta\boldsymbol{k}} + \Big\{ \frac{\gamma_1 + \gamma_2}{2} \alpha_{\boldsymbol{k}} \alpha_{-\boldsymbol{k}} + \frac{\gamma_1 - \gamma_2}{2} \beta_{\boldsymbol{k}} \beta_{-\boldsymbol{k}} - \gamma_3 \alpha_{\boldsymbol{k}} \beta_{-\boldsymbol{k}} + \text{h.c.} \Big\} \Big], \quad (7.73)$$

where $\gamma_1 = (c_x + c_y + c_z)/3$, $\gamma_2 = \sqrt{3}(c_y - c_x)/6$, $\gamma_3 = (2c_z - c_x - c_y)/6$, and $c_\alpha = \cos k_\alpha$. These expressions refer to the state I (that is, quadrupole ordering). For the angular momentum ordering state II, one finds an orbital wave Hamiltonian of the same form, with the only difference that γ_2 and γ_3 are interchanged in (7.73). The latter does not affect the spectrum of the elementary excitations, however. Indeed, diagonalizing Hamiltonian (7.73) we obtain for the orbiton dispersions $\omega_\pm(\boldsymbol{k}) = \sqrt{1 - (\gamma_1 \pm \kappa)^2}$, where $\kappa = \sqrt{\gamma_2^2 + \gamma_3^2}$.

Using the excitation spectrum $\omega_\pm(\boldsymbol{k})$ for the (α, β) doublet that describes orbital fluctuations, one may calculate the various expectation values within a linear orbital wave approximation. First, we obtain energy gain due to the quantum, orbital wave fluctuations. It is the same for both quadrupole and magnetic ordered phases:

$$E_0 = \frac{1}{2} \sum_{\boldsymbol{k}} (\omega_+(\boldsymbol{k}) + \omega_-(\boldsymbol{k})) - 1 = -0.214 \,. \quad (7.74)$$

Next, the number of bosons not in the condensate, that is the population of α and β orbitons, is given by

$$\langle n_\alpha + n_\beta \rangle = -1 + \sum_{\boldsymbol{k}} \frac{1}{\omega_+(\boldsymbol{k})} = 0.54 \,. \quad (7.75)$$

Thus, the order parameters are strongly reduced by fluctuations. We obtain $Q = 0.19$ ($m_l = 0$, of course) in the quadrupole ordered state I, and the angular magnetic moment $m_l = 0.19$ accompanied by a small quadrupole moment $Q = -0.095$ in the state II. Taking into account these orbital fluctuations, the electron density at site 1 is given by $\rho_1(\boldsymbol{r}) = n_{\tilde{c}} \psi_{\tilde{c}}^2 + n_{\tilde{a}} \psi_{\tilde{a}}^2 + n_{\tilde{b}} \psi_{\tilde{b}}^2$. This can be expressed via quadrupole order parameter Q as follows:

$$\rho_1(\boldsymbol{r}) = \frac{1}{3}(d_{yz}^2 + d_{xz}^2 + d_{xy}^2) + \frac{2}{3} Q (d_{yz} d_{xz} + d_{yz} d_{xy} + d_{xz} d_{xy}). \quad (7.76)$$

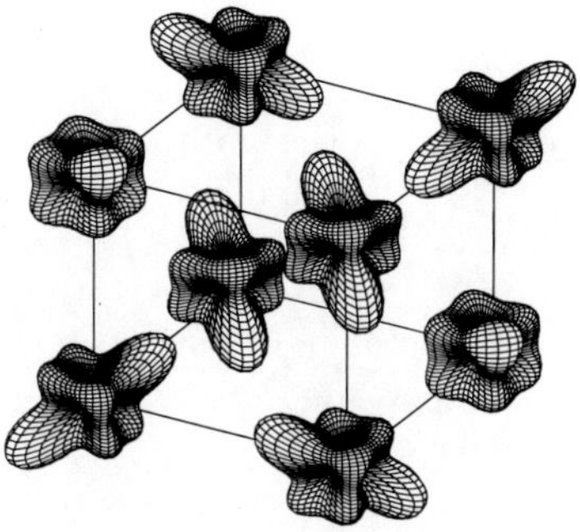

Fig. 7.11. The t_{2g}-electron density in the quadrupole ordered state, calculated from (7.76). (After [51]).

Similar expressions can easily be obtained for sublattices 2,3,4 by redefinitions of the quantization axes according to Fig. 7.10. To visualize the resulting orbital patterns, we show in Fig. 7.11 the electron density calculated including quantum fluctuations.

The anomalous reduction of the order parameters is due to the highly frustrated nature of the interactions in (7.54). A special, non spin-like feature of all orbital models is that orbitals are bond selective, resulting in a pathological degeneracy of classical states. This leads to soft modes (observe that $\omega_{\pm}(\boldsymbol{k})$ is flat along $(0, 0, \pi)$ and equivalent directions). These soft modes have their origin in special symmetry properties of the t_{2g} orbital model (7.54), which result in conservation laws with an important consequences, as discussed in [13]. Namely, the total number of orbitals of each "color" (a, b, c) are conserved during superexchange process, as can easily be seen from (7.55). Moreover, as t_{2g}-orbitals can hop only along two directions (say, xy-orbital motion is restricted to (ab) planes), the orbital number is conserved on each plane separately. Formally, these conservation rules are reflected by a possibility of a uniform phase transformation of orbiton operators, e.g. $a \to a \exp(\mathrm{i}\phi_a)$, etc., which leaves the orbital Hamiltonian invariant. These continuous symmetries are spontaneously broken in above ordered states. The breaking of a continuous symmetry is usually associated with gapless Goldstone modes. This is precisely what happens in t_{2g} orbital degenerate models. In fact, the soft modes obtained above have a two dimensional feature (stemming from 2D geometry of the t_{2g} orbital hoppings). As a result, long range orbital order is possible only at zero temperature, as is the case for 2D Heisenberg models. Formally, this is manifested as a divergence in the number of thermally excited orbitons, $\langle n_\alpha + n_\beta \rangle$, if one attempts to calculate this quantity at any finite temperature, including Bose population factor in the (7.75). The soft modes of the t_{2g} orbital model should be contrasted with gapful excitations of the e_g orbital models. This difference is due to a different structure of electron hopping matrices, which results in different symmetry properties of the superexchange interactions.

A remarkable feature of all the above phases is they have the same spin couplings in all the cubic directions due to high symmetry of the ordering patterns. Being consistent with spinwave data by Ulrich et al. [36], this may indicate the relevance of the superexchange mechanism in the orbital orderings of the titanates. On the other hand, it is also clear that realistic models should include also many other interactions such as the relativistic spin-orbital coupling, deviations of real lattices from an ideal perovskite structure etc., as has been discussed in previous chapters. In particular, the bending of the exchange bond due to octahedra tiltings has been suggested as being driving mechanism for the ferromagnetic state of $YTiO_3$ [13, 52].

Finally, it is interesting to compare the quantum energy gain obtained in ferro- and antiferromagnetic phases of t_{2g} superexchange model: E_0 is equal to -0.214 and -0.33 (in units of J), respectively. This reflects again the fact that t_{2g} superexchange for a given pair of spins 1/2 has no unique spin value in the ground state: the spin singlet/orbital triplet and spin triplet/orbital singlet states are degenerate in that case. Though such a degeneracy is absent in a cubic crystal, ferro- and antiferromagnetic states have very close ground state energies, which stems in both cases from orbital quantum fluctuations.

7.3 High Spin Systems with t_{2g} Orbital Degeneracy

The behavior of t_{2g} orbitals, when as in vanadium oxides, they are filled by two electrons forming $S = 1$ is puzzling. On one hand, the occupied t_{2g} orbitals are known to order in non-cubic vanadium compounds, such as $LiVO_2$ [53] and V_2O_3 [54, 55], while on the other hand, the situation in cubic systems might be very different since all the bonds are *a priori* magnetically equivalent, and, in this case, the quantum fluctuations among orbitals are expected to play an important role.

7.3.1 Formation of One-Dimensional Orbital Chains

The superexchange interactions between the $S = 1$ spins of V^{3+} ions arise from the virtual excitations $d_i^2 d_j^2 \rightarrow d_i^3 d_j^1$ on a given bond $\langle ij \rangle$, with the hopping t allowed only between two out of three t_{2g} orbitals, see (7.36). The d_i^3 excited state may be either a high-spin $^4A_2$ state with correlation energy $U - 3J_H$, or one of three low-spin states: ^2E, $^2T_1$ or $^2T_2$ with energies U or $U + 2J_H$, respectively [37]. Second order perturbation theory leads to the spin-orbital Hamiltonian [34]:

$$H = \sum_{\langle ij \rangle} \left[(\boldsymbol{S}_i \cdot \boldsymbol{S}_j + 1) \hat{J}_{ij}^{(\gamma)} + \hat{K}_{ij}^{(\gamma)} \right] . \tag{7.77}$$

In terms of the operators $A_{ij}^{(\gamma)}$, $B_{ij}^{(\gamma)}$ and $n_i^{(\gamma)}$ introduced previously by (7.40)-(7.42), the orbital operators $\hat{J}_{ij}^{(\gamma)}$ and $\hat{K}_{ij}^{(\gamma)}$ are as follows:

$$\hat{J}_{ij}^{(\gamma)} = \frac{J}{4}\left[(1+2\eta R)A_{ij}^{(\gamma)} - \eta r B_{ij}^{(\gamma)} - \eta R(n_i+n_j)\right]^{(\gamma)}, \tag{7.78}$$

$$\hat{K}_{ij}^{(\gamma)} = \frac{J}{2}\left[\eta R A_{ij}^{(\gamma)} + \eta r B_{ij}^{(\gamma)} - \frac{1}{2}(1+\eta R)(n_i+n_j)\right]^{(\gamma)}. \tag{7.79}$$

Here $J = 4t^2/U$, as usual. The coefficients $R = 1/(1-3\eta)$ and $r = 1/(1+2\eta)$ originate from the multiplet structure of the t_{2g}^3 excited states via $\eta = J_H/U$.

The structure of the orbital exchange operators (7.78,7.79) is very similar to that for the titanates since both are determined by the symmetry of t_{2g} electron hopping Hamiltonian (7.36). Of course, there are some differences which originate from different multiplet structure of the excited virtual states. This is reflected in coefficients R and r relevant here and which have replaced the r_1, r_2, r_3 defined earlier. If we neglect the Hund's coupling splitting of the excited states ($\eta \to 0$ limit), the Hamiltonian (7.77) reduces to:

$$H = J\sum_{\langle ij\rangle} \frac{1}{2}(\boldsymbol{S}_i\cdot\boldsymbol{S}_j+1)\Big(\boldsymbol{\tau}_i\cdot\boldsymbol{\tau}_j + \frac{1}{4}n_in_j\Big)^{(\gamma)}, \tag{7.80}$$

where a constant energy of $-2J$ per V^{3+} ion is neglected. This result should be compared with corresponding limit in d^1 case, (7.46). There is a different spin structure so that $\frac{1}{2}(\boldsymbol{S}_i\cdot\boldsymbol{S}_j+1)$ is obtained for vanadium ions while it is $2(\boldsymbol{S}_i\cdot\boldsymbol{S}_j+\frac{1}{4})$ which is relevant for spin one-half Ti^{3+} ions. The difference in spin values can, in fact, be accounted for via $(\boldsymbol{S}_i\cdot\boldsymbol{S}_j+S^2)/2S^2$. It is important also to notice that we have two electrons per V^{3+} ion; one therefore has a different constraint equation for orbiton densities $n_{ia}+n_{ib}+n_{ic}=2$.

Mean-field phase diagram for cubic vanadates has been discussed in rather detail in [56] (see also Chap. 5). We focus here on possible quantum effects. It is instructive to start again with a single bond along c-axis. A crucial observation is that the lowest energy of $-J/2$ is obtained when the spins are *ferromagnetic*, and the orbitals a and b form a singlet, with $\langle\boldsymbol{\tau}_i\cdot\boldsymbol{\tau}_j\rangle^{(c)} = -\frac{3}{4}$. The spin singlet×orbital triplet level is higher (at $-J/4$). This is in sharp contrast to the $S=1/2$ case, where the spin singlet×orbital triplet and the spin triplet×orbital singlet configurations are degenerate, resulting in a strong quantum resonance. Thus, one finds a novel mechanism by which to obtain ferromagnetic interactions which works because of the local orbital singlet made of the a and b orbitals. The dominance of the high spin configuration simply reflects the fact that spin part of the interaction, that is $(\boldsymbol{S}_i\cdot\boldsymbol{S}_j+S^2)/2S^2$ is equal to one for ferromagnetic configuration, but vanishes in spin-singlet sector (as $-1/S$) in the limit of large spins. In order to form ab orbital singlets along c-axis, the condition $n_i^{(c)} = n_j^{(c)} = 1$ must be fulfilled (no $\boldsymbol{\tau}^{(c)}$ pseudospin can be formed otherwise). This implies that the second electron on both sites has to go to an inactive (that is xy) orbital. Thus we arrive a following picture for the superexchange bond in c direction: (i) spins are aligned ferromagnetically, (ii) one electron on each site occupies either an a or b orbital forming SU(2) invariant orbital pseudospins that bind into an orbital singlet, (iii) the xy orbital has a stabilization energy of about $-J/2$ (the energy that is required to break a ab orbital singlet) and accommodates remaining, second, electron.

If the high spin state of a given pair is so stable, why is then that the whole crystal does not form a uniform ferromagnet? This is because each site has two electrons, and an orbital that is inactive in a particular (ferromagnetic bond) direction, in fact induces an antiferromagnetic coupling in the other two directions. Thus spin interactions are strongly ferromagnetic (supported by orbital singlets) in one direction, while the other bonds are antiferromagnetic. As all directions are *a priori* equivalent in a cubic lattice, we again arrive at the problem of "cubic frustration" common to all spin-orbital models with perovskite lattices. The solution of this problem is as follows. Since the spin-orbital resonance found for the titanates is suppressed in the present case of large spin $S = 1$ at d^2 ions, the quantum energy is gained mainly from the orbital sector. This implies that a particular classical spin configuration, which maximizes the energy gain from orbital fluctuations is preferred. Indeed, orbital singlets (with $n_{ia} + n_{ib} = 1$) may form on bonds parallel to c-axis, exploiting fully the SU(2) symmetry of the a, b orbital interactions in one direction, with the second electron occupying the third t_{2g} orbital ($n_{ic} = 1$) thereby controlling the spin interactions in the ab-planes. In this way one arrives at spin order of the C-type [ferromagnetic chains along c-axis which stagger within ab-planes], which is the best comprise which mixes both the ferromagnetic (driven by the orbital singlets) and antiferromagnetic (induced by the electron residing on the static orbital) interactions.

Given the spontaneous breaking of the cubic symmetry (fluctuating a, b orbitals, and stable c orbital located at lower energy) resulting in a C-type spin structure the superexchange Hamiltonian can be simplified. We may now set $n_{ic} = 1, n_{ia} + n_{ib} = 1$, and introduce the pseudospin $\boldsymbol{\tau}$ which operates within the (a, b) orbital doublet exclusively.

We focus first on orbital sector as the quantum dynamics of the system are dominated by the orbital pseudospins $\tau = \frac{1}{2}$ rather than by large, more classical spins $S = 1$. In the classical C-type spin state, $(\boldsymbol{S}_i \cdot \boldsymbol{S}_j + 1)$ is equal to two along the c-direction while it is zero on ab-plane bonds. In this approximation, the orbital interactions in the model (7.77) are given by $(2\hat{J}_{ij}^{(c)} + \hat{K}_{ij}^{(c)})$ on c-axis bonds, while on ab-plane bonds only the $\hat{K}_{ij}^{(a,b)}$ term contributes. Expressing the $A_{ij}^{(\gamma)}$ and $B_{ij}^{(\gamma)}$ operators in (7.78,7.79) via pseudospins $\boldsymbol{\tau}$, one arrives at following orbital Hamiltonian

$$H_{\mathrm{orb}} = J_{\mathrm{orb}} \sum_{\langle ij \rangle \| c} (\boldsymbol{\tau}_i \cdot \boldsymbol{\tau}_j) + J_{\mathrm{orb}}^{\perp} \sum_{\langle ij \rangle \| (a,b)} \tau_i^z \tau_j^z \,, \tag{7.81}$$

where $J_{\mathrm{orb}} = JR$ and $J_{\mathrm{orb}}^{\perp} = J\eta(R+r)/2$. As their ratio is small, $J_{\mathrm{orb}}^{\perp}/J_{\mathrm{orb}} = \eta(1 - 5\eta r/2)$ (that is about 0.1 only for the realistic values of the parameter $\eta = J_H/U$ for vanadates), we obtain one-dimensional orbital pseudospin chains coupled only weakly to each other. Orbital excitations of the model

(7.81) propagate mostly along c-chain directions. Their spectrum can be calculated within the linear spin-wave approximation, assuming a weak orbital order due to interchain coupling $J_{\rm orb}^{\perp}$. One indeed obtains a one-dimensional orbital-wave spectrum:

$$\omega_{\boldsymbol{k}} = J\sqrt{\Delta^2 + R^2 \sin^2 k_z}, \tag{7.82}$$

with a gap $J\Delta = J\{\eta(R+r)[2R+\eta(R+r)]\}^{1/2}$ at $k_z = \pi/2$. The orbital gap $J\Delta$ is small and grows with increasing Hund's coupling as $\propto J\sqrt{\eta}$.

The underlying orbital fluctuations have important consequences on spin interactions which control spinwave dispersions. In the spin sector, we obtain interactions $J_c(\boldsymbol{S}_i \cdot \boldsymbol{S}_j)$ and $J_{ab}(\boldsymbol{S}_i \cdot \boldsymbol{S}_j)$ on c- and (a,b)-bonds, respectively. The spin exchange constants follow from (7.77)–(7.78). The result involves orbital pseudospin correlations, i.e.,

$$J_c = \frac{J}{2}\Big[(1+2\eta R)\Big\langle \boldsymbol{\tau}_i \cdot \boldsymbol{\tau}_j + \frac{1}{4}\Big\rangle - \eta r\Big\langle \tau_i^z \tau_j^z + \frac{1}{4}\Big\rangle - \eta R\Big], \tag{7.83}$$

$$J_{ab} = \frac{J}{4}\left[1 - \eta(R+r) + (1+2\eta R - \eta r)\Big\langle \tau_i^z \tau_j^z + \frac{1}{4}\Big\rangle\right]. \tag{7.84}$$

While the in-plane antiferromagnetic couplings are mostly determined by the classical contribution of xy orbitals (first term in J_{ab}), the exchange constant along the c-axis has a substantial quantum contribution represented by the first term in J_c. This contribution to J_c is of negative sign due to the orbital singlet correlations along chains. The pseudospin expectation values in (7.83,7.84) can be estimated using either the Bethe ansatz results for a 1D Heisenberg chain of disordered a and b orbitals, or the orbital-wave approximation [34]. In both approximations, one observes that the ferromagnetic coupling along the c-axis is strongly enhanced by orbital fluctuations. For a realistic value of $\eta \sim 0.12$, one obtains $-J_c \sim J_{ab} \sim J/5$. This result is qualitatively different from that expected from the Goodenough-Kanamori rules. Indeed, in that classical picture one would obtain instead the smaller value $-J_c \sim 2\eta R J_{ab} \sim J_{ab}/2$.

Comparing all the coupling constants in both the spin and orbital sectors, one observes that the Heisenberg-like orbital dynamics has the largest energy scale ($J_{\rm orb} = JR$), thus dominating the physics of the present spin-orbital model. The overall picture is that "cubic frustration" is resolved here by the formation of orbital chains with Heisenberg dynamics for which a large quantum energy is gained. This is in fact very similar to the order from disorder mechanism in the case of e_g spin-orbital model (7.1), where classical $(3z^2 - r^2)$-type orbital order results in quasi one-dimensional spin chains. In the present t_{2g} orbital model with large classical spins the role of spins, and orbitals are just interchanged.

The above scenario may explain the C-AF type spin order in $LaVO_3$ [57]. A structural transition that follows the onset of the magnetic order in this compound is a natural consequence of the superexchange model. Indeed,

C-type spin ordering and formation of the pseudospin orbital chains are intimately connected and support each other. Thus the quadrupole ordering in orbital sector (stabilization of the xy orbital and weak interchain orderings among a, b orbitals) can occur only below the magnetic transition.

7.3.2 Spin-Orbital Dimerization

We shown above that C-type spin order results in quasi-one dimensional orbital chains which accumulate a large quantum energy, thereby stabilizing the state. In addition, orbitals show weak three dimensional staggered ordering driven by small interchain couplings in the orbital Hamiltonian (7.81). It is well known, that the Heisenberg chain may show another kind of ordering – Peierls order – that is the chain dimerizes, breaking the translational symmetry. Normally, one needs some extra degrees of freedom (e.g., phonons) to trigger the dimerization instability. It turns out that the spin-orbital coupling, always present in the original superexchange model (7.77), may in fact induce dimerization of the spin-orbital system even without electron-phonon coupling.

Consider the limit of $\eta = 0$. Since the second term in the orbital Hamiltonian (7.81) vanishes in this limit, we are left with a pure Heisenberg orbital Hamiltonian. Now, let us go one step back to the full spin-orbital model (7.77) (or its $\eta = 0$ version (7.80)). It is easy to see that we can actually gain more orbital quantum energy by choosing the spin structure shown in Fig. 7.12. That is, we replace C-type spin order by an alternating structure, in which the spins on every second bond are oriented antiparallel, giving $(\boldsymbol{S}_i \cdot \boldsymbol{S}_j + 1) = 0$. This implies that orbitals are completely decoupled on these bonds, see (7.80). As a result, the expectation value $\langle \boldsymbol{\tau}_i \cdot \boldsymbol{\tau}_j \rangle$ vanishes on these "weak" bonds, thus the spin exchange constant $J_{c1} = J/8$ (consider (7.83) for uncorrelated orbitals and the $\eta = 0$ limit) turns out to be positive, consistent with antiferromagnetic spin alignment on these bonds. On the other hand, the orbital singlets are fully developed on ferromagnetic bonds ($\langle \boldsymbol{\tau}_i \cdot \boldsymbol{\tau}_j \rangle = -3/4$ and $J_{c2} = -J/4$), gaining a quantum energy $-J/2$. We obtain the energy $-J/4$ per ion in the dimer state, which

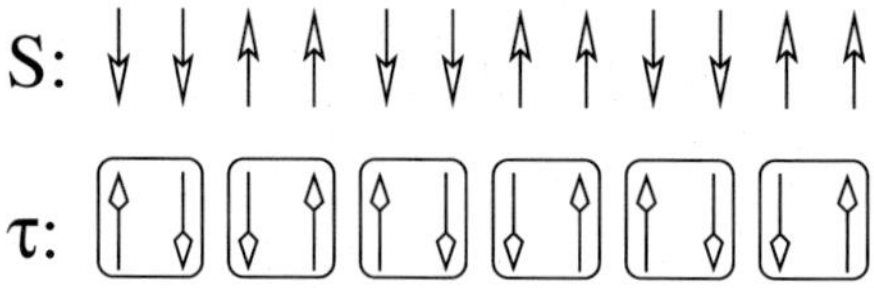

Fig. 7.12. A classical representation of the dimerized spin-orbital chain along the c-axis. Orbitals form perfect dimers indicated by the boxes and spins show a Néel order, but with 4 site periodicity due to the alternating ferromagnetic (on orbital dimer bonds) and antiferromagnetic (between the dimer bonds) spin exchange couplings. For a cubic crystal, dimers on different chains are correlated in-phase through the staggered spin correlations within the (ab)-planes.

is lower than the Bethe result $(1/2 - \ln 2)J \simeq -0.19J$ for uniform Heisenberg orbital chains with C-type spin structure. Thus, the spin-orbital chain has a dimerized ground state in the limit of small η values, as has been noticed by several authors [62–64, 66, 67]. The building blocks here are the decoupled orbital dimers with total spin 2. This construction is supported by an antiferromagnetic spin correlation between the dimers. Now, let us increase the Hund's coupling parameter η. As is easily understood from (7.83), an additional, orbital free term ηR tends to stabilize a ferromagnetic spin order along the c-axis, thus the C-type spin state with weak orbital ordering will be restored for sufficiently large values of η.

Subsequently, there is a zero temperature quantum phase transition between the dimerized state and the quasi one-dimensional ordering of orbitals. The precise value of the critical coupling strength η_{cr} at which the transition occurs, and the behavior of different correlation functions has been investigated in [64] using numerical methods. Though this method is restricted to the one dimensional version of the model (which is justified to some extend by quasi one dimensionality of the orbital sector), it gives clear and rigorous results. The transfer matrix renormalization group method used in Ref. [64] permits the study of finite temperature properties which are hard to access by analytical methods. We briefly discuss here the results obtained in Ref. [64].

The starting point is again the model (7.77) but the summation is restricted now to c-axis bonds only:

$$H = \sum_i \left[(\boldsymbol{S}_i \cdot \boldsymbol{S}_{i+1} + 1)\, \hat{J}_{i,i+1} + \hat{K}_{i,i+1} \right] . \tag{7.85}$$

Assume that the stabilization of xy orbital has already occurred by the formation of orbital chains as discussed at the beginning of previous paragraph. The orbital interactions (7.78,7.79) are then,

$$\begin{aligned} \hat{J}_{i,j} &= \frac{J}{2} \left[(1 + 2\eta R) \Big(\boldsymbol{\tau}_i \cdot \boldsymbol{\tau}_j + \frac{1}{4} \Big) - \eta r \Big(\boldsymbol{\tau}_i \otimes \boldsymbol{\tau}_j + \frac{1}{4} \Big) - \eta R \right] , \\ \hat{K}_{i,j} &= J \left[\eta R \Big(\boldsymbol{\tau}_i \cdot \boldsymbol{\tau}_j + \frac{1}{4} \Big) + \eta r \Big(\boldsymbol{\tau}_i \otimes \boldsymbol{\tau}_j + \frac{1}{4} \Big) \right] , \end{aligned} \tag{7.86}$$

with $\boldsymbol{\tau}$ acting on (a, b) orbital doublet. (An unessential energy shift stemming from the last term in $\hat{K}_{i,j}$ is not shown here). The symbol $\otimes$ denotes that $(\tau_i^+ \tau_j^+ + \tau_i^- \tau_j^-)$ is used in the pseudospin representation of operator $B_{ij}^{(c)}$, instead of the conventional $(\tau_i^+ \tau_j^- + \tau_i^- \tau_j^+)$ transverse part of a simple scalar product.

In Fig. 7.13 the temperature dependence of the nearest neighbor spin-spin correlation function $\langle \boldsymbol{S}_i \cdot \boldsymbol{S}_{i+1} \rangle$ and the orbital-orbital correlation function $\langle \boldsymbol{\tau}_i \cdot \boldsymbol{\tau}_{i+1} \rangle$ for different η values is shown. First, we want to discuss the case $\eta = 0$. Note, that in the zero temperature limit the nearest neighbor orbital correlation approaches a value near $-3/8$, the value for perfect dimerization ($\langle \boldsymbol{\tau}_i \cdot \boldsymbol{\tau}_{i+1} \rangle = -3/4$ on every second bond). In the high temperature limit the

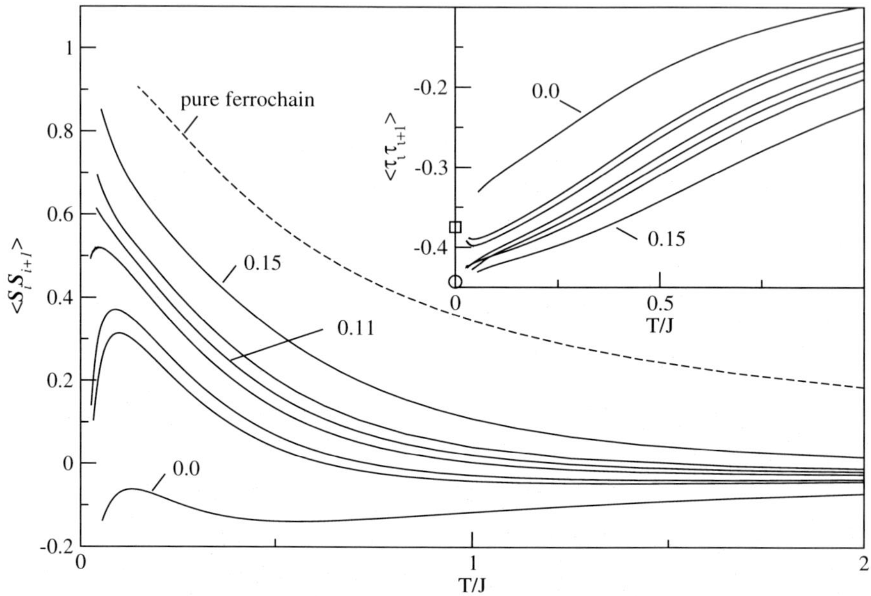

Fig. 7.13. The main figure (inset) shows the temperature dependence of nearest neighbor correlations $\langle \boldsymbol{S}_i \cdot \boldsymbol{S}_{i+1} \rangle$ ($\langle \boldsymbol{\tau}_i \cdot \boldsymbol{\tau}_{i+1} \rangle$) for η values 0.0, 0.07, 0.08, 0.1, 0.11, 0.12, 0.15. For comparison the spin only case is shown where $J_{i,j}$ is fixed by its zero temperature expectation value for $\eta = 0.15$. The circle (square) at $T = 0$ in the inset denotes $-\ln 2 + 1/4$ ($-3/8$). (After Ref. [64]).

orbitals are completely disordered and the spin pair correlation is therefore negative. Upon lowering the temperature, orbital singlets are formed and $\langle \boldsymbol{S}_i \cdot \boldsymbol{S}_{i+1} \rangle$ increases due to the ferromagnetic spin interaction within an orbital singlet. The curve reaches a maximum at $T/J \approx 0.13$ corresponding to the energy scale of the orbital gap, and antiferromagnetic quantum fluctuations of spins in neighboring orbital singlets are strongly enhanced and are dominant. This emphasizes the differences between a quantum model with $S = 1$ and the classical limit adopted in Fig. 7.12: in that picture, where fluctuations in the antiferromagnetic bonds are neglected, one would obtain that $\langle \boldsymbol{S}_i \cdot \boldsymbol{S}_{i+1} \rangle$ alternates between ± 1 with $\langle \boldsymbol{S}_i \cdot \boldsymbol{S}_{i+1} \rangle = 0$ on average.

Consider now the effects of a finite Hund's coupling. As seen from the first equation in (7.86), the term ηR tends to stabilize a ferromagnetic spin order. Numerical data for $\eta = 0.15$ indeed shows that $\langle \boldsymbol{S}_i \cdot \boldsymbol{S}_{i+1} \rangle$ is monotonously increasing and reaches 1, the value expected for an uniform ferromagnetic chain, in the zero temperature limit (see Fig. 7.13). This leads to an exact cancellation of the orbital anisotropy terms $\boldsymbol{\tau}_i \otimes \boldsymbol{\tau}_j$ in the ground state (see (7.85,7.86)), resulting in an isotropic Heisenberg antiferromagnetic coupling of the orbitals. From the Bethe ansatz result for the Heisenberg chain one expects $\langle \boldsymbol{\tau}_i \cdot \boldsymbol{\tau}_{i+1} \rangle = -\ln 2 + 1/4$ consistent with the numerical result (see Fig. 7.13).

The evolution of correlations depending on parameter η is interesting. The low temperature asymptotics of the spin and orbital expectation values (see Fig. 7.13) suggests a first order phase transition between two possible ground states: The spin/orbital dimer phase at small η, and the spin saturated ferromagnetic phase for large Hund's coupling. The critical coupling η_c for the zero temperature phase transition is somewhere in the range 0.10-0.11. This is just slightly below realistic values for the vanadium oxides. Indeed, using a representative values of $J_H \sim 0.68$ eV and $(U - \frac{20}{9} J_H) \sim 4.5$ eV [65], one estimates $\eta = 0.113$ for cubic vanadates. This suggests that the two phases are actually very close in energy and thus are strongly competing in real systems.

Another manifestation of the competition between the uniform and dimerized states in the present spin-orbital model is an anomalous behavior of spin correlations at finite temperature, as evident from Fig. 7.13. In comparison to the spin only case, where $J_{i,j}$ is fixed by its zero temperature expectation value $\eta = 0.15$ and therefore for which all dimerization effects are omitted, $\langle \boldsymbol{S}_i \cdot \boldsymbol{S}_{i+1} \rangle$ shows an unusual rapid decay with temperature. This is because the dimer state is very close in energy to the C-type ground state even above η_{cr}, and strongly affects thermodynamic properties for entropy reasons. The point is that the dimer state contains "weak" spin bonds: the spin interaction between different dimers is small when η is near the critical value (above the critical η, the spin interaction is of ferromagnetic sign but still weaker than strong ferromagnetic interaction within the dimers). Therefore, the spin entropy of an individual dimer with total spin 2, that is $\ln 5$, is released at rather low temperature, introducing substantial spin disorder.

The dynamical dimerization for $\eta > \eta_c$ could in fact be relevant to YVO_3, which shows C-type magnetic order as an intermediate, finite temperature phase with rather unusual magnetic properties [59]. Namely, the gain of spin entropy due to dimerization lowers the free energy $F = \langle H \rangle - TS$ at finite temperature and may stabilize the dimer state with alternating weak and strong ferromagnetic bonds along c-axis. In other words, a dimerization of the spin-orbital chains occurs due to the *orbital Peierls effect*, in which thermal spin fluctuations play the role of lattice degrees of freedom. This results in the modulation of the bond couplings in the spin and orbital sectors. The effect is dynamical in the present 1D case. However, a weak but finite coupling between the spin-orbital dimers on different c-chains is present in real crystals. Dimers on different chains will therefore correlate and may form a coherent dimer crystal in the intermediate phase of YVO_3. Doubling of the magnetic unit cell along the c-axis is then expected. In fact, a modulation of the ferrocouplings and the related splitting of the magnon branches has been observed by Ulrich et al. [66] in YVO_3 at finite temperature. As the dimerization is of electronic origin, concomitant lattice distortions are expected to be small because the t_{2g} orbitals are not bond oriented and are only weakly coupled to the lattice.

We may conclude that the t_{2g} spin-orbital model with high spin values shows an intrinsic tendency towards the formation of weakly coupled orbital chains followed by dimerization of these low dimensional spin-orbital structures. The dimerization is driven by a critical behavior of the orbital sector. Large Hund's coupling can stabilize uniform ferromagnetic chains in the ground state with weak orbital order, but strong dimer correlations develop again at finite temperature due to the orbital Peierls effect. Alternatively, one can say that this phenomena occurs due to a large spin entropy released by the dimer states.

7.4 Summary

In general, quantum fluctuations of orbital degrees of freedom in transition metal oxides are important and have to be taken into account in the explanation of various properties of these systems. Despite (or even due to) the absence of the rotational symmetry in the orbital sector, profound nonclassical behavior of the orbitals can result from the frustrated nature of the interactions within the spin-orbital systems and by the strong coupling of orbitals to doped holes.

References

1. J. B. Goodenough: Magnetism and Chemical Bond, Interscience Publ., New York–London, 1963
2. J. B. Goodenough: Phys. Rev. **100**, 564 (1955)
3. J. Kanamori: J. Phys. Chem. Solids **10**, 87 (1959)
4. K. I. Kugel, D. I. Khomskii: Sov. Phys. Usp. **25**, 231 (1982)
5. L. F. Feiner, A. M. Oleś, J. Zaanen: Phys. Rev. Lett. **78**, 2799 (1997)
6. For a discussion of the order from disorder phenomena in frustrated spin systems, see A. M. Tsvelik: *Quantum Field Theory in Condensed Matter Physics* (Cambridge University Press, Cambridge, 1995), Chap. 17, and references therein.
7. L. F. Feiner, A. M. Oleś, J. Zaanen: J. Phys. Condens. Matter **10**, L555 (1998)
8. G. Khaliullin, V. Oudovenko: Phys. Rev. B **56**, R14 243 (1997)
9. G. Khaliullin, R. Kilian: J. Phys. Condens. Matter **11**, 9757 (1999)
10. J. van den Brink, P. Horsch, F. Mack, A. M. Oleś: Phys. Rev. B **59**, 6795 (1999)
11. See the discussion in Ref. [4], p.253.
12. G. Khaliullin: Phys. Rev. B **64**, 212405 (2001)
13. G. Khaliullin, S. Okamoto: Phys. Rev. B **68**, 205109 (2003)
14. G. Khaliullin, P. Horsch: unpublished.
15. K. Kubo: J. Phys. Soc. Jpn. **71**, 1308 (2002)
16. R. Kilian, G. Khaliullin: Phys. Rev. B **60**, 13458 (1999)
17. S. Ishihara, J. Inoue, S. Maekawa: Phys. Rev. B **55**, 8280 (1997)

18. L. N. Bulaevskii, E. L. Nagaev, D. I. Khomskii: Sov. Phys.–JETP **27**, 836 (1968)
19. W. Koshibae, S. Maekawa: Physica C **317-318**, 205 (1999)
20. J. van den Brink, P. Horsch, A. M. Oleś: Phys. Rev. Lett. **85**, 5174 (2000)
21. Y. Endoh, K. Hirota, S. Ishihara, S. Okamoto, Y. Murakami, A. Nishizawa, T. Fukuda, H. Kimura, H. Nojiri, K. Kaneko, S. Maekawa: Phys. Rev. Lett. **82**, 4328 (1999)
22. S. Uhlenbrusk et al. Phys. Rev. Lett. **82**, 185 (1999)
23. S. Saitoh et al. Phys. Rev. B **51**, 13942 (1995)
24. J. M. De Teresa, M. R. Ibarra, P. A. Algarabel, C. Ritter, C. Marquina, J. Blasco, J. Garcia, A. del Moral, Z. Arnold: Nature (London) **386**, 256 (1997)
25. A. J. Millis, B. I. Shraiman, R. Mueller: Phys. Rev. Lett. **77**, 175 (1996)
26. T. Mizokawa, D. I. Khomskii, G. A. Sawatzky: Phys. Rev. B **63**, 024403 (2001)
27. Y. Okimoto, T. Katsufuji, T. Ishikawa, A. Urushibara, T. Arima, Y. Tokura: Phys. Rev. Lett. **75**, 109 (1995)
28. S. Ishihara, M. Yamanaka, N. Nagaosa: Phys. Rev. B **56**, 686 (1997)
29. H. Shiba, R. Shiina, A. Takahashi: J. Phys. Soc. Jpn. **66**, 941 (1997)
30. R. Kilian, G. Khaliullin: Phys. Rev. B **58**, R11841 (1998)
31. G. Khaliullin, R. Kilian: Phys. Rev. B **61**, 3494 (2000)
32. B. Keimer et al. Phys. Rev. Lett. **85**, 3946 (2000)
33. G. Khaliullin, S. Maekawa: Phys. Rev. Lett. **85**, 3950 (2000)
34. G. Khaliullin, P. Horsch, A. M. Oleś: Phys. Rev. Lett. **86**, 3879 (2001)
35. T. Katsufuji, Y. Taguchi, Y. Tokura: Phys. Rev. B **56**, 10145 (1997)
36. C. Ulrich et al. Phys. Rev. Lett. **89**, 167202 (2002)
37. J. S. Griffith, *The Theory of Transition Metal Ions* (Cambridge University Press, Cambridge, 1971)
38. K. I. Kugel, D. I. Khomskii: Sov. Phys. Solid State **17**, 285 (1975)
39. S. Ishihara, T. Hatakeyama, S. Maekawa: Phys. Rev. B **65**, 064442 (2002)
40. Y. Q. Li, M. Ma, D. N. Shi, F. C. Zhang: Phys. Rev. Lett. **81**, 3527 (1998)
41. B. Frischmuth, F. Mila, M. Troyer: Phys. Rev. Lett. **82**, 835 (1999)
42. F. Mila, B. Frischmuth, A. Deppeler, M. Troyer: Phys. Rev. Lett. **82**, 3697 (1999)
43. B. Sutherland: Phys. Rev. B **12**, 3795 (1975)
44. T. Itakura, N. Kawakami: J. Phys. Soc. Jpn. 64, 1321 (1995)
45. I. Affleck, J. B. Marston: Phys. Rev. B **37**, 3774 (1988)
46. C. L. Kane, P. A. Lee, N. Read: Phys. Rev. B **39**, 6880 (1989)
47. G. Martinez, P. Horsch: Phys. Rev. B **44**, 317 (1991)
48. A. Abragam, B. Bleaney: *Electron Paramagnetic Resonance of Transition Ions* (Oxford University Press, New York, 1970)
49. G. Khaliullin, P. Horsch: Phys. Rev. B **47**, 463 (1993)
50. A. B. Harris et al. Phys. Rev. Lett. **91**, 087206 (2003)
51. G. Khaliullin, S. Okamoto: Phys. Rev. Lett. **89**, 167201 (2002)
52. M. Mochizuki, M. Imada: J. Phys. Soc. Jpn. **70**, 1777 (2001)
53. H. F. Pen et al. Phys. Rev. Lett. **78**, 1323 (1997)
54. C. Castellani, C. R. Natoli, J. Ranninger: Phys. Rev. B **18**, 4945 (1978); **18**, 4967 (1978); **18**, 5001 (1978)
55. F. Mila et al. Phys. Rev. Lett. **85**, 1714 (2000)
56. T. N. De Silva, A. Joshi, M. Ma, F. C. Zhang: Phys. Rev. B **68**, 184402 (2003)
57. S. Miyasaka, T. Okuda, Y. Tokura: Phys. Rev. Lett. **85**, 5388 (2000)
58. H. Kawano, H. Yoshizawa, Y. Ueda: J. Phys. Soc. Jpn. **63**, 2857 (1994)
59. Y. Ren et al. Phys. Rev. B **62**, 6577 (2000)
60. T. Mizokawa, D. I. Khomskii, G. A. Sawatzky: Phys. Rev. B **60**, 7309 (1999)
61. H. Sawada and K. Terakura: Phys. Rev. B **58**, 6831 (1998)

62. Shun-Qing Shen, X. C. Xie, F. C. Zhang: Phys. Rev. Lett. **88**, 027201 (2002)
63. A. M. Oleś, P. Horsch, G. Khaliullin: Acta Physica Polonica B, **34**, 857 (2002)
64. J. Sirker, G. Khaliullin: Phys. Rev. B **67** , 100408(R) (2003)
65. T. Mizokawa, A. Fujimori: Phys. Rev. B **54**, 5368 (1996)
66. C. Ulrich et al.: Phys. Rev. Lett. **91**, 257202 (2003)
67. P. Horsch, G. Khaliullin, A. M. Oleś: Phys. Rev. Lett. **91**, 257203 (2003)

Appendices

A. Optical Conductivity

In this appendix expressions for the optical conductivity and related quantities are derived for interacting fermionic tight-binding models, by following a method mentioned in [1].

In continuum models with electric and magnetic fields, the momentum $\boldsymbol{p}$ for an electron of mass m and charge $-e$ ($e > 0$) is replaced by $\boldsymbol{p} + (e/c)\,\boldsymbol{A}$, where c is the speed of light, $\boldsymbol{A}$ is the vector potential, and where Gaussian units are used. This is reflected in the kinetic energy as

$$H_{\mathrm{kin}} = \int \mathrm{d}\boldsymbol{r}\psi^{\dagger}(\boldsymbol{r})\frac{1}{2m}\left(-\mathrm{i}\hbar\nabla + \frac{e}{c}\boldsymbol{A}\right)^{2}\psi(\boldsymbol{r}), \tag{A.1}$$

where ψ is a fermionic field operator. The same result can be obtained through a local gauge transformation

$$\psi(\boldsymbol{r}) \to \psi(\boldsymbol{r})\mathrm{e}^{\mathrm{i}\phi(\boldsymbol{r})}, \tag{A.2}$$

given that $\hbar\nabla\phi = (e/c)\boldsymbol{A}$ [2]. In lattice models with a discrete bases set, the gauge transformation (A.2) become

$$c_{i,\alpha,\sigma} \to c_{i,\alpha,\sigma}\mathrm{e}^{\mathrm{i}\phi(\boldsymbol{R}_i)}, \tag{A.3}$$

where $c_{i,\alpha,\sigma}$ is the annihilation operator of an electron with spin σ in orbital α at the lattice position $\boldsymbol{R}_i$. This phase is sometimes called the Peierls phase. The transformation (A.3) effectively modifies the hopping matrix element from the γ orbital at site j to the α orbital at site i

$$t_{ij}^{\alpha\gamma} \to t_{ij}^{\alpha\gamma}\exp\left[-\mathrm{i}(e/c\hbar)\int_{\boldsymbol{R}_j}^{\boldsymbol{R}_i}\boldsymbol{A}(\boldsymbol{r})\cdot\mathrm{d}\boldsymbol{r}\right]. \tag{A.4}$$

We consider a spatially homogeneous and time-dependent electric field with frequency ω, $\boldsymbol{E}(t) = \boldsymbol{E}(\omega)\mathrm{e}^{-\mathrm{i}\omega t}$, in the absence of magnetic field. The spatial homogeneity implies the wave vector $\boldsymbol{q} \to 0$. This limit is appropriate for the calculation of the optical response. The phase ϕ is written as $\phi(\boldsymbol{R}_i, t) = (e/c\hbar)\boldsymbol{A}(t)\cdot\boldsymbol{R}_i$ with $\boldsymbol{A}(t) = \boldsymbol{A}(\omega)\mathrm{e}^{-\mathrm{i}\omega t} = (c/\mathrm{i}\omega)\boldsymbol{E}(t)$ (remember $\boldsymbol{E}(t) = -(1/c)\partial\boldsymbol{A}(t)/\partial t$). Needless to say, $\boldsymbol{A}$ is the same at all lattice sites. Assuming that $\boldsymbol{A}$ is small, the Hamiltonian of the system with the electric field can be expressed as

$$H(\boldsymbol{A}) = \sum_{\langle i,j\rangle,\alpha,\gamma,\sigma} t_{ij}^{\alpha\gamma} \left(\mathrm{e}^{-\mathrm{i}(e/c\hbar)\boldsymbol{A}(t)\cdot\boldsymbol{R}_{ij}} c_{i,\alpha,\sigma}^{\dagger} c_{j,\gamma,\sigma} + \mathrm{h.c.} \right) + H_{\mathrm{int}}$$

$$\approx H_0 - \frac{e}{c}\boldsymbol{A}(t)\cdot\boldsymbol{j} - \frac{1}{2}\left(\frac{e}{c}\right)^2 \boldsymbol{A}(t)\cdot\boldsymbol{\tau}\boldsymbol{A}(t) \tag{A.5}$$

where with $\boldsymbol{R}_{ij} = \boldsymbol{R}_i - \boldsymbol{R}_j$, $H_0 = H(\boldsymbol{A} = 0)$, H_{int} denotes the interaction terms independent of $\boldsymbol{A}$, and the summation $\langle i, j\rangle$ runs over pairs of sites. The particle-current operator

$$\boldsymbol{j} = \mathrm{i}\hbar^{-1} \sum_{\langle i,j\rangle,\alpha,\gamma,\sigma} t_{ij}^{\alpha\gamma} \boldsymbol{R}_{ij} \left(c_{i,\alpha,\sigma}^{\dagger} c_{j,\gamma,\sigma} - \mathrm{h.c.} \right), \tag{A.6}$$

and the stress-tensor operator

$$\boldsymbol{\tau} = \hbar^{-2} \sum_{\langle i,j\rangle,\alpha,\gamma,\sigma} t_{ij}^{\alpha\gamma} \boldsymbol{R}_{ij} \otimes \boldsymbol{R}_{ij} \left(c_{i,\alpha,\sigma}^{\dagger} c_{j,\gamma,\sigma} + \mathrm{h.c.} \right), \tag{A.7}$$

where $\otimes$ represents the tensor product. The x component of the current operator (2.42) for the three-band Hubbard model can be obtained from (A.6).

The linear-response electrical current operator $\boldsymbol{j}_{\mathrm{e}}$ is, from (A.5),

$$\boldsymbol{j}_{\mathrm{e}} = -c\frac{\partial H}{\partial \boldsymbol{A}} = e\boldsymbol{j} + \frac{e^2}{c}\boldsymbol{\tau}\boldsymbol{A}(t), \tag{A.8}$$

where the first and second terms are called paramagnetic and diamagnetic current, respectively. The total paramagnetic current $\langle \boldsymbol{j}\rangle$ induced by $\boldsymbol{A}(\omega)$ is expressed within the linear response theory (Kubo formula) [3] as

$$\langle \boldsymbol{j}\rangle(\omega) = \boldsymbol{\chi}(\omega)\frac{e}{c}\boldsymbol{A}(\omega) \tag{A.9}$$

with the current-current correlation tensor

$$\boldsymbol{\chi}(\omega) = \frac{\mathrm{i}}{\hbar}\int_0^{\infty} \mathrm{d}t \mathrm{e}^{\mathrm{i}(\omega+i\eta)t} \langle [\boldsymbol{j}(t), \boldsymbol{j}(0)]\rangle, \tag{A.10}$$

where $\boldsymbol{j}(t) = \mathrm{e}^{\mathrm{i}H_0 t/\hbar}\boldsymbol{j}\mathrm{e}^{-\mathrm{i}H_0 t/\hbar}$, and $\langle\cdots\rangle = \mathrm{Tr}(\mathrm{e}^{-\beta H_0}\cdots)/\mathrm{Tr}\mathrm{e}^{-\beta H_0}$, with $\beta = 1/(k_{\mathrm{B}}T)$. An infinitesimal parameter η with positive sign is introduced in order for the external field to be switched on adiabatically, starting at $t = -\infty$. The spectral representation of $\boldsymbol{\chi}(\omega)$ reads

$$\boldsymbol{\chi}(\omega) = \frac{1}{\hbar Z}\sum_{n,m} \langle m|\boldsymbol{j}|n\rangle \otimes \langle n|\boldsymbol{j}|m\rangle \frac{\mathrm{e}^{-\beta E_n} - \mathrm{e}^{-\beta E_m}}{(E_m - E_n)/\hbar - \omega - \mathrm{i}\eta}, \tag{A.11}$$

where $|n\rangle$ is the eigenstate of H_0 with the eigenenergy E_n, and where Z is the partition function defined as $Z = \sum_n \mathrm{e}^{-\beta E_n}$. Here note that, since the diagonal terms $E_n = E_m$ in the above summation vanish automatically, one can add a constraint $E_n \neq E_m$ in the summation.

The total electric current $\langle \boldsymbol{j}_{\mathrm{e}}\rangle$ thus reads

$$\langle \boldsymbol{j}_{\mathrm{e}}\rangle(\omega) = \frac{e^2}{c}\left[\boldsymbol{\chi}(\omega) + \langle\boldsymbol{\tau}\rangle\right]\boldsymbol{A}(\omega). \tag{A.12}$$

Using the relation $\boldsymbol{A}(\omega) = c\boldsymbol{E}(\omega)/[\mathrm{i}(\omega+\mathrm{i}\eta)]$, the complex optical conductivity tensor defined as $\langle \boldsymbol{j}_\mathrm{e}\rangle(\omega) = N\boldsymbol{\sigma}(\omega)\boldsymbol{E}(\omega)$ is given by

$$\boldsymbol{\sigma}(\omega) = \boldsymbol{\sigma}'(\omega) + \mathrm{i}\boldsymbol{\sigma}''(\omega) = \frac{1}{N}\frac{e^2}{\mathrm{i}(\omega+\mathrm{i}\eta)}\left[\boldsymbol{\chi}(\omega) + \langle\boldsymbol{\tau}\rangle\right], \tag{A.13}$$

where N is the total number of sites. The complex $\boldsymbol{\chi}(\omega) = \boldsymbol{\chi}'(\omega) + \mathrm{i}\boldsymbol{\chi}''(\omega)$ and taking the limit $\eta \to 0$, the real part of $\boldsymbol{\sigma}(\omega)$ reads

$$\boldsymbol{\sigma}'(\omega) = \frac{e^2}{N}\left(-\pi\left(\langle\boldsymbol{\tau}\rangle + \boldsymbol{\chi}'(\omega)\right)\delta(\omega) + \frac{1}{\omega}\boldsymbol{\chi}''(\omega)\right), \tag{A.14}$$

where the delta function is given by $\delta(\omega) = \pi^{-1}\eta/(\omega^2+\eta^2)$. We find that $\boldsymbol{\sigma}'(\omega)$ consists of the delta function at the origin $\omega = 0$ (singular part: $\boldsymbol{\sigma}'_\mathrm{sing}$) and finite-frequency contributions in the second term (regular part: $\boldsymbol{\sigma}'_\mathrm{reg}$). Using the relation $1/(x \pm \mathrm{i}\eta) = \mathrm{P}/x \mp \mathrm{i}\pi\delta(x)$ (P is the principal part), the regular part is expressed in the spectral representation from (A.11)

$$\begin{aligned}\boldsymbol{\sigma}'_\mathrm{reg}(\omega) = {} & \frac{\pi e^2}{NZ}\frac{1-\mathrm{e}^{-\beta\hbar\omega}}{\hbar\omega} \\ & \times \sum_{\substack{n,m \\ E_m \neq E_n}} \mathrm{e}^{-\beta E_n}\langle m|\boldsymbol{j}|n\rangle \otimes \langle n|\boldsymbol{j}|m\rangle\,\delta\left(\omega - (E_m - E_n)/\hbar\right).\end{aligned} \tag{A.15}$$

The condition $E_n \neq E_m$ in the summation discussed above ensures no delta function at $\omega = 0$.

The weight of the singular part, i.e., the strength of the delta function in (A.14), is called the Drude weight, which corresponds to a well-known expression $\pi e^2 n/m^*$ for free electron systems with the electron density n and the effective mass m^*. The singular part is related to the charge stiffness tensor $\boldsymbol{D}$ through

$$\boldsymbol{\sigma}'_\mathrm{sing}(\omega) = -\frac{\pi e^2}{N}\left(\langle\boldsymbol{\tau}\rangle + \boldsymbol{\chi}'(\omega)\right)\delta(\omega) = 2\pi e^2\boldsymbol{D}\delta(\omega). \tag{A.16}$$

Therefore, one can find

$$\begin{aligned}\boldsymbol{D} &= -\frac{1}{2N}\left(\langle\boldsymbol{\tau}\rangle + \boldsymbol{\chi}'(0)\right) \\ &= -\frac{1}{2N}\langle\boldsymbol{\tau}\rangle - \frac{1}{NZ}\sum_{\substack{n,m \\ E_m \neq E_n}} \mathrm{e}^{-\beta E_n}\frac{\langle m|\boldsymbol{j}|n\rangle \otimes \langle n|\boldsymbol{j}|m\rangle}{E_m - E_n} \\ &= -\frac{1}{2N}\langle\boldsymbol{\tau}\rangle - \frac{1}{\pi e^2}\int_0^\infty \boldsymbol{\sigma}'_\mathrm{reg}(\omega)\mathrm{d}\omega.\end{aligned} \tag{A.17}$$

Using the imaginary part of $\boldsymbol{\sigma}(\omega)$, $\boldsymbol{D}$ can also be expressed as

$$\boldsymbol{D} = \frac{1}{e^2}\lim_{\omega\to 0}\omega\boldsymbol{\sigma}''(\omega). \tag{A.18}$$

We find from (A.14) the optical sum rule for the real part $\boldsymbol{\sigma}'(\omega)$

$$\begin{aligned}\int_0^\infty \boldsymbol{\sigma}'(\omega)\mathrm{d}\omega &= \frac{1}{2}\int_{-\infty}^\infty \boldsymbol{\sigma}'(\omega)\mathrm{d}\omega \\ &= \frac{\pi e^2}{2N}\left(-\langle \boldsymbol{\tau}\rangle - \boldsymbol{\chi}'(0) + \frac{\mathrm{P}}{\pi}\int_{-\infty}^\infty \frac{\boldsymbol{\chi}''(\omega)}{\omega}\mathrm{d}\omega\right) \\ &= -\frac{\pi e^2}{2N}\langle \boldsymbol{\tau}\rangle\,. \end{aligned} \tag{A.19}$$

Here, in order to obtain the third line in (A.19) the Kramers-Kronig relation [2]

$$\boldsymbol{\chi}'(\omega) = \frac{1}{\pi}\mathrm{P}\int_{-\infty}^\infty \frac{\boldsymbol{\chi}''(\omega')}{\omega'-\omega}\mathrm{d}\omega', \tag{A.20}$$

has been used. Note that, in lattice models with hoppings only between nearest-neighbor sites such as the t-J model (2.38), the diagonal component of $\langle \boldsymbol{\tau}\rangle$ in (A.7) is directly related to the kinetic energy if the nearest-neighbor distance is taken to be unity: $\langle \boldsymbol{\tau}_{\alpha\alpha}\rangle = \hbar^{-2}\langle (H_{\mathrm{kin}})_\alpha\rangle$, leading to the relation between the optical sum rule and the kinetic energy [4].

The charge stiffness $\boldsymbol{D}$ was introduced by Kohn [5] and reflects the conductive state of the system. For $T=0$, a non-zero charge stiffness implies an ideal conductor since the non-zero value means the presence of unscattered, freely accelerating charges at $\omega = 0$ in the system. On the other hand, for an insulator it must be that $\boldsymbol{D} = 0$. Kohn pointed out that the charge stiffness measures the sensitivity of the ground-state energy of the system to the change in boundary conditions, thereby being different for an ideal conductor and an insulator. The criterion has been later applied to correlated fermion systems [6] and used to characterize the superconducting state [7]. The concept has also been extended to the conductance at finite temperature and its relation to integrability of the system [8]. We show below the connection between the charge stiffness and the boundary conditions.

Here let us take the vector potential $\boldsymbol{A}$ to be constant in space and time. This means that both the electric and magnetic fields are null. The role of $\boldsymbol{A}$ is to induce a gauge transformation of the fermionic operators according to (A.3), $c_{i,\alpha,\sigma}(\boldsymbol{k}) = c_{i,\alpha,\sigma}\mathrm{e}^{\mathrm{i}\boldsymbol{k}\cdot\boldsymbol{R}_i}$ with $\boldsymbol{k} = (e/c\hbar)\boldsymbol{A}$. For a periodic system with periods $L\boldsymbol{e}_\alpha$, $\boldsymbol{e}_\alpha$ being the unit vector along the α direction, this transformation introduces a twist at the boundary and thus imposes the change of boundary condition for the wavefunction of the system, i.e.,

$$\Psi(\ldots,\boldsymbol{R}_i,\ldots) = \mathrm{e}^{\mathrm{i}L\boldsymbol{k}\cdot\boldsymbol{e}_\alpha}\Psi(\ldots,\boldsymbol{R}_i + L\boldsymbol{e}_\alpha,\ldots)\,. \tag{A.21}$$

Therefore, the boundary conditions of the system can be continuously varied by changing $\boldsymbol{A}$. Let us calculate the energy of the system with this vector potential. Using the standard perturbation theory, we evaluate the energy of the eigenstate $|n\rangle$ up to the second order of $\boldsymbol{A}$. The energy is calculated from (A.5) is then

$$E_n(\boldsymbol{A}) \approx E_n - \frac{e}{c}\boldsymbol{A}\cdot\langle n\,|\boldsymbol{j}|\,n\rangle - \frac{1}{2}\left(\frac{e}{c}\right)^2 \boldsymbol{A}\cdot\langle n\,|\boldsymbol{\tau}|\,n\rangle\,\boldsymbol{A} - \left(\frac{e}{c}\right)^2 \boldsymbol{A}\cdot \sum_{\substack{m \\ E_m \neq E_n}} \frac{\langle n\,|\boldsymbol{j}|\,m\rangle \otimes \langle m\,|\boldsymbol{j}|\,n\rangle}{E_m - E_n}\boldsymbol{A}. \tag{A.22}$$

By rewriting $\boldsymbol{D}$ in (A.17) as a thermal average

$$\boldsymbol{D} = \frac{1}{Z}\sum_n \mathrm{e}^{-\beta E_n}\,\boldsymbol{D}_n \tag{A.23}$$

with

$$\boldsymbol{D}_n = -\frac{1}{2N}\langle n\,|\boldsymbol{\tau}|\,n\rangle - \frac{1}{N}\sum_{\substack{m \\ E_m \neq E_n}} \frac{\langle m\,|\boldsymbol{j}|\,n\rangle \otimes \langle n\,|\boldsymbol{j}|\,m\rangle}{E_m - E_n}, \tag{A.24}$$

we find that the charge stiffness for individual energy level is given by

$$\boldsymbol{D}_n = \frac{1}{2N}\left(\frac{c}{e}\right)^2 \left.\frac{\partial^2 E_n(\boldsymbol{A})}{\partial \boldsymbol{A}\partial \boldsymbol{A}}\right|_{\boldsymbol{A}=0}. \tag{A.25}$$

The charge stiffness is therefore directly related to the curvature of the energy $E_n(\boldsymbol{A})$ at $\boldsymbol{A} = 0$, that is, the variation of an energy level to quadratic order with respect to the applied vector potential.

In an insulator at zero temperature, the ground state $|0\rangle$ should be insensitive to small changes of the boundary conditions because of the localized nature of the electron wave function. This leads to $\boldsymbol{D} = 0$ from (A.25) at $T = 0$. On the other hand, in an ideal conductor at $T = 0$, the wave function can be extended through the system and thus the electrons can feel the twist, therefore $\boldsymbol{D} \neq 0$. For finite-size lattices, one may encounters unphysical negative value of $\boldsymbol{D}$ at $T = 0$. In such a case, one can improve the results by changing the boundary conditions for $\boldsymbol{A} = 0$ or by averaging over different boundary conditions [9].

B. Exact Diagonalization Technique: The Lanczos Method

The exact diagonalization technique based on the Lanczos algorithm [10–12] is one of the most efficient methods for numerically calculating various physical quantities of finite-size systems in lattice models with strongly correlated electrons. The physical quantities are, for example, the ground-state energy, static and dynamical correlation functions. Since the discovery of high-T_c superconductivity, this technique has been widely used for the studies of one- and two-dimensional Hubbard and t-J models [13].

The Hamiltonian matrix for a finite-size system can be constructed from a set of the orthonormal basis vectors. For a N-site lattice of the Hubbard model (2.39) with $N_{e\uparrow}$ up-spin electrons and $N_{e\downarrow}$ down-spin electrons, the basis states $|i\rangle$ can be given by

$$|i\rangle = \prod_{m=1}^{N_{e\downarrow}} c^{\dagger}_{l_m,\downarrow} \prod_{n=1}^{N_{e\uparrow}} c^{\dagger}_{l_n,\uparrow} |\text{vac.}\rangle \,, \tag{B.1}$$

where $c^{\dagger}_{l,\sigma}$ is the creation operator of an electron with σ spin at site l, and $|\text{vac.}\rangle$ is the vacuum state. The l_m denotes the site of the m-th electron, satisfying $l_1 \neq l_2 \neq \cdots \neq l_{N_{e\downarrow}}$ and depends on the state $|i\rangle$. The total number of the basis states $|i\rangle$ is $N_{\text{st}} = {}_N\text{C}_{N_{e\downarrow}} \cdot {}_N\text{C}_{N_{e\uparrow}}$, where ${}_N\text{C}_M = N!/[M!(N-M)!]$. For the t-J model (2.38), $c^{\dagger}_{l\sigma}$ in (B.1) is replaced by $c^{\dagger}_{l,\sigma}(1-c^{\dagger}_{l,-\sigma}c_{l,-\sigma})$, and the total number of the basis states can be evaluated to be $N_{\text{st}} = {}_{N_e}\text{C}_{N_{e\downarrow}} \cdot {}_N\text{C}_{N-N_e}$ with $N_e = N_{e\downarrow} + N_{e\uparrow}$.

In order to obtain the ground state of such systems, one has to numerically diagonalize a $N_{\text{st}} \times N_{\text{st}}$ hermitian matrix. (If one can divide the total space into subspaces by making use of various symmetries contained in the system, the size of the matrices to be diagonalized is very usefully reduced.) Since N_{st} increases rapidly with N as well as N_e, the standard diagonalization technique that requires full storage of the matrix elements is of no practical use due to the limitation of the memory size in present computers. Instead of the full diagonalization, the Lanczos algorithm is widely used for obtaining the ground state (and a few excited states). One of the advantages of the method is that it is only necessary to store non-zero matrix elements rather than the full matrices in memory or on the hard-drive.

Let us review the Lanczos algorithm. Consider a unitary matrix U that transforms a $N_{\rm st} \times N_{\rm st}$ hermitian matrix to a real symmetric tridiagonal matrix T:

$$U^\dagger H U = T, \qquad U^\dagger U = I \tag{B.2}$$

and

$$T = \begin{pmatrix} a_1 & b_2 & 0 & \dots & 0 & 0 \\ b_2 & a_2 & b_3 & \dots & 0 & 0 \\ 0 & b_3 & a_3 & \dots & 0 & 0 \\ \vdots & \vdots & \vdots & \ddots & \vdots & \vdots \\ 0 & 0 & 0 & \dots & a_{N_{\rm st}-1} & b_{N_{\rm st}} \\ 0 & 0 & 0 & \dots & b_{N_{\rm st}} & a_{N_{\rm st}} \end{pmatrix}. \tag{B.3}$$

Taking $U = (|u_1\rangle, |u_2\rangle, |u_3\rangle, \dots, |u_{N_{\rm st}}\rangle)$, $|u_i\rangle$ being orthonormal vectors with $N_{\rm st}$ components, one can find from (B.2) and (B.3) a set of recurrence relations

$$\begin{aligned} H|u_1\rangle &= a_1|u_1\rangle + b_2|u_2\rangle \\ H|u_i\rangle &= b_i|u_{i-1}\rangle + a_i|u_i\rangle + b_{i+1}|u_{i+1}\rangle \qquad (2 \le i \le N_{\rm st}-1) \\ H|u_{N_{\rm st}}\rangle &= b_{N_{\rm st}}|u_{N_{\rm st}-1}\rangle + a_{N_{\rm st}}|u_{N_{\rm st}}\rangle . \end{aligned} \tag{B.4}$$

In the Lanczos algorithm, the vectors $|u_i\rangle$ and the coefficients a_i and b_i in (B.5) are numerically obtained in the following procedures: First prepare a normalized vector $|u_1\rangle = \sum_{i=1}^{N_{\rm st}} \alpha_i |i\rangle$ with randomly chosen α_i. Then set $|v_1\rangle = H|u_1\rangle$ and $a_1 = \langle u_1 \mid v_1\rangle$, then repeat the following process for $i = 1, 2, 3, \dots, M-1$ ($M \ll N_{\rm st}$).

$$\begin{cases} |w_i\rangle = |v\rangle_i - a_i|u_i\rangle \\ b_{i+1} = \sqrt{\langle w_i \mid w_i\rangle} \\ |u_{i+1}\rangle = |w_i\rangle / b_{i+1} \\ |v_{i+1}\rangle = H|u_{i+1}\rangle - b_{i+1}|u_i\rangle \\ a_{i+1} = \langle u_{i+1} \mid v_{i+1}\rangle . \end{cases} \tag{B.5}$$

After $M-1$ iterations, obtained is a real symmetric tridiagonal matrix T with dimension M. Approximate eigenvalues $\tilde{E}_i$ ($i = 0, 2, \dots, M-1$) for H are obtained by diagonalizing T, using standard numerical library routines. The virtues of this algorithm are that extreme (smallest and largest) eigenvalues including the ground-state energy $\tilde{E}_0$ converge rapidly for small M to the correct eigenvalues of H. In fact $M = 100$ to 200 is usually sufficient to achieve convergence for the ground-state energy, i.e., $\tilde{E}_0 \to E_0$.

By using the eigenvector $|x_0\rangle$ for $\tilde{E}_0$ obtained by diagonalizing the $M \times M$ matrix T, the ground-state wave function is given by

$$|0\rangle = \sum_{i=1}^{M} [|x_0\rangle]_i \, |u_i\rangle , \tag{B.6}$$

where $[|x_0\rangle]_i$ represents the i-th component of $|x_0\rangle$. Since in the numerical procedure (B.5) the vectors $|u_i\rangle$ are usually overwritten with $|u_{i+1}\rangle$ and thus not preserved, one has to repeat (B.5) to create them. If one would need a very accurate wave function, it might be better to use the inverse iteration method, in which $|x_0\rangle$ is obtained by repeating $|y_{i+1}\rangle = (\tilde{E}_0 I - H)^{-1} |y_i\rangle$, staring from an arbitrary vector $|y_1\rangle$. (See appropriate textbooks for this method.)

Here let us comment on the problem encountered when one uses the Lanczos algorithm. The Lanczos procedure (B.5) is numerically subject to roundoff errors introduced by the finite-precision arithmetics. This leads to the loss of orthogonality of the vectors $|u_i\rangle$. Reorthogonalization procedures for each step might be effective to recover orthogonality. However, such procedures require huge number of numerical operations, thereby being impractical. If one performs the Lanczos procedure (B.5) and calculates $\tilde{E}_i$ step by step with increasing M, one will find that the effect of the roundoff errors appear as *ghost* eigenvalues [12]. The ghost moves with increasing M. When the ghost is very close to the real eigenvalues, one must be cautious of the accuracy of the corresponding eigenvectors. In order to check the accuracy of the eigenvectors obtained from (B.5), it may be important to examine the norm of the vectors as well as physical quantities whose operators commute with Hamiltonian, such as the total spin, since they reflect well the accuracy of the vectors.

Let us move onto the calculation of the dynamical correlation functions at zero temperature. For an operator A the spectral representation is given by

$$I(\omega) = \sum_i |\langle \nu_i| A |0\rangle|^2 \, \delta\left(\omega - (E_i - E_0)/\hbar\right), \tag{B.7}$$

where $|\nu_i\rangle$ is the i-th final state with energy E_i. Because it converges rapidly for low-energy eigenstates as mentioned above, the Lanczos algorithm (B.5) may be useful for the calculation of $I(\omega)$ if one only wishes to know low-energy excitations accurately. First, prepare the initial vector

$$|u_1\rangle = A|0\rangle / \sqrt{C} \tag{B.8}$$

with the normalization factor $C = \langle 0| A^\dagger A |0\rangle$. After performing the procedure (B.5) and diagonalizing the $M \times M$ tridiagonal matrix T, one may obtain an approximate spectral representation

$$\begin{aligned} I(\omega) &\approx C \sum_{i=0}^{M-1} |\langle x_i \mid u_1\rangle|^2 \, \delta\left(\omega - (\tilde{E}_i - E_0)/\hbar\right) \\ &= C \sum_{i=0}^{M-1} [|x_i\rangle]_1^2 \, \delta\left(\omega - (\tilde{E}_i - E_0)/\hbar\right). \end{aligned} \tag{B.9}$$

Since the energies $\tilde{E}_i$ are discrete for finite-size systems, an artificial broadening with Lorentzian type is usually introduced into the delta function when spectral-weight distributions are discussed (see, for instance, Fig. 2.12). It is important to emphasize again that low-energy excitation spectra show good convergence if M is taken to be $M = 200$ to 300 in (B.9). For high-energy spectra, the convergency is not good, but weight distributions for a certain-energy range can be reproduced very well.

The application of the Lanczos method to finite-temperature physical quantities have also been realized [1] and various quantities have been calculated for the t-J and Hubbard models. This is called the finite-temperature Lanczos method. For details, please see the review article [1].

Finally let us comment on the shapes of finite-size systems for two-dimensional Hubbard and t-J models. We consider N-site square lattices (sometimes written as $\sqrt{N} \times \sqrt{N}$) with periodic boundary conditions, N being even number. The number sites N should satisfy $N^2 = l^2 + m^2$, l and m being integer, and the allowed numbers are N $(l, m) = 4$ $(2, 0)$, 8 $(2, 2)$, 10 $(3, 1)$, 16 $(4, 0)$, 18 $(3, 3)$, 20 $(4, 2)$, 26 $(5, 1)$, 32 $(4, 4)$, 34 $(5, 3)$, 36 $(6, 0)$, ... Note that wave vectors defined on the lattices are discrete in the Brillouin zone, and all of the lattices contain the wave vectors $\boldsymbol{k} = (0, 0)$ and (π, π). The $(\pi, 0)$ and $(0, \pi)$ points are defined in the lattices satisfying $N = 4n$ (n: integer). Note that the $N = 20$ lattice is used in Figs. 2.29 and 2.33b. The maximum sizes of square lattices examined up until now are $N = 20$, $N = 32$, and $N = 36$ for the Hubbard, t-J, and Heisenberg models, respectively.

C. Projection Method (Memory Function Method, Composite Operator Method)

The projection method, which is also termed the memory function method, the composite operator method and Liouville operator method, is one of the non-perturbational methods widely used in many body physics. Consider the calculation of the retarded Green's function:

$$G^R_{AA}(t) = \theta(t)\langle[A^\dagger(t), A(0)]\rangle, \tag{C.1}$$

where A and B are boson operators and $[\cdots]$ indicates a commutation relation. The method is also applicable for fermion operators. A thermal average is implied by $\langle\cdots\rangle$. The Fourier transform of $G^R_{AA}(t)$ is involves the spectral function $\sigma_{AA}(\omega)$, i.e.,

$$G^R_{AA}(\omega) = \int \mathrm{d}\kappa \frac{\sigma_{AA}(\kappa)}{\omega - \kappa + \mathrm{i}\eta}, \tag{C.2}$$

where η is an infinitesimal. Instead of the retarded Green function, it is convenient to introduce the relaxation (Kubo) function at temperature $T(\equiv 1/\beta)$ defined by

$$C_{AA}(t) = \theta(t)T\int_0^\beta \mathrm{d}\lambda\langle A^\dagger(t)A(\mathrm{i}\lambda)\rangle \equiv \theta(t)\langle A^\dagger(t)A(0)\rangle_\lambda. \tag{C.3}$$

The spectral representation of the relaxation function is

$$C_{AA}(\omega) = \int \mathrm{d}\kappa \frac{\sigma_{AA}(\kappa)}{\kappa(\omega - \kappa + \mathrm{i}\eta)}. \tag{C.4}$$

Then, we have a relation:

$$\mathrm{Im}G_{AA}(\omega) = \beta\omega\mathrm{Im}C_{AA}(\omega). \tag{C.5}$$

The relaxation function is also expressed, using the Liouville operator L, by

$$C_{AB}(z) = T\int_0^\beta \mathrm{d}\lambda\Big\langle A^\dagger(0)\frac{1}{z-L}A(\mathrm{i}\lambda)\Big\rangle \equiv \Big\langle A\Big|\frac{1}{z-L}\Big|A\Big\rangle, \tag{C.6}$$

where z is defined in the complex frequency plane, and

$$\mathrm{i}LA \equiv \frac{\partial A}{\partial t}. \tag{C.7}$$

Now, $\langle A|$ and $|A\rangle$ are treated as vectors in the Hilbert space. The subspace spanned by $|A\rangle$ is projected out by the projection operator P as

$$P = |A\rangle \frac{1}{\langle A|A\rangle} \langle A|, \tag{C.8}$$

while $Q = 1 - P$ projects onto the remaining space. Using the identity:

$$\frac{1}{z-L} = \frac{1}{z-LQ} + \frac{1}{z-LQ} LP \frac{1}{z-L}, \tag{C.9}$$

(C.6) is rewritten as

$$C_{AA}(z) = \left\langle A \middle| \frac{1}{z} \middle| A \right\rangle + \left\langle A \middle| \frac{1}{z-LQ} L \middle| A \right\rangle \frac{1}{\langle A|A\rangle} C_{AA}(z). \tag{C.10}$$

Then, we obtain

$$C_{AA}(z) = \frac{1}{z - M_0 I^{-1} - \delta M(z) I^{-1}} I, \tag{C.11}$$

where we define the normalization factor:

$$I = \langle A|A\rangle, \tag{C.12}$$

the static part of the self-energy:

$$M_0 = \langle A|L|A\rangle, \tag{C.13}$$

and the dynamical part of the self energy called "the memory function":

$$\delta M(z) = \left\langle A \middle| L \left(Q \frac{1}{z - QLQ} Q \right) L \middle| A \right\rangle. \tag{C.14}$$

This corresponds to the relaxation function for LA except that the Hilbert space is restricted. The same procedure for $C_{AA}(z)$ is applicable to $\delta M(z)$. Finally, $C_{AA}(z)$ in (C.11) is expressed as a continued fraction form.

As seen above, this method is related to the equation of motion method in the Green function formalism, and Lanczos method used for numerical calculations. Since, the operator products which arise in the equation of motion of A are treated as a single quantum variable in this scheme, this method is sometimes called the composite operator formalism. It is usually difficult to evaluate the continued fraction to higher order while maintaining the restrictions in Hilbert space. At some stage in the development of the continued fraction, it is convenient to relate the new operators to the one which might be treated as a composite operator, etc. The relaxation function for the latter be evaluated, e.g., by the loop approximation using the diagrammatic technique.

D. Thermoelectric Effects

D.1 Seebeck Effect

The thermopower (Q) relates the linear response of the gradient of electro-chemical potential ($\nabla(\mu - e\varphi)$) to the temperature gradient (∇T), i.e.,

$$\nabla\left(\mu/e - \varphi\right) = Q\nabla T. \tag{D.1}$$

The thermopower is measured using the experimental setup shown in Fig. D.1 (in the following, let us consider the thermoelectric response in one-dimension, for simplicity). The thermocouple consists of the materials A (the hatched segment) and B (the white segment). The material B is cut in two and the open ends are connected to a voltmeter via connecting wires (solid lines). The coordinates $x_1 \sim x_5$ and lengths $L_1 \sim L_5$ of each part are shown in Fig. D.1 which also indicates the definition of the junction temperatures T and $T + \varDelta T$ ($\varDelta T > 0$). The wires connected to B have the same temperature T_s. The temperature at the terminals of the voltmeter is T_p. The subscripts A, B and l are used to distinguish the materials, their thermopowers and the electro-chemical potentials. For the material A, integrating the left-hand-side of (D.1) gives

$$\int_0^{L_3}\left(\frac{1}{e}\frac{\partial\mu_A}{\partial x_3} - \frac{\partial\varphi_A}{\partial x_3}\right)\mathrm{d}x_3$$

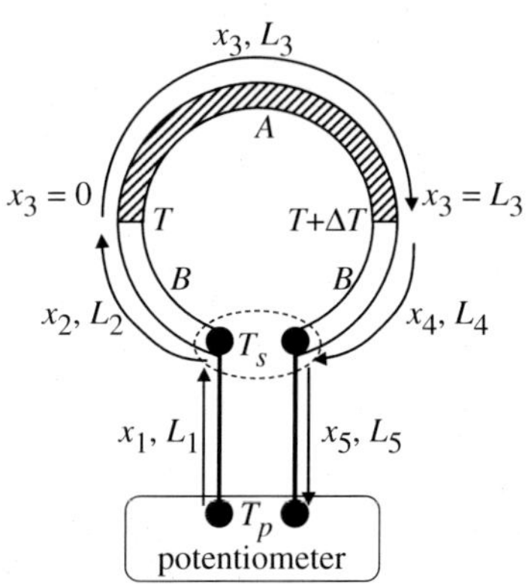

Fig. D.1. An experimental setup to measure the thermopower using thermocouple.

$$= \left(\frac{1}{e}\mu_A - \varphi_A\right)_{x_3=L_3} - \left(\frac{1}{e}\mu_A - \varphi_A\right)_{x_3=0}. \tag{D.2}$$

while from the right-hand-side is obtained

$$\int_0^{L_3} Q_A \frac{\partial T}{\partial x_3} \mathrm{d}x_3 = \int_T^{T+\Delta T} Q_A \mathrm{d}T. \tag{D.3}$$

In the same way, we find the following equations by the integration along the circuit in Fig. D.1 in clockwise direction:

$$\left(\frac{1}{e}\mu_l - \varphi_l\right)_{x_1=L_1} - \left(\frac{1}{e}\mu_l - \varphi_l\right)_{x_1=0} = \int_{T_p}^{T_s} Q_l \mathrm{d}T, \tag{D.4}$$

$$\left(\frac{1}{e}\mu_B - \varphi_B\right)_{x_2=L_2} - \left(\frac{1}{e}\mu_B - \varphi_B\right)_{x_2=0} = \int_{T_s}^{T} Q_B \mathrm{d}T, \tag{D.5}$$

$$\left(\frac{1}{e}\mu_A - \varphi_A\right)_{x_3=L_3} - \left(\frac{1}{e}\mu_A - \varphi_A\right)_{x_3=0} = \int_{T}^{T+\Delta T} Q_A \mathrm{d}T, \tag{D.6}$$

$$\left(\frac{1}{e}\mu_B - \varphi_B\right)_{x_4=L_4} - \left(\frac{1}{e}\mu_B - \varphi_B\right)_{x_4=0} = \int_{T+\Delta T}^{T_s} Q_B \mathrm{d}T, \tag{D.7}$$

$$\left(\frac{1}{e}\mu_l - \varphi_l\right)_{x_5=L_5} - \left(\frac{1}{e}\mu_l - \varphi_l\right)_{x_5=0} = \int_{T_s}^{T_p} Q_l \mathrm{d}T. \tag{D.8}$$

The measurement is performed in the steady state, i.e., there is no electric current flow, and this implies that there must be no difference in the electrochemical potentials of materials which form an iso-thermal junction, i.e.,

$$\left(\frac{1}{e}\mu_l - \varphi_l\right)_{x_1=L_1} = \left(\frac{1}{e}\mu_B - \varphi_B\right)_{x_2=0}, \tag{D.9}$$

$$\left(\frac{1}{e}\mu_B - \varphi_B\right)_{x_2=L_2} = \left(\frac{1}{e}\mu_A - \varphi_A\right)_{x_3=0}, \tag{D.10}$$

$$\left(\frac{1}{e}\mu_A - \varphi_A\right)_{x_3=L_3} = \left(\frac{1}{e}\mu_B - \varphi_B\right)_{x_4=0}, \tag{D.11}$$

$$\left(\frac{1}{e}\mu_B - \varphi_B\right)_{x_4=L_4} = \left(\frac{1}{e}\mu_l - \varphi_l\right)_{x_5=0}. \tag{D.12}$$

From (D.4)–(D.12) one has,

$$\begin{aligned} &\left(\frac{1}{e}\mu_l - \varphi_l\right)_{x_5=L_5} - \left(\frac{1}{e}\mu_l - \varphi_l\right)_{x_1=0} \\ &= \int_{T_p}^{T_s} Q_l \mathrm{d}T + \int_{T_s}^{T} Q_B \mathrm{d}T + \int_{T}^{T+\Delta T} Q_A \mathrm{d}T + \int_{T+\Delta T}^{T_s} Q_B \mathrm{d}T + \int_{T_s}^{T_p} Q_l \mathrm{d}T \\ &= \int_{T}^{T+\Delta T} \left(Q_A - Q_B\right) \mathrm{d}T. \end{aligned} \tag{D.13}$$

Since the connecting wires are chemically identical, $\mu_l(x_5 = L_5) = \mu_l(x_1 = 0)$ and the difference of electro-static potential $\varphi_l(x_5 = L_5) - \varphi_l(x_1 = 0)$ measured by the voltmeter is

$$\varphi_l(x_5 = L_5) - \varphi_l(x_1 = 0) = \int_T^{T+\Delta T} (Q_B - Q_A)\,\mathrm{d}T. \tag{D.14}$$

The difference $Q_{AB} \equiv Q_B - Q_A$ is called the Seebeck coefficient of the thermocouple AB.

When the material B is a superconductor and $T + \Delta T$ is lower than the transition temperature of the superconductivity, (D.14) becomes

$$\begin{aligned}\varphi_l(x_5 = L_5) - \varphi_l(x_1 = 0) &= -\int_T^{T+\Delta T} Q_A \mathrm{d}T \qquad \text{(D.15)}\\ &= \left(\frac{1}{e}\mu_A - \varphi_A\right)_{x_3=L_3} - \left(\frac{1}{e}\mu_A - \varphi_A\right)_{x_3=0}.\end{aligned}$$

Thus, the gradient of the electro-chemical potential is measured by the voltmeter and the *absolute* Seebeck coefficient, i.e. the thermopower is determined. Measurement of the thermopower at higher temperatures use platinum (Pt) or lead (Pb), for which thermopower is known [14], as the reference material B.

D.2 Peltier Effect

At a junction consisting of different materials, the electric current is continuous. However, the heat current can be discontinuous because the thermoelectric system exchanges heat with a reservoir. This discontinuity causes the Peltier effect.

Let us consider an electric current through a junction consisting of materials A and B without a temperature gradient as shown in Fig. D.2. If $\partial T/\partial x = 0$, the linear dynamical law (6.9) reduces to

$$\begin{cases} j_1 = -M^{11}\dfrac{1}{T}\dfrac{\partial(\mu - e\varphi)}{\partial x}, \\ j_2 = -M^{21}\dfrac{1}{T}\dfrac{\partial(\mu - e\varphi)}{\partial x}, \end{cases} \tag{D.16}$$

and as a result,

$$j_2 = \frac{M^{21}}{M^{11}} j_1. \tag{D.17}$$

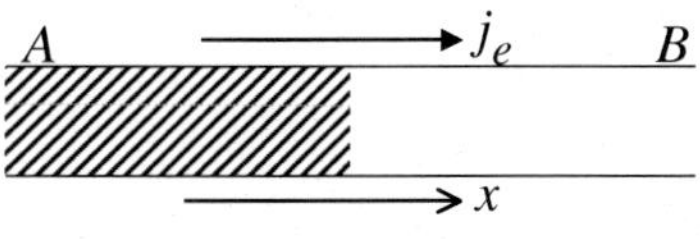

Fig. D.2. Electric current through a junction of the materials A and B without temperature gradient. j_e is the electric current density. x axis is taken along the current.

Using (D.17), the heat currents in the materials are given by

$$j_{heat,A} = j_{2,A} - \mu_A j_{1,A} = \left(\frac{M_A^{21}}{M_A^{11}} - \mu_A \right) j_{1,A}, \tag{D.18}$$

$$j_{heat,B} = j_{2,B} - \mu_B j_{1,B} = \left(\frac{M_B^{21}}{M_B^{11}} - \mu_B \right) j_{1,B}. \tag{D.19}$$

The subscripts A and B are used to distinguish the materials. Since the particle current is continuous at the junction, $j_{1,A} = j_{1,B} = j_e/(-e)$ and so finally, we obtain

$$\begin{aligned} j_{heat,B} - j_{heat,A} &= \left\{ \left(-\frac{1}{e} \frac{M_B^{21}}{M_B^{11}} + \frac{\mu_B}{e} \right) - \left(-\frac{1}{e} \frac{M_A^{21}}{M_A^{11}} + \frac{\mu_A}{e} \right) \right\} j_e \\ &\equiv \Pi_{AB} j_e. \end{aligned} \tag{D.20}$$

This is the definition of the Peltier coefficient Π_{AB}. The electric current flowing across the junction of A and B is expressed as $j_e S_c$ where S_c denotes the cross sectional area of the materials in Fig. D.2. Thus the junction of A and B absorbs the heat $\Pi_{AB} j_e S_c$ per unit time. This is the Peltier effect. The Peltier effect cannot occur in a uniform single material.

Using the Onsager reciprocity $M^{12} = M^{21}$, the definition of the thermopower (6.19) and (D.20) give

$$TQ_{AB} = \Pi_{AB}, \tag{D.21}$$

which is known as the second Kelvin relation. Measurements of the thermopower and the Peltier coefficient can be used to verify Onsager reciprocity.

D.3 Thomson Effect

Thomson (Lord Kelvin) discovered a third effect in addition to those of Seebeck and Peltier, and is known as the Thomson effect. Figure D.3 shows an electric current in the presence of a temperature gradient in a single material. The electric current generates a quantity of heat q given by

$$q\Delta x = \tau j_e S_c \Delta T, \tag{D.22}$$

where S_c is the cross sectional area of the material and τ is the Thomson coefficient. The heat $q\Delta x$ is that required to change the temperature of the electric current $j_e S_c$ and so Thomson described τ as *"specific heat of electricity"* [15].

Consider an electric current flowing in a single material as shown in Fig. D.3. Using the electrical resistivity ρ, thermopower Q, and thermal conductivity κ, the linear dynamical law (6.9) can be written as

$$j_1 = -\frac{1}{e^2} \rho^{-1} \frac{\partial}{\partial x} (\mu - e\varphi) + \frac{1}{e} \rho^{-1} Q \frac{\partial T}{\partial x}, \tag{D.23}$$

$$j_2 = -eTQ j_1 - \kappa \frac{\partial T}{\partial x} + \mu j_1. \tag{D.24}$$

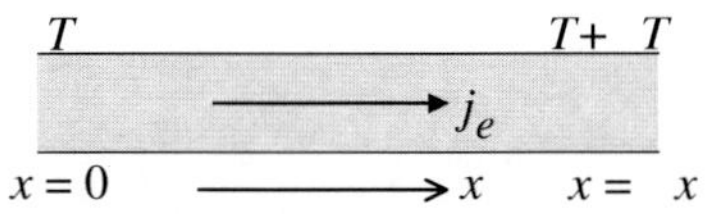

Fig. D.3. Electric current in a single material (the gray area) with temperature gradient. The quantity j_e is the electric current density. The x-axis is taken along the material.

Since the electric and energy currents flow in the x direction, the vector particle and energy fluxes are $\boldsymbol{j}_1 = (j_1, 0, 0)$ and $\boldsymbol{j}_2 = (j_2, 0, 0)$ in a Cartesian coordinate system. The conservation of energy with heat and particle currents currents and an electro-static field (φ) takes the form

$$\frac{\partial n_{energy}}{\partial t} = \mathcal{S}\,\mathrm{div}\,(\boldsymbol{j}_2\,\mathcal{S}\,e\varphi\boldsymbol{j}_1) = \mathcal{S}\frac{\partial}{\partial x}(j_2\,\mathcal{S}\,e\varphi j_1)\,, \tag{D.25}$$

where n_{energy} is the local energy-density. Inserting (D.24) into (D.25), the right-hand-side of (D.25) gives

$$\begin{aligned}\frac{\partial}{\partial x}(j_2\,\mathcal{S}\,e\varphi j_1) &= \mathcal{S}e\frac{\partial}{\partial x}(TQj_1)\,\mathcal{S}\,\frac{\partial}{\partial x}\left(\kappa\frac{\partial T}{\partial x}\right) + \frac{\partial}{\partial x}\left[(\mu\,\mathcal{S}\,e\varphi)\,j_1\right] \\ &= \mathcal{S}e\frac{\partial T}{\partial x}(Qj_1)\,\mathcal{S}\,eT\left(\frac{\partial Q}{\partial x}\right)j_1\,\mathcal{S}\,eTQ\left(\frac{\partial j_1}{\partial x}\right) \\ &\quad\mathcal{S}\,\frac{\partial}{\partial x}\left(\kappa\frac{\partial T}{\partial x}\right) \\ &\quad+\left[\frac{\partial(\mu\,\mathcal{S}\,e\varphi)}{\partial x}\right]j_1 + (\mu\,\mathcal{S}\,e\varphi)\left(\frac{\partial j_1}{\partial x}\right).\end{aligned} \tag{D.26}$$

Using the conservation of particle number, $\partial j_1/\partial x (= \mathrm{div}\boldsymbol{j}_1) = 0$, combining (D.23) and (D.26) then gives

$$\begin{aligned}\frac{\partial}{\partial x}(j_2\,\mathcal{S}\,e\varphi j_1) &= \mathcal{S}e\frac{\partial T}{\partial x}(Qj_1)\,\mathcal{S}\,eT\left(\frac{\partial Q}{\partial x}\right)j_1\,\mathcal{S}\,\frac{\partial}{\partial x}\left(\kappa\frac{\partial T}{\partial x}\right) \\ &\quad\mathcal{S}e^2\rho\,(j_1)^2 + e\,(j_1Q)\,\frac{\partial T}{\partial x} \\ &= \mathcal{S}eT\left(\frac{\partial Q}{\partial x}\right)j_1\,\mathcal{S}\,\frac{\partial}{\partial x}\left(\kappa\frac{\partial T}{\partial x}\right)\mathcal{S}\,e^2\rho\,(j_1)^2\,.\end{aligned} \tag{D.27}$$

Since the thermopower Q is not an explicit function of x,

$$\frac{\partial Q}{\partial x} = \frac{\partial Q}{\partial T}\frac{\partial T}{\partial x}. \tag{D.28}$$

Finally (D.25), corresponding to the conservation of energy, becomes

$$\begin{aligned}\frac{\partial n_{energy}}{\partial t} &= eT\frac{\partial Q}{\partial T}\left(j_1\frac{\partial T}{\partial x}\right)\mathcal{S}\,\frac{\partial}{\partial x}\left(\mathcal{S}\kappa\frac{\partial T}{\partial x}\right) + e^2\rho\,(j_1)^2 \\ &= \mathcal{S}T\frac{\partial Q}{\partial T}\left(j_e\frac{\partial T}{\partial x}\right)\mathcal{S}\,\frac{\partial}{\partial x}\left(\mathcal{S}\kappa\frac{\partial T}{\partial x}\right) + \rho\,(j_e)^2\,.\end{aligned} \tag{D.29}$$

The meaning of the terms on the right-hand-side is transparent, i.e., the last and second terms are the Joule heat and the divergence of the Fourier heat, respectively. The first term is the Thomson heat, i.e., the heat generated or absorbed in a single material where the electric current flows along the temperature gradient. The Thomson coefficient τ is the heat per unit current and per unit temperature gradient:

$$\tau = T\frac{\partial Q}{\partial T}. \tag{D.30}$$

The equation (D.30) might be compared with the relation between the specific heat c and entropy s:

$$c = T\frac{\partial s}{\partial T}. \tag{D.31}$$

As argued in the paragraph following (6.23), thermopower Q is the entropy of the electric current. Thomson's description of τ as the specific heat of electricity is therefore quite reasonable. About 150 years ago, (D.30) and (D.21) were derived by Thomson using thermodynamic principles. This historic and interesting application of thermodynamic ideas is presented in the next section.

D.4 The First and Second Laws of Thermodynamics and Thermoelectric Response in Solids: Thomson's Considerations

Let us consider the balance of the thermodynamic quantities in the circuit of interest, Fig. D.4. Following phenomenological laws are well established experimentally:

i) An electro-motive force ΔE occurs due to the Seebeck effect:

$$\Delta E = Q_{AB}\Delta T, \tag{D.32}$$

where Q_{AB} is the Seebeck coefficient of the thermocouple.

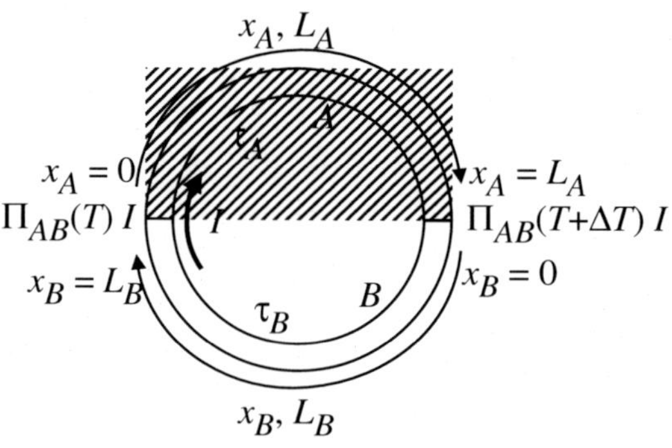

Fig. D.4. Thermocouple of the materials A and B. The temperature of the left (right) junction is T $(T + \Delta T)$.

ii) The Peltier heat emitted per unit time at the left junction

$$\Delta q_l = \Pi_{AB}(T)\, I. \tag{D.33}$$

At the right junction, the absorbed Peltier heat per unit time

$$\Delta q_r = \Pi_{AB}(T + \Delta T)\, I, \tag{D.34}$$

where Π_{AB} is the Pertier coefficient of the thermocouple and where $\Pi_{AB} = -\Pi_{BA}$ has been used .

iii) In order to maintain a temperature gradient $\Delta T/L_A$ the quantity of heat which must be supplied from the reservoir per unit time to the material A is

$$\Delta q_A = \tau_A I \Delta T, \tag{D.35}$$

which involves the Thomson coefficient τ_A for the material A. In the same way, the heat removed from B per unit time is

$$\Delta q_B = \tau_B I \Delta T, \tag{D.36}$$

where τ_B is the Thomson coefficient for the material B.

The energy of the electricity supplied by the emf ΔE of i) must balance the heats identified in ii) and iii), i.e., the conservation energy leads to

$$I Q_{AB} \Delta T = \Pi_{AB}(T + \Delta T)\, I - \Pi_{AB}(T)\, I + \tau_A I \Delta T - \tau_B I \Delta T. \tag{D.37}$$

There are two solutions to (D.37):

a) One possibility is that $I = 0$. In this case, (D.37) is an identity.
b) However when $I \neq 0$, (D.37) reduces to

$$Q_{AB} = \frac{\mathrm{d}\Pi_{AB}}{\mathrm{d}T} + \tau_A - \tau_B. \tag{D.38}$$

This equation is a simple consequence of the first law of thermodynamics and is known as the "first Kelvin relation".

The Seebeck, Peltier and Thomson effects are considered to be reversible in the sense that a change of the sign of the temperature difference ΔT and a change in the direction of the electric current I causes a change of the sign of the heat flows. Thomson assumed the thermoelectric effects are reversible and may be considered separately from the irreversible phenomena (e.g. a generation of Joule heat). The second law of thermodynamics can then be used to insist that the total change of entropy is zero. Equating zero to the various entropy changes per unit time results in

$$0 = -\frac{\Pi_{AB}(T)}{T} I + \frac{\Pi_{AB}(T + \Delta T)}{T + \Delta T} I + I \int_T^{T+\Delta T} \frac{\tau_A - \tau_B}{\tilde{T}} \mathrm{d}\tilde{T}. \tag{D.39}$$

There are again two solutions for (D.39):

c) If $I = 0$, (D.39) is an identity.

d) When $I \neq 0$, (D.39) in the limit $\Delta T \to 0$ reduces to

$$0 = \frac{1}{T}\frac{\mathrm{d}\Pi_{AB}}{\mathrm{d}T} - \frac{\Pi_{AB}}{T^2} + \frac{\tau_A - \tau_B}{T}. \tag{D.40}$$

Inserting (D.38) into (D.40), we have the "second Kelvin relation":

$$Q_{AB} = \frac{\Pi_{AB}}{T}. \tag{D.41}$$

Even for the insulators, some electric current flows at finite temperature. However, the solutions a) and c) are relevant for insulators and semiconductors since their electrical resistivity diverges as $\rho \sim \exp(\Lambda/T)$ as $T \to 0$ where $\Lambda > 0$. Since ρ diverges Ohm's law implies zero electric current at least in the absence of a temperature gradient. It might be imagined that the second term in (D.23), i.e. a temperature gradient might drive an electric current. However, it is well confirmed that Q/ρ goes to zero when $T \to 0$ in insulators and semiconductors, so that the electric current *does* disappear. This follows from the third law of thermodynamics. Let us examine this issue in the next section.

D.5 The Third Law of Thermodynamics and Thermoelectric Response in Solids

Let us discuss the behavior of the thermopower in the low temperature limit $T \to 0$. Figure D.5 shows a closed circuit composed of two materials A and B with a source of emf of magnitude V. The electric current is I and there is no temperature gradient. The emitted heat by the left junction per unit time is $\Pi_{AB}(T)I$, where Π_{AB} is the Peltier coefficient of the thermocouple. The Peltier effect causes a change of the entropy $\Pi_{AB}(T)I/T$, per unit time, at this junction. The third law of thermodynamics requires;

$$\lim_{T \to 0} \frac{\Pi_{AB}(T)I}{T} = 0. \tag{D.42}$$

From Ohm's law the electrical current $I = V/R_{AB}$ where R_{AB} is the resistance of the circuit. Because the applied voltage V is arbitrary, it must be that

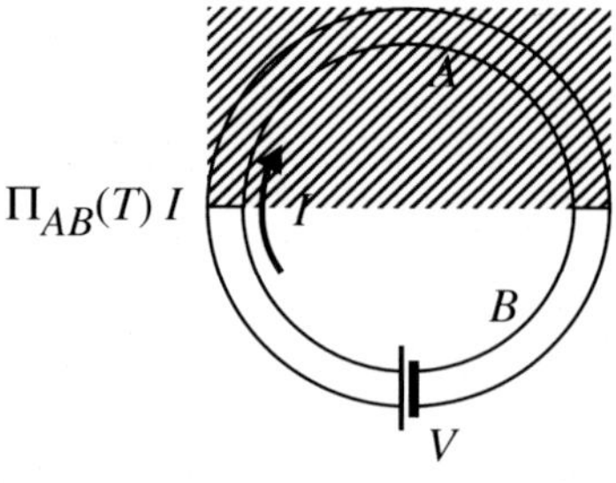

Fig. D.5. The Peltier effect. The temperature of the circuit is T.

$$0 = \lim_{T \to 0} \frac{\Pi_{AB}(T)}{T}(1/R_{AB}) = \lim_{T \to 0}(Q_{AB}/R_{AB}), \tag{D.43}$$

where, the second Kelvin relation, $\Pi_{AB}(T)/T = Q_{AB}$ is used, and where $Q_{AB}(= Q_B - Q_A)$ is the thermopower of the thermocouple. Since the physical dimensions cannot be relevant (D.43) implies

$$\lim_{T \to 0} Q/\rho = 0. \tag{D.44}$$

Then *if*

$$\lim_{T \to 0}(1/\rho) \neq 0, \tag{D.45}$$

it follows that

$$\lim_{T \to 0} Q = 0, \tag{D.46}$$

i.e., the thermopower of conductors must be zero when $T \to 0$ and this follows from the third law.

On the other hand, the electrical conductivity $1/\rho$ of insulators and semi-conductors goes to zero exponentially as $T \to 0$. Therefore even though the thermopower Q of insulators and semi-conductors diverges as $1/T$ at low-temperatures, the third law of thermodynamics in the form (D.44) is *not* violated [16, 17].

References for Appendices

1. J. Jaklič, P. Prelovšek: Adv. Phys. **49**, 1 (2000); J. Jaklič: Dynamics of Strongly Correlated Electrons at Finite Temperature. Thesis, University of Ljubljana, Ljubljana (1996)
2. C. Kittel: *Introduction to Solid State Physics* (John Wiley & Sons, Inc., New York 1995)
3. R. Kubo: J. Phys. Soc. Jpn. **12**, 570 (1957)
4. P. F. Maldague: Phys. Rev. B **16**, 2437 (1977)
5. W. Kohn: Phys. Rev. **133**, A171 (1964)
6. B. S. Shastry, B. Sutherland: Phys. Rev. Lett. **65**, 243 (1990)
7. D. J. Scalapino, S. R. White, S. Zhang: Phys. Rev. B **47**, 7995 (1993)
8. X. Zotos, P. Prelovšek: 'Transport in One Dimensional Quantum Systems'. In: *Interacting Electrons in Low Dimensions.* (Kluwer Academic Publishers, 2003); cons-mat/0304630
9. D. Poilblanc: Phys. Rev. B **44**, 9562 (1991)
10. C. Lanczos: J. Res. Natl. Bur. Stand. **45**, 255 (1950)
11. J. H. Wilkinson: *The Algebraic Eigenvalue Problem* (Calrendon, Oxford 1965)
12. C. C. Paige: J. Inst. Maths. Applics **10**, 373 (1972)
13. E. Dagotto: Rev. Mod. Phys. **66**, 763 (1994)
14. The thermopower of the reference materials can be determined by Thomson effect. See the following sub-sections.
15. Sir William Thomson, Mathematical and Physical Papers Vol. 1 Cambridge University Press (1882)
16. M. Kohler, Abhandl. Brauns. Wiss. Ges. **3**, 49 (1951)
17. J. Tauc, *Photo and thermoelectric effects in semiconductors*, (Pergamon Press, New York, 1962)

Index

AMO_3 4
A-type antiferromagnet 18
angle-resolved photoemission spectroscopy (ARPES) 63, 81, 133, 135, 142, 146
anisotropic superexchange interaction 25, 27
anomalous Hall effect 20
anti-symmetric exchange-interaction 25
anyon 120, 127
atomic scattering factor 201
auxiliary particle 102, 126, 128, 137, 163
azimuthal angle dependence 199, 203

$Ba_2CuGe_2O_7$ 34
$BaTiO_3$ 4
BCS theory 101, 112
Berry phase 180
Bethe ansatz 59, 70
(Bi,Pb)-Sr-Co-O 246
Bose-Einstein condensation 120, 124, 126, 134, 163

$Ca_2Co_2O_5$ 245
$CaMnO_3$ 15, 19, 20
charge and spin velocities 59
charge inhomogeneity 89
charge-transfer energy 13, 40
charge-transfer insulators 14
coherent state 120, 143
colossal magnetoresistance (CMR) 19, 167
cooperative Jahn-Teller effect 178, 180
corner-sharing cuprates 43
crystal field 1, 3
crystal-field splitting 6
cyclic four-spin interaction 49

d-p model 42
d-wave 133, 163
d-wave gap 123
d-wave pairing 105, 106
d-wave singlet 102
direct exchange interaction 11, 12
double exchange interaction 177
doublon 69, 77, 129
Drude weight 91, 92, 313
dynamical charge correlation function 71
Dzyaloshinski-Moriya interaction 25, 27

edge-sharing cuprates 47
electro-chemical potential 249, 323
electro-modulation spectroscopy 76
electron energy-loss spectroscopy 72
Emery model 42
exact diagonalization method 84

Fick's law 248
first Kelvin relation 329
first law of thermodynamics 329
flux state 136, 141, 142, 156
flux tube 116, 120, 149
Fourier's law 248

$GdFeO_3$-type distortion 226
gossamer superconductivity 158, 164
Grassman variable 120, 121, 163
Gutzwiller projection 132, 136, 158

hard core boson 116, 120, 126, 143
Heisenberg model 114, 116, 122, 140
high-spin state of Co ion 242
holon 59, 61, 65, 69, 77, 129, 134
holon-doublon model 68, 78
Holstein-Primakoff theory 103
Holstein-Primakoff transformation 111, 163
Hubbard band 128, 131, 132, 157, 161, 162

Hubbard model 11, 13, 102
Hund coupling 11, 12, 16, 171

intermediate-spin state of Co ion 242

Jahn-Teller effect 178, 236
Jordan-Wigner formulation 137
Jordan-Wigner transformation 103, 116, 118, 122, 138, 142, 163

K_2CuF_4-type structure 5
Kanamori-Goodenough rule 13
kinetic coefficient 248
Kramers-Kronig relation 314

$(La_{1.8}Eu_{0.2})CuO_4$ 34
La_2CuO_4 5, 29, 39, 51
La_2NiO_4 25
$La_{1-x}Ca_xMnO_3$ 21
$La_{1-x}Sr_xCoO_3$ 241
$La_{1-x}Sr_xMnO_3$ 4
$La_{1.2}Sr_{1.8}Mn_2O_7$ 8
$La_{1.6-x}Nd_{0.4}Sr_xCuO_4$ 88
$La_{2-x}Ba_xCuO_4$ 88
$La_{2-x}Sr_xCuO_4$ 38, 91
$LaCoO_3$ 242
$LaMnO_3$ 18, 19
Lanczos method 57, 62, 317
$LaTiO_3$ 282, 283, 290
$LaVO_3$ 301
Li_2CuO_2 38, 39, 57, 58
Lie algebra 151, 152, 164
linear dynamical law 248
linear response theory 312
Lord Kelvin 326
low-spin state of Co ion 242
lower and upper Hubbard bands 52
Luther-Emery model 62, 67

Madelung potential 2, 6, 40, 193
magnetic anisotropy 22
magnetic polaron 116
magnetic Raman scattering 15
Manganite with Layered Structure 190
manganites 167
Meissner effect 141
Mott gap 64, 68
Mott insulator 12, 67, 72, 73, 102, 111, 163
Mott-Hubbard insulators 14, 40

$Na_{0.35}CoO_2{\cdot}1.3H_2O$ 241
$NaCo_2O_4$ 241, 245
$Nd_{2-x}Ce_xCuO_4$ 90, 91
NiF_2 24, 25
nonlinear optical response 74

O(3) 144
Ohm's law 248
Onsager reciprocity 247, 326
optical conductivity 55, 281
orbital $t-J$ Hamiltonian 272
orbital bond 282, 287
orbital chain 298
orbital degree of freedom 168
orbital excitation 211
orbital gap 266
orbital liquid 186, 279, 282
orbital order 169, 182, 227
orbital polaron 276
orbital singlet 299
orbital wave 18, 211, 266
orbital-charge coupling 272, 277
orbiton 18, 281, 289

Pauli exclusion principle 13
Peierls phase 311
Peltier effect 325
perovskite structure 3, 4, 15
phase separation 89, 157
phase string 67
phonon 107, 108
polaron 115
Pr_2CuO_4 57
pre-formed pair 124
projection method 209
pseudo-gap 135
pseudodipolar interaction 25
pseudogap 87
pump and probe spectroscopy 76

$R_{1-x}A_xMnO_3$ 167
$R_{1-x}A_{1+x}MnO_4$ 167
$R_{1-x}A_xTiO_3$ 225
$R_{1-x}A_xVO_3$ 225
$R_{2-2x}A_{1+2x}Mn_2O_7$ 167
$R_{2-2x}A_{1+2x}Mn_2O_7$ 190
Racah parameters 172
Raman scattering 214
resonant inelastic x-ray scattering 73, 214
resonant valence bond (RVB) 102, 122, 123, 132, 134, 140, 156, 163
resonant x-ray scattering 197, 230

s-wave 133
s-wave pairing 105

s-wave singlet 102
second Kelvin relation 330
second law of thermodynamics 329
Seebeck effect 248, 323
self-consistent Born approximation 82, 115, 273, 289
single ion anisotropy 22
single-particle spectral function 61
slave boson 65, 126, 146
slave-boson representation 281
SO(3) 150
SO(5) 102, 111, 144, 151, 156, 164
spin and charge correlation functions 60
spin charge separation 130
spin wave 112, 142
spin-charge separation 58, 59, 68, 126
spin-gap 87
spin-orbit 235
spin-orbit interaction 22, 26
spin-orbital dimerization 302
spin-orbital resonance 286
spinon 59, 61, 65, 133, 135, 138, 139
$Sr_2CuO_2Cl_2$ 82
Sr_2CuO_3 38, 39, 57, 58, 73, 76, 77, 79, 80
$SrCu_2O_3$ 38, 39
$SrCuO_2$ 65, 73, 74
string gauge 116, 118
stripes 88
SU(2) 144
SU(3) 143, 144
superexchange interaction 12, 13, 16, 43

t-J model 55, 60, 64, 102, 115, 163
t-t'-t''-J model 82, 91, 92, 95
t_{2g} -superexchange 283
thermoelectric material 245
thermopower 249, 250, 323
third harmonic generation spectroscopy (THG) 76
third law of thermodynamics 330, 331
Thomson effect 326
three-band Hubbard model 42
titanates 225
tolerance factor 182, 193
Tomonaga-Luttinger liquid 58, 67
Tomonaga-Luttinger model 62
two photon absorption (TPA) 76

U(1) 136, 140, 144, 150

V_2O_3 225
vanadates 225

Wannier orbital 41, 46

XY model 116, 122, 163

$YBa_2Cu_3O_7$ 5
$YBa_2Cu_3O_{6.5}$ 5
$YTiO_3$ 283, 298
YVO_3 305

Zhang-Rice singlet state 50

Springer Series in Solid-State Sciences

Editors: M. Cardona P. Fulde K. von Klitzing H.-J. Queisser

1 **Principles of Magnetic Resonance** 3rd Edition By C. P. Slichter
2 **Introduction to Solid-State Theory** By O. Madelung
3 **Dynamical Scattering of X-Rays in Crystals** By Z. G. Pinsker
4 **Inelastic Electron Tunneling Spectroscopy** Editor: T. Wolfram
5 **Fundamentals of Crystal Growth I** Macroscopic Equilibrium and Transport Concepts By F. E. Rosenberger
6 **Magnetic Flux Structures in Superconductors** 2nd Edition By R. P. Huebener
7 **Green's Functions in Quantum Physics** 2nd Edition By E. N. Economou
8 **Solitons and Condensed Matter Physics** Editors: A. R. Bishop and T. Schneider
9 **Photoferroelectrics** By V. M. Fridkin
10 **Phonon Dispersion Relations in Insulators** By H. Bilz and W. Kress
11 **Electron Transport in Compound Semiconductors** By B. R. Nag
12 **The Physics of Elementary Excitations** By S. Nakajima, Y. Toyozawa, and R. Abe
13 **The Physics of Selenium and Tellurium** Editors: E. Gerlach and P. Grosse
14 **Magnetic Bubble Technology** 2nd Edition By A. H. Eschenfelder
15 **Modern Crystallography I** Fundamentals of Crystals Symmetry, and Methods of Structural Crystallography 2nd Edition By B. K. Vainshtein
16 **Organic Molecular Crystals** Their Electronic States By E. A. Silinsh
17 **The Theory of Magnetism I** Statics and Dynamics By D. C. Mattis
18 **Relaxation of Elementary Excitations** Editors: R. Kubo and E. Hanamura
19 **Solitons** Mathematical Methods for Physicists By. G. Eilenberger
20 **Theory of Nonlinear Lattices** 2nd Edition By M. Toda
21 **Modern Crystallography II** Structure of Crystals 2nd Edition By B. K. Vainshtein, V. L. Indenbom, and V. M. Fridkin
22 **Point Defects in Semiconductors I** Theoretical Aspects By M. Lannoo and J. Bourgoin
23 **Physics in One Dimension** Editors: J. Bernasconi and T. Schneider
24 **Physics in High Magnetics Fields** Editors: S. Chikazumi and N. Miura
25 **Fundamental Physics of Amorphous Semiconductors** Editor: F. Yonezawa
26 **Elastic Media with Microstructure I** One-Dimensional Models By I. A. Kunin
27 **Superconductivity of Transition Metals** Their Alloys and Compounds By S. V. Vonsovsky, Yu. A. Izyumov, and E. Z. Kurmaev
28 **The Structure and Properties of Matter** Editor: T. Matsubara
29 **Electron Correlation and Magnetism in Narrow-Band Systems** Editor: T. Moriya
30 **Statistical Physics I** Equilibrium Statistical Mechanics 2nd Edition By M. Toda, R. Kubo, N. Saito
31 **Statistical Physics II** Nonequilibrium Statistical Mechanics 2nd Edition By R. Kubo, M. Toda, N. Hashitsume
32 **Quantum Theory of Magnetism** 2nd Edition By R. M. White
33 **Mixed Crystals** By A. I. Kitaigorodsky
34 **Phonons: Theory and Experiments I** Lattice Dynamics and Models of Interatomic Forces By P. Brüesch
35 **Point Defects in Semiconductors II** Experimental Aspects By J. Bourgoin and M. Lannoo
36 **Modern Crystallography III** Crystal Growth By A. A. Chernov
37 **Modern Chrystallography IV** Physical Properties of Crystals Editor: L. A. Shuvalov
38 **Physics of Intercalation Compounds** Editors: L. Pietronero and E. Tosatti
39 **Anderson Localization** Editors: Y. Nagaoka and H. Fukuyama
40 **Semiconductor Physics** An Introduction 6th Edition By K. Seeger
41 **The LMTO Method** Muffin-Tin Orbitals and Electronic Structure By H. L. Skriver
42 **Crystal Optics with Spatial Dispersion, and Excitons** 2nd Edition By V. M. Agranovich and V. L. Ginzburg
43 **Structure Analysis of Point Defects in Solids** An Introduction to Multiple Magnetic Resonance Spectroscopy By J.-M. Spaeth, J. R. Niklas, and R. H. Bartram
44 **Elastic Media with Microstructure II** Three-Dimensional Models By I. A. Kunin
45 **Electronic Properties of Doped Semiconductors** By B. I. Shklovskii and A. L. Efros
46 **Topological Disorder in Condensed Matter** Editors: F. Yonezawa and T. Ninomiya

Springer Series in Solid-State Sciences

Editors: M. Cardona P. Fulde K. von Klitzing H.-J. Queisser

47 **Statics and Dynamics of Nonlinear Systems**
Editors: G. Benedek, H. Bilz, and R. Zeyher

48 **Magnetic Phase Transitions**
Editors: M. Ausloos and R. J. Elliott

49 **Organic Molecular Aggregates**
Electronic Excitation and Interaction Processes
Editors: P. Reineker, H. Haken, and H. C. Wolf

50 **Multiple Diffraction of X-Rays in Crystals**
By Shih-Lin Chang

51 **Phonon Scattering in Condensed Matter**
Editors: W. Eisenmenger, K. Laßmann, and S. Döttinger

52 **Superconductivity in Magnetic and Exotic Materials** Editors: T. Matsubara and A. Kotani

53 **Two-Dimensional Systems, Heterostructures, and Superlattices**
Editors: G. Bauer, F. Kuchar, and H. Heinrich

54 **Magnetic Excitations and Fluctuations**
Editors: S. W. Lovesey, U. Balucani, F. Borsa, and V. Tognetti

55 **The Theory of Magnetism II** Thermodynamics and Statistical Mechanics By D. C. Mattis

56 **Spin Fluctuations in Itinerant Electron Magnetism** By T. Moriya

57 **Polycrystalline Semiconductors**
Physical Properties and Applications
Editor: G. Harbeke

58 **The Recursion Method and Its Applications**
Editors: D. G. Pettifor and D. L. Weaire

59 **Dynamical Processes and Ordering on Solid Surfaces** Editors: A. Yoshimori and M. Tsukada

60 **Excitonic Processes in Solids**
By M. Ueta, H. Kanzaki, K. Kobayashi, Y. Toyozawa, and E. Hanamura

61 **Localization, Interaction, and Transport Phenomena** Editors: B. Kramer, G. Bergmann, and Y. Bruynseraede

62 **Theory of Heavy Fermions and Valence Fluctuations** Editors: T. Kasuya and T. Saso

63 **Electronic Properties of Polymers and Related Compounds**
Editors: H. Kuzmany, M. Mehring, and S. Roth

64 **Symmetries in Physics** Group Theory Applied to Physical Problems 2nd Edition
By W. Ludwig and C. Falter

65 **Phonons: Theory and Experiments II**
Experiments and Interpretation of Experimental Results By P. Brüesch

66 **Phonons: Theory and Experiments III**
Phenomena Related to Phonons
By P. Brüesch

67 **Two-Dimensional Systems: Physics and New Devices**
Editors: G. Bauer, F. Kuchar, and H. Heinrich

68 **Phonon Scattering in Condensed Matter V**
Editors: A. C. Anderson and J. P. Wolfe

69 **Nonlinearity in Condensed Matter**
Editors: A. R. Bishop, D. K. Campbell, P. Kumar, and S. E. Trullinger

70 **From Hamiltonians to Phase Diagrams**
The Electronic and Statistical-Mechanical Theory of sp-Bonded Metals and Alloys By J. Hafner

71 **High Magnetic Fields in Semiconductor Physics**
Editor: G. Landwehr

72 **One-Dimensional Conductors**
By S. Kagoshima, H. Nagasawa, and T. Sambongi

73 **Quantum Solid-State Physics**
Editors: S. V. Vonsovsky and M. I. Katsnelson

74 **Quantum Monte Carlo Methods in Equilibrium and Nonequilibrium Systems** Editor: M. Suzuki

75 **Electronic Structure and Optical Properties of Semiconductors** 2nd Edition
By M. L. Cohen and J. R. Chelikowsky

76 **Electronic Properties of Conjugated Polymers**
Editors: H. Kuzmany, M. Mehring, and S. Roth

77 **Fermi Surface Effects**
Editors: J. Kondo and A. Yoshimori

78 **Group Theory and Its Applications in Physics**
2nd Edition
By T. Inui, Y. Tanabe, and Y. Onodera

79 **Elementary Excitations in Quantum Fluids**
Editors: K. Ohbayashi and M. Watabe

80 **Monte Carlo Simulation in Statistical Physics**
An Introduction 4th Edition
By K. Binder and D. W. Heermann

81 **Core-Level Spectroscopy in Condensed Systems**
Editors: J. Kanamori and A. Kotani

82 **Photoelectron Spectroscopy**
Principle and Applications 2nd Edition
By S. Hüfner

83 **Physics and Technology of Submicron Structures**
Editors: H. Heinrich, G. Bauer, and F. Kuchar

84 **Beyond the Crystalline State** An Emerging Perspective By G. Venkataraman, D. Sahoo, and V. Balakrishnan

85 **The Quantum Hall Effects**
Fractional and Integral 2nd Edition
By T. Chakraborty and P. Pietiläinen

86 **The Quantum Statistics of Dynamic Processes**
By E. Fick and G. Sauermann

87 **High Magnetic Fields in Semiconductor Physics II**
Transport and Optics Editor: G. Landwehr

88 **Organic Superconductors** 2nd Edition
By T. Ishiguro, K. Yamaji, and G. Saito

89 **Strong Correlation and Superconductivity**
Editors: H. Fukuyama, S. Maekawa, and A. P. Malozemoff

Springer Series in Solid-State Sciences

Editors: M. Cardona P. Fulde K. von Klitzing H.-J. Queisser

90 **Earlier and Recent Aspects of Superconductivity** Editors: J. G. Bednorz and K. A. Müller

91 **Electronic Properties of Conjugated Polymers III** Basic Models and Applications Editors: H. Kuzmany, M. Mehring, and S. Roth

92 **Physics and Engineering Applications of Magnetism** Editors: Y. Ishikawa and N. Miura

93 **Quasicrystals** Editors: T. Fujiwara and T. Ogawa

94 **Electronic Conduction in Oxides** 2nd Edition By N. Tsuda, K. Nasu, F. Atsushi, and K. Siratori

95 **Electronic Materials** A New Era in Materials Science Editors: J. R. Chelikowsky and A. Franciosi

96 **Electron Liquids** 2nd Edition By A. Isihara

97 **Localization and Confinement of Electrons in Semiconductors** Editors: F. Kuchar, H. Heinrich, and G. Bauer

98 **Magnetism and the Electronic Structure of Crystals** By V.A. Gubanov, A.I. Liechtenstein, and A.V. Postnikov

99 **Electronic Properties of High-T_c Superconductors and Related Compounds** Editors: H. Kuzmany, M. Mehring, and J. Fink

100 **Electron Correlations in Molecules and Solids** 3rd Edition By P. Fulde

101 **High Magnetic Fields in Semiconductor Physics III** Quantum Hall Effect, Transport and Optics By G. Landwehr

102 **Conjugated Conducting Polymers** Editor: H. Kiess

103 **Molecular Dynamics Simulations** Editor: F. Yonezawa

104 **Products of Random Matrices** in Statistical Physics By A. Crisanti, G. Paladin, and A. Vulpiani

105 **Self-Trapped Excitons** 2nd Edition By K. S. Song and R. T. Williams

106 **Physics of High-Temperature Superconductors** Editors: S. Maekawa and M. Sato

107 **Electronic Properties of Polymers** Orientation and Dimensionality of Conjugated Systems Editors: H. Kuzmany, M. Mehring, and S. Roth

108 **Site Symmetry in Crystals** Theory and Applications 2nd Edition By R. A. Evarestov and V. P. Smirnov

109 **Transport Phenomena in Mesoscopic Systems** Editors: H. Fukuyama and T. Ando

110 **Superlattices and Other Heterostructures** Symmetry and Optical Phenomena 2nd Edition By E. L. Ivchenko and G. E. Pikus

111 **Low-Dimensional Electronic Systems** New Concepts Editors: G. Bauer, F. Kuchar, and H. Heinrich

112 **Phonon Scattering in Condensed Matter VII** Editors: M. Meissner and R. O. Pohl

113 **Electronic Properties of High-T_c Superconductors** Editors: H. Kuzmany, M. Mehring, and J. Fink

114 **Interatomic Potential and Structural Stability** Editors: K. Terakura and H. Akai

115 **Ultrafast Spectroscopy of Semiconductors and Semiconductor Nanostructures** 2nd Edition By J. Shah

116 **Electron Spectrum of Gapless Semiconductors** By J. M. Tsidilkovski

117 **Electronic Properties of Fullerenes** Editors: H. Kuzmany, J. Fink, M. Mehring, and S. Roth

118 **Correlation Effects in Low-Dimensional Electron Systems** Editors: A. Okiji and N. Kawakami

119 **Spectroscopy of Mott Insulators and Correlated Metals** Editors: A. Fujimori and Y. Tokura

120 **Optical Properties of III–V Semiconductors** The Influence of Multi-Valley Band Structures By H. Kalt

121 **Elementary Processes in Excitations and Reactions on Solid Surfaces** Editors: A. Okiji, H. Kasai, and K. Makoshi

122 **Theory of Magnetism** By K. Yosida

123 **Quantum Kinetics in Transport and Optics of Semiconductors** By H. Haug and A.-P. Jauho

124 **Relaxations of Excited States and Photo-Induced Structural Phase Transitions** Editor: K. Nasu

125 **Physics and Chemistry of Transition-Metal Oxides** Editors: H. Fukuyama and N. Nagaosa

The New Springer Global Website
Be the first to know
Benefit from new practice-driven features.
Search all books and journals – now faster and easier than ever before.
Enjoy big savings through online sales.
springeronline.com – the innovative website with you in focus.
springeronline.com
The interactive website for all Springer books and journals
Springer
010048x

Printing: Mercedes-Druck, Berlin
Binding: Stein + Lehmann, Berlin

MIX
Papier aus verantwortungsvollen Quellen
Paper from responsible sources
FSC® C105338

Printed by Books on Demand, Germany